Agricultural Marketing in India

SEVENTH EDITION

Agricultural Marketing in India

Seventh Edition

ISBN: 978-93-89688-06-1

© 2021, SS Acharya and NL Agarwal

Seventh Editon: 2021
Reprint: 2022, 2024, **2025**
First Edition: 1987
Second Edition: 1992
Third Edition: 1999
Fourth Edition: 2004
Fifth Edition: 2011
Sixth Edition: 2016
Reprint: 2017, 2018, 2020

OXFORD & IBH

New Delhi
(A Unit of CBS Publishers & Distributors Pvt Ltd)

Published by **Satish Kumar** Jain and Produced by **Varun Jain** for

CBS Publishers & Distributors Pvt Ltd
4819/XI Prahlad Street, 24 Ansari Road, Daryaganj, New Delhi 110 002, India.

Ph: 011-23289259, 23266861, 23266867   Fax: 011-23243014   Website: www.cbspd.com
e-mail: delhi@cbspd.com;
cbspubs@airtelmail.in.

Corporate Office: 204 FIE, Industrial Area, Patparganj, Delhi 110 092, India
Ph: 011-4934 4934   Fax: 011-4934 4935   e-mail: publishing@cbspd.com;
publicity@cbspd.com

Branches

- **Bengaluru:** Seema House 2975, 17th Cross, KR Road, Banasankari 2nd Stage, Bengaluru 560 070, Karnataka, India
  Ph: +91-80-26771678/79   Fax: +91-80-26771680   e-mail: bangalore@cbspd.com
- **Chennai:** 18/8B, Subbaraya Street, Shenoy Nagar, Chennai 600 030, Tamil Nadu, India
  Ph: +91-044-42032115, 044-26681266   e-mail: chennai@cbspd.com
- **Kochi:** 42/1325, 1326, Power House Road, Opp KSEB, Power House, Ernakulum Kochi 682 018, Kerala, India
  Ph: +91-484-4059061-65, 67   Fax: +91-484-4059065   e-mail: kochi@cbspd.com
- **Kolkata:** 147, Hind Ceramics Compound, 1st Floor, Nilgunj Road, Belghoria, Kolkata-700056, West Bengal, India
  Ph: +033-25633055, 033-25633056   e-mail: kolkata@cbspd.com
- **Lucknow:** Basement, Khushnuma Complex, 7 Meerabai Marg (Behind Jawahar Bhawan), Lucknow-226001, UP, India
  Ph: +0522-4000032   e-mail: tiwari.lucknow@cbspd.com
- **Mumbai:** PWD Shed, Gala no 25/26, Ramchandra Bhatt Marg, Next to JJ Hospital Gate no. 2, Opp. Union Bank of India, Noorbaug, Mumbai-400009, Maharashtra, India
  Ph: 022-66661880/89   e-mail: mumbai@cbspd.com

Representatives.

Gujarat	0-9879558667	Hyderabad	0-9885175004	Jharkhand	0-9811541605
Nagpur	0-8692091830	Patna	0-9334159340	Pune	0-9664372571
Uttarakhand	0-9716462459				

Printed at: Mudrak, Noida, UP, India

Agricultural Marketing in India

SEVENTH EDITION

S.S. Acharya
Former Chairman
Commission for Agricultural Costs and Prices, Government of India

and

N.L. Agarwal
Former Professor of Agricultural Economics
Rajasthan Agricultural University

Oxford & IBH Publishing Co. Pvt. Ltd.
New Delhi
(*A Unit of* CBS Publishers & Distributors Pvt Ltd)

CBSPD

CBS Publishers & Distributors Pvt Ltd

New Delhi • Bengaluru • Chennai • Kochi • Kolkata • Lucknow • Mumbai
Gujarat • Hyderabad • Jharkhand • Nagpur • Patna • Pune • Uttarakhand

Preface to the Seventh Edition

I am happy to note that the sixth edition of the book was found very useful by the students, teachers, agricultural economists, agribusiness professionals, development thinkers and policy makers in India as well as other countries, both in developing and the developed world. Since the publication of the sixth edition in 2016, the macro-economic environment has further undergone a sea change. Indian agriculture is increasingly becoming commercialized and market orientation of farmers has further gone up. The farmers are shifting towards high-value crops and nutritionally rich livestock products. The main drivers are changing demand preferences of consumers towards nutritive and value-added farm products, better market prices for farmers, availability of new and appropriate technologies, increased investment in agriculture and allied sectors, expansion of export opportunities and support from favourable policies and programmes, including incentives. Indian agriculture is also gaining increasing corporate attention. India has embarked on 'Ever Green' revolution and shifted its focus from 'production growth' to 'sustainable growth' as well as enhancing the farmers income that makes the role of an efficient agricultural marketing system even more critical. In addition, diversification of rural economy, increase in rural household incomes and consequent shift in demand patterns of even rural households underline the new role of marketing system, both in terms of its cause and effect. While revising the book for seventh edition, I have tried to capture these dimensions, especially in terms of the response of government in the form of marketing policies and programmes, and of other stakeholders, including farmers, market functionaries and agribusiness managers. However, I continue to remain conscious of the importance of retaining some of the previous marketing policies, programmes and practices, which have been critical part of the evolution of agricultural marketing system in India during the last seven and half decades. As was true with the preceding six editions, my intention and dedication continue to be towards guiding the readers to the evolution and frontiers of agricultural marketing system in India.

As has been our practice, in this edition also, I have re-written many portions and added a large number of new sections/sub-sections. As I make extensive use of factual data to support the observations, statements and analysis, all the relevant tables have been updated with latest available data and also added some new tables. New references have been mentioned where necessary and the bibliography has been enriched. However, I have retained the structure of the book to 11 chapters.

In the first chapter on Definition and Scope of Agricultural Marketing, I have added the recent departure in India's approach to agricultural development from production to farmers' income in the section on new role of agricultural marketing. In the second chapter, latest estimates of income and price elasticities of demand for important farm products have been added. We have also updated the factual information at several places, including relevant tables. Further, a few new concepts in agricultural marketing have been added. In the third chapter on Agricultural Marketing and Development, I have updated the estimates of marketed surplus, contribution of different size group of farms to marketed surplus and projections of demand and supply of farm products. There are quite a few changes and alterations in chapter 4 on Marketing Functions. The factual information, wherever appearing, has been updated throughout the chapter. Additions in this chapter include new AGMARK logo, relative roles of rail and road transport, new sub-section on warehousing industry, gaps in cool chains and cold storage and options trading. This chapter also includes illustration of as to how farmers can use futures trade to reduce their price risks.

In the fifth chapter on Marketing Agencies, Institutions and Channels, I have added marketing channels for fruits and vegetables and milk. Further, some details of New Model Contract Farming Act have been added. The sixth chapter on Marketing of Farm Inputs has also been revised by updating the factual information and including additional developments. These include recent steps to improve fertilizer marketing; National Seeds Reserve and SAARC Seed Bank; major achievement in electricity generation and supply in rural areas; and trend in purchased farm inputs. There are several additions and changes in chapter 7 on Government Intervention and Role in Agricultural Marketing. The additions in this chapter include provisions of new APLM Act, 2017; E-NAM; suggestions related to agricultural marketing by the committee on DFI; my suggestions related to speeding up of marketing reforms; updates on the progress of Food Security (Right to Food) Act; updates on FSSAI regulations; changes in MSP policy (level of floor) and its implementation; direct income support to farmers; and Goods and Service Tax. These apart, there is considerable updating of factual information in various tables.

I have also revised the eighth chapter on Cooperation and Cooperatives in Agricultural Marketing. Apart from updating the factual information, a new section on size of Indian cooperative sector and another on National Cooperative Union of India have been added. In the ninth chapter on Market Integration and Efficiency, I have added a new section on steps in co-integration analysis and updated data on terms of trade for agriculture and that for farmers. In chapter 10 on Training, Research, Extension and Statistics, I have updated the information on training facilities and emphasized the need for establishment of a system for generation of crop price forecasts on regular basis as a part of agricultural marketing research system. The last chapter (11) on External Trade in Farm Products has been considerably changed. Apart from updating the trade related factual information, some of the sections/sub-sections have been restructured and latest situation described. The section on promotion of agricultural exports has been redrafted. New sections include Foreign Trade policy 2015–20, recent trade facilitation reforms, Agricultural Trade Policy 2018, FTAs/RTAs and Tariff Rate Quotas. In the appendix, the lists of abbreviations, marketing institutions and selected readings have been updated.

In making these revisions, my constant endeavour has been to keep the presentation so simple that the students as well as teachers, researchers, scholars, marketing specialists, marketing managers, policy makers, agribusiness entrepreneurs and farmers or farmers' leaders are able to understand the realities and complexities of agricultural marketing system and benefit from the improved understanding.

While working on the revision of the book for seventh edition, I was frequently reminded of the contributions of my colleague and co-author Dr N.L. Agarwal in conceiving and bringing out such a book in 1987 and its subsequent revisions till the 6th Edition in 2016. I had the privilege of being his Ph.D. thesis supervisor in 1970s. He superannuated from the post of University Professor in 1999. I pay homage to Dr Agarwal, who left for heavenly abode on 28th January, 2017.

I am grateful to all the readers of first six editions of this book for providing me feedback and useful comments, which have helped in revising the book and making this complex subject simple and easily understandable by all. The debts can be only warmly acknowledged but never fully recompensed. My greatest debts are due to the reviewers of earlier editions of the book, for their pains-taking reading and invaluable comments. I also gratefully acknowledge the suggestions received from teachers, students, scholars, businessmen, entrepreneurs, marketing professionals, development thinkers, and policy makers in India and other countries during the last 33 years since the publication of the first edition of the book. I sincerely hope that this enlarged and revised seventh edition will inspire even greater response from the readers.

January, 2021   S.S. Acharya

Preface to the First Edition

The agricultural situation in India has undergone a rapid change in last two decades. Investment in agricultural sector, both in public and private sectors, has risen. Agricultural production, in general has achieved reasonable growth rate. But the growth rate has not only to be maintained, but accelerated and fluctuations in agricultural production are to be minimised. The efforts are already under way to evolve location-specific technologies, transfer them to farmers' fields and assure input supply to farmers in right time, place and quality. The rate at which new technology, and yield increasing inputs are adopted by farmers, is affected by the prices of inputs and output. Simultaneously, consumers also expect the availability of goods at reasonable prices. For achieving these conflicting objectives, marketing system for agricultural commodities and inputs has to play a very crucial role. This necessitates an understanding of the marketing system and market structure for agricultural products and inputs by all the sections of the population viz., farmers, traders, consumers, extension workers, scientists, sociologists, administrators, planners and politicians. Therefore, it was felt necessary to analyse all dimensions of agricultural marketing scene in a single volume.

The book is based on the experience of teaching courses on agricultural marketing, prices and related subjects to undergraduate and postgraduate students, guiding and conducting research studies and extensive visits of agricultural markets by the authors during last more than 20 years. The material in the book has been synthesised from the available research findings of the authors and others, data from different published sources, practical insights of the authors and critical appraisal of the policies of the government. The book evaluates the performance of the existing marketing system, institutions and policy in accelerating agricultural development of the country. It contains an incisive analysis with special emphasis on marketing functions, institutions, efficiency, costs and margins, government efforts in the improvement of agricultural marketing and market research. Although the main emphasis is on the performance of India's agricultural marketing system, the methodology and policy implications are applicable to the other developing countries.

The book runs in eight chapters. Chapter 1 deals with the meaning, scope and subject-matter of agricultural marketing. Theory of markets and market structure, development of markets, market forces and market models for price determination have been spelt in Chapter 2. In Chapter 3, importance of agricultural marketing in economic development of the country, history and growth of agricultural marketing, producers' surplus and characteristics of an ideal system of marketing have been discussed at length. These three chapters provide the theoretical base for better understanding of the problems in marketing of farm products and inputs. Next three chapters deal with the marketing system of farm products and inputs along with the steps taken by the government for combating the problems confronting the farmers. Chapter 4 describes the marketing functions, marketing agencies and marketing channels. Marketing of farm inputs necessary for the technological breakthrough in Indian agriculture has been presented in Chapter 5. Chapter 6 describes the role played by the government in improvement of agricultural marketing system. This includes the critical study of the traditional agricultural marketing system, forms of government intervention viz., establishment of Directorate of Marketing and Inspection, regulation of markets, state trading, procurement and distribution operations of foodgrains, quality control, price support, price controls and rationing of foodgrains. Cooperative marketing has also been dealt in this Chapter. Evaluation of market integration, marketing efficiency and costs and margins in marketing of different farm products has been presented in Chapter 7. The detailed analysis in this Chapter highlights the areas necessitating improvement in the field of agricultural marketing. Last chapter deals with the training facilities, research and statistics in agricultural marketing. This chapter also describes the approaches for conducting marketing studies and the institutions engaged in agricultural marketing research. At the end, available literature on agricultural marketing has also been reviewed for the benefit of readers. The subject-matter in the book is so graded that undergraduate and postgraduate students as well as teachers and research scholars, specialist in marketing, policy-planners and those interested in the welfare of the farmers could benefit from this book.

We sincerely acknowledge the help taken in writing of this book from various books, reports, journals and magazines. We are thankful to Dr. K.N. Nag, Vice-Chancellor, Sukhadia University, Udaipur; Dr. R.C. Mehta, Dean, S.K.N. College of Agriculture, Jobner; Dr. M.S. Manohar, Dean, Rajasthan College of Agriculture, Udaipur and Dr. B.S. Rathore, Ex-Professor and Head, Department of Agricultural Economics, Rajasthan College of Agriculture, Udaipur for their encouragement in writing the book. The authors also feel obliged to the National Book Trust, New Delhi for providing the financial assistance under their programme of production of low-cost textbooks. Shri Mohan Primlani of Oxford and IBH deserves our special thanks in bringing out this work in a short period.

April, 1986

S.S. Acharya
N.L. Agarwal

Contents

1

Agricultural Marketing (Definition and Scope)

Marketing is as critical to better performance in agriculture as farming itself. Therefore, market reform and marketing system improvement ought to be an integral part of policy and strategy for agricultural development. Although a considerable progress has been achieved in technological improvements in agriculture by the use of high-yielding variety seeds and chemical fertilisers, and by the adoption of plant protection measures, the rate of growth in farming in developing countries has not attained the expected levels. This has been largely attributed to the fact that not enough attention was devoted to the facilities and services which must be available to farmers if agriculture is to develop.

Agricultural marketing was, for many decades, not fully accepted as an essential element in agricultural development in the countries of Asia and Africa. Although opinions differ as to the extent and precedence, there was general agreement till 1970 that the question of markets for agricultural commodities had been neglected[1]. Agricultural marketing occupied a fairly low place in agricultural development policies of developing countries. The National Commission on Agriculture (1976) and Farmers Commission (2007) had emphasised that it is not enough to produce a crop or an animal product; it must be satisfactorily marketed.

CONCEPT AND DEFINITION

The term *agricultural marketing* is composed of two words—agriculture and marketing. Agriculture, in the broadest sense, means activities aimed at the use of natural resources for human welfare, i.e. it includes all the primary activities of production. But, generally, it is used to mean growing and/or raising crops and livestock. Marketing connotes a series of activities involved in moving the goods from the point of production to the point of consumption. It includes all activities involved in the creation of time, place, form and possession utility.

Philip Kotler[2] has defined marketing as a human activity directed at satisfying the needs and wants through exchange process. American Marketing Association defined marketing as the performance of business activities that directs the flow of goods and services from producers to users.

The interest of scholars and development thinkers in marketing of agricultural commodities began with a focus on food or foodgrains marketing and on marketing of agricultural raw material (cotton, jute etc.). In early 1970s, Moore, Johl and Khusro[3] conceptualized foodgrains marketing as business activities involved in moving foodgrains from farmers to consumers through time (storage), space (transport), form (processing), and transfer of ownership at various stages in the marketing channel. Later, Kohls and Uhl[4] defined foodgrains marketing as the performance of all business activities involved in the flow of food products and services from the point of initial agricultural production until they are in the hands of consumers.

According to Thomsen[5], the study of agricultural marketing comprises all the operations, and the agencies conducting them, involved in the movement of farm-produced foods, raw materials and their derivatives, such as textiles, from the farms to the final consumers, and the effects of such operations on farmers, middlemen and consumers. This definition does not include the input side of agriculture.

Agricultural marketing system in developing countries including India can be understood to compose of two major sub-systems viz., product marketing and input (factor) marketing. The actors in the product marketing sub-system include farmers, village/primary traders, wholesalers, processors, importers, exporters, marketing cooperatives, regulated market committees and retailers. The input sub-system includes input manufacturers, distributors, related associations, importers, exporters and others who make available various farm production inputs to the farmers.

Agricultural marketing is the study of all the activities, agencies and policies involved in the procurement of farm inputs by the farmers and the movement of agricultural products from the farms to the consumers. The agricultural marketing system is a link between the farm and the non-farm sectors. It includes the organisation of agricultural raw materials supply to processing industries, the assessment of demand for farm inputs and raw materials, and the policy relating to the marketing of farm products and inputs.

According to the National Commission on Agriculture (XII Report, 1976), agricultural marketing is a process which starts with a decision to produce a saleable farm commodity, involves all the aspects of market structure or system, both functional and institutional, based on technical and economic considerations, and includes pre- and post-harvest operations, assembling, grading, storage, transportation and distribution.

However, as Acharya[6] has described, in a dynamic and growing agricultural sector, the agricultural marketing system ought to be understood and developed as a link between the farm and the non-farm sectors. A dynamic and growing agricultural sector, requires fertilisers, pesticides, farm equipments, machinery, diesel, electricity, packing material and repair services which are produced and supplied by the industry and non-farm enterprises. The expansion in the size of farm output stimulates forward linkages by providing surpluses of food and natural fibres which require transportation, storage, milling or processing, packaging and retailing to the consumers. These functions are variously performed by non-farm enterprises. Further, if the increase in agricultural production is accompanied by a rise in real incomes of farm families, the demand of these families for non-farm consumer goods goes up as the proportion of income spent on non-food consumables and durables tends to rise with the increase in real per capita income. Several industries, thus find new markets for their products in the farm sector.

Agricultural marketing, therefore, can be defined as comprising of all activities involved in supply of farm inputs to the farmers and movement of agricultural products (food, feed and fibre) from the individual farms to the ultimate consumers. According to this definition, agricultural marketing system includes the assessment of demand for farm-inputs and their supply, post-harvest handling of farm products, performance of various activities required in transferring farm products from farm gate to processing industries and/or ultimate consumers, assessment of demand for farm products and public policies and programmes relating to the pricing, handling, and purchase and sale of farm inputs and agricultural products.

NEED FOR UNDERSTANDING

A decision on an appropriate strategy, the evolution of a proper policy and a choice of policy instruments calls for a continuous flow of advice, information and assessment of the existing system. Every system generates impulses as a result of environmental changes. These impulses have to be observed, recorded, analysed and interpreted for the benefit of the policy-makers.

A study of the agricultural marketing system is necessary for an understanding of the complexities involved and the identification of bottlenecks with a view to providing efficient services in the transfer of farm products and inputs from producers to consumers. An efficient marketing system minimizes costs, and benefits all the sections of the society.

The expectations from the system vary from group to group; and, generally, the objectives of various groups are in conflict. The efficiency and success of the system depends on how best these conflicting objectives are reconciled.

- *Producers:* Producer-farmers want the marketing system to purchase their produce without loss of time and provide the maximum share in the consumer's rupee. They want the maximum possible price for their surplus produce from the system. Similarly, they want the system to supply them the inputs at the lowest possible price.
- *Consumers:* The consumers of agricultural products are interested in a marketing system that can provide food and other items in the quantity and of the quality required by them at the lowest possible price. However, this objective of marketing for consumers is contrary to the objective of marketing for the farmer-producers.
- *Market Middlemen and Traders:* Market middlemen and traders are interested in a marketing system which provides them a steady and increasing income from the purchase and sale of agricultural commodities. This objective of market middlemen may be achieved by purchasing the agricultural products from the farmers at low prices and selling them to consumers at high prices.
- *Agro-Processors:* Those who are engaged in processing of agricultural commodities are interested in farm products (raw material) which are suitable for processing and able to give higher output of processed products with least wastage. They want these products at lower prices.
- *Agri-Exporters:* Exporters of agricultural commodities expect the marketing system to provide them the farm products which meet the international quality standards at minimum and reasonable prices.
- *Government:* The objectives and expectations of all the three-groups of society— producers, consumers and market middlemen—conflict with one another. All the groups are indispensable to society. The government has to act as a watch-dog to safeguard the interests of all the groups associated in marketing. It tries to provide

the maximum share to the producer in the consumer's rupee; food and other farm products of the required quality to consumers at the lowest possible price; and enough margin to market middlemen so that they may remain in the trade and not think of going out of trade and jeopardise the whole marketing mechanism. Thus, the government wants that the marketing system should be such as may bring about the overall welfare to all the segments of society.

The overall objective of agricultural marketing system in a developing country like India should be to help the primary producers viz., the farmers in getting remunerative prices for their produce on the one hand and to provide right type of goods at the right place, in the right quantity and quality at a right time and at right prices to the processors and/or ultimate consumers on the other.

SCOPE AND SUBJECT MATTER

Agricultural marketing in a broader sense is concerned with the marketing of farm products produced by farmers and of farm inputs and services required by them in the production of these farm products. Thus, the subject of agricultural marketing includes product marketing as well as input marketing.

The subject of output marketing is as old as civilization itself. The importance of output marketing has become more conspicuous in the recent past with the increased marketable surplus of the crops and other agricultural commodities following the technological breakthrough. The market orientation of farming has increased. Input marketing is a comparatively new subject. Farmers in the past used such farm sector inputs as local seeds and farmyard manure. These inputs were available with them; the purchase of inputs for production of crops from the market by the farmers was almost negligible. The importance of farm inputs—improved seeds, fertilizers, insecticides and pesticides, farm machinery, implements and credit—in the production of farm products has increased over time. The new agricultural technology is input-responsive. Thus, the scope of agricultural marketing must include both product marketing and input marketing. In this book, the subject-matter of agricultural marketing has been dealt with; both from the theoretical and practical points of view. It covers what the system is, how it functions, and how the given methods or techniques may be modified to get the maximum benefits.

Specially, the subject of agricultural marketing includes marketing functions, agencies, channels, efficiency and costs, price spread and market integration, producer's surplus, government policy and research, training and statistics on agricultural marketing and imports/exports of agricultural commodities.

NEW ROLE OF AGRICULTURAL MARKETING

The new role of agricultural marketing system is evident from the fact that Indian agriculture is increasingly becoming more commercialized and market-orientation has gone up. This is a very positive transformation of Indian agriculture, in which agricultural marketing system and policies have contributed in the past but have to play an increasing role in times to come. The farmers are producing for the market and marketed surpluses have gone up. There is a greater use of purchased inputs and services by farmers. They are shifting towards high value crops and nutritionally rich products. The main drivers of increasing commercialization of Indian agriculture are better market prices, changing demand preferences of consumers towards high-value nutritive farm products, availability of new agricultural technologies, increased investment in agriculture, expansion of export opportunities and support from favourable policies and programmes, including incentives.

Indian agriculture is also gaining increasing corporate attention. Traditionally seen as unorganized and fragmented, Indian agri-business has now become a hot sector. As the market for processed foods with better quality is expanding, it is being considered as hot entrepreneurial activity. Agribusiness in areas like farm inputs, logistics, warehousing, processing and marketing of dairy and food products is scaling-up and attracting investment from a plethora of risk capital investors. Branded food products are particularly attracting the investors. Western food chains are also flocking to India.

These apart, India has embarked on second Green Revolution which is reflected in accelerated target growth of agriculture (4% during 12th Five Year Plan), increased attention to eastern India, increasing the availability of high value crops and nutritionally-rich livestock products, and enhancing the output of oilseeds and pulses for reducing dependence on imports. The role of an efficient agricultural marketing system is quite critical in achieving the goals of second green revolution. In addition, there is one other factor that underlines (both as a cause and effect) the role of agricultural marketing system is the increase in rural household incomes, diversification of rural economy, increase in rural wages and shifts in demand patterns of rural households (towards safe and quality food, processed and branded products and more nutritionally healthy diets).

In 2016, Government of India launched a mission to double the farmers' income within six years. This is a distinct departure, in the approach to agricultural and rural development, from production-enhancing to farmers' income-enhancing approach. Farmers income is directly related to the size of marketed surplus (out of the harvested produce) and the prices farmers are able to realize. In this connection, the importance of the access of the farmers to an efficient marketing system hardly needs any emphasis. Without appropriate market linkages, higher production by the farmers does not automatically translate into higher returns to farmers and neither are the benefits of higher production fully passed on to the consumers.

Therefore, the framework under which agricultural produce markets function and the factors which influence the prices received by the farmers now need to be understood in a different perspective compared to that in the past. The role of marketing now starts right from the time of decision relating to what to produce, which variety to produce and how to prepare the product for marketing rather than limiting it to when, where and to whom to sell.

It may be noted here that there is a considerable difference among various groups of agricultural commodities as regards the marketing requirement. Agricultural commodities from this perspective can be understood to compose of foodgrains (cereals and pulses), oilseeds, fibres (cotton, jute, mesta etc.), sugar crops (sugarcane etc.), plantation crops (tea, coffee, rubber etc.), horticultural crops (fruits, vegetables, flowers), and livestock products (milk, meats, eggs, and fisheries). The perishability of horticultural and livestock products is very high as compared to other agricultural commodities. The post-harvest losses are substantial for these commodities. According to the Working Group on Agricultural Marketing for the 12th Five Year Plan (2012–17), the post harvest losses are 3.9% (sorghum) to 6% (wheat) for cereals, 4.3% (chickpea) to 6% (gram) for pulses, 2.8% (cotton seed) to 10.1% (groundnut) for oilseeds, 5.8% (sapota) to 18% (guava) for fruits, 6.8% (cauliflower) to 12.4% (tomato) for vegetables, 3.9% (black pepper) to 7.4% (turmeric) for spices, and 0.8% (milk) to 6.9% (inland fisheries) for livestock products. But these estimates are only up to the first point of marketing by the farmers. If all the losses, till the produce reaches the final consumer, i.e. the losses in the entire value chain, are considered, these are considerably higher than the Working Group estimates mentioned above. These are around 30% of the total production in the case of fruits and vegetables. Our estimate is that total value of agricultural products lost in the entire marketing chain is currently at around ₹ 1 lakh crore (₹ 1 trillion). Inadequacy of technology-aided farm-to-market logistics contributes to high food losses in the case of perishables.

Agricultural marketing has to play a special role in handling perishable farm products, which are also high-value commodities. The demand for horticultural and livestock products is increasing and farmers in several areas are shifting to the production of such commodities, which is helping in increasing the farmers' incomes. But these products require special facilities to handle and market them. Horticultural and livestock products have some special characteristics, which are distinctly different than other farm products.

DIFFERENCES IN MARKETING OF AGRICULTURAL AND MANUFACTURED GOODS

The marketing of agricultural commodities is different from the marketing of manufactured commodities because of the special characteristics of the agricultural sector (demand and supply) which have a bearing on marketing. Because of these characteristics, the subject of agricultural marketing has been treated as a separate discipline—and this fact makes the subject somewhat complicated. These special characteristics of the agricultural sector affect the supply and demand of agricultural products in a manner different from that governing the supply and demand of manufactured commodities. The special characteristics which the agricultural sector possesses, and which are different from those of the manufactured sector, are as follows:

(i) Perishability of the Product

Most farm products are perishable in nature; but the period of their perishability varies from a few hours to a few months. To a large extent, the marketing of farm products is virtually a race with death and decay. Their perishability makes it almost impossible for producers to fix the reserve price for their farm-grown products. The supply of agricultural products is irregular; the price of the crop therefore fluctuates both during the year and from year to year. The extent of perishability of farm products may be reduced by the processing function; but they cannot be made non-perishable like manufactured products. Nor can their supply be made regular.

(ii) Seasonality of Production

Farm products are produced in a particular season of the year; they cannot be produced throughout the year. It leads to intra-year seasonality in prices. In the harvest season, prices of farm products fall. But the supply of manufactured products can be adjusted or made uniform throughout the year. Their prices therefore remain almost the same throughout the year.

(iii) Bulkiness of Products

The characteristic of bulkiness of most farm products makes their transportation and storage difficult and expensive. This fact also restricts the location of production to somewhere near the place of consumption or processing. The price spread in bulky products is higher because of the higher costs of transportation, handling and storage.

(iv) Variation in Quality of Products

There is a large variation in the quality of agricultural products, which makes their grading and standardization somewhat difficult. There is no such problem in manufactured goods, for they can be produced of uniform quality.

(v) Irregular Supply of Agricultural Products

The supply of agricultural products is uncertain and irregular because of the dependence of agricultural production on natural conditions. With the varying supply, the demand remaining almost constant, the prices of agricultural products fluctuate substantially more than that of manufactured goods.

(vi) Small Size of Holdings and Scattered Production

Farm products are produced throughout the length and breadth of the country and most of the producers are of small size. This makes the estimation of supply difficult and also creates problems in marketing as well as in discovery of prices.

(vii) Product Pricing

Apart from the problem in estimation of total supply in a small-farm agriculture, an individual farmer faces a typical marketing situation. As his share in total supply is very small, he cannot influence the market supply. Further, owing to the inelastic nature of demand for most of the farm products, the market price for his product is determined independent of his supply. It is in this context that an individual farmer is supposed to be operating in a buyer's market. Contrary to this, most of the manufacturing firms, owing to their large share in the market, can control, to some extent, the supply and thus influence the price of the product they sell. All the manufacturers who sell their products in the packaged form, are required, by the law, to interalia, display the maximum retail price (MRP) on the packages. However, it does not mean that producers of non-farm goods can fix the prices of their produce at will. Even a monopoly seller cannot do that. His choice of selling price is determined by the nature of demand curve faced by him. If he fixes a higher price, he can sell only a lower quantity of his product and vice versa. He can decide either the quantity he wants to produce and sell or the price he wants to charge and not both of them simultaneously.

(viii) Processing

Most of the farm products need some kind of processing before their consumption by the ultimate consumers. This processing function increases the price spread of agricultural commodities. Processing firms enjoy the advantage of monopsony, oligopsony or duopsony in the market. This situation sometimes creates disincentives for the producers and may have an adverse effect on production in the next year.

The characteristics of agricultural commodities mentioned above make their marketing system complex and different than that for manufactured goods. These are illustrated in Chart 1.1.

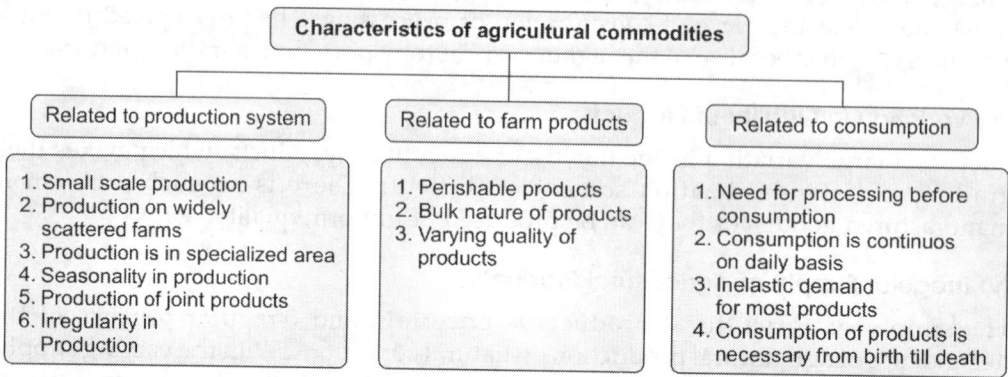

Chart 1.1: Some distinguishing characteristics of agricultural commodities that make their marketing system complex and different

PLAN OF THE BOOK

The subject-matter of agricultural marketing in this book has been presented in eleven chapters. The first chapter covers the concept, definition, need for understanding, scope and subject-matter of agricultural marketing and the major differences in the marketing features of agricultural and manufactured goods. The basic theory and some concepts of market and market structure, which are necessary for an understanding of agricultural marketing, have been presented in Chapter 2. Some common terms used in the field of marketing have also been explained in the chapter. The relevance of agricultural marketing in economic development has been described in Chapter 3. It covers the importance, history and growth of agricultural marketing, the issues related to marketed surplus, the characteristics of an ideal marketing system, and the need for the modernization of agricultural markets in India.

Fourth chapter gives a description of different marketing functions with special reference to farm products. This chapter also includes marketing infrastructure. Marketing agencies, institutions and channels, involved in marketing of farm products are discussed in Chapter 5. Marketing of agricultural inputs has been given a special treatment in Chapter 6. The government intervention in agricultural marketing has been analysed in Chapter 7. This chapter covers market regulation, role of marketing organizations, procurement and public distribution of foodgrains, quality control, the price policy, government sponsored national organizations, salient features of the legislative measures, new initiatives for improving agricultural marketing in India, alternative marketing systems and second phase of agricultural marketing reforms. Basic concepts of cooperation and the role of cooperatives in agricultural marketing and processing have been dealt in Chapter 8. The issues related to market integration, marketing efficiency, costs, margins and price spread in marketing of farm products have been analysed in Chapter 9. Tenth chapter deals with the status and requirements of training, research, extension and statistics on agricultural marketing in India. Some aspects of external trade (imports and exports) in agricultural commodities and India's share in the world trade have been brought out in Chapter 11. International trade regime for agricultural commodities has also been dealt in this chapter, which includes GATT, WTO and implications for Indian agriculture. A broad survey of the literature on agricultural marketing has been presented in the Bibliography at the end. Some common abbreviations in agricultural marketing have been also given at the end.

REFERENCES

1. Spinks, G.R., "Attitude Towards Agricultural Marketing in Asia and the Far East", Agricultural Development in Developing Countries—Comparative Experience, The Indian Society of Agricultural Economics, Bombay, 1972, p. 205.
2. Kotler, Philip, Marketing Management: Analysis, Planning and Control, Prentice Hall of India Pvt. Ltd., New Delhi, 1972.
3. Moore, J.R., S.S. Johl, and A. M. Khusro, Indian Foodgrains Marketing, Prentice Hall of India Private Limited, New Delhi, 1973, p.1.
4. Kohls, R.L. and J.N. Uhl, Marketing of Agricultural Products, Macmillan Publishing Company, Inc, New York, 1980, p. 8.
5. Thomsen, F.L., *Agricultural Marketing*, McGraw-Hill Book Company, Inc., New York, 1951, p.1.
6. Acharya, S.S., "Agriculture-Industry Linkages: Public Policy and Some Areas of Concern", Agricultural Economics Research Review, Vol. 10, No. 2, July–Dec. 1997, p. 162.

2

Markets and Market Structure

This chapter deals with the concept and components of a market, classification of markets based on 12 criteria, and growth of markets. It also covers market structure, conduct and performance. The concept of market forces and simplest theory of price determination have also been presented in this chapter.

▌MARKET

MEANING

The word *market* comes from the latin word 'marcatus' which means merchandise or trade or a place where business is conducted.

Word 'market' has been widely and variedly used to mean: (a) a place or a building where commodities are bought and sold, e.g. a cluster of shops or supermarket; (b) potential buyers and sellers of a product, e.g. wheat market and cotton market; (c) potential buyers and sellers of a country or region, e.g. Indian market and Asian market; (d) an organization which provides facilities for exchange of commodities, e.g. Bombay stock exchange; and (e) a phase or a course of commercial activity, e.g. a dull market or bright market.

There is an old English saying that two women and a goose may make a market. However, in common parlance, a market includes any place where persons assemble for the sale or purchase of commodities intended for satisfying human wants. Other terms used for describing markets in India are *Haats, Painths, Shandies and Bazar*.

The word *market* in the economic sense carries a broad meaning. Some of the definitions of *market* are given as follows:

1. A *market* is the sphere within which price determining forces operate.[1]
2. A *market* is the area within which the forces of demand and supply converge to establish a single price.[2]
3. The term *market* means not a particular market place in which things are bought and sold but the whole of any region in which buyers and sellers are in such a free intercourse with one another that the prices of the same goods tend to equality, easily and quickly.[3]
4. *Market* means a social institution which performs activities and provides facilities for exchanging commodities between buyers and sellers.[4]

5. Economically interpreted, the term *market* refers, not to a place but to a commodity or commodities and buyers and sellers that are in free intercourse with one another.[5]
6. The American Marketing Association has defined a market as the aggregate demand of the potential buyers for a product/service, while Kotler defined market as an area for potential exchanges.[6]

A market exists when buyers wishing to exchange the money for a good or service are in contact with the sellers who are willing to exchange goods or services for money. Thus, a market is defined in terms of the existence of fundamental forces of supply and demand and is not necessarily confined to a particular geographical location. The concept of a market is basic to most of the contemporary economies, since in a free market economy, this is the mechanism by which resources are allocated.

COMPONENTS OF A MARKET

For a market to exist, certain conditions must be satisfied. These conditions should be both necessary and sufficient. They may also be termed as the components of a market.

1. The existence of a good or commodity for transactions (physical existence is, however, not necessary);
2. The existence of buyers and sellers;
3. Business relationship or intercourse between buyers and sellers; and
4. Demarcation of area such as place, region, country or the whole world.

The existence of perfect competition or a uniform price is not necessary.

DIMENSIONS OF A MARKET

There are various dimensions of any specified market. These dimensions are:

1. Location or place of operation
2. Area or coverage
3. Time span
4. Volume of transactions
5. Nature of transactions
6. Number of commodities
7. Degree of competition
8. Nature of commodities
9. Stage of marketing
10. Extent of public intervention
11. Type of population served
12. Accrual of marketing margins.

Any individual market may be classified in a twelve-dimensional space.

CLASSIFICATION OF MARKETS

Markets may be classified on the basis of each of the twelve dimensions already listed.

1. On the Basis of Location or Place of Operation

On the basis of the place of location or place of operation, markets are of the following types:

(a) *Village Market*: A market which is located in a small village, where major transactions take place among the buyers and sellers of a village, is called a village market.

(b) *Primary Markets*: These markets are located in towns near the centres of production of agricultural commodities. In these markets, a major part of the produce is brought for sale by the producer-farmers themselves. Transactions in these markets usually take place between the farmers and primary traders.

(c) *Secondary Wholesale Markets*: These markets are located generally at district headquarters or important trade centres or near railway junctions. The major transactions in commodities in these markets take place between the village traders and wholesalers. The bulk of the arrivals in these markets is from other markets. The produce in these markets is handled in large quantities. There are, therefore, specialized marketing agencies performing different marketing functions, such as commission agents, brokers and weighmen in these markets.

(d) *Terminal Markets*: A terminal market is one where the produce is either finally disposed of to the consumers or processors, or assembled for export. In these markets, merchants are well organized and use modern methods of marketing. Commodity exchanges exist in these markets which provide facilities for forward trading in specific commodities. Such markets are located either in metropolitan cities or at seaports. Delhi, Mumbai, Chennai, Kolkatta and Cochin are terminal markets for many commodities.

(e) *Modern Terminal Markets:* The Department of Agriculture and Cooperation, Ministry of Agriculture, Government of India has taken the initiative to promote modern terminal markets for fruits, vegetables and other perishable commodities in important urban centres of the country to provide the state-of-the art infrastructure facilities of electronic auction, cold chain and logistics. These terminal markets are envisaged to operate on a 'hub and spoke' format where the terminal market (the hub) is linked to a number of collection centres (the spokes), conveniently located in key production centres to allow easy access to farmers for marketing of their produce. These terminal markets are built, owned and operated by either a corporate, private, or cooperative entity. This entity could be a consortium of entrepreneurs from agri-business, cold chain, logistics, warehousing, agri-infrastructure and related background. The central/State Government lend support to the initiative by providing financial support to the project by equity participation up to a maximum of 49% determined through a competitive bidding process.

The operational guidelines have been circulated by the Ministry to the States for setting up of terminal markets. Accordingly some states have amended their Agricultural Produce Marketing Regulation Acts. The states of Andhra Pradesh, Bihar, Madhya Pradesh, Maharashtra, Orissa, Punjab, Rajasthan, Tamil Nadu, West Bengal and Nagaland and Union Territory of Chandigarh have identified land for setting up of such terminal markets. Government of India has launched a scheme under which subsidy is available up to 25% for capital investment in agricultural marketing infrastructure for terminal markets with a ceiling limit for private agencies.

(f) *Seaboard Markets*: Markets which are located near the seashore and are meant mainly for the import and/or export of goods are known as seaboard markets. These are generally seaport towns. Examples of these markets in India are Mumbai, Chennai, Kolkata and Cochin (Kochi).

2. On the Basis of Area/Coverage

On the basis of the area from which buyers and sellers usually come for transactions, markets may be classified into the following four classes:

(a) *Local or Village Markets*: A market in which the buying and selling activities are confined among the buyers and sellers drawn from the same village or nearby villages. The village markets exist mostly for perishable commodities in small lots, e.g. local milk market or vegetable market.

(b) *Regional Markets*: A market in which buyers and sellers for a commodity are drawn from a larger area than the local markets. Regional markets in India usually exist for foodgrains.

(c) *National Markets*: A market in which buyers and sellers are spread at the national level. Earlier national markets existed for only durable goods like jute and tea. But with the expansion of roads, transport and communication facilities, the markets for most of the agriculture products have taken the form of national markets.

(d) *World or International Market*: A market in which the buyers and sellers are drawn from more than one country or the whole world. These are the biggest markets from the area point of view. These markets exist in the commodities which have a worldwide demand and/or supply, such as coffee, tea, machinery, gold, silver, etc. International trade in agricultural commodities has gone up since 1994 when AoA was signed and WTO was created (details in Chapter 11).

3. On the Basis of Time Span

On this basis, markets are of the following types:

(a) *Short-period Markets*: The markets which are held only for a day or few hours are called short-period markets. The products dealt within these markets are of a highly perishable nature, such as fish, fresh vegetables, liquid milk and ready-to-eat foods. In these markets, the prices of commodities are governed mainly by the extent of demand for, rather than by the supply of, the commodity.

(b) *Periodic Markets:* The periodic markets are congregation of buyers and sellers at specified places either in villages, semi-urban areas or some parts of urban areas on specific days and time. Major commodities traded in these markets is the farm produce grown in the hinterlands. The periodic markets are held weekly, biweekly, fortnightly or monthly according to the local traditions. These are similar to 'spontaneous markets' in several developed countries.

(c) *Long-period Markets*: These markets are held for a longer period than the short-period markets. The commodities traded in these markets are less perishable and can be stored for some time; like foodgrains and oilseeds. The prices of the products in these markets are governed both by the supply and demand forces.

(d) *Secular Markets*: These are markets of a permanent nature. The commodities traded in these markets are durable in nature and can be stored for many years. Examples are markets for machinery and manufactured goods.

4. On the Basis of Volume of Transactions

There are two types of markets on the basis of volume of transactions at a time.

(a) *Wholesale Markets*: A wholesale market is one in which commodities are bought and sold in large lots or in bulk. These markets are generally located in either towns or cities. The economic activities in and around these markets are so intense that over time the population tends to get concentrated around these markets. These markets occupy an extremely important link in the marketing chain of all the commodities including farm products. Apart from balancing the supply and demand and discovery of the prices of a commodity, these markets and functionaries in them serve as a link between the production system and consumption system. The wholesale markets for farm products in India can be classified as primary, secondary and terminal wholesale markets. The primary wholesale markets are in the nature of assembling centres located in and around producing regions. The transactions in primary wholesale markets take place mainly between farmers and traders. Secondary wholesale markets are generally located between primary wholesale and terminal markets. The transactions in these markets take place between primary wholesalers and traders of terminal market. The terminal markets are generally located at the large urban metropolitan cities or export centres catering to the large consuming population around them or in the overseas markets.

(b) *Retail Markets*: A retail market is one in which commodities are bought by and sold to the consumers as per their requirements. Transactions in these markets take place between retailers and consumers. The retailers purchase the goods from wholesale market and sell in small lots to the consumers in retail markets. These markets are very near to the consumers.

The distinction between the wholesale and retail market can be made mainly on the basis of buyer. A retail market means that the buyers are generally ultimate consumers, whereas in the wholesale market the buyers can be wholesalers or retailers. But sometimes-bulk consumers also purchase from the wholesale markets. The quantity transacted in retail markets is generally smaller than that in the wholesale markets.

5. On the Basis of Nature of Transactions

The markets which are based on the types of transactions in which people are engaged are of two types:

(a) *Spot or Cash Markets*: A market in which goods are exchanged for money immediately after the sale is called the spot or cash market.

(b) *Forward Markets*: A market in which the purchase and sale of a commodity takes place at time t but the exchange of the commodity takes place on some specified date in future, i.e. time $t + 1$. Sometimes even on the specified date in the future ($t + 1$), there may not be any exchange of the commodity. Instead, the differences in the purchase and sale prices are paid or taken.

6. On the Basis of Number of Commodities in which Transaction Takes Place

A market may be general or specialized on the basis of the number of commodities in which transactions are completed:

(a) *General Markets*: A market in which all types of commodities, such as foodgrains, oilseeds, fibre crops, gur, etc., are bought and sold is known as general market. These markets deal in a large number of commodities.

(b) *Specialized Markets*: A market in which transactions take place only in one or two commodities is known as a specialized market. For every group of commodities, separate markets exist. The examples of specialized markets are foodgrain markets, vegetable markets, wool market and cotton market.

7. On the Basis of Degree of Competition

Each market can be placed on a continuous scale, starting from a perfectly competitive point to a pure monopoly or monopsony situation. Extreme forms are almost non-existent. Nevertheless, it is useful to know their characteristics. In addition to these two extremes, various midpoints of this continuum have been identified. On the basis of competition, markets may be classified into the following categories:

(a) *Perfect Markets*: A perfect market is one in which the following conditions hold good:
  (i) There is a large number of buyers and sellers;
  (ii) All the buyers and sellers in the market have perfect knowledge of demand, supply and prices;
  (iii) Prices at any one time are uniform over a geographical area, plus or minus the cost of getting supplies from surplus to deficit areas;
  (iv) The prices are uniform at any one place over periods of time, plus or minus the cost of storage from one period to another;
  (v) The prices of different forms of a product are uniform, plus or minus the cost of converting the product from one form to another.

(b) *Imperfect Markets*: The markets in which the conditions of perfect competition are lacking are characterized as imperfect markets. The following situations, each based on the degree of imperfection, may be identified:
  (i) *Monopoly Market*: Monopoly is a market situation in which there is only one seller of a commodity. The seller exercises sole control over the quantity or price of the commodity. In this market, the price of a commodity is generally higher than in other markets. Indian farmers operate in monopoly market when purchasing electricity for irrigation. When there is only one buyer of a product, the market is termed as a monopsony market. The sugarcane farmers in the catchment area of a sugar factory, generally face a monopsony market situation.
  (ii) *Duopoly Market*: A duopoly market is one which has only two sellers of a commodity. They may mutually agree to charge a common price which is higher than the hypothetical price in a common market. The market situation in which there are only two buyers of a commodity is known as the duopsony market.
  (iii) *Oligopoly Market*: A market in which there are more than two but still a few sellers of a commodity is termed as an oligopoly market. A market having a few (more than two) buyers is known as oligopsony market.
  (iv) *Monopolistic Competitive Market*: When a large number of sellers deal in heterogeneous and differentiated form of a commodity, the situation is called monopolistic competition. The difference is made conspicuous by different trade marks on the product. Different prices prevail for the same basic product. Examples of monopolistic competition faced by farmers may be drawn from the input markets. For example, they have to chose between various makes of insecticides, pumpsets, fertilizers and equipments. Market for diesel and petrol in India is a case of this type of market.

8. On the Basis of Nature of Commodities

On the basis of the type of goods dealt in, market may be classified into the following categories:

(a) *Commodity Markets*: A market which deals in goods and raw materials, such as wheat, barley, cotton, fertilizer, seed, etc., are termed as commodity markets. Commodity markets are the markets where a particular commodity is transacted in large quantity and all necessary facilities exist in that market for efficient sale of that commodity. In India, there are some markets which are famous for marketing of a particular commodity. Some of these are jute market at Kolkata, cotton market at Mumbai, wheat market at Hapur and Khanna, cumin market at Unzha (Gujarat), sugar market at Kolkata, Kanpur, Meerut and Muzzaffarnagar, and hides and skin market at Kanpur and Chennai.

(b) *Capital Markets*: The market in which bonds, shares and securities are bought and sold are called capital markets; for example, money markets and share markets.

9. On the Basis of Stage of Marketing

On the basis of the stage of marketing, markets may be classified into two categories:

(a) *Producing Markets*: Those markets which mainly assemble the commodity for further distribution to other markets are termed as producing markets. Such markets are located in producing areas.

(b) *Consuming Markets*: Markets which collect the produce for final disposal to the consuming population are called consumer markets. Such markets are generally located in areas where production is inadequate, or in thickly populated urban centres. The urban areas, including cities, are consuming markets for agricultural commodities.

10. On the Basis of Extent of Public Intervention

Based on the extent of public intervention, markets may be placed in any one of the following two classes:

(a) *Regulated Markets*: These are those markets in which business is done in accordance with the rules and regulations framed by the statutory market organization representing different sections involved in markets. The marketing costs in such markets are standardized and, marketing practices are regulated.

(b) *Unregulated or Informally Regulated Markets*: These are the markets in which business is conducted without any set rules and regulations. Traders frame the rules for the conduct of the business and run the market. These markets suffer from many ills, ranging from unstandardised charges for marketing functions to imperfections in the determination of prices. However, the situation is changing in these markets. The association of traders or other functionaries are prescribing rules of the game, which are binding on the traders and sellers. There are several cases where functioning of such markets is better than formally regulated markets.

11. On the Basis of Type of Population Served

On the basis of population served by a market, it can be classified as either urban or rural market:

(a) *Urban Market*: A market which serves mainly the population residing in an urban area is called an urban market. The nature and quantum of demand for agricultural products arising from the urban population is characterised as urban market for farm products.

(b) *Rural Market*: The word rural market usually refers to the demand originating from the rural population. There is considerable difference in the nature of embedded services required with a farm product between urban and rural demands.

Rural markets generally have poor marketing facilities as compared to urban markets. According to the survey of the Directorate of Marketing and Inspection (DMI) of Government of India[7], in mid-1980s only 46% of rural primary markets of the country had the facility of market yards; 6.4% had office buildings, 3.2% had cattle shed, 3% had canteen, 4.9% had storage facilities, 5.1% had auction platforms, 12.9% had drinking water facility and 5.2% markets had electricity facility. Marketing support services such as godowns, cleaning, price information and extension services were found completely non-existent in most of these rural markets.

However, the situation improved considerably in the later period. According to DMI (1999)[8], out of the total primary regulated markets, all have office buildings, 64 to 67% have auction platforms, 74% have storage godowns, 85% have weighing equipments, 61% have farmers' rest houses, and more than 84% have internal roads, boundary walls and public telephone systems. Several of the market yards now have banks, post offices, agri-input shops and drainage systems. Neverthless, as reported by Acharya[9], the amenities in several market yards are considerably less than the requirements for making these yards farmers' friendly.

12. On the Basis of Market Functionaries and Accrual of Marketing Margins

Markets can also be classified on the basis of as to who are the market functionaries and to whom the marketing margins accrue. Over the years, there has been a considerable increase in the producers or consumers cooperatives or other organizations handling marketing of various products. Though private trade still handles bulk of the trade in farm products, the cooperative marketing has increased its share in the trade of some agricultural commodities like milk, fertilizers, sugarcane and sugar. In the case of marketing activities undertaken by producers or consumers cooperatives, the marketing margins are either negligible or shared amongst their members. In some cases, farmers themselve work as sellers of their produce to the consumers. On this basis, the market can be (a) farmers markets, (b) cooperative markets or (c) general markets.

It must be noted that each market or market place can be classified on the basis of the 12 criteria mentioned above. A 12-dimensional classification of markets is shown in Chart 2.1.

GROWTH OF MARKETS

Following the economic development of society, there is a tendency among the markets to grow. This tendency is termed as market development or growth. The growth of a market may be natural or induced. Market development takes place, both qualitatively and quantitatively. Two of the important dimensions of the growth of a market are functional and geographical.

Functional Growth

Initially, a market is generally a multi-commodity market characterized by a large number of traded commodities, each with a low volume of business. The sale and purchase take place in the physical presence and handling of the produce. Such a

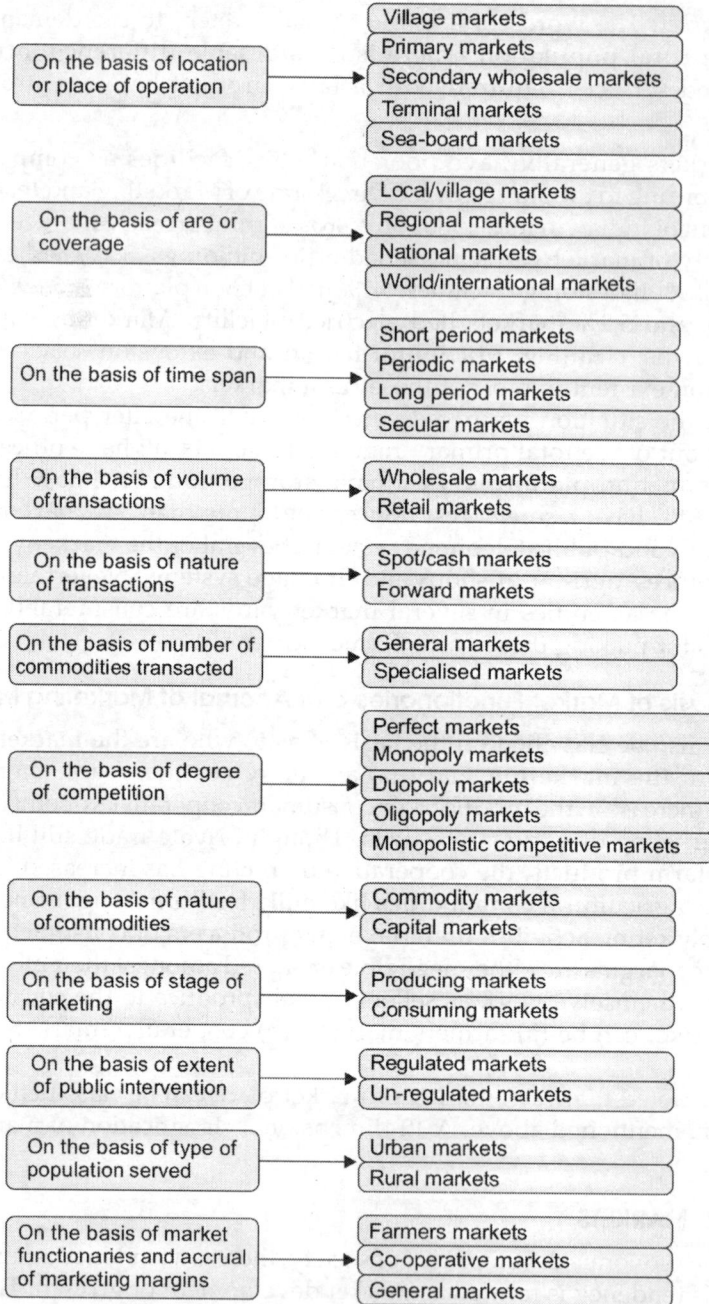

On the basis of location or place of operation	Village markets
	Primary markets
	Secondary wholesale markets
	Terminal markets
	Sea-board markets

On the basis of are or coverage	Local/village markets
	Regional markets
	National markets
	World/international markets

On the basis of time span	Short period markets
	Periodic markets
	Long period markets
	Secular markets

| On the basis of volume of transactions | Wholesale markets |
| | Retail markets |

| On the basis of nature of transactions | Spot/cash markets |
| | Forward markets |

| On the basis of number of commodities transacted | General markets |
| | Specialised markets |

On the basis of degree of competition	Perfect markets
	Monopoly markets
	Duopoly markets
	Oligopoly markets
	Monopolistic competitive markets

| On the basis of nature of commodities | Commodity markets |
| | Capital markets |

| On the basis of stage of marketing | Producing markets |
| | Consuming markets |

| On the basis of extent of public intervention | Regulated markets |
| | Un-regulated markets |

| On the basis of type of population served | Urban markets |
| | Rural markets |

On the basis of market functionaries and accrual of marketing margins	Farmers markets
	Co-operative markets
	General markets

Chart 2.1: 12-Dimensional Classification of Markets

market is called a general market. Over time, there is a tendency among buyers and sellers to specialize in some commodities, though the volume of business may still be low. Markets thus reveal the tendency of becoming specialised markets. Specialization leads to increase in the volume of business; and deals take place through inspection of samples rather than complete lots. The next stage in the functional growth of a market is one when grades of commodities are standardized, and deals are struck in the name of the specific grades of a commodity. In this way, at least four stages in the functional growth of a market have been identified. These are:

(a) First stage: General markets
(b) Second stage: Specialized markets
(c) Third stage: Dealings with samples
(d) Fourth stage: Dealings with grades

Spatial or Geographical Growth

Markets also grow in terms of their area. Initially, the buyers and sellers in a limited area interact to buy and sell a commodity without linkages with outsiders. Such markets are termed as local markets. An increase in the quantity of the commodities to be handled, communication and transportation facilities, and better exchange of information lead to the spatial growth of a market from local to regional, national and international. This dimension of market growth has often been termed as the spatial or geographical development of a market.

Factors Affecting Rate of Market Development/Growth

Markets for some commodities and in some countries have developed at a faster rate than for others. Some of the factors affecting the rate of market development are given here.

1. *Nature of Demand*: Markets for goods which have a relatively regular demand develop faster than the markets for goods for which the demand is seasonal.
2. *Nature of Products*: Markets for durable goods develop at a faster rate than for perishable commodities.
3. *Transportation and Communication Facilities*: The markets in areas well connected by national highways, state highways, railways and post and telecommunication facilities develop faster than in other areas. There is a lot of variation in the availability of these facilities from area to area; and hence the variation in the development of markets.
4. *Quantum of Supply and Demand:* Markets in high producing or consuming areas develop at a faster rate than in other areas.
5. *Public Policies*: Public policy is a very important determinant of the rate of market development. Public support for the development of markets has led to a faster development of markets in some countries than in others.
6. *Banking Facilities*: In areas where banking infrastructure is well developed, markets develop at a faster rate than in areas where such infrastructure is poorly developed.
7. *Peace and Security*: In countries where political and social disturbances are infrequent, business runs smoothly and the markets develop at a faster rate than in disturbed areas. Political stability is very essential before business can flourish; and this leads to the expansion and development of markets.
8. *Economic Growth*: The rate of economic growth is the most important factor which affects the development of market.[10] While the growth of agricultural sector has a multiplier effect on the growth of the economy, via expansion in trade and services required to handle the agricultural surpluses and supply of essential farm inputs, the development of markets play an important role in triggering the growth process. Thus, the rate of economic growth not only affects the market development but is also conditioned by it.

MARKET STRUCTURE

MEANING

The term *structure* refers to something that has organization and dimension—shape, size and design; and which is evolved for the purpose of performing a function. A function modifies the structure, and the nature of the existing structure limits the performance of functions.

By the term *market structure* we refer to the size and design of the market. It also includes the manner of the operation of the market. Some of the expressions describing the market structure are:

1. Market structure refers to those organizational characteristics of a market which influence the nature of competition and pricing, and affect the conduct of business firms,[11]
2. Market structure refers to those characteristics of the market which affect the traders' behaviour and their performances.[12]
3. Market structure is the formal organization of the functional activity of a marketing institution.[13]

An understanding and knowledge of the market structure is essential for identifying the imperfections in the performance of a market.

COMPONENTS OF MARKET STRUCTURE

The components of the market structure, which together determine the conduct and performance of the market, are:

1. Concentration of Market Power

The concentration of market power is an important element determining the nature of competition and consequently of market conduct and performance. This is measured by the number and size of firms existing in the market. The extent of concentration represents the control of an individual firm or a group of firms over the buying and selling of the produce. A high degree of market concentration restricts the movement of goods between buyers and sellers at fair and competitive prices, and creates an oligopoly or oligopsony situation in the market.

2. Degree of Product Differentiation

Whether or not the products are homogeneous affects the market structure. If products are homogeneous, the price variations in the market will not be wide. When products are heterogeneous, firms have the tendency to charge different prices for their products. Everyone tries to prove that his product is superior to the products of others.

3. Conditions for Entry of Firms in the Market

Another dimension of the market structure is the restriction, if any, on the entry of firms in the market. Sometimes, a few big firms do not allow new firms to enter the market or make their entry difficult by their dominance in the market. There may also be some government restrictions on the entry of firms.

4. Flow of Market Information

A well-organized market intelligence information system helps all the buyers and sellers to freely interact with one another in arriving at prices and striking deals.

5. Degree of Integration

The behaviour of an integrated market will be different from that of a market where there is no or less integration either among the firms or of their activities. Firms plan their strategies in respect of the methods to be employed in determining prices, increasing sales, co-ordinating with competing firms and adopting predatory practices against rivals or potential entrants. The structural characteristics of the market govern the behaviour of the firms in planning strategies for their selling and buying operations.

DYNAMICS OF MARKET STRUCTURE—CONDUCT AND PERFORMANCE

The market structure determines the market conduct and performance. The term *market conduct* refers to the patterns of behaviour of firms, specially in relation to pricing and their practices in adapting and adjusting to the market in which they function. Specifically, market conduct includes:

(a) Market sharing and price setting policies;
(b) Policies aimed at coercing rivals; and
(c) Policies towards specification of the quality of products.

The term *market performance* refers to the economic results that flow from the industry as each firm pursues its particular line of conduct[14]. Society has to decide the criteria for satisfactory market performance. Some of the criteria for measuring market performance and of the efficiency of the market structure are:

1. Efficiency in the use of resources, including real cost of performing various marketing functions;
2. The existence of monopoly or monopoly profits, including the relationship of margins with the average cost of performing various functions;
3. Dynamic progressiveness of the system in adjusting the size and number of firms in relation to the volume of business, in adopting technological innovations and in finding and/or inventing new forms of products so as to maximize general social welfare.
4. Whether or not the system aggravates the problem of inequalities in inter-personal, inter-regional, or inter-group incomes. For example, inequalities increase under the following situations:
   (a) A market intermediary may pocket a return greater than its real contribution to the national product;
   (b) Small farmers are discriminated against when they are offered a lower return because of the low quantum of surplus;
   (c) Inter-product price parity is substantially disturbed by new uses for some products and wide variations and rigidities in the production pattern between regions.

The market structure, therefore, has always to keep on adjusting to changing environment if it has to satisfy the social goals. A static market structure soon becomes obsolete because of the changes in the physical, economic, institutional and technological factors. For a satisfactory market performance, the market structure should keep pace with the following changes:

1. *Production Pattern:* Significant changes occur in the production pattern because of technological, economic and institutional factors. The market structure should be re-oriented to keep pace with such changes.

2. *Demand Pattern:* The demand for various products, specially in terms of form and quality, keeps on changing because of change in incomes, the pattern of distribution among consumers, and changes in their tastes and habits. The market structure should be re-oriented to keep it in harmony with the changes in demand.

3. *Costs and Patterns of Marketing Functions:* Marketing functions such as transportation, storage, financing and dissemination of market information, have a great bearing on the type of market structure. Government policies with regard to purchases, sales and subsidies affect the performance of market functions. The market structure should keep on adjusting to the changes in costs and government policy.

4. *Technological Change in Industry:* Technological changes necessitate changes in the market structure through adjustments in the scale of business, the number of firms, and in their financial requirements.

MARKET FORCES

The key function of a market is to determine the price of the "lot" at which the product should change hands. This process goes on continually at all times and all the places between buyers and sellers. The forces which affect the process of price determination, either directly or indirectly, may be termed as market forces. These forces may be tangible, like the quantity of arrivals at a particular point of time in the market, or heavy rainfall; or they may be intangible, like the announcement of a particular government policy. All these forces affect price determination by affecting either the demand behaviour of buyers or the supply behaviour of sellers.

DEMAND

Meaning

The word *demand* usually refers to the quantity of a product or service which the buyers are likely to purchase at different prices in a given market at a given time. It must be understood that demand represents the willingness and ability to buy under specified conditions. Even if no actual transaction takes place, the demand for a product may exist.

The law of demand formalizes the relationship between the quantities purchased and their prices. The law states that the price and quantity demanded are inversely related, other things remaining the same. The usual law of demand may be depicted either through a demand schedule or a demand curve on a two dimensional space. Table 2.1 shows the hypothetical demand schedule of milk at a given time and place.

The graphic presentation of the demand schedule gives a curve, which is known as the demand curve. The demand curve slopes downward from left to right, as is shown in Fig. 2.1. The demand schedule as well as the demand curve illustrate the nature of the relationship between quantity and price. Thus the word, *demand* connotes the whole schedule of price-quantity demanded pairs. The following concepts of demand should be understood thoroughly:

(a) *Effective Demand*: Effective demand is the desire of the consumer for the commodity backed up by his purchasing power. Therefore, the pertinent question in marketing is: "how much will be bought at a price?" And not: "How much will be needed or desired?" Sometimes, a distinction is made between demand and desire by defining demand as the desire which is backed by purchasing power or the capacity to purchase.

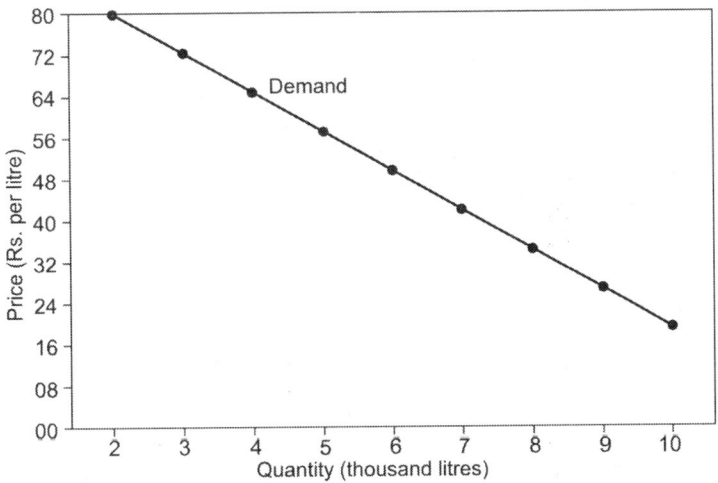

Table 2.1: Hypothetical market demand schedule for Milk in 'A' market	
Price of milk (₹ per litre)	**Quantity demanded (thousand litres)**
16	10
24	9
32	8
40	7
48	6
56	5
64	4
72	3
80	2

Fig. 2.1: Hypothetical demand curve for milk

(b) *Derived Demand*: The demand for some commodities exists only because they are used in producing other commodities which satisfy human wants. The demand for such commodities is termed as a *derived demand*. For example, farmyard manure and fertilizers are demanded not because they satisfy any human want directly but because these are used in producing goods—crops which are directly demanded by consumers. In a strict sense, therefore, the demand for farm inputs is a derived demand, for it is derived from the demands for goods produced with their help.

(c) *Reservation Demand and Price*: The term *reservation demand* refers to the quantity of a product a seller would like to retain (rather than sell) at a given price. At each price, the seller himself has a demand to keep a certain quantity with himself for later sale. He may not like to sell a particular "lot" if the price offered for it is lower than some preconceived price, which is known as the *reservation price*.

At any point in time, a schedule showing the quantity to be retained or, for that matter, the quantity to be sold at each price may be developed for each stockist. Such a schedule may be termed as *reservation demand and price schedule*.

(d) *Demand Function*: The demand for a commodity is not affected by price alone, whether one thinks of an individual consumer or a group of consumers. Factors such as income, tastes, habits, weather, the prices of substitutes and incentives for savings affect demand in the sense that they shift the whole range of price-quantity relationship. A demand schedule is usually expressed as

$$Q = f(P_q)$$

where Q = Quantity demanded; P_q = Price per unit of Q.

The demand function for the same commodity is expressed as

$$Q = f(P_q, P_s, P, Y, W, ...)$$

where P_S = Price of substitute or complementary goods; P = Population or family size; Y = Income; W = Weather

and the dots show that some more factors may be identified, depending upon the purpose of analysis, the group of consumers and the commodity that is considered.

Factors Affecting Demand for Farm Products

Farm products are demanded by consumers to satisfy their nutritional wants, and by cattle raisers to feed their livestock, by traders for sale and by manufacturers for converting them into processed foods or other goods.

At the macro or national level, aggregate demand for farm products is determined by the size of the population and per capita income. Changes in tastes, processing technology and income distribution affect the aggregate, as well as the composition of the demand for farm products.

The effect of an increase in population and per capita incomes on the demand for foodgrains may be expressed in simple terms as

$$D = P + eY$$

where D = Rate of increase in aggregate demand for foodgrains; P = Rate of growth in population; e = Income elasticity of demand for foodgrains; Y = Rate of growth in per capita income.

In other words, if the population of a country increases at a rate of 2.2% per year and if the per capita income increases at a rate of 3% per year, assuming an income elasticity of demand as 0.6, the demand for foodgrains in the country would grow at a rate of 4.0% [2.2 + (0.6) 3.0] per year.

The estimates of the rates of growth of population and per capita incomes are readily available. The coefficient of income elasticity of demand is very important in determining the pressure of demand. The coefficient varies from commodity to commodity, from area to area, and between income levels.

Estimates have been made from time to time to study the relationship between consumer income and consumption of various farm products. Most of the estimates are based on the cross-sectional data obtained from the National Sample Surveys (NSS) and show the relationship between total consumer expenditure and expenditure on individual items rather than that between income and purchased quantity. These are, therefore, sometimes expressed as expenditure elasticities of demand. For illustration the estimates of income elasticities of demand for foodgrains and edible oils as worked out by Murty[15] from the data for seventies are shown in Table 2.2. The estimates for a large number of commodities generated by Bhalla[16] from the data for the eighties are shown in Table 2.3. For policy formulation, one should always look at the estimates based on the latest data. A comparison of data in Tables 2.2 and 2.3 shows that there

Table 2.2: Income and direct price elasticities of demand for selected foodgrains and edible oils in India (1983)

Commodity groups	Income elasticity		Direct price elasticity	
	Rural	Urban	Rural	Urban
Wheat and rice	0.81	0.48	−0.69	−0.66
Sorghum	0.40	0.53	−0.68	−1.95
Bajra	0.43	0.33	−0.74	−1.29
Gram	0.47	0.25	−0.81	−1.01
Edible oils	1.03	1.01	−0.46	−0.57

All pertain to middle expenditure class (34–55).

Source: Murty, K.N., Consumption and Nutritional Patterns of ICRISAT Mandate Crops in India, ICRISAT, Economics Programme, Report No. 53, August 1983, pp. 10–11.

Table 2.3: Income elasticities of demand in India (1995)

Commodity	Rural population	Urban population
Cereals	0.33	0.18
Pulses	0.62	0.47
Milk	1.47	1.01
Edible oils	0.77	0.63
Egg/meat	1.04	0.75
Vegetables	0.60	0.60
Fruits	1.33	1.16
Sugar	0.95	0.55
Total food	0.67	0.60
Clothing	1.81	1.62

Source: Bhalla, G.S. (1995). Globalisation and Agricultural Policy in India. Indian Journal of Agricultural Economics, Vol. 50, No. 1, Jan–March, 1995, pp. 1–26.

has occurred a considerable change in the income elasticity of demand for farm products in India.

The income elasticity of demand differs between rural and urban areas. In the case of cereals, the income elasticity of demand in rural areas (as per Bhalla) is 0.33 and that in urban areas is 0.18. Another important point to be noted is that the income elasticity of demand for all the agricultural commodities of urban population is generally lower than that of rural population.

The estimates of price elasticities of demand for a later period generated by Kumar and Mathur[17] are shown in Table 2.4. The price elasticities are obviously negative for all the commodities included in the table. An important point that needs to be noted is that the direct price elasticities are generally lower for higher income classes than that for lower income classes meaning

The latest available estimates of consumer demand elasticities for India (Table 2.5) are quite revealing. The own price elasticities are much higher than income elasticities. For major staple foodgrains (rice and wheat), own price elasticities are more than one. Cross price elasticities are positive for many pairs indicating substitution across cereals.

Table 2.4: Price elasticities of demand for foodgrains and edible oils in India (1996)

Expenditure groups	% in total population	Rice	Wheat	Coarse cereals	Pulses	Oilseeds
RURAL						
Very poor	23	−0.472	−0.400	−0.389	−0.775	−0.832
Moderately poor	20	−0.360	−0.317	−0.308	−0.686	−0.740
Non-poor low	29	−0.245	−0.227	−0.214	−0.545	−0.600
Non-poor high	28	−0.133	−0.140	−0.111	−0.335	−0.386
All groups	100	−0.282	−0.242	−0.286	−0.524	−0.567
URBAN						
Very poor	12	−0.464	−0.319	−0.451	−0.784	−0.798
Moderately poor	13	−0.402	−0.312	−0.392	−0.738	−0.757
Non-poor low	26	−0.302	−0.216	−0.281	−0.597	−0.622
Non-poor high	49	−0.205	−0.143	−0.166	−0.406	−0.422
All groups	100	−0.288	−0.217	−0.309	−0.516	−0.522

Source: Kumar, P. and V.C. Mathur, Agriculture in Future: Demand-Supply Perspective for the Ninth Five Year Plan, Economic and Political Weekly, Vol. 31 (39), Sept. 28, 1996, pp. A–137.

Table 2.5: Income elasticity, direct price elasticity and cross price elasticity of demand for cereals in India (2014 estimates)

Elasticity with respect to	Rice	Wheat	Maize	Other coarse cereals
Income	−0.03	−0.05	−0.06	−0.04
Price				
Rice	−1.30	0.45	0.04	0.0
Wheat	0.57	−1.81	−0.09	−0.03
Maize	2.22	−3.85	−0.45	0.26
Other coarse cereals	−0.09	−0.20	0.07	−0.77

Source: Kumar, P. and P.K. Joshi, Input Subsidy vs. Farm Technology—Which is More Important for Agricultural Development, Agricultural Economics Research Review, Vol. 27 (1), June 2014, p. 1–18.

The income elasticity of demand for both major staple cereals viz. wheat and rice has turned negative. The direct price elasticities for both rice and wheat are obviously negative, but very high (−1.81 for wheat and −1.30 for rice). This is mainly because wheat and rice have become good substitutes. Their cross elasticities are +ve and quite high. This means that a rise in price of any one of these increases the demand for the other grain. A 10% increase in price of wheat increases the demand for rice by 5.7%. Similarly, ten% increase in price of rice increases the demand for wheat by 4.5%. This reflects a distinct change in the pattern of staple food consumption in the country.

The latest estimates of income (expenditure) elasticities of demand, separately for rural and urban India, for six important food groups, used by the Working Group for Projections of Demand and Supply in India, are shown in Table 2.6. Note that for cereals, as a group, the income elasticity of demand is negative, though a small figure (−0.13 for rural and −0.04 for urban areas). It implies that for an average income household, the increase in income may not lead to increased expenditure on cereals.

Table 2.6: India: Expenditure elasticities of demand for food products (2018)

Food groups	Rural	Urban
Cereals	−0.13	−0.04
Pulses	0.55	0.36
Edible oils	0.88	0.37
Fruits and vegetables	0.85	0.42
Milk	0.82	0.40
Meat	0.82	0.40

Source: Demand and Supply Projections Towards 2033, Working Group Report, NITI Aayog, GoI, 2018.

For all other food groups, the income elasticities are positive. The other point to be noted is that for other food groups, the income elasticities for rural consumers are substantially higher than that for urban consumers.

As mentioned earlier, the income and price elasticities of demand vary among income groups of consumers. The latest estimates of elasticities of demand by Kumar (2019) for major food items in India are shown in Table 2.7. The main observations are (a) with 10% increase in income of poor families, the demand for rice increases by 1.5%; (b) with increase in income of all income households, the demand for nutri cereals decreases; (c) own price elasticities are negative for all food commodities for all income households; and (d) except for nutri cereals and rice for higher income groups, the income (expenditure) elasticities of demand for all food commodities are positive.

The demand for individual farm products is also affected by the availability of substitutes. Foodgrains are the basic items of necessity, and there are no substitutes

Table 2.7: India: Expenditure and own price elasticities of demand for food by Income groups (2019)

Food commodity	Expenditure elasticity				Own price elasticity			
	Poor	Middle income	High income	All h.h.	Poor	Middle income	High income	All h.h.
Rice	0.15	0.03	−0.02	0.03	−0.47	−0.31	−0.20	−0.29
Wheat	0.10	0.07	0.08	0.08	−0.51	−0.40	−0.30	−0.38
Nutri cereals	−0.17	−0.18	−0.10	−0.15	−0.38	−0.24	−0.14	−0.25
Pulses	0.50	0.27	0.10	0.21	−0.70	−0.53	−0.35	−0.46
Edible oils	0.63	0.35	0.14	0.26	−0.75	−0.58	−0.39	−0.49
Sugar	0.27	0.10	0.00	0.06	−0.59	−0.41	−0.25	−0.34
Vegetables	0.57	0.33	0.14	0.26	−0.73	−0.57	−0.40	−0.50
Fruits	0.70	0.50	0.28	0.37	−0.80	−0.69	−0.56	−0.62
Spices and beverages	1.13	0.92	0.62	0.74	−0.96	−0.94	−0.91	−0.92
Milk	0.78	0.54	0.28	0.38	−0.83	−0.71	−0.53	−0.60
Meat, fish, eggs	1.01	0.80	0.52	0.65	−0.91	−0.87	−0.80	−0.84

Figures rounded to two decimal points.

Source: Kumar P., Food and Nutrition Security-Empowering the Farmers and Consumers to fight Hunger and Poverty, Presidential Address, IJAM, Vol. 33 (3), 2019.

for them as a group. However, within the group, one commodity may be substituted for another.

Consumers' tastes affect the demand for individual food products. Their tastes are related to the area, weather, religion, occupation, social status, urbanisation and levels of development. Changes in the proportion of the population in the various groups in the country may be expected to affect the overall demand for various food products over time.

SUPPLY

Meaning

The term *supply* refers to schedule or quantities of a product that will be offered for sale at different prices at a given time and in a given market. There is a logical relationship between supply and price. The higher the price, the larger the quantity that is offered for sale, and vice versa. Thus, supply indicates a relationship between the quantity and price of a commodity from the seller's viewpoint. A hypothetical supply schedule for milk is given in Table 2.8.

Table 2.8: Hypothetical market supply schedule for Milk in 'A' market

Price (₹ per litre)	Quantity offered for sale (thousand litres)
80	10
72	9
64	8
56	7
48	6
40	5
32	4
24	3
16	2

The graphical presentation of the supply schedule is known as the supply curve. The supply curve is positively sloped; and it always slopes upward left to right on a graph, as shown in Fig. 2.2.

Time is a very important consideration in supply analysis. The given supply exists at a given time. The time periods can be of different lengths.

(a) *Short Run:* This means that the existing production is already on hand and that the cost incurred on its production does not influence its price. The response of the quantity offered for sale to a change in price is very low.

(b) *Intermediate Run:* This refers to the time during which goods can be produced only with the existing production facilities. The existing capacity puts an upper limit on the quantity that can be offered for sale:

(c) *Long Run:* The term long run refers to the time during which production facilities themselves may be expanded or contracted. In *the long run, supply* is more responsive to prices and other incentives. The time element greatly complicates the process of the analysis of agricultural supply.

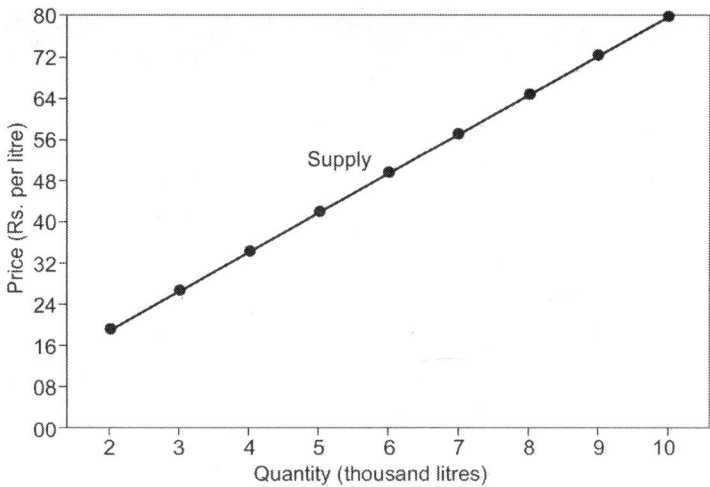

Fig. 2.2: Hypothetical supply curve for milk

Factors Affecting Supply of Farm Products

There are two *main sources* of supply *of farm* products at the national level: one is the production on the farms and the other is import from other countries.

The factors which affect the domestic production of farm products are weather, technology, irrigation facilities, land suitable for cultivation, acreage under various crops, availability of inputs and relative inter-crop and input-output prices. The effect of prices and other factors on supply or production through the acreage may be studied by using the supply response models suggested by Nerlove. The basic supply response models have been presented in Appendix 2.1.

When the objective of the study is short-run forecasting, single one-stage econometric procedure, such as the Nerlovian model can be employed to directly estimate the functions using market level time series data. However, for long term forecasting, single equation approaches may yield misleading projections.

When crop-wise projections are to be made, it is advisable to use comprehensive approach involving simultaneity between output supply and factor demand functions. Using the comprehensive approach, own price and fertilizer price elasticities for supply/production of rice, wheat, pulses, oilseeds and sugarcane are shown in Table 2.9. The latest estimates of Kumar and Joshi (2014) show that supply response elasticity wrt own price for rice is 0.2742 and for wheat is 0.2667.

Domestic supply of a product can be either augmented by imports or restricted by allowing exports. The use of such instruments to affect the domestic supply depends on: (a) the difference between the domestic price and the international price after allowing for the shipment costs; (b) the objective of the national policy in terms of protection of the domestic producers and safeguarding the interests of the consumers in the immediate future; (c) the ability of the country to resort to imports (i.e. the availability of the foreign exchange); and (d) the need of the country to earn foreign exchange by resorting to exports. The relative importance of these factors varies from country to country and even within the same country, from time to time.

Apart from the current production and net imports/exports, the carryover stocks determine the quantity which is actually available to the consumers during a given year. For example, the net availability of cereals in a country during any given year is

Table 2.9: Supply response elasticities in India		
Crop	Own price	Fertilizer price
Rice	0.2357	0.0001
Wheat	0.2164	−0.0095
Nutri grains	0.5333	0.2791
Pulses	0.1695	−0.0013
Edible oilseeds	0.5079	0.0062
Sugarcane	0.1216	0.0045

Source: Praduman Kumar, P. Sinoj, S.S. Raju, Anjani Kumar, Karl M. Rich and Siwa Msangi, "Factor Demand, Output Supply Elasticities and Supply Projections for Major Crops in India", Agricultural Economics Research Review, Vol. 23 (1), January–June 2010, P.1–14, Kumar P. (2019).

equal to the carryover stocks plus current year's net production plus imports minus exports minus balance at the end of the year. Net availability, defined in this way, is the quantity which is actually available in the domestic market during a given period. This has been illustrated by using a case of cereals in India in Table 2.10. The net availability of pulses and total foodgrains along with per capita availability have also been shown.

SIMPLE MARKET MODEL AND PRICE DETERMINATION

The simplest of the simple market models is one where it is assumed that the quantity demanded and quantity supplied are affected only by the price of the commodity. The price and quantity which satisfy both the buyer(s) and seller(s) are called the equilibrium price and equilibrium quantity. Price determination in the case of this simplified model is illustrated below by three alternative approaches.

TABULAR APPROACH

The hypothetical demand and supply schedules given, respectively in Tables 2.1 and 2.8 have been reproduced in Table 2.11. It is evident from the table that at the price of ₹ 48 per litre, the quantity demanded by consumers (6,000 litres) is equal to the quantity offered for sale by the sellers. The equilibrium price, therefore, is ₹ 48 per litre and the equilibrium quantity is 6,000 litres. At a price below ₹ 48, the quantity demanded exceeds the quantity supplied; there is thus a pressure on the price to rise. At a price higher than ₹ 48, the quantity supplied exceeds the quantity demanded, there is thus a downward pressure on price.

GRAPHICAL APPROACH

Graphs of demand and supply schedules drawn in Figs 2.1 and 2.2 have been reproduced in Fig. 2.3. The intersection of the demand and supply curves indicates the equilibrium price and quantity. Since the demand and supply curves represent the same demand and supply schedules as were used in the tabular analysis, the equilibrium price and equilibrium quantity obtained by the graphical analysis are the same as those obtained from the tabular analysis, i.e. ₹ 48 per litre and 6,000 litres, respectively.

Table 2.10: Production, imports and net availability of foodgrains in India

Year	Population (million persons)	Cereals (million tonnes)				Pulses net availability (million tonnes)	Foodgrains net availability (million tonnes)	Per capita net availability of foodgrains (gms per day)
		Net production	Net imports	Change in government stocks	Net availability			
1951	363.2	40.10	4.10	+0.6	43.6	8.00	52.30	394.9
1956	397.3	50.43	1.39	−0.6	52.42	10.23	62.65	430.9
1961	442.4	60.89	3.49	−0.17	64.55	11.14	75.69	468.7
1966	493.2	54.6	10.3	+0.1	64.8	8.7	73.5	408.1
1971	551.3	84.53	2.03	+2.57	83.99	10.32	94.31	468.7
1976	617.2	94.50	0.66	+10.74	84.42	11.41	95.83	424.3
1981	688.5	104.09	0.52	−0.24	104.85	9.44	114.29	454.8
1986	766.1	119.94	(−)0.06	−1.58	121.46	12.30	133.76	478.1
1991	851.7	141.90	(−)0.60	(−)4.40	145.70	12.90	158.60	510.1
1996	939.5	147.1	(−)3.5	(−)8.5	152.0	11.3	163.4	476.2
2001	1033.2	162.5	(−)4.5	(+)12.3	145.6	11.3	156.9	416.2
2006	1119.8	170.8	(−)3.8	(−)1.8	168.8	13.30	182.1	445.3
2011	1201.9	198.0	(−)9.6	(+)8.3	180.1	18.9	199.0	453.6
2016	1309.1*	205.8	(−)8.7	(−)9.2	206.3	20.3	226.6	474.2

Note:

1. Population figures relate to mid-year.
2. Net production is 87.5% of total production, remaining 12.5% being provided for seed, feed and wastage.
* UNO projections

Source: Government of India, Economic Survey 2014–15 and 2017–18, Ministry of Finance, Government of India.

Table 2.11: Equilibrium price and quantity of milk		
Price (₹ per litre)	Quantity demanded ('000 litres)	Quantity supplied ('000 litres)
16	10	2
24	9	3
32	8	4
40	7	5
48	6	6 (equilibrium)
56	5	7
64	4	8
72	3	9
80	2	10

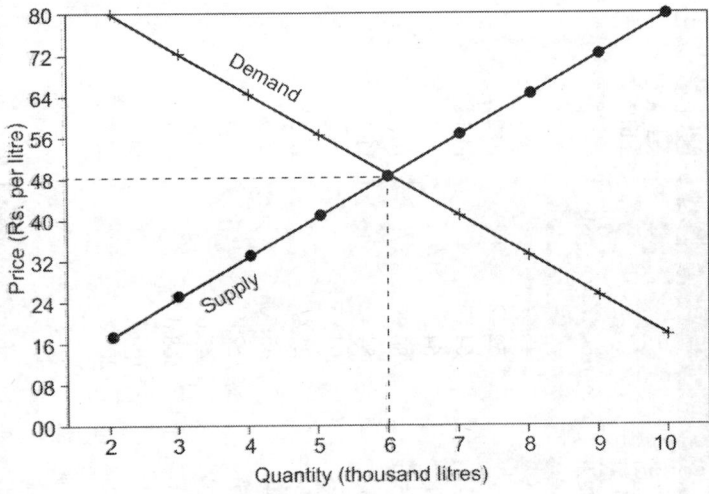

Fig. 2.3: Equilibrium price and quantity

SIMULTANEOUS EQUATIONS APPROACH

The simplified market model may also be expressed in three equations:

Demand	$Q_d = f(p_q)$
Supply	$Q_s = f(P_q)$
Equilibrium conditions	$Q_d = Q_s$

Assuming a linear and exact demand and supply relationship (for simplification), the hypothetical market model may be written as follows:

General		Specific
$Q_d = a + bP$	$b < 0$	$Q_d = 12 - 0.125P$
$Q_s = c + dP$	$d > 0$	$Q_s = 0 + 0.125P$
$Q_d = Q_s$		$Q_d = Q_s$

where Q_d = Quantity demanded in '000 litres; Q_s = Quantity supplied in '000 litres; P = Price per unit of milk in ₹ per litre.

From the equilibrium condition stated in the last equation, the equilibrium value of P may be calculated as follows:

General	Specific
$a + b\overline{P} = c + d\overline{P}$	$12 - 0.125\overline{P} = 0 + 0.125\overline{P}$

or $\quad b\overline{P} - d\overline{P} = -a + c \qquad -0.125\overline{P} - 0.125\overline{P} = 0 - 12$

or $\quad \overline{P} = \dfrac{-a+c}{b-d} \qquad\qquad \overline{P} = \dfrac{-12}{-0.25} = 48.0$

By substituting the equilibrium value of P (₹ 48) in either the demand or the supply relation, the equilibrium value of Q may be calculated as follows:

General Specific

$$\overline{Q} = a + b\left(\dfrac{-a+c}{b-d}\right) \qquad \overline{Q} = 12 - 0.125\,(48) = 6$$

or $\quad \overline{Q} = c + d\left(\dfrac{-a+c}{b-d}\right) \qquad \overline{Q} = 0 + 0.125\,(48) = 6$

The mechanism of determination at the macro level may now be understood. One may think of a demand schedule for, say, foodgrains at the national level.

Over the years, the demand schedule shifts to the right because of the growth in population and incomes and other factors stated earlier. If aggregate supply expands so that the aggregate supply curve also shifts appropriately, the price level may not rise and the equilibrium may be established at the same level of price and at higher quantity. But this rarely happens. At least in·India, it has not happened.

To illustrate this point, let us consider Fig. 2.4. Let D_1 and S_1 be the demand and supply schedules respectively in period t_1. The equilibrium price and quantity during this period are P_1 and Q_1 respectively. Assume that, in period t_2. the demand schedule shifts to D_2. Now if there is no shift in the supply schedule, the equilibrium will be established at a price of P_2 (higher than P_1) and the quantity of Q_2 (higher than Q_1). If the supply curve in period t_2 shifts to S_2, the equilibrium will be established at the old price level (P_1) and higher quantity (Q_4). But as the experience in India shows that the supply of farm products does not shift to the extent shown by the supply curve S_2. Nevertheless, barring a few exceptionally bad crop years, the supply schedule does shift to the right, say, as S_3, resulting in an increase in the price level from P_1 to P_3, and

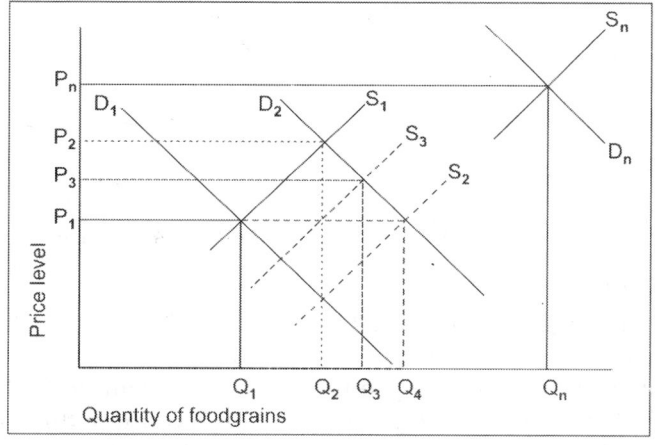

Fig. 2.4: Determination of Price Level of Foodgrains over time at Macro Level

the higher equilibrium supply and demand Q_3. Every year, the process continues; and one observes larger quantities of supply and purchases at higher and higher levels of prices. The magnitude of the rise in prices from year to year depends on the relative shifts in, and the relative slopes of, the demand and supply schedules. In the nth year if the demand curve shifts to D_n and the supply curve shifts to S_n the equilibrium price would be P_n and the equilibrium quantity would be Q_n.

Having understood the theoretical concept of interaction of demand and supply schedules to arrive at equilibrium quantity and price at a point of time, it needs to be recognized that it is the interaction of all factors that affect the supply and demand, which finally determines the actual quantity bought and consumed by consumers and sold by the producers or suppliers. In a large country like India, billions of such transactions take place daily, for different agricultural products. The National Sample Survey Organization (NSSO) has estimated the per capita per month consumption of different agricultural products in India during 2011–12 as shown in Table 2.12.

Table 2.12: India: Per capita per month consumption during 2011–12 (kg)

Commodity	Rural	Urban
Rice and its products	6.13	4.66
Wheat and its products	4.43	4.32
Nutri cereals	0.69	0.33
All cereals	11.25	9.31
Pulses and products	0.79	0.90
Edible oils	0.68	0.85
Fruits	1.53	2.36
Vegetables	6.96	7.12
Sugar	0.78	0.86
Milk and products	4.38	5.55
Eggs, meat and fish	0.64	0.80

Source: NSSO, Report No. 558 (68th round), quoted in Working Group Report, NITI Aayog, GoI, 2016.

SOME GENERAL CONCEPTS USED IN MARKETING

For a better understanding of agricultural marketing, it is necessary to familiarise with some common concepts which are used in the field of general marketing. Some of these are given in this section.

Balance of Trade

The balance of trade is the difference between the value of exports and imports. The value of nation's exports less its imports is called balance of trade.

Brand and Branding

Brand is any name, sign, symbol and/or design used to identify the product of one firm which set them apart from competitor's offerings. Branding is a practice of giving a specified name to a product or group of products from one seller. The specified name creates individuality for the product and on account of this the specified brand of the product can be easily distinguished or recognized in the market from the rival products offered by other sellers. The sole purpose of the branding is to distinguish the branded product from those sold by the competitors.

Cartel

An association of firms that attempts to regulate industry output and prices through mutual agreement is termed as cartel.

Consumer Franchise

The consumer franchise connotes the consumer loyalty to a branch or to a firm as a result of past experience or promotional efforts.

Consumerism

Consumerism can be defined as a movement or social force designed to protect consumers' interest in the market. It is in a sense organized consumers' pressure on the marketing system. Consumerism is an attempt to redress the existing imbalance in the exchange transactions between sellers and buyers. Consumerism is also a form of protest of consumers against unfair business practices and business injustices. The basic aim of consumerism is to remove injustices and eliminate unfair marketing practices which put the consumers to loss. Such practices could be in the nature of mis-branding of products; sale of spurious and unsafe products; sale of obsolete products; adulteration of products; price collusion among sellers; deceptive packaging of products; misleading advertisement; defective warranties; hoarding, profiteering and black marketing of the products; and use of short weights and measures.

In the normal marketing system, producer has the right of power to design the product, distribute, advertise and price it; while the consumer has only the power of not buying it. As such the consumer often feels that while he/she has the power of veto but is not always fully equipped to exercise that power in his/her best interest. This situation may be the result of lack of information or misinformation from one or several competing producers. All these situations have led to the movement of consumerism.

Consumer organizations could provide united and organized efforts to fight against unfair marketing practices and to secure consumers protection. Commendable work has been done by the government as well as voluntary organizations in extending the rights of the consumers by keeping the producers conscious of consumer rights and interests (see more in chapter 7). These consumer organizations/agencies help the consumers by providing information and knowledge in following areas:

1. Consumer education: They provide information about the availability of consumer goods, their prices, standard trade practices, consumer expectations and consumer's rights.
2. Product rating: Consumer organizations also guide the consumers in their choice of the products by carrying tests for products and provide knowledge about the rating scale.
3. Maintaining liaison with government and producer companies of the products to protect the interests of consumers in providing good quality goods at reasonable prices.

Departmental Store

It is a retail organization that carries a wide variety of product lines like grocery, fresh farm products, clothing, home furnishings, personal care products, accessories and household goods; each line is operated as a separate department managed by specialist merchandisers. The merchandises are displayed with fixed prices. These are not the places to haggle about prices. The department store was invented in the 19th century and these are still surviving despite many competing retail outlets, both physical and online. (The first department store was established in Paris in 1852).

Differentiated Marketing

The practice whereby firms develop separate marketing programmes for different target consumer groups is known as differentiated marketing.

Farmers' Terms of Trade

The farmers' terms of trade connote the ratio of the indices of prices received to the indices of prices paid by the farmers. This ratio indicates whether farmers as a group are gainers or losers in the price situation prevailing in the country.

Franchise and Franchising

It is a trading agreement between a supplier (called franchiser) and a retail outlet (franchisee), often in the service sector, where the retail outlet is given the support of the supplier in return for the use of the franchiser's brand and various marketing tools. Franchising is a marketing and sales process in which the franchiser contracts to the franchisee to market and sell his products and service, with strict controls on supply of product, use of the franchiser's logo and quality control over product and services. The franchiser provides the franchisee with marketing services; and in return the franchisee purchases equipment and supplies and pays the franchiser a percentage of revenues and a franchise fee. The franchisee has exclusive marketing and retail sales rights in a designated area or location. It is being considered as a lower risk form of business expansion. The franchiser sells the right to market a product under its name to a franchisee. The franchiser and the franchisee remain separate entities but with interdependent business.

Free Market

A market with no direct involvement of government in market decisions is called a free market.

Game Theory

It is a study of competitive strategies among rival groups with incomplete information about each other's intentions, but in which the participants' intentions are in conflict with each other and the path is determined by individual choices rather than measurable cost variables. This was developed by John Von Neumann, a mathematician in 1944 and is often used by sophisticated marketers to devise competitive strategies and to assist with decision-making.

Hypermarket

It is a large retail outlet combining a supermarket and a departmental store, typically in the range of merchandise under one roof and typically built away from the main town or city thoroughfare to enable easy access and parking.

Inflation

Persistent and appreciable rise in general price level is denoted as inflation. It is the process of continuous rising prices and not a situation of sudden rise in prices. Based on the rate of rise in prices, the inflationary phase is characterized as either creeping inflation, running inflation or galloping/hyperinflation.

Law of One Price (LOOP)

The law of one price connotes a marketing principle which holds that under perfectly competitive market conditions, all prices within a market will be uniform after the costs of place, time and form utilities are taken into consideration.

Marketing Mix

The marketing mix connotes the way in which a firm or industry combines its pricing, promotional and distribution strategies which appeal the consumers. A successful marketing strategy must have a marketing mix as well as a target market for which the marketing mix is prepared. There are four elements of marketing mix viz., (a) product mix; (b) price mix; (c) distribution mix; and (d) promotional mix. Marketing manager is a mixer of all marketing ingredients and creates a mix of all the marketing elements and resources. The marketing mix offers an optimum combination (least cost combination) of all marketing ingredients to maximize company's objectives/goals viz., profit, return on investment, sales volume and market share. Marketing mix has to change according to the changes in marketing conditions and environmental factors which affect the market.

Marketing Process

Marketing process is the sequence of events and actions that coordinate the flow of goods and the value adding activities in the marketing system. Marketing is a matching process by which a producer provides a marketing-mix that meets consumer demand of a target market. The marketing process brings together producers and consumers for the exchange of the product. Each producer or seller has certain goals in marketing. The exchange takes place when market offering is acceptable to the consumer. In the process of exchange, both parties, i.e. producer-sellers and consumers derive some gains. The producer gets the surplus value in the form of profit and the consumer gets the surplus in the form of utility or individual satisfaction. The marketing process is influenced by competition, government laws and policy, media of communication and consumer advocates.

Marketing Strategy

Marketing strategy is a particular procedure used by the seller to achieve a marketing goal. For example, one firm may use the strategy of low prices to attract the consumers, while the other might employ the strategy of selling quality products. The marketing strategy is a functional strategy. Thus, marketing strategy is a comprehensive plan of action designed to meet the needs of a certain enterprise operating in a particular environment. The marketing strategy of a firm can be made unique by emphasizing certain elements of the marketing mix. Formulation of a marketing strategy involves the following steps:

1. Determination of marketing objectives;
2. Generating alternative marketing mix options;
3. Selecting the most profitable marketing mix option; and
4. Creating conditions for implementing the chosen marketing mix.

Market Segmentation

The marketing technique of developing separate products and marketing programmes to appear to different consumer classes is called market segmentation. Market segmentation is a method for achieving maximum market response from limited marketing characteristics of various parts of the market. As such the market segmentation is the strategy of divide and conquer, i.e. dividing markets to conquer them. Market segmentation enables the marketers to give better attention to the selected customers and offer an appropriate marketing mix for each chosen segment or group of buyers having homogenous demand. Each sub-division or segment is selected as a market target to be reached with a distinct marketing mix. In the imperfect competition, generally two types of marketing strategies are adopted viz., (a) product differentiation through branding; and (b) market segmentation. Segmentation strategy is a customer oriented philosophy and offers following benefits:

(a) Marketers are in a better position to locate and compare marketing opportunities;
(b) Marketers can effectively formulate and implement marketing programmes which will be tuned with the demand of the particular segment of the market;
(c) Marketers can make finer adjustments in their products and marketing communication;
(d) Marketers can avoid fierce competition by assessing the strengths and weaknesses of the competitors and can use resources more profitably by catering to customers demand; and
(e) Segmentation leads to a more effective utilization of marketing resources because customer is the focus of marketing effort and in this only target markets are served.

Maximum Retail Price (MRP)

All packaged items in India carry MRP on their label. The compulsion to mention MRP was introduced in 1990 by an amendment to the Standards of Weights and Measures Act (Packaged Commodity Rules) 1976. The objective was to prevent tax evasion and protect consumers from profiteering by the retailers. Before this amendment, the manufacturers could print either 'retail price inclusive of all taxes' or the 'retail price (local taxes extra)'. In the latter case, retailers were found to charge more in the guise of local taxes. Having observed the practice for 25 years, many scholars are raising the question whether this has really helped the consumers.

Nutri Cereals

Grains like jowar (sorghum), bajra (pearl millet), ragi and other millets, which were earlier termed as coarse cereals, have now been renamed (since April 2018) as nutri cereals. The objective is to dispense with a lingering perception that these grains are inferior to rice and wheat even when their health benefits are larger.

Supermarket

It is a medium-sized store that sells groceries and packaged goods, usually in urban setting.

Superstore

It is a large store which sells everything from groceries to packaged goods to clothing, usually outside an urban setting.

Terms of Trade

The terms of trade connote the relationship of prices of one country or a sector with the other. In international trade, it is the rate at which a country exchanges its exports with imports. Within a country, this term is used for inter-sectoral exchanges reflected through relative prices. In a two sector economy (agriculture and industry), it can be defined as the ratio of the index of agricultural prices to the index of industrial prices with reference to a common base year.

Trade Mark

Trade mark is a legal term given to a particular branded product duly registered under the Trade Names and Trade Marks Act. Thus, a branded product enjoys legal protection. The English letter 'R' in a circle on each package indicates that this brand of the product is duly registered.

Value Chain

Conceptually, a value chain involves a series of value generating activities, through which a farm or firm develops competitive advantage and creates value. A fully commercial value chain is comprised of primary activities and support activities. The primary activities include (a) inbound logistics (receiving and distribution of inputs); (b) operations (transforming inputs into products); (c) outbound logistics (storage and distribution of products), marketing and sales; and (d) after-sales services. The secondary activities include infrastructure (organization, control, business culture, finance, legal, quality management etc.); human resource management; technology development; and procurement (inputs, equipments, etc.).

REFERENCES

1. Hibbard, B.H., *Marketing Agricultura! Products*, D. Appleton and Company, Inc., New York, 1921, pp. 13–15.
2. *Encyclopaedia of Social Sciences*, Vol. 10, 1933, p. 133.
3. Cournot, Recher Sur les Principles Mathematics de la Theorie des Richesses, Chap. IV.
4. Gupta, A.P. *Marketing of Agricultura! Produce in India*, Vora and Co. Publishers Pvt. Limited, Bombay, 1975, p. 15.
5. Chapman, *Economics*, Quoted in Jain S.C. *Principles and Practice of Agricultural Marketing and Prices*, Vora and Co. Publishers Pvt. Limited, Bombay, 1971, p. 2.
6. Sherleker, S. A., *Marketing Management*, Himalayan Publishing House, Mumbai, 1981, p.1.
7. Directorate of Marketing and Inspection (DMI), Survey of Planning of Rural Markets in India, MPDC, Report No. 23, July 1987, p. 39.
8. Directorate of Marketing and Inspection (DMI), Infrastructural Facilities provided by the APMCs and their Utilization, MRPC-25, Ministry of Agriculture, Government of India, Nagpur, 1999.
9. Acharya, S. S., Agricultural Marketing, State of the Indian Farmer-A Millennium Study, Volume 17, Ministry of Agriculture, Government of India, Academic Foundation, New Delhi, 2004.
10. Acharya S.S., *"Agricultural-Industry Linkages: Public Policy and Some Areas of Concern"*, Agricultural Economics Research Review, Vol. 10, No. 2, July–Dec. 1997.
11. George, M.V. and A.J. Singh, "Structure, Conduct and Performance of Wholesale Vegetable Market in Punjab," Proceedings of the Second All-India Agriculture Marketing Research Conference held at Jabalpur on June 8–9, 1970, pp. 80–81.
12. Stifel, L.D., *Teaching and Research Forum*, Series No. 3, Agricultural Development Council, New York, October, 1976.
13. Amarchand, D. and B. Varadharajan. *An Introduction to Marketing*, Vikas Publishing House Pvt. Ltd., New Delhi, 1980, p. 18.

14. George, M.V. and A.J. Singh, *op. cit.*, pp. 80–81.
15. Murty, K.N., Consumption and Nutritional Patterns of ICRISAT Mandate Crops in India, ICRISAT, Progress Report No. 53, Aug. 1983, pp. 10–11.
16. Bhalla, G.S., *"Globalisation and Agricultural Policy in India"*, Indian Journal of Agricultural Economics, Vol. 50, No. 1, Jan–March 1995, pp. 1–26.
17. Kumar, P. and V.C. Mathur, *"Agriculture in Future: Demand-Supply Projections for the Ninth Five Year Plan"* ; Economic and Political Weekly, Vol. 31 (39), Sept. 28, 1996, pp A-131 to 139.
18. Kumar, P. and P.K. Joshi, *"Input Subsidy vs. Farm Technology–which is more important for Agricultural Development,"* Agricultural Economics Research Review, Vol. 27(1), June 2014, p. 1–18.
19. Kumar, P. *"Food and Nutrition Security"*, Presidential Address, Indian Journal of Agricultural Marketing, Vol. 33 (3), 2019.

APPENDIX 2.1

BASIC NERLOVIAN PRICE EXPECTATION AND ACREAGE ADJUSTMENT MODELS

These models were initially suggested and used by Nerlove to estimate the effect of price on the acreage under crops. In the price expectation model, it was hypothesized that farmers would react not only to the last year's prices but to the price they expected, and that this expected price would depend, but only to a limited extent, on last year's price. This model is known as the Nerlovian Price Expectation Model and is of the following form:[1]

$$A_t = a + bP_t^* + U_t \qquad \qquad ...(1)$$

$$(P_t^* - P_{t-1}^*) = \beta(P_{t-1} - P_{t-1}^*) \qquad \qquad ...(2)$$

$$0 < \beta < 1$$

where A_t = actual acreage under the crop in year t; p_t^* = expected price of the crop in year t; p_{t-1}^* = expected price of the crop in the preceding year; P_{t-1} = actual price of the crop in year $t-1$; U_t = error term; a and b = constants; β = coefficient of expectation.

The Nerlovian model of price expectation was modified by other scholars. Rainfall, relative yield and total irrigated area in the first equation of basic Nerlovian model were included as the explanatory variables to explain the variation in acreage.[2] Further, the price variability as one of the independent variables in the basic Nerlovian model was added as a measure of uncertainty in the price by Behrman[3].

In the acreage adjustment model, the hypothesis is that the desired acreage (A_t^*) depends on last year's price (P_{t-1}) and that the acreage adjustment from one year to another depends on the difference between last year's acreage (A_{t-1}) and the desired area for the current year (A_t^*). The two equation acreage adjustment model can be expressed as:

$$A_t^* = a + bP_{t-1} + U_t \qquad \qquad ...(3)$$

$$A_t - A_{t-1} = \gamma(A_t^* - A_{t-1}) \qquad \qquad ...(4)$$

$$0 < \gamma < 1$$

where a, b and the coefficient of adjustment (γ) are the parameters to be estimated.

Both the price expectation and acreage adjustment models lead to the same form of the estimating equation, which is given below:

$$A_t = \pi_0 + \pi_1 P_{t-1} + \pi_2 A_{t-1} + W_t$$

where $\pi_0 = a\beta$ or $a\gamma$; $\pi_1 = b\beta$ or $b\gamma$; $\pi_2 = 1 - \beta$ or $1 - \gamma$; W_t = new disturbance term.

The relationship of the new disturbance term in the estimating (reduced form) equation and the structural equations, and the corresponding assumptions about the behaviour of these terms, specially in respect of auto-correlation, differ between the price expectation and acreage adjustment models.

REFERENCES

1. Nerlove, Marc. "Distributed Lags and Estimation of Long-run Supply and Demand Elasticities: Theoretical Considerations," *Journal of Farm Economics*, Vol. 40, 1958, pp. 302–11.
2. Rajkrishna. Farm Supply Response in India-Pakistan: A Case Study of Punjab Region, *The Economic Journal*, Vol. 73, 1963, pp. 447–87
3. Behrman, J.R. *Supply Response in Underdeveloped Agriculture*, North Holland Publishing Co., Amsterdam, 1968.

3

Agricultural Marketing and Development

One of the most important goal of development is eliminating hunger, food insecurity and malnutrition. Orderly and efficient marketing of foodgrains plays an important role in solving the problem of hunger. Most of those who go hungry do so because they cannot get food at affordable prices. If marketing system is not efficient, price signals arising at the consumers' level are not adequately transferred to the producers, as a result farmers do not get sufficient price incentive to increase the production of the commodities which are in short supply. Further, considerable quantity of food gets lost in the marketing chain. Thus, an inefficient marketing system adversely affects the living standards of both the farmers and consumers. In agricultural-oriented developing countries like India, agricultural marketing plays a pivotal role in fostering and sustaining the tempo of rural and economic development. Markets trigger the process of development.

The development of an efficient marketing system is important in ensuring that scarce and essential commodities reach different classes of consumers. Marketing is not only an economic link between the producers and the consumers; it maintains a balance between demand and supply. The objectives of price stability, rapid economic growth and equitable distribution of goods and services cannot be achieved without the support of an efficient marketing system.

IMPORTANCE OF AGRICULTURAL MARKETING

Agricultural marketing plays an important role not only in stimulating production and consumption, but also in accelerating the pace of economic development. Its dynamic functions are of primary importance in promoting economic development. For this reason, it has been described as the most important multiplier of agricultural development.

India's age-old farming practices have taken a turn in recent decades. There has been a technological breakthrough—the evolution of high-yielding variety seeds, increasing use of fertilizers, insecticides, pesticides, the installation of pumping sets, and tractorization. This technological breakthrough has led to a substantial increase in production on the farms and to the larger marketable and marketed surplus. To maintain this tempo and pace of increased production through technological development, an assurance of remunerative prices to the farmer is a prerequisite, and this assurance can be given to the farmer by developing an efficient marketing system.

The agricultural marketing system plays a dual role in economic development in countries whose resources are primarily agricultural. Increasing demands for money with which to purchase other goods leads to increasing sensitivity to relative prices on the part of the producers, and specialization in the cultivation of those crops on which the returns are the greatest, subject to socio-cultural, ecological and economic constraints. It is the marketing system that transmits the crucial price signals. On the other hand, and in order to sustain the growth of the non-agricultural sector, resources have to come from the agricultural sector—physical resources to guarantee supplies of food and raw materials for the agro-industry and financial resources for investment in non-farm economy as well as for re-investment in agriculture.

On the basis of IADP experience, Kiehl[1] has shown that the "marketing problem" begins to emerge in the process of shifting from traditional to modern agriculture because of production surpluses generated by the shift. Indeed, the term *modern agriculture* implies a *market-oriented agriculture*. The scope for moving towards modern agriculture must include market dimensions if the momentum of production transformation is to be sustained.

The importance of agricultural marketing in economic development is revealed from the following:

1. Optimization of Resource Use and Output Management

An efficient agricultural marketing system leads to the optimisation of resource use and output management. An efficient marketing system can also contribute to an increase in the marketable surplus by scaling down the losses arising out of inefficient processing, storage and transportation. A well-designed system of marketing can effectively distribute the available stock of modern inputs, and thereby sustain a faster rate of growth in the agricultural sector.

2. Increase in Farm Income

An efficient marketing system ensures higher levels of income for the farmers by reducing the number of middlemen or by restricting the cost of marketing services and the malpractices in the marketing of farm products. An efficient system guarantees the farmers better prices for farm products and induces them to invest their surpluses in the purchase of modern inputs so that productivity and production may increase. This again results in an increase in the marketed surplus and income of the farmers. If the producer does not have an easily accessible market-outlet where he can sell his surplus produce, he has little incentive to produce more. The need for providing adequate incentives for increased production is, therefore, very important, and this can be made possible only by streamlining the marketing system. Farmer's income is directly related to the quantity sold by the farmer and the price at which it is sold.

3. Widening of Markets

An efficient and well-knit marketing system widens the market for the products by taking them to remote corners both within and outside the country, i.e. to areas far away from the production points. The widening of the market helps in increasing the demand on a continuous basis, and thereby guarantees a higher income to the producer farmer. At the same time it helps a large number of consumers in satisfying their demand for various farm products.

4. Growth of Agro-based Industries

An improved and efficient system of agricultural marketing helps in the growth of agro-based industries and stimulates the overall development process of the economy. Many industries like cotton, sugar, edible oils, food processing and jute depend on agriculture for the supply of raw materials.

5. Price Signals

An efficient marketing system helps the farmers in planning their production in accordance with the needs of the economy. This work is carried out by the marketing system through transmitting price signals at different stages of marketing, which helps, apart from the farmers, market functionaries in planning their buying and selling activities.

6. Adoption and Spread of New Technology

The marketing system helps the farmers in the adoption of new scientific and technical knowledge. New technology requires higher investment and farmers would invest only if they are assured of market clearance of their surplus production at remunerative prices.

7. Employment Creation

The marketing system provides employment to millions of persons engaged in various activities, such as assembling, packaging, transportation, storage and processing. Persons like commission agents, brokers, traders, retailers, weighmen, *hamals*, packagers and regulating staff are directly employed in the marketing system. This apart, several others find employment in supplying goods and services required by the marketing system.

8. Addition to National Income

Marketing activities add value to the product thereby increasing the nation's gross national product and net national product.

9. Better Living

The marketing system is essential for the success of the development programmes which are designed to uplift the population as a whole. Any plan of economic development that aims at diminishing the poverty of the agricultural population, reducing consumer food prices, earning more foreign exchange or eliminating economic waste has, therefore, to pay special attention to the development of an efficient marketing for food and agricultural products. In the absence of an efficient marketing system, farmers receive lower prices, consumers pay higher prices, and lots of agricultural products are wasted.

10. Creation of Utility

Marketing is equally productive, and is as necessary as the farm production. It is, in fact, a part of production itself, for production is complete only when the product reaches a place in the form and at the time required by the consumers. Marketing adds cost to the product, but, at the same time, it adds utilities to the product. The following four types of utilities of the product are created by marketing:

(a) *Form Utility:* The processing function adds form utility to the product by changing the raw material into a finished form. With this change, the product becomes more useful than it is in the form in which it is produced by the farmer. For example, through processing, oilseeds are converted into oil, sugarcane into sugar, cotton into cloth and wheat into flour and bread. The processed forms are more useful than the original raw materials.

(b) *Place Utility:* The transportation function adds place utility to products by shifting them to a place of need from the place of plenty. Products command higher prices at the place of need than at the place of production because of the increased utility of the product.

(c) *Time Utility:* The storage function adds time utility to the products by making them available at the time when they are needed.

(d) *Possession Utility*: The marketing function of buying and selling helps in the transfer of ownership from one person to another. Products are transferred through marketing to persons having a higher utility from persons having a low utility.

11. Contribution to Food and Nutritional Security

The food marketing system can contribute immensely in improving food and nutritional security on the following counts:

(a) Foodgrains, edible oilseeds (cooking medium) and horticultural crops account for around 95% of the gross cropped area in the country. Marketing of these products provides income to most of the Indian farmers, so that they are able to buy the required inputs as well as other items of food which they do not produce themselves.

(b) The food marketing business provides livelihood to millions of traders, processors, retailers and other market functionaries.

(c) An efficient food marketing system helps in reaching the food items to the consumers at minimum possible cost and saving the wastage of food products that occurs in the marketing chain.

(d) The food marketing system, which influences the food prices at different stages of marketing, is politically very sensitive. An efficient food marketing system, can thus contribute towards political stability in the country. With the increase in consciousness towards nutritive foods, the importance of marketing and pricing of fruits, vegetables, milk and other products has considerably gone up.

HISTORY AND GROWTH OF AGRICULTURAL MARKETING

Marketing had its beginning in agriculture. It developed only after man was able to produce more food than he needed for himself, and only after a way of exchanging the products of his labour for those of others had been found. This transition from production for consumption to production for exchange came about slowly. About a century ago, farmers used to consume most of what they produced; but, now, most of what the farmers produce is exchanged for other things which they require. To reach this stage, farmers became business-minded. This tendency increased their dependence on marketing, which has resulted in the overall development of the market mechanism and the entire economy.

The early pioneers of our country did not face much of the marketing problem. Producers and consumers, if not actually the same individuals, lived next door to each other. The following factors have led to the growth of agricultural marketing in India.

1. Specialization

The tendency towards increasing specialization by farmers and regions in certain crops or livestock has resulted in an increase in their efficiency and the breakdown in the self-sufficiency of the family unit. Specializatlon, thus, has resulted in increased production, which is the base for the growth of marketing and, in turn, of the economy. This has also resulted in improved use efficiency of natural resources like land and water.

2. Urbanization

Urban people are the main buyers of agricultural surpluses. The urban population of India had increased from 33.5 million in 1931, to 62.4 million in 1951, 109.1 million in 1971, to 217.6 million in 1991, 285.5 million in 2001 and 377.1 million in 2011. This has necessitated a faster growth of agricultural marketing activities. The rate of growth of urban population is much higher than rural population(due to rural-urban migration), which has further increased the importance of marketing system for farm products.

3. Transportation and Communication

The increases in transportation and communication facilities have widened the market for farm products. The length and breadth of the market to which a product is taken from the production areas have increased. In the absence of these facilities, the movement of produce from one area to another was limited, and the consumption of a product was restricted only to the areas of production or, at the most, to nearby areas. The scope of marketing has, thus increased manifold.

4. Technological Change in Agriculture

Technological developments in agriculture, such as the evolution of high-yielding varieties of seeds, increased use of modern inputs and cultivation practices in the agricultural sector, have resulted in substantial increase in farm production. The marketed surplus of the agricultural produce has therefore increased. Production-conscious farmers have also become income/price conscious. This has resulted in the growth of the marketing system.

The importance of an efficient marketing system as a vital link between the farmer and the consumer was recognized way back in 1928 by the Royal Commission on Agriculture. Since then, a good deal of progress has been made in organizing agricultural marketing by the adoption of the various administrative and legislative measures by the Government from time to time. The establishment of the Directorate of Marketing and Inspection in 1935, the enactment of the Act for the grading and standardization of agricultural commodities in 1937, the conduct of commodity market surveys, and the establishment of regulated markets in the country—these are some of the measures which have been taken up to improve the marketing situation and to make agricultural marketing as efficient as possible.

During the First and Second Five Year Plans, agricultural marketing did not receive importance. Whatever development that took place in the sphere of marketing was due to the gradual progress towards the commercialization of agriculture, as a result of its own dynamic nature, and not because of any specific government efforts.

The National Commission on Agriculture[2] (the first commission which suggested measures for the development of agriculture in the post-independence period) remarked: "There is an increasing awareness that it is not enough to produce a crop or animal product, it must be marketed well. Increased production, resulting in a greater percentage increase in the marketable surplus accompanied by the increase in demand from urban population, calls for a rapid improvement in the existing marketing system." This statement emphasized the increasing importance of marketing of agricultural commodities and the need for the adoption of measures to increase production.

Since then, lots of efforts were made in various Five Year Plans to promote organized marketing of agricultural commodities through a network of regulated markets. Most of the States and Union Territory governments enacted legislations (Agricultural Produce Market Regulation Acts), and the number of regulated markets went up. These have helped mitigate the market handicaps of producers/sellers at the wholesale assembling level. During the 10th Five Year Plan, the Union Ministry of Agriculture formulated another modified Model Law on agriculture marketing, which provides for establishment of private markets/yards, direct purchase centres, consumers or farmers' markets for direct sale, and promotion of public-private partnership in the management and development of agricultural markets in the country. Many States and UTs have amended their Acts.

During the last five years, several new initiatives have been taken to improve the agricultural marketing system in India. These include (a) circulation of yet another new Model Act titled "Agricultural Produce and Livestock Marketing (Promotion and Facilitation) Act in 2017; (b) creation and launch of electronic National Agricultural Market (e-NAM); (c) a new Model Contract Farming Act; (d) re-definition of the floor for fixation of minimum support prices; (e) strengthening of market intervention scheme for such crops as potato, onion and tomato by creating a Price Stabilization Fund; and (f) launch of several schemes for expansion of marketing infrastructure and agro-processing facilities (see details in subsequent chapters).

▌ PRODUCER'S SURPLUS OF AGRICULTURAL COMMODITIES

In any developing economy, the producer's surplus of agricultural product plays a significant rote. This is the quantity which is actually made available to the non-producing population of the country. From the marketing point of view, this surplus is more important than the total production of commodities. The arrangements for marketing and the expansion of markets have to be made only for the surplus quantity available with the farmers, and not for the total production.

The rate at which agricultural production expands determines the pace of agricultural development, while the growth in the marketable surplus determines the pace of economic development. An increase in production must be accompanied by an increase in the marketable surplus for the economic development of the country. Though the marketing system is more concerned with the surplus which enters or is likely to enter the market, the quantum of total production is essential for this surplus. The larger the production of a commodity, the greater the surplus of that commodity and vice versa. The knowledge of marketed and marketable surplus helps the policy-makers as well as the traders in the following areas:

1. *Framing Sound Price Policies:* Price support programmes are an integral part of agricultural policies necessary for stimulating agricultural production. The knowledge of quantum of marketable surplus helps in framing these policies.
2. *Developing Proper Procurement and Purchase Strategies:* The procurement policy for feeding the public distribution system has to take into account the quantum and

behaviour of marketable and marketed surplus. Similarly, the traders, processors and exporters have to decide their purchase strategies on the basis of marketed quantities of different farm products.

3. *Checking Undue Price Fluctuations:* A knowledge of the magnitude and extent of the surplus helps in the minimization of price fluctuations in agricultural commodities because it enables the government and the traders to make proper arrangements for the movement of produce from one area, where these are in surplus, to another area which is deficient. The knowledge of marketed surplus also helps the traders and the Government in decisions related to storage.

4. *Export/Import Decisions:* Advanced estimates of the surpluses of such commodities which have the potential of external trade are useful in decisions related to the export and import of the commodity. If surplus is expected to be less than what is necessary, the country can plan for imports and if surplus is expected to be more than what is necessary, avenues for exporting such a surplus can be explored.

5. *Development of Transport and Storage Systems:* The knowledge of marketed surplus helps in developing adequate capacity of transport and storage system to handle it.

MEANING AND TYPES OF PRODUCER'S SURPLUS

The producer's surplus is the quantity of produce which is, or can be, made available by the farmers to the non-farm population. The producer's surplus is of two types:

1. Marketable Surplus

The marketable surplus is that quantity of the produce which can be made available to the non-farm population of the country. It is a theoretical concept of surplus. The marketable surplus is the residual left with the producer-farmer after meeting his requirements for family consumption, farm needs for seeds and feed for cattle, payment to labour in kind, payment to artisans—carpenter, blacksmith, potter and mechanic— payment to landlord as rent, and social and religious payments in kind. This may be expressed as follows:

$$MS = P - C$$

where MS = Marketable surplus; P = Total production; and C = Total requirements (family consumption, farm needs, payment to labour, artisans, landlord and payment for social and religious work).

2. Marketed Surplus

Marketed surplus is that quantity of the produce which the producer-farmer actually sells in the market, irrespective of his requirements for family consumption, farm needs and other payments. The marketed surplus may be more, less or equal to the marketable surplus.

Whether the marketed surplus increases with the increase in production has been under continuous theoretical scrutiny. It has been argued that poor and subsistence farmers sell that part of the produce which is necessary to enable them to meet their cash obligations. This results in distress sale on some farms. In such a situation, any increase in the production of marginal and small farms should first result in increased on-farm consumption.

An increase in the real income of farmers also has a positive effect on on-farm consumption because of positive income elasticity. Since the contribution of this group to the total marketed quantity is not substantial, the overall effect of increase in production must lead to an increase in the marketed surplus.

Bansil[3] writes that there is only one term—*marketable surplus*. This may be defined subjectively or objectively. Subjectively, the term *marketable surplus* refers to theoretical surplus available for sale with the producer-farmer after he has met his own genuine consumption requirements and the requirements of his family, the payment of wages in kind, his feed and seed requirements, and his social and religious payments. Objectively, the marketable surplus is the total quantity of arrivals in the market out of the new crop.

RELATIONSHIP BETWEEN MARKETED SURPLUS AND MARKETABLE SURPLUS

The marketed surplus may be more, less or equal to the marketable surplus, depending upon the condition of the farmer and type of the crop. The relationship between the two terms may be stated as follows:

$$\text{Marketed surplus} \gtreqless \text{Marketable surplus}$$

1. The marketed surplus is more than the marketable surplus when the farmer retains a smaller quantity of the crop than his actual requirements for family and farm needs. This is true especially for small and marginal farmers, whose need for cash is more pressing and immediate. This situation of selling more than the marketable surplus is termed as distress or forced sale. Such farmers generally buy the produce from the market in a later period to meet their family and/or farm requirements. The quantity of distress sale increases with the fall in the price of the product. A lower price means that a larger quantity should be sold to meet some fixed cash requirements.
2. The marketed surplus is less than the marketable surplus when the farmer retains some of the surplus produce. This situation holds true under the following conditions:
   (a) Large farmers generally sell less than the marketable surplus because of their better retention capacity. They retain extra produce in the hope that they would get a higher price in the later period. Sometimes, farmers retain the produce even up to the next production season.
   (b) Farmers may substitute one crop for another crop either for family consumption purpose or for feeding their livestock because of the variation in prices. With the fall in the price of the crop relative to a competing crop, the farmers may consume more of the first and less of the second crop.
3. The marketed surplus may be equal to the marketable surplus when the farmer neither retains more nor less than his requirement. This holds true for perishable commodities and for the average farmer.

FACTORS AFFECTING MARKETABLE SURPLUS

The marketable surplus differs from region to region and, within the same region, from crop to crop. It also varies from farm to farm. On a particular farm, the quantity of marketable surplus depends on the following factors:

1. *Size of Holding:* There is positive relationship between the size of the holding and the marketable surplus.
2. *Production*: The higher the production on a farm, the larger will be the marketable surplus, and vice versa.
3. *Price of the Commodity*: The price of the commodity and the marketable surplus have a positive as well as a negative relationship, depending upon whether one considers the short and long run or the micro and macro levels.
4. *Size of Family*: The larger the number of members in a family, the smaller the surplus on the farm.
5. *Requirement of Seed and Feed*: The higher the requirement for these uses, the smaller the marketable surplus of the crop.
6. *Nature of Commodity*: The marketable surplus of non-food crops is generally higher than that for food crops. For example, in the case of cotton, jute and rubber, the quantity retained for family consumption is either negligible or very small part of the total output. For these crops, a very large proportion of total output is marketable surplus. Even among food crops, for such commodities like sugarcane, spices and oilseeds which require some processing before final consumption, the marketable surplus as a proportion of total output is larger than that for other food crops.
7. *Consumption Habits*: The quantity of output retained by the farm family depends on the consumption habits. For example, in Punjab, rice forms a relatively small proportion of total cereals consumed by farm-families compared to those in southern or eastern states. Therefore, out of a given output of paddy/rice, Punjab farmers sell a greater proportion than that sold by rice eating farmers of other states.

The functional relationship between the marketed surplus of a crop and factors affecting the marketed surplus may be expressed as:

$$M = f(x_1, x_2, x_3, x_4, ..., x_n)$$

where M = Total marketed surplus of a crop in quintals; x_1 = Size of holding in hectares; x_2 = Size of family in adult units; x_3 = Total production of the crop in quintals; x_4 = Price of the crop in ₹ per quintal. The other factors may be specified.

In a study of wheat in Rajasthan[4] during 1969–70 and 1970–71 (pre-green revolution period), the elasticity of marketed surplus with respect to total production was estimated at more than unity (1.09). Some later studies also indicate that the marketed surplus-output elasticity of wheat and rice in India is more than one. This implies that as the production of these commodities expands, the market sales increase proportionately more than the increase in output and if the output of these commodities falls, the decrease in the market sales is proportionately more than the fall in output.

RELATIONSHIP BETWEEN PRICES AND MARKETED SURPLUS

Two main hypotheses have been advanced to explain the relationship between prices and the marketed surplus of foodgrains in subsistence agriculture.

1. Inverse Relationship

There is an inverse relationship between prices and the marketed surplus. This hypothesis was presented by P.N. Mathur and M. Ezekiel. They postulate that the farmers' cash requirements are nearly fixed, and given the price level, the marketed portion of the output is determined. This implies that the farmers' consumption is a residual, and that the marketed surplus is inversely proportional to the price level. This behaviour assumes that farmers have inelastic cash requirements.

The argument is that, in the poor economy of underdeveloped countries, farmers sell that quantity of the output which gives them the amount of money they need to satisfy their cash requirements; they retain the balance of output for their own consumption purpose. With a rise in the prices of foodgrains, they sell a smaller quantity of foodgrains to get the cash they need, and vice versa. In other words, with a rise in price, farmers sell a smaller, and with the fall in price, they sell a larger quantity. Olson and Krishnan have argued that the marketed surplus varies inversely with the market price on a different basis. They contend that a higher price for a subsistence crop may increase the producer's real income sufficiently to ensure that the income effect on demand for the consumption of the crop outweighs the price effect on production and consumption.

2. Positive Relationship

V.M. Dandekar and Rajkrishna put forward the case for a positive relationship between prices and the marketed surplus of foodgrains in India. This relationship is based on the assumption that farmers are price conscious. With a rise in the prices of foodgrains, farmers are tempted to sell more and retain less. As a result, there is increased surplus. The converse, too, holds true.

Rajkrishna has pointed that the elasticity of the marketable surplus is not negative so long-as the substitution effect is non–zero.

Three models to indirectly investigate the size and magnitude of the elasticity of the marketed surplus of a subsistence crop are given below. The detailed derivation is given in Appendix 3.1.

(a) Rajkrishna Model [5]

$$M = Q - C$$

$$e = rb - (r - 1)(g + mkh)$$

where M = Marketed surplus of the crop; Q = Total production of the crop; C = On-farm consumption of the crop; e = Elasticity of marketed surplus with respect to price of the crop; r = Reciprocal of sales ratio (M/Q); b = Output price elasticity of the crop; g = Consumption price elasticity of the crop; m = Sales ratio (M/Q); h = Consumption income elasticity; $k = (PQ)/I$ = Ratio of the total value of production to the total net income of the producers.

(b) Behrman Model [6]

$$M_1 = Q_1 - C_1$$

$$e = rb_1 - (r - 1)[q + kh(1 + b_1)] - (r - 1)hb_2(1 - k)$$

where M_1 = Marketed surplus of Q_1; b_1 = Price elasticity of Q_1 with respect to P_1/P_2; b_2 = Price elasticity of Q_2 with respect to P_1/P_2 and all other notations having the same meaning as defined earlier.

(c) T.N. Krishnan Model[7]

$$e = -(\beta - \alpha)\frac{r}{1-r}$$

$$0 < r < 1$$

where r = Proportion of the output consumed on the farm; b = Consumption income elasticity; a = Consumption price elasticity.

In terms of Behrman's notations, this model is

$$e = -(h - g)\frac{1-m}{m}$$

▌ESTIMATION OF MARKETABLE AND MARKETED SURPLUS

In such countries as India where the production activity is carried out by millions of farmers which are spatially scattered throughout the length and breadth of the vast country, the estimates of the marketable/marketed surplus of agricultural products at the national level is not easy specially of the foodgrains and other food items such as milk which are consumed by the producing families also.

Estimation of marketed and marketable surplus at the level of individual farmer is easy. There has been a large number of studies which provides the estimates of marketed and marketable surplus at micro level. Apart from the use of such estimates for ascertaining the marketed surplus at the national level, such micro studies have been used to bring out the nature and extent of distress sale of foodgrains by small and marginal farmers.

Consider the following data of a case farm of Mr. Jagdish for the year 2014–15 (July to June).

There are six adult units in the family of Mr. Jagdish. He also maintains two milch animals. Mr. Jagdish sold 120 quintals of wheat, 15 quintals of barley, 48 quintals of mustard, 40 quintals of gram, 60 quintals of bajra and 30 quintals of guar at different times between July, 2014 and June 2015. Given this information, the marketable and marketed surplus of the different crops of this farmer can be worked out. Total production and the requirements of the farm and family are shown below, followed by the computation of marketable and marketed surplus of his farm:

Crops	Area under crop (ha)	Productivity (Qtl/ha)	Seed requirement (Qtl)	Consumption requirement per adult (Qtl)	Requirement for livestock and artisans unit (Qtl)
Wheat	8	20	6.0	2.00	2.00 for artisans
Barley	2	12	1.0	0.50	–
Mustard	5	10	0.4	0.10	–
Gram	5	10	2.5	0.25	1.00 for artisans
Bajra	15	6	3.0	1.00	–
Guar	5	8	2.0	–	2.50 per milch cow

Crops	Total production (Area × Productivity)	Seed requirement	Consumption requirement No. of units × Requirement per unit	Other requirements	Total requirement
Wheat	8 × 20 = 160	6.0	6 × 2.0 = 12.0	2.0	20.0
Barley	2 × 12 = 24	1.0	6 × 0.5 = 3.0	–	4.0
Mustard	5 × 10 = 50	0.4	6 × 0.1 = 0.6	–	1.0
Gram	5 × 10 = 50	2.5	6 × 0.25 = 1.5	1.0	5.0
Bajra	15 × 6 = 90	3.0	6 × 1.00 = 6.0	–	9.0
Guar	5 × 8 = 40	2.0	–	5.0	7.0

The marketable and marketed surplus of different products for this farm are as follows:

Crops	Total production (Qtls)	Total requirement (Qtls)	Marketable surplus (Qtls)	Marketed surplus (Qtls)	As percentage of production	
					Marketable surplus	Marketed surplus
Wheat	160	20	140	120	87.50	75.00
Barley	24	4	20	15	83.33	62.50
Mustard	50	1	49	48	98.00	96.00
Gram	50	5	45	40	90.00	80.00
Bajra	90	9	81	60	90.00	66.67
Guar	40	7	33	30	82.50	75.00

As per this exercise, the marketed quantity is less than the marketable surplus on this case farm and there is no distress sale. Rather the farmer has retained a part of the produce for a later sale.

The variation in the marketed and marketable surplus of some cereals, oilseeds and pulse crops across farm size groups can be seen from the estimates available from some sample studies during the 1990s given in Tables 3.1 and 3.2.

In the case of staple food like cereals, the marketable surplus-output ratio on marginal farms in some states is negative which means that such farmers are net buyers of cereals. The marketed or marketable surplus as a proportion of the total output of oilseeds and pulses is considerably higher than that of cereals. Among cereals, the marketed surplus-output ratio for paddy/rice and wheat is generally higher than that for maize. This is mainly owing to higher per farm output of rice and wheat. Another important point to be noted is that marketed surplus-output ratio for paddy/rice in Punjab and Haryana is higher than that in Orissa, West Bengal, and Tamil Nadu. This is on account of both higher per farm output in Punjab and Haryana than other states and differences in consumption pattern of the farm families in these two groups of states.

The situation has considerably changed in the recent years. Even in the case of cereals, including nutri cereals, both marketable and marketed surplus ratios have considerably gone up. Even marginal and small farmers are selling more than half of their production of cereals. It is a clear indication of a shift from subsistence to market-oriented agriculture for a majority of the farmers (Table 3.3).

Table 3.1: Marketable and marketed surplus of wheat, rice and maize on different size groups of farms in India (1990s)

(% of production)

Commodity/State/Particulars		Marginal	Small	Medium	Large	Overall
WHEAT						
Uttar Pradesh	ML	−11.0	48.1	52.7	64.0	45.7
	MS	17.1	22.7	48.6	68.3	42.2
Haryana	ML	32.0	50.0	65.0	63.0	62.0
	MS	38.9	60.8	76.9	69.1	70.7
Punjab	ML	61.3	70.1	80.7	88.1	83.1
	MS	54.2	62.3	91.8	91.1	87.2
Rajasthan	ML	−31.2	0.8	47.1	41.4	32.9
	MS	18.5	13.7	53.2	49.7	49.1
All India	ML	4.1	49.6	61.4	70.8	59.7
	MS	23.5	32.3	62.3	73.0	58.9
PADDY/RICE						
Haryana	ML	77.5	75.0	92.7	84.7	88.0
	MS	74.2	72.3	83.8	61.3	73.5
Punjab	ML	81.8	87.1	94.2	98.0	96.0
	MS	92.6	79.1	93.8	87.1	89.2
Orissa	ML	4.4	15.7	74.3	49.6	56.1
	MS	18.4	19.8	5.5	21.8	12.1
West Bengal	ML	−27.3	68.0	70.4	74.4	53.6
	MS	31.5	55.8	62.9	86.7	59.4
Tamil Nadu	ML	13.3	31.1	26.4	52.8	30.1
	MS	32.5	33.0	46.1	31.0	35.3
All India	ML	−9.0	23.9	61.7	76.4	46.5
	MS	27.6	34.8	43.8	51.7	41.7
MAIZE						
Uttar Pradesh	ML	16.1	44.7	46.2	52.0	40.3
	MS	22.5	32.7	51.9	52.7	40.8
Rajasthan	ML	16.8	20.6	31.0	34.7	24.0
	MS	4.1	9.6	10.5	21.9	10.9
All India	ML	5.8	28.1	41.6	50.5	32.3
	MS	12.8	20.6	38.1	45.0	28.9

ML = Marketable surplus; MS = Marketed surplus

Source: Directorate of Marketing and Inspection, Production, Utilization, Marketable and Marketed Surpluses of Wheat, Rice and Maize, Government of India, Faridabad, 1995, pp. 53–73.

Due to low level of production on marginal and small farms coupled with higher proportion of output being consumed by the farm families, the contribution of these size groups of farms to the total marketed quantity is considerably lower than medium and large farmers. The estimates given in Table 3.4 reveal that marginal and small farmers contribute 12.6% to the total marketed quantity of wheat. In the case of rice and maize, the shares of marginal and small farmers in total marketed quantities are 16.8% and 30.5% respectively. As regards milk, marginal and small farmers contribute 53.2 and 22.4% to the total marketed surplus.

Table 3.2: Marketed and marketable surplus of oilseeds and pulses across different size groups of farms

(% of production)

Commodity/State/Particulars	Farm size			Overall
	Small	Medium	Large	
Groundnut (Gujarat)				
Marketable surplus	70.33	78.47	80.01	78.56
Gram (Rajasthan)				
Marketable surplus	71.1	75.7	79.7	76.6
Marketed Surplus	78.7	81.2	86.3	83.6
Mustard (Rajasthan)				
Marketed surplus	91.88	93.29	93.89	92.88

Source:
1. Acharya, S.S., Agricultural Production, Marketing and Price Policy in India, Mittal Publications, Delhi, 1988, p. 268.
2. Patel, G.N., Price Behaviour and Marketing of Groundnut in Gujarat, Ph.D. Thesis, Rajasthan Agricultural University, 1991.
3. Hari Om, Marketing of Rapeseed and Mustard in Bharatpur District of Rajasthan, M.Sc. (Ag. Eco.) Thesis, Rajasthan Agricultural University, 1988.

Table 3.3: Marketable and marketed surplus ratios of foodgrains on different farm size groups (2011–12)

Farm size		Wheat	Paddy/Rice	Maize	Bajra
Marginal	ML	64.8	64.7	86.8	67.5
	MS	61.2	63.0	79.9	60.2
Small	ML	72.2	73.6	88.7	71.8
	MS	69.4	74.0	83.7	66.2
Semi-medium	ML	79.9	81.0	91.6	74.7
	MS	77.7	80.9	91.1	68.7
Medium	ML	84.7	90.3	93.4	72.8
	MS	82.6	83.2	93.3	67.9
Large	ML	88.1	99.1	81.1	80.4
	MS	86.0	72.1	88.2	74.1
All farms	ML	83.0	85.5	90.5	73.5
	MS	80.7	78.0	88.3	67.7

Source: V.P. Sharma and Harsh Wardhan, Assessment of Marketable and Marketed Surplus of Major Foodgrains in India, IIM, Ahmedabad, April 2015.

Table 3.4: Percentage contribution of different size group of farms to total marketed surplus in India

Farm size	Wheat	Rice	Maize	Bajra	Gram	Tur/Arhar	Milk
Marginal	4.1	5.0	10.4	11.5	4.2	3.9	53.2
Small	8.5	11.8	20.1	21.4	9.3	10.3	22.4
Semi-medium	18.4	23.9	28.8	20.4	15.0	16.6	
Medium	34.9	38.3	24.2	29.8	34.0	33.0	15.7
Large	34.1	21.0	16.5	17.0	37.5	36.2	8.7
Total	100.0	100.0	100.0	100.0	100.0	100.0	100.0

Source: Sharma V.P and Harsh Wardhan, Assessment of Marketable and Marketed Surplus of Major Foodgrains in India, IIM, Ahmedabad, 2015. (Kumar P., et al., 2018 for milk)

The quantum of market arrivals in main wholesale assembling markets also provides an assessment of at least that part of the marketed surplus which the marketing system has to handle. But these are often under estimates of the total marketed surplus originating at the farm level because a part of the marketed surplus reaches the consumers specially of the rural areas either directly from the farmers or through village traders. This apart, there is also some under reporting of arrivals in the wholesale assembling markets. The market arrival figures reported in various official publications should, therefore, be used with care. However, there is ample evidence to show that there has occurred a considerable increase in the marketed surplus and market arrivals of farm products. The increase in the market arrivals has been due to the following factors:

1. Increase in the productivity per unit of land area and production per farm;
2. Increase in the inter-regional specialization in crop production which requires larger quantities to be moved from one area to the other;
3. Increases in the monetisation of the farm economy;
4. Migration of population from rural to urban areas as also from farm to the non-farm jobs which decreases the on-farm consumption requirements;
5. Expansions in area under higher yielding and improved seeds which reduces the on-farm requirements of grains for seed; and
6. Increase in the processing facilities which decreased on-farm processing and consequently the consumption of the unprocessed or on-farm processed grains by the farm families.

Planning for marketing system improvement requires estimates of quantities of various farm products entering the market network. While the production statistics is regularly generated by various official and non-officials agencies, the information of marketed quantities is derived from the estimates of marketed surplus-output ratios. Such ratios are essentially based on sample surveys. However, the sample surveys are neither regular nor cover all agricultural products at a time.

Dharam Narain[8] estimated that in 1950–51 considering all agricultural commodities as a group, marketed surplus as a percentage of gross value of output was around 33.4%. There are no such comprehensive estimates available for the recent period. The Task Force[9] on Terms of Trade appointed by the Ministry of Agriculture, Government of India for the purpose of working out the weights for constructing the index of prices received by the agricultural sector estimated the value of marketed surplus in triennium ending (TE) 1990–91. According to these estimates, the marketed surplus output ratio during TE 1990–91 was 64.1%. This ratio has gone up further to around 80% in the recent period.

The marketed surplus-output ratio for various agricultural commodities for 1950–51, 1997, 2007 and 2014–15 are given in Table 3.5.

The marketed surplus-output ratio for all agricultural commodities has shown an upward rise overtime. This increased from around 30% in rice and wheat in 1950–51 to 84.4 and 73.8% in 2014–2015. Similarly, for coarse cereals (bajra, jowar and maize), the marketed surplus-output ratio went up from 27% in 1950–51 to 68.4% in bajra and 88.1% in maize in 2014–2015. Among the crops, there existed considerable difference in the marketed surplus-output ratios. The ratio is:

(a) Lower for coarse grains among cereals;
(b) Lower for cereals among foodgrains;

Table 3.5: Marketed surplus-output ratios of important agricultural commodities in India at different points of time

				(% of production)
Commodity	1950–51	1997	2007	2014–15
Rice	30.0	60.1	79.2	84.4
Wheat	30.0	61.4	66.3	73.8
Bajra	27.0	47.2	72.2	68.4
Maize	24.0	59.1	78.6	88.1
Jowar	24.0	55.0	61.0	66.1
Barley	–	55.8	58.9	77.7
Other cereals	18.0	–	30.0	48.9
Total cereals	29.2	–	73.5	79.2
Gram	35.0	–	76.8	91.1
Arhar	50.0	47.2	83.6	88.2
Urad	–	53.2	78.4	92.2
Moong	–	76.5	80.3	90.6
Other pulses	55.0	72.1	79.0	94.4
Total pulses	45.3	–	79.7	91.2
Total foodgrains	30.3	72.4	73.9	79.3
Groundnut	68.3	–	91.6	91.6
Mustard	84.3	82.4	87.7	90.9
Sesamum	–	69.3	91.3	95.4
Sunflower	–	86.9	97.2	100.0
Soyabean	–	93.8	95.8	97.6
Other oilseeds	86.3	94.1	94.1	100.0
Total oilseeds	73.6	86.3	93.0	94.5
Sugarcane	100.0	92.5	99.6	99.6
Cotton	100.0	100.0	96.2	98.6
Jute and mesta	100.0	96.5	97.4	98.6
Spices	–	–	92.5	94.0
Total fruits	–	97.0	89.9	97.0
Total vegetables	–	83.0	79.2	94.0
Total F and V	70.0	–	82.9	95.1
Flowers	–	–	100	100.0
Tobacco	–	–	96.8	98.0
Fodder	–	–	23.3	30.0
Milk	50.0	60.0	70.0	70.0
Meat	98.0	100.0	100.0	99.0
Fish	98.0	98.0	98.0	99.0
Eggs	98.0	88.2	90.0	98.0
Raw wool, skin, hides	–	100.0	100.0	100.0

Source:

1. Dharam Narain (1961), Distribution of Marketed Surplus of Agricultural Produce by Size Level of Holdings in India, Institute of Economic Growth, Occasional Paper No. 2, pp. 33–38 for 1950–51.
2. Acharya, S.S., Agricultural Marketing in India, Millennium Study of Indian Farmers, Ministry of Agriculture, CoI, 2003 for 1997.
3. For 2007 and 2014–2015, Weighted averages calculated by us based on crop data given in Agricultural Statistics at a Glance, 2009, 2014 and 2018, Ministry of Agriculture, Government of India, New Delhi.

(c) Lower for foodgrains than for other crops;

(d) The ratio is almost equal to one for non-food crops like cotton and jute.

(e) For edible oilseeds this ratio is estimated to be more than 94%.

The marketed surplus-output ratios also differ across size group of farmers. Small sized farmers with lower output per farm usually sell a smaller proportion of their output than the larger sized farmers. The marketed surplus output ratios also increases with the increase in per farm output.

The estimated quantity of marketed surplus of individual agricultural products and product groups and changes therein during the last 65 years are given in Table 3.6.

Table 3.6: Marketed surplus of agricultural commodities

(million tonnes)

Commodities	1951			2014–15		
	Production	MS ratio	MS	Production	MS ratio	MS
Rice	20.58	30.0	6.2	105.5	84.4	89.0
Wheat	6.46	30.0	1.9	86.6	73.8	63.9
Bajra	2.60	27.0	0.7	9.2	68.4	6.3
Maize	1.73	24.0	0.4	24.2	88.1	21.3
Jowar	5.50	24.0	1.3	5.4	66.1	3.6
Others	5.54	18.0	1.0	4.1	48.9	2.0
Total cereals	42.41	29.2	11.5	234.9	79.2	186.1
Gram	3.65	35.0	1.3	7.3	91.1	6.7
Arhar	1.72	50.0	0.9	2.8	88.2	2.5
Other pulses	3.04	55.0	1.7	7.0	91.4	6.4
Total pulses	8.41	45.3	3.9	17.1	91.2	15.6
Total foodgrains	50.82	30.2	15.4	252.0	79.2	199.5
Groundnut	3.48	68.3	2.4	7.4	91.6	6.8
Mustard	0.76	84.3	0.6	6.3	90.9	5.7
Soyabean	–	–	–	10.4	97.6	10.2
Other oilseeds	0.92	86.3	0.8	3.4	97.1	3.3
Total oilseeds (9)	5.16	73.6	3.8	27.5	94.5	26.0
Sugarcane	57.05	100.0	57.0	362.3	99.6	360.8
Cotton*	3.04	100.0	3.0	34.8	98.8	34.4
Jute and mesta*	3.31	100.0	3.3	11.1	98.6	10.9
Fruits	NA	NA	NA	90.2	97.0	87.5
Vegetables	NA	NA	NA	169.2	94.0	159.0
Total F and V	54.52**	70.0	38.2	259.3	95.1	246.5
Milk	17.0	50.0	8.5	146.3	70.0	102.4
Meat	NA	98.0	NA	6.7	99.0	6.6
Fish	0.75	98.0	0.7	10.3	99.0	10.2
Eggs***	1.83	98.0	1.8	78.5	98.0	76.9

* Million bales; ** Pertain to 1981–82; *** Billion number

The marketed surplus of cereals is estimated to have gone up from 11.5 million tonnes during 1950–51 to 186.1 million tonnes during 2014–15. This has happened on account of increase in both output and marketed surplus output ratio. The marketed surplus of rice is estimated to have gone up from 6.2 to 89.0 million tonnes, wheat from 1.9 to 63.9 million tonnes and of maize from 0.4 to 21.3 million tonnes. In the case of pulses, the marketed quantity is estimated to have gone up from 3.9 million tonnes during 1950–51 to 15.6 million tonnes during 2014–15. For foodgrains as a whole, the marketed surplus is estimated as 199.5 million tonnes for 2014–15 as against 15.4 million tonnes for 1950–51.

The marketed surplus of oilseeds went up from 3.8 million tonnes to 26.0 million tonnes and of sugarcane from 57.0 million tonnes to 360.8 million tonnes during this period. In the case of cotton, almost whole of the output is marketed. There has been considerable increase in the marketed quantities of vegetables and fruits. The marketed surplus of both of these crop groups taken together is estimated to have gone up from 38.2 million tonnes during 1981–82 to 246.5 million tonnes during 2014–15. Marketed surplus of livestock products also went up considerably during this period. It increased from 8.5 million tonnes to 102.4 million tonnes of milk, from 0.7 million tonnes to 10.2 million tonnes of fish and from 1.8 billion to 76.9 billion eggs during this period.

PATTERN OF DISPOSAL OF MARKETED SURPLUS

For understanding the marketing system, it is important to know the pattern of disposal of marketed surplus according to place and time of sale.

1. Place of Sale

Based on the place of disposal, the sale by farmers may be divided into village sale and market sale. The quantity of the produce marketed by the farmers in the villages and markets varies with the commodity and the size group of the farmers. The extent of the sales made in the villages and in the markets by the farmers in Rajasthan during the 1970s is shown in Table 3.7.

Table 3.7: Extent of sale within villages and in markets				
				(percentage of total sales)
Commodity	Place of sale	Size group of farms		
		Small	Medium	Large	Overall
Wheat	Within village	39.90	32.60	16.80	23.70
	In market	60.10	67.40	83.20	76.30
Maize	Within village	46.93	36.19	23.81	30.65
	In market	53.07	63.81	76.19	69.35
Bajra	Within village	24.26	26.90	21.65	23.36
	In market	75.74	73.10	78.35	76.64

Source: Research Report of the Department of Agricultural Economics, University of Udaipur, Rajasthan, Campus Jobner, (Jaipur), 1976.

About 70 to 75% of the marketed surplus of foodgrains was marketed by the farmers in the market, and only 25 to 30% in the village. The percentage of the quantity marketed in the villages decreases with the increase in the size of the farm or the quantity of the produce available for marketing.

There is one more important factor which affects the choice of place of sale by the farmers. The farmers living in a village located away from the market sell more in the village as compared to those living in the villages located near the market. Similarly, the farmers of those villages which are not connected to the markets by roads usually sell a higher percentage of their marketed surplus in the villages as compared to the farmers of the villages which are connected to the markets by roads.

The advantages of market sales in receiving a higher price as compared to the village sales is slowly disappearing with the changes in the market structure and purchase behaviour of traders or processors. The system of pre-harvest contracts for fruits is prevalent since long. Contract farming arrangements, facilitated by amendments in State Agricultural Marketing Acts, help in linking-up the farmers with contractors for, inter alia, purchase of farmers produce. These new developments have blurred the distinction between village and market sales to a large extent.

2. Time of Sale

The producer's decision to sell his produce at a particular time is of vital importance, for it may bring about a glut or scarcity in the market, which ultimately affects the price of the produce.

Earlier, as a normal practice nearly half of the marketed surplus of cereals used to reach the markets during the first three months after harvest and other half was spread over the remaining nine months; with farmers or traders performing the function of carrying stocks from post-harvest season to the later part of the season. But as a consequence of increase in the involvement of public agencies in the foodgrains trade and increase in the cost of carrying stocks, some significant changes in the pattern of market arrivals have taken place as has been brought out by Tyagi[10]. Though Tyagi's results pertain to mid-eighties, no significant change occurred since then in the time pattern of sale of wheat and rice by the farmers.

As shown in Table 3.8 out of the total market arrivals of wheat, the proportion arriving in the first quarter after harvest has increased from 51.2% in 1961–62 to 63.6% in 1983–84 at the all-India level. In major wheat producing States of Punjab and Haryana, the concentration of market arrivals in the first quarter after harvest has sharply increased from 52.5% in 1961–62 to more than 88% in 1983–84. In the case of rice, the concentration of arrivals in the post-harvest season is even more than that in wheat. This phenomenon of very high concentration of market arrivals during a short period has been termed by Tyagi as 'Markets Getting Choked'.

In the case of rice and wheat, owing to the large size of public distribution system, quite often the inta-year rise in prices is not sufficient to cover the storge costs. Moreover, many farmers do not have adequate storage facilities. Therefore, tendency of large sales during first three months of harvest should not be considered as irrational behaviour of farmers.

▌ COMMERCIALIZATION OF INDIAN AGRICULTURE

There are two simple ways to assess the degree of commercialization of Indian agriculture. One is the degree of output commercialization and second relates to the use of purchased inputs. The growth in marketed surplus, arising from both increase in production and increase in marketed surplus-output ratios, as illustrated in the preceding section, clearly indicates the step-up in the commercialization of output side of Indian agriculture. Input side of commercialization of Indian agriculture has been discussed in Chapter 6.

Table 3.8: Changes in time pattern of market arrivals of wheat and rice in some states of India

(% of total arrivals)

Crop/State	Year	Quarters after harvest			
		I	II	III	IV
WHEAT					
All India	1961–62	51.2	17.1	16.4	15.3
	1970–71	56.8	21.5	12.6	9.1
	1983–84	63.6	13.0	13.0	10.4
Punjab	1961–62	52.5	22.3	15.6	9.6
	1970–71	70.1	19.6	7.6	2.7
	1983–84	93.2	4.7	1.3	0.8
Haryana	1961–62*	52.5	22.3	15.6	9.6
	1970–71	67.0	21.7	7.3	4.0
	1983–84	88.7	6.0	4.3	1.1
RICE					
All India	1961–62	28.6	33.5	23.1	14.8
	1970–71	36.4	23.3	20.4	10.9
	1983–84	51.4	22.2	17.4	9.0
Punjab	1961–62	71.6	26.6	1.8	–
	1970–71	87.2	9.3	2.2	1.3
	1983–84	96.6	2.0	0.3	1.2
Haryana	1961–62*	71.6	26.6	1.8	–
	1970–71	91.6	7.8	0.5	0.1
	1983–84	95.7	3.6	0.7	–

* Same as for Punjab as at that time they were parts of one State, First quarter is April–June for wheat and October–December for rice.
Source: Tyagi, D.S., Managing India's Food Economy, Sage Publications, New Delhi, 1990, pp. 114–16.

PROJECTIONS OF DEMAND AND SUPPLY OF FARM PRODUCTS

Projections of demand and supply are continuously made by traders and agribusiness organizations, but these are mostly informal projections and not available to others. On the other hand, the government and government organizations need such projections as a part of perspective planning of various development programmes. Projections of demand and supply of food are crucial for attaining food security at global, national and regional levels. Demand projections are needed for planning of production programmes as also for decisions relating to exports and imports. The interest in projections of demand and supply started with the launch of Five Year Plans, and it increased considerably during the last seven decades.

The demand projections for agricultural commodities are made by both the government as well as by researchers and academicians. In all the Five Year Plan documents, the medium-term demand projections for the terminal year of the plan are made with a view to preparing programmes to achieve the required levels of production or supply. Long-term demand projections are made by governments, researchers and academicians or international agencies for strategic or perspective planning.

In making projections of demand for food commodities, to be used as an input in planning for production programmes, three essential components need to be kept in view. One is the consumption demand which depends inter alia on population growth, growth of per capita income, and income elasticity of demand. The second component is the demand for seed, feed and the part of the grains that are wasted at different stages of marketing. The second component is usually expressed as a percent of gross production, which for long has been 12.5% (5% for seed, 5% for feed, and 2.5% wastage). These two components together constitute the domestic demand. The third component is the export demand, which relates to commitments, if any, made to other countries for export. India does not have any such formal firm commitment. The predictions of consumption demand can be made by following three different approaches. First is the simplest approach to project the demand by using projected population and base year consumption parameters (assuming the static behavior of consumption). Second is the normative approach that is based on the normative nutritional requirement as recommended by Nutritionists. And the third is the behaviouristic approach, which is based on the growth of population, changing behaviour of consumption owing to changes in per capita income and life styles (income elasticities). The projections of domestic demand made by different organizations vary widely mainly due to the differences in assumptions about (a) income elasticity of demand, (b) rate of growth of per capita income, (c) rate of growth of population, and (d) demand for seed, feed and wastage.

Supply projections are usually based on the predictions of area under crops or number of animals, growth of critical inputs, productivity growth, and output-input price ratios.

Medium-term projections of demand and supply of key agricultural commodities for the terminal year of 11th Five Year Plan (2011–12) are available from the Planning Commission's documents (Table 3.9) The demand for food grains (including 2 million tonnes for augmenting buffer stocks and export of 8 million tonnes) was projected as 244 million tonnes. As against this, the projected production was 214 to 240 million tonnes. In the case of oilseeds, while the demand was projected at 53 million tonnes, the supply was likely to be 45 million tonnes. In order to compare the projections of supply with the actual production during the year 2011–12, we have given the production levels of these crops achieved during 2011–12 in the last column of Table 3.9. It can be seen that country could achieve the projected/targeted production levels

Table 3.9: India: Demand and supply projections for 2011–12 and actual production

(million tonnes/bales)

Crops	Projected demand	Projected supply	Actual production
Foodgrains	244	214–240	259.3
Oilseeds	53	45	29.8
Sugarcane	340	278–334	361.0
Cotton	29	16–50	35.2
Jute and mesta	10	11	11.4

Source: Planning Commission, Government of India, Report of the concerned Working Group, Dec. 2006 and Agricultural Statistics at a Glance, 2014.

for foodgrains, sugarcane, cotton and jute-Mesta. It was only the case of oilseeds where the actual output level was far short of the projected supply.

Projections of demand and supply of cereals in India have been made for the year 2020 by Kumar, Rosegrant and Hazell[11] as a part of IARI–IFPRI study. Their projections are shown in Table 3.10. According to these projections, total demand for cereals in 2020 is likely to be 293.4 million tonnes. As regards supply prospects, if the current productivity growth is sustained, the production would exceed the demand by 15.9 million tonnes. However, if productivity growth declines, India may have to import 23 million tonnes of cereals by the year 2020.

Table 3.10: India: Projected cereal supply, demand and net trade (2020 AD)

(million tonnes)

Particulars	Rice	Wheat	Coarse cereals	Total
Declining productivity Growth				
Supply	120.5	107.6	42.3	270.4
Demand	124.5	111.0	57.9	293.4
Net exports	–4.0	–3.4	–15.6	–23.0
Sustained productivity Growth				
Supply	134.0	127.3	48.0	309.3
Demand	124.5	111.0	57.9	293.4
Net exports	9.5	16.3	–9.9	15.9

Source: Kumar, P., Mark Rosegrant and Peter Hazell, Cereals Prospects in India to 2020: Implications for Policy, IFPRI Vision 2020, Brief 23 June, 1995.

Kumar et al.[12] have also come with projections of demand for some key agricultural commodities for the year 2025 (Table 3.11). According to these projections, the domestic demand for rice will be 117.3 million tonnes and for wheat 104 million tonnes. In the case of pulses, the demand during 2025 will go up to 23.9 million tonnes, and of oilseeds 57.6 million tonnes. The derived demand for sugarcane is projected at 382.7 million tonnes. They have also given projections of supply of these key agricultural commodities. The supply projections are based on the assumption that the trend in area growth, total factor productivity growth and input-output prices as observed during 1981–2005 will continue.

Table 3.11: India: Projected domestic demand and supply of key agricultural commodities

(million tonnes)

Crop	TE 2005		TE 2025	
	Demand	Supply	Demand	Supply
Rice	93.16	87.78	117.34	151.88
Wheat	70.84	70.04	104.01	125.68
Pulses	13.96	13.88	23.87	17.34
Oilseeds	30.29	25.94	57.62	73.27
Sugarcane	255.67	252.78	382.66	476.91

Source: Kumar et al. (2010).

The comparison of projections of demand and supply imply that the demand for pulses would far exceed the potential supply by the year 2025. As regards rice, wheat, oilseeds and sugarcane, the domestic supplies may be sufficient to meet the domestic demand, if assumptions used in the study hold good.

Projections of demand for livestock products were made by Dastagiri[13] (2002) using simple growth rate model. His projections are shown in Table 3.12. It may be mentioned here that against a projected demand of 94 million tonnes, India produced 108 million tonnes of milk in 2010. It should be noted that under the new Dairy Development Plan launched in 2012 (being implemented by NDDB), the targeted milk production by 2020 is 210 million tonnes.

Table 3.12: Projections of demand for livestock products in India

(million tonnes)

Product	Area	2010	2020
Milk		94.21	147.21
	Rural	52.66	70.19
	Urban	41.55	77.02
Mutton and goat meat		3.80	12.72
	Rural	0.65	0.77
	Urban	3.15	11.95
Beef and buffalo meat		0.84	1.14
	Rural	0.46	0.55
	Urban	0.38	0.59
Chicken		0.52	0.81
	Rural	0.28	0.39
	Urban	0.24	0.42
Fish		1.79	1.63
	Rural	1.29	1.18
	Urban	0.50	0.45
Eggs*		24.79	44.05
	Rural	10.87	16.67
	Urban	13.92	27.38

* Billion numbers

Source: Dastagiri, M.B., Demand for Livestock Products in India—Current Status and Projections for 2020, Agricultural Economics Research Review, 2002, pp. 176–82.

The latest projections of demand and supply of agricultural commodities for the year 2021–22 and 2032–33, by a working group of NITI Aayog, Government of India, are shown in Table 3.13. Except for oilseeds, the projected supply is either more or nearly equal to the anticipated demand. In edible oils, India would continue to depend on imports because demand will exceed the supply. The Government is making efforts to reduce the dependence on imports for edible oils.

CHARACTERISTICS OF DEVELOPED MARKETS AND IDEAL SYSTEM OF AGRICULTURAL MARKETING

A developed market is the *sine qua non* of any developing economy. It should satisfy the objectives of marketing system for all the persons associated with marketing in the process of the movement of produce from producer to the consumer.

Table 3.13: India: Projections of demand and supply of agricultural commodities (million tonnes)				
Commodity	**2021–22**		**2032–33**	
	Demand	Supply	Demand	Supply
Rice	109.3	122.2	120.8	151.7
Wheat	97.1	109.5	113.5	138.8
Other cereals	46.9	47.9	67.5	61.8
Total cereals	253.3	279.6	301.8	352.3
Pulses	26.7	24.4	35.2	34.0
Foodgrains	280.0	304.0	337.0	386.3
Oilseeds (edible oils)	61.4 (17.2)	39.2	99.6 (27.9)	60.0
Sugar cane (sugar)	(40.3)	360.1 (36.7)	(46.4)	435.7 (44.4)
Fruits	126.7	120.8	203.6	202.7
Vegetables	224.3	221.0	360.8	362.9
Total (Hort)	NA	383.5*	NA	659.4*
Nuts	2.4	NA	3.9	NA
Milk and products	181.9	202.7	292.2	329.7
Eggs (E)**	NA	116.2	–	215.9
Meat (M)	NA	13.2	–	42.3
Fish (F)	NA	14.3	–	23.6
Total (EMF)	18.6	NA	35.5	NA
Cotton***	NA	49.2	NA	113.9
Jute and mesta	NA	11.1	NA	12.4
Wool	NA	48.8	NA	50.2

* includes flowers, ** billion number, *** million bales
Source: D&S Projections Towards 2033, Working Group Report, NITI Aayog, GoI, Feb. 2018

A good developed market should possess the following characteristics:

1. A good developed market should provide commodities which the consumers want and are ready to pay for.
2. It should provide a wide variety of products to consumers so that they may easily choose for themselves. The varieties should not be so wide as to create a confusion for them.
3. No harmful products should be offered for sale in the market. Precautions should be taken to protect consumers.
4. The information on the presence of goods in the market and their relative merits should be available to all the prospective consumers.
5. There should not be any sort of pressure on the consumers to buy products from a particular trader or class of traders.
6. The retailing services should be available in the market (together with wholesale facilities) for small consumers.
7. Prices should be fair and uniform for the products for all categories of consumers.
8. There should not be any wastage of the products in the market.
9. The producer should be able to sell his surplus quickly and get a price which is consistent with the demand and supply situation.

10. Adquate and efficient storage, transportation and processing facilities should be available.
11. Proper grading facilities should be available.
12. The methods of packaging should be as per the requirements of different farm products as well as the consumer choices.

IDEAL MARKETING SYSTEM

In agriculture-oriented developing countries like India, agricultural product marketing services play a pivot role in fostering and sustaining the tempo of agricultural and rural development. The ideal marketing system has been defined by Moore, Johl and Khusro[14] in the following words:

"The ideal marketing system is one that maximizes the long run welfare of society. To do this, it must be physically efficient; otherwise the same output could be produced with fewer resources; and it must be allocatively efficient; otherwise a change in allocation could increase the total welfare where income distribution is not a consideration."

An ideal marketing system should operate with maximum physical and allocative efficiency. For maximum physical efficiency, such basic physical functions as transportation, storage, and processing should be carried on in such a way as to achieve the highest output per unit of cost incurred on them. Similarly, an ideal marketing system must allocate agricultural products in time, space and form to intermediaries and consumers in such proportions and at such prices as to ensure that no other allocation would make consumers better off. To achieve these conditions, prices throughout the marketing system must be efficient and must, at the same time, be equal to the marginal costs of production and marginal consumer utility.

The pricing system is efficient when it fulfils the following conditions:

1. The prices through space (geographically) should vary on the basis of the cost of transportation from one point to another. This can be judged by observing the price differential between the markets and comparing it with the cost of transportation, as well as by studying the correlation coefficient between the prices obtaining in the various markets. Higher correlation coefficients indicate that the prices in different markets move closely with one another.
2. The prices through time should vary no more than the cost of storage from one period to another.
3. The prices of different forms of products should vary no more than the differences in the costs of processing.

CHARACTERISTICS OF A GOOD MARKETING SYSTEM

The following characteristics should exist in a good marketing system:

1. One of the conditions of an efficient agricultural marketing system is that there should not be unnecessary government interference in free market transactions. The methods of intervention include restrictions on foodgrain movements, restrictions on the quantity to be processed or on the construction of processing plant, price supports, rationing, price ceiling, entry of persons in the trade, etc. When these conditions are violated, the inefficiency in the marketing system creeps in, and commodities pass into the black market. They are not then easily available at fair prices to the consumers.

2. The marketing system should operate on the basis of the independent, but systematic and orderly, decisions of the millions of the individual consumers and producers whose lives are affected by it.

3. The marketing system should be capable of developing into an intricate and far-flung marketing system in view of the rapid development of the urban-industrial economy.

4. The marketing system should bring demand and supply together and should establish an equilibrium between the two.

5. The marketing system should be able to generate employment by ensuring the development of processing industries and convincing the people to consume more processed foods, consistent with their tastes, habits and income levels.

6. The marketing system should eliminate wastage of farm products along the supply chain, which helps both farmers and consumers.

SCIENTIFIC MARKETING OF FARM PRODUCTS

The tendency among the farmers to market their produce has been increasing. Production is complete only when the product is marketed at a price remunerative to the farmer. Increasing specialization in production, higher marketable/marketed surplus of the produce and alternative channels of marketing have increased the importance of the marketing decisions and related activities for the farmers.

COMMANDMENTS OF SCIENTIFIC MARKETING

The farmers can gain more if they follow the following commandments of scientific marketing:

1. *Bring the Produce for Sale after Cleaning.* The produce brought by the farmers for sale in the market must be clean and free from such impurities as dirt, sand and pebbles, and should be unmixed with produce of another quality. Impurities, when present, lower the price offered by the trader-buyer in the market. The fall in price is more than the extent of the impurity present in the produce would warrant. Many buyers are interested in buying only clean produce, and are ready to pay a higher price for it. Clean produce attracts more buyers and in turn higher prices to the seller.

2. *Sell Different Qualities of Products Separately.* The produce of different varieties should be marketed separately. Many farmers mix the lots because of their small quantity, and get a low price for the mixed lot. It has been observed that when different varieties of products are marketed separately, the farmers get a higher average price because of the buyer's preference for specific varieties.

3. *Sell the Produce after Grading.* It is always advantageous for the farmer to market the produce after grading. Graded produce is sold off quickly. Studies have shown that sellers get a higher total income when they sell the produce after grading it. The additional income generated by the adoption of grading is more than the cost incurred on the process of grading. This shows that there is an incentive for the farmers for the production of good quality products and selling after grading.

4. *Keep Abreast of Market Information.* The farmer must keep in touch with market news in order to know the prices that prevail in different markets. Price information helps him to take decisions about when and where to sell the produce so that a better price may be obtained. Farmers who go to the market for sale without prior knowledge of the prevailing prices repent later. By and large, they agree to sell at

a price which is lower than the price expected by them because it is difficult and cumbersome for them to postpone the sale or take the produce to another market. They should know in advance whether the crop is covered under price support programme and what is the support price. They should never sell below this price.

5. *Carry Bags/Packs of Standard Weighs*. Farmers should weigh their produce and fill each bag with fixed quantity before moving to the market for sale. A majority of the farmers do not weigh their produce before taking it for sale and suffer loss by way of a possible malpractice in weighing; or they may have to make excess payments in transit (octroi, transport costs).

6. *Avoid Immediate Post-harvest Sales*. Farmers should avoid sales immediately after a harvest. The prices of the produce touch the lowest level in the peak marketing season. Farmers can get better prices by availing themselves of the warehousing facilities existing in their area. Farmers can meet their cash needs by pledging the warehouse receipts to nationalized banks. However, before taking such decisions, they should carefully compare the expected rise in the prices from the post-harvest season to the later months with cost of storage and interest on the value of the produce.

7. *Patronize Cooperative or Group Marketing*. Farmers can get better prices by sales through a cooperative marketing society or a marketing group and can avoid the possibility of being cheated in the process of marketing their produce. The cost of marketing, particularly the transportation cost for farmers having a small quantity of marketable surplus, is minimized, for transportation is arranged cooperatively by the society or the group, and the profit earned by the society or group is shared among its members.

8. *Sell the Produce in Regulated Markets*. The farmers should take their produce for sale to the nearby regulated markets rather than selling them in villages or unregulated markets. In regulated markets, farmers are not required to bear, many of the marketing charges. They get the sale slips in the regulated markets, which show the quantity of the produce marketed and the amount of charges deducted from the value of the produce. Sale slips protect farmers against the malpractice of deliberate erroneous accounting or unauthorized deductions.

9. *Choose the Right Varieties*. Different varieties fetch different prices in the market. Certain varieties enjoy a price premium in the market. For example, basmati varieties of paddy command a higher price than others. Scented varieties are priced higher than non–scented varieties. In the case of cotton, long staple varieties fetch better prices. This is also true in the case of vegetables. The farmers should, therefore, ought to be selective and look for varieties which can be sold at relatively higher prices.

10. *Minimise Chemical Residues*. Over the years, the consumers are becoming increasingly conscious of chemical residues in the products. So much so that the products grown on land without the application of chemical fertilizers and insecticidal sprays/dustings (organic products) are bought at premium prices. Given this change in consumer preferences, the growers would need to avoid the indiscriminate use of chemicals in farming for realizing higher prices of their products.

11. *Avail the Advantages of Contract Farming Arrangement*. The corporate sector (processors or bulk traders) is increasingly coming forward to make advance agreements with the farmers for purchase of the farmers' produce in advance of even the sowing season. This apart, the seeds of the desired variety and other inputs, including guidance on agronomic practices, are provided by the contractors. The farmers should avail of such arrangements, which can reduce their price risks.

MECHANIZATION/MODERNIZATION OF FOODGRAIN MARKETS

Mechanization has become popular in almost every human activity. In recent years, the pace of mechanization in all stages of farm production, including harvesting and threshing, has gained momentum. Increased mechanization in agriculture has reduced the time taken in performing various farm operations as well as the cost of these operations. It has also saved farmers from the drudgery of farm operations. Mechanization has led to increased production. As a result, the marketable and marketed surplus has also gone up. This has created overcrowding and congestion in the markets, specially those of foodgrains.

Indian grain markets are intensive users of human labour in all the marketing processes, from the unloading of the produce to the final sale. This results in a higher cost of performing marketing functions, a high spread of marketing margin/price, and inconvenience to the sellers because more time is taken in loading, unloading, weighing and preparing the sale slips of the produce. At present, in many market yards virtually no mechanical device is used either to unload, clean, dry, weigh, bag or load the grains for moving out of the markets. As a result, the present market yards are congested and call for an enlargement of market capacity. This market capacity can be enlarged either by enlarging the size of the market yard or by building a new market yard. The building of a new market yard is very costly because of the high cost of land and building materials. Moreover, the construction of a new market yard necessitates the simultaneous development of such infrastructural facilities as roads, telephones, light, water, godowns and banking; and all these are capital intensive in nature. Therefore, the only alternative to the enlargement of market capacity is the enlargement of the size of the market yard; and this can be done by the introduction of mechanization for various marketing operations. And this is precisely what has been suggested by many.

Mechanization may be introduced at three stages of the *marketing process*, i.e. movement of goods from the production point to the market; buying and selling operations; and movement of the produce from the market to the consumers.

The introduction of mechanization in the following marketing operations will enlarge the market capacity:

1. *Drying of Produce:* The produce, after harvesting/threshing, has to be dried to the desired moisture level to prevent losses during storage and to prolong its keeping quality. The presence of excess moisture makes milling operations difficult and unsatisfactory. Mostly the drying of produce is done by placing it in the sun. This takes a long time, and at times, the produce gets spoiled, particularly when the weather is bad. Quick drying is necessary if the price advantage is to be taken. The drying of the produce up to the desired level of moisture can be done by mechanical driers. Mechanical driers of different capacity are available, and can easily dry up to 20 quintals produce in an hour. The cost of mechanical drying is not very high. These driers reduce the moisture content in the produce to the extent needed.

   The Central Storage Institute, Chandigarh, has developed a special electronic device for the measurement of moisture in the grains. The principle behind the device is that grains have a dielectric constant, which varies with the moisture content of the grain. The variation in the dielectric constant is used to vary the effective capacitance of a suitable capacitor assembly. By measuring this capacitance, it is possible to indicate the moisture content of the grains. The device can measure the moisture over a range of 5–30%. Such moisture meters should be made available in all the markets.

2. *Cleaning the Produce:* The second stage, at which mechanization may be introduced in the marketing process, is in the cleaning of the produce to make it free from extraneous matter and admixture of other grains. This operation is presently manually performed. The grain is passed over a set of screens (sieves) and then subjected to oscillations. A single person is able to clean only about one quintal of grains per hour. As a result, the market process is slowed down. The process of the cleaning of the produce can be mechanized by using the available electric or diesel-operated cleaners with air blowers. Scalper cleaners are also available. Power-operated cleaners comprise either oscillating or rotating screens and aspiration arrangements. In these units, grain is lifted to the hopper by a conveyor; and as it passes from the hopper to the screens, it is subjected to aspirations. Almost all extraneous matter, except that having the same size and weight as the grain, is removed by this cleaner. The capacity of these cleaners equipped with electricity or diesel-driven engines ranges between 25 and 100 quintals per hour. However, for a small quantity of the produce, manual cleaners, with some fixed screen and a blow fan, and having a capacity of 10 to 15 quintals per hour per pair of workers, may be used.

The Central Institute of Agricultural Engineering, Bhopal has developed a hand-operated double screen grain cleaner. The cleaner assembly consisting of three sieves is suspended at an elevated point with the help of four ropes and is operated by oscillating it to and fro with hands. Grain is fed to the topmost sieve in batches of 5 to 10 kg. Separation takes place on the basis of difference in the size of the foreign matter and grains. The grain cleaner can clean 1.5 to 2 quintals of grain per hour. For such markets as in Punjab and Haryana, where arrivals are very high, automatic grain cleaning machines are needed. The Punjab State Agricultural Marketing Board is arranging to provide movable cleaning units to the farmers to ease the problem of labour faced in major markets during the peak marketing seasons.

3. *Grading of Produce:* The importance of the grading and standardization of grains in the marketing process has been increasingly realized. The produce of one grade can be auctioned at one time instead of auctioning each lot separately. But this procedure calls for the grading of the produce on commercial lines. Mechanical graders are available and can be used. Oscillating screens for the "shape" separation of the produce, vibratory boards and airblasts for gravity separation, and the photo-electric or magnetic eye for reflection separation—all these be used.

4. *Bagging and Stitching:* This is another marketing operation which requires a lot of manual labour. In the peak season, it is difficult for the available palledars in the market to bag and stitch the total grain arriving on a particular day. Mechanical baggers and automatic stitching machines can help in completing the work in time.

5. *Weighing:* Weighment of the produce in the market is done either by beam scales or platform scales. Both the methods are labour-intensive and slow. A mechanical device to weigh the produce has become necessary to facilitate the bulk handling of increased production and marketed surplus. A weighbridge can weigh a full cartload or truckload. This reduces time as well as cost. Weighbridges have been installed in several areas and markets of the country. Electronic balances are common in wholesale markets. These balances are more accurate in weighing.

6. *Grain Sampling:* The grain stored in bags is sampled with the help of *purkhi* or a tubular sampler. This sampling technique is difficult when grain is stored in bulk or is moving over a conveyor. Vacuum samplers operated by power can be employed for drawing samples.

7. *Loading and Unloading:* The facilities for bulk handling of grain in India are very limited. Grain is, therefore, carried in jute bags on the backs of labourers. The process of unloading or loading a truck takes very long time even when two to three *palledars* are employed at a time. Often, because of the availability of a limited number of *palledars* in the market, trucks have to wait for a long period of time till loading/unloading operations are completed. The installation of a mechanical conveying system is inevitable in the present times because of the non-availability of labourers and high labour cost. A mechanical conveying system is in use in rice mills, flour mills and pulse mills. Four different types of conveyors are available for this work. These are:

   (a) *Bucket Elevators:* These are similar to Persian wheels and are suitable for vertical lifting up to heights ranging from 2 to 50 metres, with their capacity between 5 to 50 tonnes per hour.

   (b) *Belt Conveyors:* They are used for horizontal conveying of grains. Belts mounted on pulleys are used for long distances; trolley-mounted small portable belts are used for short distances.

   (c) *Screw Conveyors:* These are horizontal or vertical or inclined, and are used for loading grain into transport vehicle or from vehicle into storage godowns. Their capacity is up to 10 tonnes per hour.

   (d) *Pneumatic Conveyors:* These are used for moving the grain upward, downward and sideways. These are best suited where specialized multi-purpose handling is involved. Their capacity ranges from 5 to 100 tonnes per hour. Their use is confined to areas where silo storage structures are used for grain storage. In India, at Hapur (U.P.), Moga (Punjab), and some other locations, silo structures have been erected and these pneumatic conveyors are in use.

8. *Transportation:* This is one area where mechanization in marketing has become most popular. The produce is transported by trucks/tractor trolleys to a large extent. Even in areas where the produce is transported by bullock/camel carts, there has been an improvement, which has been effected by the introduction of pneumatic tyres. Pneumatic tyres have increased the capacity and speed of the bullock and camel carts and reduced the per quintal transport cost of the grain.

9. *Bulk Handling System*: There are considerable losses in foodgrains in the marketing chain starting from harvest stage to storage, handling, transportation and distribution to the final consumer. As per some estimates, about 10% of the foodgrains production is lost at various stages of marketing which comes to a huge quantity of around 20 million tonnes annually. The post-harvest storage and transportation losses can be substantially curtailed by adopting bulk handling, storage and transportation facilities.

   The bulk handling (movement and storage) system minimizes the throughput time, overall cost of handling and transportation and results in lesser damage or contamination by rodents, insects and micro-organisms. This keeps foodgrains in better quality for longer period in storage. This is being done by the use of specially designed wagons with either top filling or bottom or side discharge.

The mechanical handling system also leads to low operating cost as by its adoption the cost involved towards gunny bags, cost of stitching the bags and handling (man power) is eliminated and thus the overall cost of handling is lower.

The bulk handling, storage and transportation technology is common in Australia, Canada, USA and many other countries. This system, to some extent, exists in India with the Food Corporation of India. The present system and technology available in India requires up-scaling and up-gradation in terms of world standards.

10. *Electronic Auctioning*: Auctioning is another area where mechanization in marketing of agricultural commodities is necessary. Over the years, the produce brought by the farmers for sale/auction has increased manifold. Prospective buyers have also increased many times in the markets. Recording of bids manually by the auction clerk in the auction register is a cumbersome process and time consuming. The electronic auction system facilitates the recording of bids and hastens the process of auction. This also avoids legal problems. The system ensures efficiency and transparency in sale proceeds. It does not require the physical presence of buyers at the time of auction.

In the natural course, mechanization in marketing operations is bound to increase. But there are social costs of mechanizing marketing operations—the costs in terms of increase in demand for non-renewable sources of energy like electricity and diesel and the displacement of human labour, accentuating the problem of under-employment and unemployment. The level of mechanization, though is still at a growing stage, has helped and will help in improving the efficiency of the marketing system.

REFERENCES

1. Kiehl, Elmer R. *Agricultural Marketing in India—Role, Strategies and Implications*, The Ford Foundation, New Delhi, January 1969, pp. 1–2.
2. National Commission on Agriculture, Report No. XII, Government of India, Ministry of Agriculture and Irrigation, New Delhi, 1976; p.110.
3. Bansil, P.C., "Problems of Marketable Surplus", *Indian Journal of Agricultural Economics*, Vol. XVI, No. l, January–March, 1961, pp. 26–32.
4. Chauhan, K.K.S. and R.V. Singh., *Marketing of Wheat in Rajasthan*, University of Udaipur, SKN College of Agriculture, Jobner, 1973, pp. 96–97.
5. Rajkrishna, "A Note on the Elasticity of the Marketable Surplus of a Subsistence Crop", *Indian Journal of Agricultural Economics*, Vol. XVII, No. 3, July–September 1962, pp. 79–85.
6. Behrman, J.R., "Price Elasticity of the Marketed Surplus of a Subsistence Crop", *Journal of Farm Economics*, Vol. XLVIII, 4 Part I, November 1966, pp. 875–893, and *Supply Response in Underdeveloped Agriculture*, North Holland Publishing Company, 1968, pp. l86–90.
7. Krishnan, T.N., "The Marketed Surplus of Foodgrains—Is it Inversely Related to Price?", *Economic Weekly*, XVII, February 1965, pp. 325–28.
8. Dharam Narain, "Distribution of the Marketed Surplus of Agricultural Produce by Size Level of Holdings in India", Occasional Paper No. 2, Institute of Economic Growth, Delhi, p. 33–38.
9. Directorate of Economics and Statistics, Indian Agriculture at a Glance, Ministry of Agriculture, Government of India, New Delhi, 2001, 2002.
10. Tyagi, D.S., Managing India's Food Economy, Sage Publications, New Delhi, 1990, pp. 113–16.
11. Kumar, P., Rosegrant and P. Hazell, "Cereals Prospects in India to 2020: Implications for Policy", IFPRI Vision 2020, Brief 23, June 1995.
12. Kumar, Praduman, P.Sinoj, S.S. Raju, Anjani Kumar, Karl M. Rich and Siwa Msangi, "Factor Demand, Output Supply Elasticities and Supply Projections for Major Crops in India", Agricultural Economics Research Review, Vol. 23, No. 1, January–June 2010, p. 1–14.

13. Dastagiri, M.B., "Demand for Livestock Products in India — Current Status and Projections for 2020", Agricultural Economics Research Review, Conference Proceedings 2002, pp. 176–82.
14. Moore, J.R., S.S. Johl and A.M. Khusro. *Indian Foodgrain Marketing*, Prentice Hall of India Pvt. Limited, New Delhi, 1973, p. 18.

APPENDIX 3.1

DERIVATION OF ELASTICITY OF MARKETED SURPLUS OF SUBSISTENCE CROPS

Rajkrishna's Model[1,2]

Only one subsistence crop is considered. The marketed surplus (M) is defined as the difference between production (Q) and consumption (C) by the farm family as follows:

$$M = Q - C \qquad \qquad ...(1)$$

By differentiation with respect to price (P) of the crop concern, we get:

$$\frac{dM}{dP} = \frac{dQ}{dP} - \frac{dC}{dP} \qquad \qquad ...(2)$$

By multiplying all the terms with $\dfrac{P}{M}$, we get:

$$\frac{dM}{dP} \cdot \frac{P}{M} = \frac{dQ}{dP} \cdot \frac{P}{M} - \frac{dC}{dP} \cdot \frac{P}{M} \qquad \qquad ...(3)$$

or

$$\frac{dM}{dP} \cdot \frac{P}{M} = \frac{dQ}{dP} \cdot \frac{P}{Q} \cdot \frac{Q}{M} - \frac{dC}{dP} \cdot \frac{P}{C} \cdot \frac{C}{M} \qquad \qquad ...(4)$$

If we denote

$e = \dfrac{dM}{dP} \cdot \dfrac{P}{M}$, i.e. Marketed surplus-price elasticity

$b = \dfrac{dQ}{dP} \cdot \dfrac{P}{Q}$, i.e. Output-price elasticity

$g = \dfrac{dC}{dP} \cdot \dfrac{P}{C}$, i.e. Consumption-price elasticity

$r = \dfrac{Q}{M}$, i.e. Reciprocal of the sales ratio

(Sales ratio is M/Q)

Then $\dfrac{C}{M}$ may be expressed as ($r - 1$) and equation (4) may be expressed as:

$$e = rb - (r - 1)g \qquad \qquad ...(5)$$

This may be further extended by the introduction of the effect of a price change on consumption via the income effect because a price change affects the income of producer.

Thus, equation (2) may be rewritten as

$$\frac{dM}{dP} = \frac{\partial Q}{\partial P} - \left[\frac{\partial C}{\partial P} + \frac{\partial C}{\partial I} \cdot \frac{\partial I}{\partial P} \right] \qquad \qquad ...(6)$$

where I = Total (net) income of the producer. The increase in the income of the producer by one unit increase in price may be taken as equal to Q, i.e. $\dfrac{dI}{dP} = Q$, if the farmer is only the producer. But if the farmer is producer as well as consumer, the increase in income per unit increase in price is only M.

i.e. $$\frac{dI}{dP} = M \qquad \qquad ...(7)$$

By substituting (7) in (6), multiplying all the terms by P/M, and rearranging the terms, we get:

$$\frac{dM}{dP} \cdot \frac{P}{M} = \left[\frac{dQ}{dP} \cdot \frac{PQ}{QM} \right] - \left[\frac{dC}{dP} \cdot \frac{PC}{CM} \right] - \left[\frac{dC}{dI} \cdot \frac{ICMPQ}{CMQI} \right] \qquad ...(8)$$

If we denote

h = consumption–income elasticity, i.e. $\dfrac{dC}{dI} \cdot \dfrac{I}{C}$

m = sales ratio, i.e. $\dfrac{M}{Q}$

k = ratio of value of production to income, i.e. $\dfrac{PQ}{I}$,

then equation (8) may be expressed as:

or $$e = rb - (r - 1)\, g - (r - 1)\, hmk$$

$$e = rb - (r - 1)\, (g + mkh)$$

Behrman's Model[3]

Behrman in his model considered more than one commodity and introduced the concept of relative prices. He defines the marketed surplus of a subsistence crop 1 as:

$$M_1 = Q_1 - C_1 \qquad \qquad ...(1)$$

where M_1 = Marketed surplus of Crop 1; Q_1 = Production of Crop 1; C_1 = On-farm consumption of Crop 1.

If we define

P_1 = Absolute price of Crop 1
P_2 = Aggregate prices of all other crops produced
P_3 = Aggregate prices of other commodities in consumption,

then the derivative of (1) with respect to P_1 is

$$\frac{\partial M_1}{\partial P_1} = \frac{\partial Q_1}{\partial (P_1 / P_2)} \cdot \frac{\partial (P_1 / P_2)}{\partial P_1} - \left[\frac{\partial C_1}{\partial (P_1 / P_3)} \cdot \frac{\partial (P_1 / P_3)}{\partial P_1} + \frac{\partial C_1}{\partial I} \cdot \frac{\partial I}{\partial P_1} \right] \qquad ...(2)$$

Since $\dfrac{\partial(P_1/P_2)}{\partial P_1} = \dfrac{1}{P_2}$ and $\dfrac{\partial(P_1/P_3)}{\partial P_1} = \dfrac{1}{P_3}$, equation (2) may be written as:

$$\frac{\partial M_1}{\partial P_1} = \frac{\partial Q_1}{\partial(P_1/P_2)} \cdot \frac{1}{P_2} - \left[\frac{\partial C_1}{\partial(P_1/P_3)} \cdot \frac{1}{P_3} + \frac{\partial C_1}{\partial I} \cdot \frac{\partial I}{\partial P_1} \right] \qquad \ldots(3)$$

If the net income (I) is considered approximately equal to $P_1Q_1 + P_2Q_2$, then:

$$\frac{\partial I}{\partial P_1} = \frac{\partial(P_1Q_1 + P_2Q_2)}{\partial P_1}$$

or

$$\frac{\partial I}{\partial P_1} = Q_1 + P_1 \frac{\partial Q_1}{\partial(P_1/P_2)} \cdot \frac{1}{P_2} + P_2 \frac{\partial Q_2}{\partial(P_1/P_2)} \cdot \frac{1}{P_2}$$

or

$$\frac{\partial I}{\partial P_1} = Q_1 + \frac{P_1}{P_2} \cdot \frac{\partial Q_1}{\partial(P_1/P_2)} + \frac{\partial Q_2}{\partial(P_1/P_2)} \qquad \ldots(4)$$

If we substitute (4) in (3), there will be five terms on the right hand side. Multiply each term on both sides by P_1/M_1. Then multiply and divide the first term on the right hand side by Q_1, the second term by C_1, the third term by I and C_1, the fourth term by Q_1, C_1 and I, and the fifth term by Q_2, C_1, I and P_1/P_2. By a suitable adjustment and rearrangement of terms, we get an approximation of the marketed surplus–price elasticity of subsistence crop as follows:

$$e = rb_1 - (r-1)[g + kh(1 + b_1)] - (r-1)hb_2(1-k)$$

where e = Marketed surplus–price elasticity of Crop 1; r = Reciprocal of sales ratio, i.e. Q_1/M_1; g = Consumption (on farm) elasticity of Crop 1 w.r.t. relative price (P_1/P_3); k = Share of income from Q_1 in total income; h = Consumption elasticity of Crop 1 w.r.t. income; b_1 = Production elasticity of Crop 1 w.r.t. relative price (P_1/P_2); and b_2 = Production elasticity of Crop 2 w.r.t. relative price (P_1/P_2).

REFERENCES

1. Rajkrishna, "The Marketed Surplus Function for a Subsistence Crop: An Analysis with Indian Data", *The Economic Weekly*, XVII, February 1965, pp. 309–20.
2. Rajkrishna, "A Note on the Elasticity of the Marketable Surplus of a Subsistence Crop", *The Indian Journal of Agricultural Economics*, Vol. XVII(3), July–September 1962, pp. 79–84.
3. Behrman, J.R., *Supply Response in Underdeveloped Agriculture*, North Holland Publishing Company, 1968, pp. 186–190.

4

Marketing Functions

The marketing of farm products is a complex process. It includes all the functions and processes involved in the movement of the produce from the farmers (producers) to the consumers. Neither the producers nor the consumers of farm products are located at one place. They are spread all over the country and even across countries. Timewise, too, the production and consumption of farm products do not coincide. Moreover, farm products are produced in a form which is different from the one in which they are consumed. They move in different ways and at different places and times. The number and type of functions, the cost of performing these functions, the margins or profits of those who perform these functions, and the competition in the trade—all these vary from commodity to commodity, from time to time and from place to place. This chapter includes marketing functions and marketing infrastructure.

MARKETING FUNCTIONS—MEANING AND CLASSIFICATION

MEANING

Any single activity performed in carrying a product from the point of its production to the ultimate consumer may be termed as a marketing function. A marketing function may have any one or combination of four dimensions, viz., time, space, form and exchange (transfer of ownership).

The marketing functions involved in the movement of goods from the producer to its ultimate consumer vary from commodity to commodity, market to market, the level of economic development of the country or region, and the final form of the consumption. For example, the marketing of wheat may involve bagging, loading on to a bullock cart, transportation to the primary market, unloading, making heaps in the market yard, auction, weighing, sieving, deciding the price, taking ownership by the purchaser, payment of value, rebagging, loading on to the truck, transportation to the consuming centre, unloading, sale to the retailer, weighing by the retailer, and sale to the consumer. Alternatively, if a farmer sells directly to the consumer in the village itself or at the farm, only weighing, bagging, making payment to the farmer, taking possession and transportation to the consumer's home are involved.

CLASSIFICATION

The marketing functions may be classified in various ways. For example, Thomsen[1] has classified the marketing functions into three broad groups. These are:

1. Primary Functions:	• Assembling or Procurement
	• Processing
	• Dispersion or Distribution
2. Secondary Functions:	• Packing or Packaging
	• Transportation
	• Grading, Standardization and Quality Control
	• Storage and Warehousing
	• Determination or Discovery of Prices
	• Risk Taking
	• Financing
	• Buying and Selling
	• Demand Creation
	• Dissemination of Market Information
3. Tertiary Functions:	• Banking
	• Insurance
	• Communications—Posts and Telecommunication
	• Supply of Energy—Electricity

Altenatively Kohls and Uhl[2] have classified marketing functions as follows:

1. Physical Functions:	• Storage and Warehousing
	• Grading
	• Processing
	• Transportation
2. Exchange Functions:	• Buying
	• Selling
3. Facilitative Functions:	• Standardization of Grades
	• Financing
	• Risk Taking
	• Dissemination of Market Information

These apart, Converse, Huegy and Mitchell[3] have classified marketing functions in a different way. According to them, the classification is as follows:

1. Physical Movement Functions:	• Storage
	• Packing
	• Transportation
	• Grading
	• Distribution
2. Ownership Movement Functions:	• Determining Need
	• Creating Demand
	• Finding Buyers and Sellers
	• Negotiation of Price
	• Rendering Advice
	• Transferring the Title to Goods
3. Market Management Functions:	• Formulating Policies
	• Financing
	• Providing Organization
	• Supervision
	• Accounting
	• Securing Information

Irrespective of classification suggested by different scholars, 13 functions are most important in marketing of farm products. Some details of these are discussed in this chapter.

PACKAGING

Packaging is the first function performed in the marketing of agricultural commodities. It is required for nearly all the farm products at every stage of the marketing process. The type of the container used in the packing of commodities varies with the type of the commodity as well as with the stage of marketing. For example, gunny bags are used for cereals, pulses and oilseeds when they are taken from the farm to the market. For packing milk or milk products, plastic, polythene, aluminium, tin or glass containers are used. Wooden boxes with straw/ bamboo baskets and plastic trays or containers are used for packing fruits and vegetables.

MEANING OF PACKING AND PACKAGING

Packing means, the wrapping and crating of goods before they are transported. Goods have to be packed either to preserve them or for delivery to buyers. Packaging is a part of packing, which means placing the goods in small packages like bags, boxes, bottles or parcels for sale to the ultimate consumers. In other words, it means putting goods on the market in the size and pack which are convenient for the buyers.

TYPES OF PACKAGING

There are basically three kinds of packaging.

1. **Primary or Sales Packaging:** Sales packaging is the packaging of goods meant for delivery to the final consumer. The examples of this type of packaging are one litre/ half litre packs of milk, 5 kg/10 kg packs of wheat flour, 200 gm/500 gm/1 kg packets of spices or loaves of breads.
2. **Secondary or Group Packaging:** It is the packaging of a group of consumer packets. Such packaging is removed to bring out a number of consumer packets. Removal of secondary package does not affect the basic quality and quantity of the final packets meant for sale to the consumers. The examples are plastic containers in which 10 or 20 packets of one litre milk or breads are packaged. Processors undertake secondary packaging for delivery to retailers (or some times to even wholesaler or stockists).
3. **Tertiary or Transport Packaging:** This kind packaging is designed to facilitate handling and transportation of a number of secondary packages to prevent physical damage during transportation. The examples of transport packets are packaging of 10 or 20 secondary packets into big wooden cases or hundreds of secondary packages in shipping containers.

ADVANTAGES OF PACKING AND PACKAGING

Packaging is a very useful function in the marketing process of agricultural commodities. Most of the commodities are packed with a view to preserving and protecting their quality and quantity during the period of transit and storage. For some commodities, packing acts as a powerful selling tool. Packaging is designed to ensure that the product reaches the consumers in good condition. The product quality and quantity is protected, through packaging during transportation and distribution, from climatic effects, hazardous substances and contamination from infestation.

Packaging, therefore, contributes to food safety, quality and nutrition. Packaging technology has made major contributions to advancing food science and food safety and reduction of food spoilage.

The main advantages of packing and packaging are:

1. It protects the goods against breakage, spoilage, leakage or pilferage during their movement from the production to the consumption point.
2. The packaging of some commodities involves compression, which reduces the bulk like cotton, jute and wool.
3. It facilitates the handling of the commodity, specially such fruits as apples, mangoes, etc., during storage and transportation.
4. It helps in quality identification, product differentiation, branding and advertisement of the product, e.g. Amul ghee and Amul butter.
5. Packaging helps in reducing the marketing costs by reducing the handling and retailing costs.
6. It helps in checking adulteration.
7. Packaging ensures cleanliness of the product.
8. Packaging with labelling facilitates the conveying of instructions to the buyers as to how to use or preserve the commodity. The label also shows the composition of the product, which is a staututory requirement
9. Packaging prolongs the storage quality of the products by providing protection from the ill effects of weather, specially for fruits, vegetables, dairy products and other perishable goods.
10. Packaging prevents the loss of aroma or flavour and creates barriers to absorption of undesirable external odour by the product.

PACKING MATERIAL AND INVENTIONS IN PACKAGING

Over the years, there have been many new developments in the use of the materials for packaging the products. Most of these have been in the area of packaging for consumers with a view to making the commodity more attractive. The introduction of fibre board containers, polythene, pollyshell (polythene + cellofan) and multiwall paper bags are some of the innovations in this direction. The packing material is sometimes changed even during the course of the movement of the produce from the producer to the consumer.

Though packaging is advantageous, it adds to the cost of the product. Some of these are rather fanciful, and add more to the cost than to the utility of the product. In order to reduce the packaging cost, unnecessary use of fancy packing material should be avoided. In general, the material used for packaging must have the following characteristics:

1. *Protective Strength.* The material used for packaging must have enough strength to protect the goods from breakage, leakage, spoilage and pilferage.
2. *Attractive.* The material used for packaging must be attractive to tempt the onlooker to try it.
3. *Consumer Convenience.* The packing material should be used and the packets be made of such size as is convenient and suits the needs of the consumers.
4. *Economy.* It must be cheap and the material used in packaging should be useful for domestic and other purposes after the use of the contents. In case the material cannot be reused, it must be biodegradable.

5. *Free from Chemical Reaction.* Packaging material should not give rise to any adverse chemical reaction and should conform to the safety standards prescribed by the health authorities.

The material used for packaging differs from commodity to commodity depending upon the degree of perishability and according to the stage of marketing.

1. The foodgrains and oilseeds, particularly at the wholesale level, are generally packed in gunny bags made of jute. Even sugar is packed in gunny bags. However, the jute cloth used for making sugar bags is different than that used for other gunny bags. The hessian, another cloth made of jute, is used for packing raw cotton (pressed or loose). In recent years, bags made up of jute blended with synthetics are also being used for packing many commodities like cement and fertilizers.

2. For fresh fruits generally wooden crates and straw-board boxes or bamboo baskets are used. However, in recent years some non-conventional packing materials have become common mainly on cost considerations. Some of these are as follows:
   (a) polythene foam wrappers for high value mangoes;
   (b) thermoformed PVC trays fitting to the size of fruits;
   (c) wrappers using tissue papers;
   (d) plastic trays;
   (e) plastic nets with convenient handles; and
   (f) multicolour printed duplex/corrugated-board cartons.

3. Vegetables are normally packed in jute bags, bamboo baskets and expensive wooden boxes. Normally, the following material is used for packaging of vegetables:
   (a) Jute bags—these are used for packaging of less perishable vegetables like cabbage, carrot, radish, peas, beans, turnip, brinjal and cucurbits.
   (b) Bamboo baskets and wooden boxes—these are used for packaging of tomatoes, cauliflower and chillies.
   (c) CFB (Corrugated fibre board) containers—these are used for packaging of vegetables both for domestic and export markets.
   (d) Corrugated craft paper cartons—these are good alternatives for wooden boxes and bamboo baskets in packaging of vegetables.
   (e) Plastics—plastic film bags, plastic nets, corrugated-board trays, moulded plastic trays and plastic hollow boards are also used for packing vegetables.
   It is being felt that for proper transportation, there is need for designing packages of certain modular dimension which can ensure maximum utilisation of space.

4. For processed food, the packaging materials are of the following types:
   (a) Tin containers: They are easily malleable, ductile and can be put to other uses by the consumer. Such properties as thickness, tolerance and surface finish influence the efficiency in the use of tin containers.
   (b) Glass containers: They are chemically inert, impermeable, non-porous and generally more hygienic: If the cover is properly provided, glass containers protect the moisture sensitive food products against atmospheric conditions for a specified period.
   (c) Polyethylene: This material has made great inroads in the packaging of processed foods due to its low cost, light weight, flexibility and convenience in use. By extrusion or blow moulding process, this material can be easily converted into flexible semi-rigid or rigid containers in the form of wraps, bags and pouches.

(d) Aluminium foils: These foils are able to preserve freshness, flavour and texture of such sensitive products as butter and cooked foods. They protect the contents from light, odour, moisture or bacteria. The aluminium foils can be embossed, printed or coated.

(e) Polylaminate pouches: These pouches possess the property of being good barriers of moisture and volatile oils. Such packages usually have double pouches—the inner one of cellophane and outer one of 250 gauge LDPE.

The packaging industry in India is growing at a rapid rate. With the increase in the consumers awareness and improvement in the living standards of masses, the demand for packaged agricultural commodities is increasing. During 2000, the total use of all packaging material was estimated at around 9.33 million tonnes[4]. The growth rate of consumption and demand for packaging in India is envisaged at 12% per annum. For packaging of food items, the Food Safety and Standards Authority of India (FSSAI) has notified regulations for packaging in 2011 (see details in Chapter 7).

The Indian Railways have prescribed packaging conditions for transport of different agricultural products. These conditions must be taken in to account while transporting the agricultural commodities by rail. If the prescribed conditions are not followed at the time of dispatch of produce, the railway authorities usually put a remark on the railway receipt (RR) that packing conditions have not been fulfilled. Under such a situation, the Railways do not entertain any claim of the dispatching-party for any loss in quantity or quality of the produce during the transit.

MANDATORY USE OF JUTE PACKAGING

For promoting the use of eco-friendly jute packaging, there is a Jute Packaging Material (JPM) Act, 1987 that provides for mandatory use of jute in packaging of foodgrains and sugar. Nearly 40 lakh farmers are engaged in cultivation of jute and 3.7 workers are engaged in jute processing industry. From time to time, government extends the mandatory use of jute for packaging foodgrains and sugar. Government agencies also purchase jute products worth ₹ 5500 crore every year.

BAN ON SINGLE USE PLASTIC PACKAGING

In 2019, Government of India banned the use of single use plastic products as a part of the global concern about environmental damages due to extensive use of plastic packages that are non-degradable. A movement against single use plastic products has also been launched in India. Several government organizations, voluntary organizations (NGOs) and private sector companies are actively participating. Quite a few organizations have installed recycling plants in this regard.

INDIAN INSTITUTE OF PACKAGING (IIP)

The Indian Institute of Packaging (IIP) is the apex body of packaging industry in India. The Institute was established in 1966 at Mumbai to serve the packaging industry towards excellence in packaging. The Institute aims at making the existing modes of packaging to perform better and develop new packaging materials to meet the present and potential demand of the market.

The Indian Institute of Packaging is a cooperative effort of Government of India and the industries engaged in manufacture of packaging materials and ancillaries, converters of packaging materials into packages, manufacturers of packaging machinery, users of packaging materials, and transporters of goods. The IIP has been recognized as research and development centre by the Department of Science and Technology, Government of India.

The head office of the institute is at Mumbai with regional centres, for example at Kolkata, New Delhi and Chennai. These are equipped with laboratories to carry out testing of packing materials, training of the human resources, and other activities connected with packaging. The Institute has a large number of industries on its membership roll.

TRANSPORTATION

Transportation or the movement of products between places is one of the most important marketing functions at every stage, i.e. right from the threshing floor to the point of consumption. Most of the goods are not consumed where they are produced. All agricultural commodities have to be brought from the farm to the local market and from there to primary wholesale markets, secondary wholesale markets, retail markets and ultimately to the consumers. The farm inputs from the factories must be taken to the warehouses and from the warehouses to the wholesalers, retailers and finally to the consumers (farmers). Transportation adds the place utility to goods.

Transport is an indispensable marketing function. Its importance has increased with urbanization. For the development of trade in any commodity or in any area transport is a sine qua non. Trade and transport go side by side; the one reinforces and strengthens the other.

ADVANTAGES OF TRANSPORT FUNCTION

The main advantages of the transport function are:

1. *Widening of the market:* Transport helps in the development or widening of markets by bridging the gap between the producers and consumers located in different areas. Without transport, the markets would have mainly been local markets. The exchange of goods between different districts, regions or countries would be impossible in the absence of this function. The example is the market for Himachal or Kashmir apples. The producers are located mainly in the states of Himachal Pradesh and Jammu & Kashmir; but apples are consumed throughout the country. Similarly, transportation of fresh vegetables from Indian ports to Gulf countries has expanded the markets for vegetable growers of India.

2. *Narrowing Price Difference Over Space:* The transportation of goods from surplus areas to the places of scarcity helps in checking price rise in the scarcity areas and price fall in surplus areas, thus reduce the spatial differences in prices.

3. *Creation of Employment:* The transport function provides employment to a large number of persons through the construction of roads, loading and unloading, plying of the means of transportation and repair services.

4. *Facilitation of Specialized Farming:* Different areas of the country are suitable for different crops, depending on their soil and agro-climatic conditions. Farmers can go in for specialization in the commodity most suitable to their area, and exchange the goods required by them from other areas at a cheaper price than their own production cost.

5. *Transformation of the Economy:* Transportation helps in the transformation of the economy from the subsistence stage to the developed commercial stage. Industrial growth is stimulated by being fed with the raw material produced in rural areas. Manufactured goods from industries to village or rural areas, too, can be moved.

6. *Mobility of the Factors of Production:* Transport helps in increasing the mobility of capital and labour from one area to another. Entrepreneurs get opportunities for the investment of their capital in newly-opened areas of the country, where the prospects of profit are very bright. Moreover, transportation helps in the migration of people in search of better remunerative jobs.

MEANS OF TRANSPORTATION

The available means of transportation can be classified as shown in Chart 4.1.

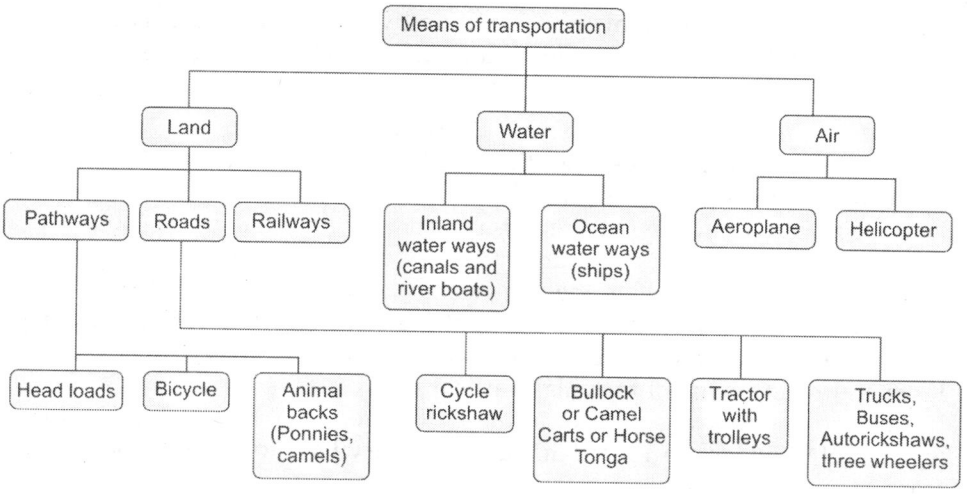

Chart 4.1

The transportation of agricultural commodities is mainly done by bullock or camel carts, tractor-trolleys and trucks, depending upon the availability, quantity and the stage of marketing. The most common means of transportation used at different stages of marketing are:

Stage of marketing	Transportation is done by	Means of transportation used
1. From the threshing floor to the village market nearest to the road or railway point	Farmer	Head loads or bullock or camel carts and tractor-trolleys
2. From the village market/railway station to primary/secondary wholesale markets	Traders	Trucks, buses or railway wagons
3. From retail or wholesale market to consumers	Consumers	Camel or bullock carts, thela, bicycle, hand carts or by head loads
4. International trade	Importer/Exporter	Ship, aircraft, other carriers.

A study by Acharya[5] on the marketing of pulses in Rajasthan during the 1980s had indicated that out of the total quantity transported by the farmers to the primary markets, 58.4% was carried by animal driven carts, 27.6% by tractor-trolleys, 5% by trucks and 9% by other modes. However, in recent years, with the expansion in the road network and increase in the use of faster means of transportation, the use of bullock or camel carts has gone down in most of the regions. According to one estimate, the bullock or camel carts now account for only 25% of the total produce arriving in the primary markets. Tractor-trolleys account for 65% and trucks around 10% of the produce arriving in primary markets. However, the bullock cart or camel cart is still an important means of transport employed by some farmers in remote areas. It is perhaps the best suited vehicle in the agricultural conditions of the country where all the villages or farms are not connected by tar roads. Bullocks or camel carts will continue to serve as the primary means of transport in rural areas in the times to come. The carrying capacity of bullock carts varies from five to ten quintals. They move slowly at a speed of 2.8 to 3.0 km/per hour. The cost of transportation by a bullock cart per quintal per km is high compared to any other mechanized means of transportation. Inspite of all these problems of transportation by a bullock cart, it is popular and indispensable mode of transportation because of the following reasons:

1. It is a cheap and easily available conveyance for the farmer having a small quantity of produce for transportation to short distances;
2. Its operational cost is low because the time is so adjusted that no opportunity cost is involved in using bullocks for this purpose;
3. It can be manufactured by village artisans from the material (wood) obtained from the farm. Farmers are not required to make a heavy initial investment. Repair facilities are also available in the villages;
4. It can be used on roads, kaccha paths and even in sandy areas. No special type of road is needed. It can operate even in muddy and stony areas;
5. It generates employment for village artisans;
6. It is multi-purpose transport, for it carries the produce to the market, inputs from the market to villages, farmyard manure from the village to the farm and passengers from one village to another.
7. It is environment-friendly mode of transportation.

TRANSPORTATION INFRASTRUCTURE

Transportation infrastructure consists of roads, railways, and transport vehicles. The status of length of roads in the country at different points of time is shown in Table 4.1.

Table 4.1: Length of roads in India

('000 km)

Year	Length of roads		National highways	State highways
	Total	Surface		
1950–51	400.0	157.0	19.8	NA
1960–61	524.5	263.0	23.8	NA
1970–71	915.0	398.0	23.8	56.8
1980–81	1485.4	684.0	31.7	94.4
1990–91	2327.4	1090.2	33.7	127.3
2000–01	3373.5	1601.7	57.7	132.1
2010–11	4690.3	2524.7	70.9	163.9
2016–17	5897.7	3501.9*	114.2	175.0

* For 2015–16

Source: Ministry of Road Transport and Highways. Taken from Economic Survey 2017–18 and 2018–19, Ministry of Finance, Government of India, New Delhi.

The length of roads has considerably gone up during the last 67 years. There has been considerable increase in national and state highways also. India now has the largest network of roads in the world, with 59 lakh km long roads. It includes 1.14 lakh km of national highways, and 1.75 lakh km of state highways. The number of vehicles (trucks, trailers, buses and tractors) running on these roads have also increased considerably in the country (Table 4.2).

Year	Registered vehicles			Railway route		Goods carried by railways (billion tonnes km)
	Total (lakh)	Goods vehicles (lakh)	Buses (lakh)	Total ('000 km)	Electrified ('000 km)	
1950–51	3.06	0.82	0.34	53.6	0.4	44.1
1960–61	6.65	1.68	0.57	56.2	0.8	87.7
1970–71	18.65	3.43	0.94	59.8	3.7	127.4
1980–81	53.91	5.54	1.62	61.2	5.4	158.5
1990–91	213.74	13.56	3.31	62.4	10.0	242.7
2000–01	549.91	29.48	6.34	63.0	14.9	315.5
2010–11	1418.66	70.64	16.04	64.4	19.6	626.5
2016–17	2533.11	122.56	18.64	68.4	29.4	693.3

Table 4.2: Number of transport vehicles in India

Source: Economic Survey 2017–18, Government of India.

However, a worrisome aspect is that the surfaced roads comprise only 60% of the total road length in the country. The status of roads is poor in the eastern states— Assam, Manipur, Nagaland, Tripura, Bihar and Orissa—where only 40% of the roads are surfaced. In the agriculturally and industrially developed states like Punjab, Haryana and Maharashtra, the proportion of surfaced roads is quite high.

The number of goods carriers went up from 0.82 lakh to 122.56 lakh, route length of electrified railways increased from 0.4 thousand to 29.4 thousand km and the haulage of goods by railways went up from 44.1 billion tonne km to 693 billion tonne km during the last 67 years in the country. Railways route availability is relatively poor in the states of Madhya Pradesh, Orissa Rajasthan and Karnataka as is clear by low density of rail route length in these states.

Rural roads are crucial in improving rural livelihoods and reducing rural poverty. Up to the mid-1990s, nearly 92% of the villages with more than 1500 population had road connectivity. The road connectivity of villages with a population in the range of 1000–1500 was only 76% and that for population less than 1000 was only 37% (Table 4.3). The connectivity of small villages or hamlets specially in hilly and tribal areas was very poor. It is in this context that a special rural roads programme called Pradhan Mantri Gram Sadak Yojna (PMGSY) was launched in December 2000. Subsequently, Bharat Nirman Programme, launched in 2005, included rural roads as an essential component. The target for rural roads under Bharat Nirman Programme was to provide all-weather roads to every habitation over 1000 population (500 in the case of tribal districts). The situation of rural connectivity in March 2007 (the beginning of XI Five Year Plan) was as shown in Table 4.4. The XI Five Year Plan aimed at providing connectivity to 60638 habitations, along with up gradation and widening of roads in the already connected habitations. The PMGSY continued during the XII five year plan also. In addition, the state govenments are also striving to provide road connectivity of all the villages to the nearest market towns.

Table 4.3: Road connectivity of villages in India

(till middle of nineties)

Group of villages with population	Villages connected with roads	Villages not connected with roads	Total number of villages
<1000	172062 (37.45)	280733 (63.55)	452795 (100)
1000–1500	44031 (75.88)	13904 (24.12)	57935 (100)
>1500	65698 (91.73)	5713 (8.27)	71411 (100)
Total	281791 (48.41)	300350 (51.59)	582141 (100)

Note: Figures in parentheses are percentages of total number of villages in the respective rows.

Table 4.4: Road connectivity of villages (till March 2007)

Population category	No. of eligible habitations	No. of habitations connected	No. to be connected
1000 and above	60030	20478	39552
500–1000	79208	13193	66015
250–500	39530	3816	35714
Total	178768	37487	141281

RELATIVE IMPORTANCE OF RAIL AND ROAD TRANSPORT

Both rail and road transport infrastructure has considerable contribution to reaching the food from production points to the consumers. As shown in Table 4.5 and 4.6, around 70 million tonnes of rice is transported by rail and roads combined, of which 68% is by roads and 32% by rail. The average distance rice is moved is 639 km. Nearly 42 million tonnes of wheat is moved to an average distance of 714 km. Fruits and vegetables are mainly moved by roads (97%). The average lead for food and vegetables is 552 km.

Table 4.5: India: Volume of some farm products transported by rail and roads

(million tonnes)

Commodity	Rail		Roads		Total	
	Quantity	%	Quantity	%	Quantity	%
Rice	22.4	32	47.1	68	69.5	100
Wheat and flour	12.3	30	29.4	70	41.7	100
Edible oils	1.1	4	25.3	96	26.4	100
Fruits and vegetables	1.9	3	69.9	97	71.8	100
Sugar and khandsari	6.0	24	18.9	76	24.9	100

Source: Report of the Committee for Doubling Farmers' Income, Volume III, MoA & FW, Aug. 2017, p. 85.

TRANSPORTATION COST

The transportation cost accounts for about 50% of the total cost of marketing. It is higher when the produce is transported by bullock or camel carts than when it is carried by other means. Further, the low capacity of the bullock or camel cart and their

Table 4.6: India: Average transport lead of some farm products (km)

Commodity	Rail	Road	Average of both
Rice	1294	327	639
Wheat and flour	1375	437	714
Edible oils	1519	538	579
Fruits and vegetables	1653	522	552
Pulses	1261	607	619
Milk products	2223	160	165
Oilseeds	1155	576	598

Source: Report of the Committee for Doubling Farmers Income, Vol.III, MoA & FW, August 2017, p. 86.

slow moving character restrict its use. In surplus producing areas tractors and trucks are being used for carrying the produce to the primary assembling markets. The truck industry can develop in rural areas for the transportation of agricultural produce if proper roads connecting the villages are available.

The efficiency of transportation depends on the speed and the care with which goods move from one place to another, the extent of the facilities provided, and the degree of care with which goods are handled en route and at terminal stations. However, there is a need for reducing the cost of transportation.

FACTORS AFFECTING THE COST OF TRANSPORTATION

Other things remaining the same, the transportation cost of a commodity depends on the following factors:

1. *Distance:* With an increase in the distance over which a commodity is transported, the total transportation cost increases; but the transportation cost per unit quantity of the produce decreases after a certain distance.
2. *Quantity of the Product:* The transportation cost per unit quantity of a commodity decreases with the increase in the volume. It will be less if a full truckload is available than if only a few quintals are transported.
3. *Mode of Transportation:* The cost of transportation varies with the mode of transportation, e.g. bullock cart, tractor, truck, railway, etc.
4. *Condition of Road:* The cost of transportation is less where metalled or tar roads have been constructed than in places where gravelled roads exist or where there are no roads at all.
5. *Nature of Products:* The cost of transportation per unit is higher for the products having the following characters:
   (a) Perishability (e.g. vegetables);
   (b) Bulkiness (e.g. straw);
   (c) Fragility (e.g. tomatoes);
   (d) Inflammability (e.g. petrol);
   (e) Requirement of a special type of facility (for example, for livestock and milk).
6. *Availability of Return Journey Consignment:* If goods are also available for transportation when a truck is to return to its starting place, the per unit cost of transportation is less.

7. *Risk Associated:* The transportation cost is less if the produce is transported at the owner's/sender's risk than when the risk is on the agency transporting the produce. The goods being transported can be insured, which, though adds to the cost in the form of insurance premium, reduces the risk arising from losses during transportation.

PROBLEMS IN TRANSPORTATION OF AGRICULTURAL COMMODITIES

The problems in the transportation of agricultural commodities are very serious because of the special factors associated with them; for example, the perishability of the produce, its bulkiness, the small quantity in which it is available, and a large number of suppliers and purchasers. The following are some of the important problems arising out of the transportation of agricultural commodities:

1. The means of transportation used are slow moving;
2. There are more losses/damages in transportation because of the use of poor packaging material, overloading of the produce and poor handling, specially of fruits and vegetables, at the time of loading and unloading;
3. The transportation cost per 100 rupees worth of the farm produce is higher than that for other goods. This is so because of its bulky character and the prevailing practice of fixing charges on the basis of weight or volume rather than on the basis of its value;
4. There is lack of co-ordination between different transportation agencies, e.g. the railways and truck companies. Some of the places are not connected by railway. The produce is often transported for a part of the distance by rail and a part by trucks or other means of transportation.
5. The multi-gauge system of railways was also a serious problem in transportation of goods by railways. However, now the country is moving towards a uni-gauge system which augurs well for marketing of farm products. Now, nearly 93% are broad gauge lines in India.

SUGGESTIONS FOR IMPROVEMENT

The following are some of the suggestions for effecting improvements in the transport function and reducing transport costs:

1. There must be full utilization of the capacity of the transportation facility in terms of the load. This would reduce the per quintal cost of transportation.
2. The transportation cost per quintal can be reduced by fixing the rate of transportation for different means. At present, each agency charges what it likes and not on the basis of any rational computation of the cost factor.
3. There should be a reduction in spoilage, damage, breakage and pilferage during the period of movement as a result of better handling, packing and the use of the proper types of wagons.
4. There should be a reduction in the barriers to inter-state movement of the produce. If this happens, the time taken in transportation and the quantity of the fuel consumed would be reduced. After the launch of Goods and Services Tax system and abolition of octroi by some states, the situation has somewhat improved.
5. A reduction in the bulk of the produce by processing can help in minimizing the transport cost. For example, milk may be processed into condensed milk, butter or ghee and fruits into juices or pulp.

6. The speed and capacity of the vehicles used in transportation should be increased. This can be done by research in respective areas. The speed and capacity of bullock carts can be increased by:
   (a) The use of pneumatic tyres instead of the existing wooden and iron wheels;
   (b) The use of springs in the axle of the cart;
   (c) The development of atleast good all-weather roads in the rural areas.
7. It must be recognised that roads and railways are important components of infrastructure, therefore, more public initiative in their expansion is called for. Nearly 10% of the villages in the country are still not connected by roads. This apart, there are sharp differences among the states. For example, in states like Orissa, Rajasthan and Madhya Pradesh large number of the villages are still to be connected by link roads. The rail transport, though capable of transporting agricultural commodities to longer distances in larger quantities with greater speed but it also suffers from multi-gauge system, shortages of wagon capacity and congestion on trunk routes. Therefore, in the overall scheme of public investment, development of this infrastructure should receive more allocation.
8. Despite construction of several over-bridges at roads and railway crossings, in some areas, level crossings continue to be a serious bottleneck and trucks have to wait for long time at railway crossings. This not only adds to the cost of transportation, but also results in lot of national wastage of precious non-renewable fuel resources. Construction of over-bridges for roads and by-passes to avoid populated areas is absolutely an urgent necessity. Several initiatives have been taken in recent past to resolve such issues.

GRADING AND STANDARDIZATION

IMPORTANCE AND MEANING

Grading is an important marketing function, to facilitate buying/selling, price discovery and movement of the produce. Grading basically means sorting of the produce into different homogenous lots on the basis of certain characteristics, reflecting quality of the produce. The characteristics vary from commodity to commodity.

Products are graded according to quality specifications. But if these quality specifications vary from seller to seller, there would be a lot of confusion about its grade. The top grade of one seller may be inferior to the second grade of another. This is when buyers lose confidence in grading. To avoid this eventuality, it is necessary to have fixed grade standards which are universally accepted and followed by all in the trade. The process of specifying the grade standards is called standardization. Without standardization the rule of caveat emptor (let the buyer beware) prevails; and there is confusion and unfairness as well. Standardization is a term used in a broader sense. Grade standards for commodities are laid down first and then the commodities are sorted out according to the accepted standards.

Standardization has been defined as the determination of the basic limits on grades or the establishment of model processes and methods of producing, handling and selling goods and services.

Standards are established on the basis of certain characteristics—such as weight, size, colour, appearance, texture, moisture content, staple length, amount of foreign matter, ripeness, sweetness, taste, and chemical content. These characteristics, on the basis of which products are standardized, are termed grade standards. Thus, standardization means making the quality specifications of the grades uniform among buyers and sellers over space and over time. As mentioned earlier, grading means the sorting of the unlike lots of the produce into different lots according to the quality specifications laid down. Each lot has substantially the same characteristics in so far as quality is concerned. It is a method of dividing products into certain groups or lots in accordance with predetermined standards. Grading follows standardization. It is a sub-function of standardization.

ADVANTAGES OF GRADING

Grading offers the following advantages to different groups of persons:

1. Grading before sale enables farmers to get a higher price for their produce. Studies during sixties and seventies revealed that on an average, the producers obtained a premium on graded tobacco at Ongole (AP) and on kapas at Hubli (Karnataka)[6]. Graded apple fetched a premium price of 11.27% over that of ungraded apple[7]. Graded dasheri and desi mango also fetched a premium over the price of ungraded mangoes[8]. Obviously, grading also serves as an incentive to producers to market a produce of better quality.

2. Grading facilitates marketing, for the size, colour, qualities and other grade designations of the product are well known to both the parties, and there is no need on the part of the seller to give any assurance about the quality of the product.

3. Grading widens the market for the product, for buying can take place between the parties located at distant places on the telephone without any inspection of the quality of the product.

4. Grading reduces the cost of marketing by minimizing the expenses on the physical inspection of the produce, minimizing storage losses, reducing its bulk, minimizing advertisement expenses and eliminating the cost of handling and weighing at every stage.

5. Grading makes it possible for the farmer—
   (a) To get easy finance when commodities are stored;
   (b) To get the claims easily settled by the railways and insurance companies;
   (c) To get storage place for the produce;
   (d) To get market information;
   (e) To pool the produce of different farmers;
   (f) To improve the "keeping" quality of the stored products by removing the inferior goods from the good lot; and
   (g) To facilitate futures trading in a commodity.

6. Grading helps consumers to get standard quality products at fair prices. It is easier for them to compare the prices of different qualities of a product in the market. It minimizes their purchasing risk, for they will not get a lower quality product at the given price.

7. Grading contributes to market competition and pricing efficiency. The product homogeneity resulting from grading can bring the market closer to perfect competition, encourages price competition among sellers, and reduces extraordinary profits.

Thus, the grading of product is beneficial to all the sections of society, i.e. the producers, traders and consumers of the product.

GENERAL CRITERIA FOR GRADE STANDARDS

The criteria which determine the adequacy of grade standards are:

1. Standards should be built on the characteristics which the users consider important, and these characteristics should be easily recognizable. More weightage should be given to the user's opinion.
2. Grade standards should be built on those factors which can be accurately and uniformly measured and interpreted. The grade standards based on subjective measurement will be difficult to apply uniformly, particularly by different graders. There will be a lot of variation in the subjective measurement of grade factors, which will reduce the usefulness of the grade itself.
3. The grade standards terminology should be uniform at all levels of the marketing channel.
4. The cost of operating the grading system must be reasonable.

The best practical test of the adequacy of grade standards is their acceptance and use by various market functionaries. If these grade standards are widely used, it means that they are fairly adequate and meaningful; but, if the large segment of the market functionaries does not use the standards, it may be assumed that some of the criteria to satisfy the consumers have not been adequately met.

TYPES OF GRADING AND CERTIFICATION

Grading and certification of agricultural commodities in India can be understood to consist of (a) grading for internal trade, (b) grading at farmers' level, and (c) grading for export.

Grading may be done on the basis of fixed standards or variable standards. It is of four types:

1. *Fixed Grading/Mandatory Grading:* This means sorting out of goods according to the size, quality and other characteristics which are of fixed standards. These do not vary over time and space. It is obligatory for a person to follow these grade standards if he wants to sell graded products. For a number of agricultural commodities, grade standards have been fixed by the Agricultural Marketing Advisor, Government of India, and it is compulsory to grade the produce according to these grade specifications. Individuals are not free to change these standards. The use of mandatory standards is compulsory for the export of the agricultural commodities to various countries.
2. *Permissive/Variable Grading:* The goods are graded under this method according to standards, which vary over time. The grade specifications in this case are fixed over time and space, but changed every year according to the quality of the produce in that year. Under this method, individual choice for grading is permitted. In India, grading by this method is not permissible.
3. *Centralised/Decentralised Grading:* Based on the degree of supervision exercised by the government agencies on grading of various farm products, the programme of voluntary grading for internal trade can be categorised into centralised and decentralised grading.

The centralized grading system is applicable to those commodities which require elaborate testing arrangements for assessment of quality and determination of grade. The commodities under centralized grading system include ghee, butter, vegetables oils, oilcakes, powdered spices, honey, wheat flour (atta) and besan (gram flour).

Under the centralised grading system, an authorised packer either sets up his own laboratory manned by qualified chemists or seeks access to an approved grading laboratory set up for the purpose by the state authorities/cooperatives/ associations/private agencies. Grading in respect of commodities such as ghee, butter and vegetable oils where elaborate testing facilities are needed for checking purity and assessing quality has been placed under centralised grading system. In this system, the Directorate of Marketing and Inspection exercises close supervision on grading work of approved chemists through periodical inspection of the grading stations and the quality of the graded produce.

The decentralized grading system is applicable to those agricultural commodities, which do not require very elaborate testing facilities. These commodities include wheat, rice, other cereals, oilseeds, edible nuts, fibre crop products, fresh fruits, fresh vegetables and eggs. The grading for such commodities is usually done on the basis of physical factors which require relatively simple tests. The grain quality is characterized into intrinsic and extrinsic parameters. Intrinsic parameters are colour, bulk density, odour, aroma, size and shape. The extrinsic parameters are broken or damaged grains, immature grains, foreign matter, infected grains and moisture content.

The decentralised grading system is implemented by State Marketing Authorities under the overall supervision and guidance of the Directorate of Marketing and Inspection.

Both these grading programmes are consumer-oriented with the aim of making available the pretested and AGMARK certified quality products to them.

4. *Grading at Producer's Level:* In addition to these, the state marketing authorities also implement a grading programme for the benefit of the producer-sellers, viz., grading at producer's level. Under this programme, free grading services are provided to the farmers for sorting the produce before offering for sale. This, in turn, enables them to realise prices commensurate with the quality of the produce. Grading units manned with grading personnel have been set up in several regulated markets and warehouses with the central assistance.

INSPECTION AND QUALITY CONTROL

To ensure the confidence of consumers, it is essential that grading is done in accordance with the standards that have been set. For this purpose, the inspection of the goods at regular intervals by a third party is essential.

Inspection involves the testing of the graded goods with a view to determining whether they conform to the prescribed standards. It ensures quality control. For purposes of inspection, samples of the product are drawn at various stages—from the manufacturers, the market middleman or the consumer at his doorstep—and are tested in the laboratory. These inspections are carried out by inspectors appointed by the government, and not by a producer or a buyer.

The network of Agmark laboratories in the country for testing the quality includes a central Agmark laboratory at Nagpur and several regional Agmark laboratories (RALs). The RALs are located at Amritsar, Bhopal, Chennai, Guntur, Jaipur, Kanpur, Kochi, Kolkata, Mumbai, Rajkot and New Delhi. Besides these, there are many other laboratories.

Regular inspection creates confidence among the buyers. Producers, too, know that there is someone who checks the standards of the produce graded by them. This avoids the temptation of adopting such malpractices in the grading as mixing of inferior grade produce, etc. After laboratory tests, if the produce is found to be below standards, the licence of the grader is cancelled and legal action is initiated against him.

The number of approved grading laboratories functioning in the country increased from 566 by the end March 1984 to 1269 in 2016. There are 100 state-owned grading laboratories, 1092 laboratories of the licencees (packers' own), 18 laboratories in cooperative sector and 59 private commercial laboratories. These approved grading and/or testing laboratories are engaged in the analysis and determination of AGMARK grades. The concentration of grading/testing laboratories is more in the states of Rajasthan, Uttar Pradesh, Haryana, Punjab, Maharashtra, Tamil Nadu and Delhi.

The Agricultural Marketing Advisor to the Government of India (AMA) is the authority empowered to implement the provisions of the Act, and suggest suitable modifications. The Central Agricultural Marketing Department (Directorate of Marketing and Inspection) maintains some staff for the inspection of the grading premises and the collection of the samples of graded products from different points in the marketing process. The collected samples are examined and analyzed either at the Central Agmark laboratory or at other laboratories set up in different parts of the country to test whether the graded products conform to the standards of quality laid down in the Act. If the sample is found below standard, necessary legal action against the party is taken, and the graded product is with drawn or removed from the market. The licence of the party, too, is cancelled.

LABELLING

The graded products, according to the standard fixed by the Agricultural Marketing Advisor, Government of India, bear the label 'AGMARK', as shown in Fig. 4.1.

Fig. 4.1: AGMARK symbols (old and new)

AGMARK is the abbreviation of Agricultural Marketing. It is a quality certification mark under the Central Agricultural Produce (Grading and Marking) Act, 1937. This label indicates the purity and quality of the product on the basis of the standards that have been laid down. The labels of different colours are used to indicate the grade of the product. The AGMARK labels are printed on special quality paper and issued by the Agricultural Marketing Advisor. They are serially numbered, and the firm is required to maintain the account of the labels, which are issued to the grader, in a register. It is a voluntary scheme. Interested traders and manufacturers are given licence to grade their products under AGMARK quality certification mark.

AGMARK label is attached to the container of the product in such a way that it will not be possible to remove the contents of the package without tampering the AGMARK labels. Each AGMARK package bears the date of packing and date of expiry of the product. AGMARK products are pretested and certified for their quality. AGMARK products are of assured quality and different from adulterated and spurious goods. If any AGMARK product purchased by the consumer is found defective, the consumer gets the product replaced or gets the money back as per the procedure laid out. There are about 14000 licencees manufacturing and marketing their products under AGMARK quality certification marks.

GRADE SPECIFICATIONS FOR AGRICULTURAL COMMODITIES

The Directorate of Marketing and Inspection, Government of India, has fixed grade standards for a number of agricultural commodities for domestic consumption as well as for export. The grade standards specified for some of the agricultural products are given in Tables 4.7 to 4.10 for illustration. These are regularly reviewed and revised from time to time by DMI. For the latest specifications, one should look at the website of DMI.

Table 4.7: Grade standards for eggs (hen)

Grade designation	Agmark label colour	Minimum weight (ounce/gm)	Other conditions
Special	White	2.00/56	1. Eggs should not have been preserved by any method.
'A' Grade	Red	1.75/49	2. Eggs should be spot-free and without blemishes.
'B' Grade	Blue	1.50/42	3. Yolk of eggs should be central.
'C' Grade	Yellow	1.25/35	4. Eggs should be solid.
			5. Eggs should be transparent.
			6. There must be air space of size less than 3/8".

Table 4.8: Grade standards for oranges

Grade designation	Agmark label colour	Minimum weight (inches)	Other conditions
Special	White	3.50	1. Oranges should be ripened so that they may not get spoiled in transportation.
Grade I	Red	3.00	
Grade II	Blue	2.75	2. The colour of oranges should be as per variety but should not be green.
Grade III	Yellow	2.50	3. There should not be wrinkles on oranges.
Grade IV	Green	2.25	4. Oranges should be free from cuts, infestation of insects and diseases.
			5. Tolerance limit is 10% of the next lower grade.

Table 4.9: Grade designations for rapeseed–mustard and taramira

Grade designation	Special characteristics				
	Foreign matters	Discoloured and damaged grains	Unriped and shriveled grains	Small atrophied seeds	Admixture of other varieties
Rapeseed and Mustard					
Special	1.0	1.0	1.0	5.0	5.0
Standard	2.0	1.5	3.0	10.0	10.0
General	3.0	2.0	4.0	20.0	15.0
Taramira					
Special	2.0	3.0	2.0	10.0	–
Standard	4.0	5.0	4.0	15.0	–
General	6.0	8.0	8.0	20.0	–

General characteristics

1. Have shape, size, colour and pungency characteristics of the variety.
2. Seeds be matured, hard, wholesome, well dried, moisture not exceeding 6%.
3. Should not have any trace of argemone seeds.
4. Free from insect damages.
5. Be in a sound merchantable condition
6. Not have the grains of other variety

Source: Reports of Directorate of Marketing and Inspection, Government of India, Faridabad/Nagpur.

Table 4.10: Grade designations for ghee

Tests	Special grade ghee	General grade ghee
1. Baudouin test	Negative	Negative
2. B.R. reading at 40°C	40–43	40–43
3. R.M. value	Not less than 28	Not less than 28
4. Polenseke value	1 to 2	1 to 2
5. Moisture content	Not more than 0.3%	Not less than 0.3%
6. Percentage of free fatty acids	Not more than 1.4%	Not more than 3.0%

PROCEDURE FOR FORMULATION OF STANDARDS FOR PROCESSED PRODUCTS

Standards for processed products are formulated in a very systematic manner. The Committee consisting of experts drawn from industry, educational institutions, research organizations, consumer bodies and government officials, connected with the field of preparing standards, is entrusted with the task of formulation of standards for the product. The members formulate the standards based on the available international standards, analytical study carried out and data on the quality of produce. The draft of the standards is circulated to interested and concerned persons in India and abroad for inviting comments. The comments received are again considered by the Committee and then finalised for printing as an Indian Standard. The standards so formulated are reviewed from time to time and necessary amendments are made as per emerging needs. The standards, in general, cover the physical, sensory, chemical, hygienic, micro-biological and packaging requirements, which vary from product to product.

PROGRESS OF GRADING IN INDIA

To improve the quality of agricultural products in India, grading and marking were introduced under an Act—The Agricultural Produce (Grading and Marking) Act, 1937. The Act authorises the Central Government to frame rules relating to the fixing of grade standards and the procedure to be adopted to grade the agricultural commodities included in the schedule. The Act of 1937 was amended from time to time to widen its scope, so that more number of commodities may be included under the changed circumstances. Initially, only 19 commodities were included for grading purposes; later their number increased to 213 and now there are 220 commodities in the schedule for which grade standards are available. The commodities included in the schedule are foodgrains, fruits and vegetables, dairy products, tobacco, coffee, oilseeds, edible oils, oilcakes, fruit products, cotton, sannhemp, edible nuts, jaggery, lac, spices and condiments, essential oils, honey, besan, suji, atta and maida.

The commodities for which AGMARK Grade Standards have been prescribed are as follows:

Group of commodities	No. of commodities
Foodgrains and allied products	30
Fruits and vegetables	57
Spices and condiments	27
Edible nuts	8
Oilseeds	17
Vegetable oils and fats	19
Oil cakes	8
Essential oils	8
Fibre crops	5
Livestock products, dairy and poultry products	10
Other products	31
Total	220

The grading and marking of goods for internal trade was first started in ghee (1938), followed by vegetable oils (1939), creamery butter (1941); gur, citrus fruits and eggs (1949), potato (1950), rice (1954), and wheat flour (1955). "Agmarking" for exportable commodities is compulsory, and was first started for sannhemp (1942), followed by tobacco (1945), bristles (1954), wool (1955), lemongrass oil (1956), sandalwood oil (1957), and goat hair (1961). Grading before export is compulsory for pepper, cardamom, cashewnuts, ginger, shellac, goat skin and essential oils.

Grading of agricultural commodities under Agmark for internal trade (Voluntary Grading) is carried out for a number of commodities like vegetable oils, ghee, creamery butter, honey, ground spices, rice, wheat flour, jaggery, potatoes, fruits, pulses, eggs, paddy, nutri cereals, oilcakes, castor seed, besan (gram flour), etc.

For domestic or internal trade, valid certificate of authorization holders (as on 31st March 2016) were 4583. Majority of these were in Tamil Nadu (981), Telangana (544), Delhi (513), Maharashtra (341), Rajasthan (339), Uttar Pradesh (334), Madhya Pradesh (326), Gujara (323), Andhra Pradesh (224) and Punjab (183).

The value of agricultural commodities/products graded for internal trade, which was only ₹ 15 lakh in 1938–39, increased to ₹ 223 crore in 1970–71, ₹ 10664 crore in 2010–11 and further to ₹ 17432 crore in 2015–16.

The grading at the producer's or farmer's level started in 1963–64. The objective was to help farmers get a price commensurate with the quality of produce. Grading units were established in regulated markets and also in the premises of marketing cooperatives. Some market committees made the grading compulsory before auction of the produce. Some examples are pulses and wheat in Jalgaon (Maharashtra), raw cotton in Hubli (Karnataka) and groundnut in Kolhapur and Sholapur (Maharashtra) markets.

The value of commodities graded at producer's level, which was ₹ 121 crore in 1970–71, increased to ₹ 4451 crore in 1990–91 and further to ₹ 6708 crore in 2015–16. However, this is considerably lower than the volume of trade in primary wholesale markets.

As on March 31, 2016, the producer's level grading units were only 1162, of which 813 were in regulated markets, 234 were operated by cooperatives and 115 by others. Most of the grading units were in Maharashtra (413), Madhya Pradesh (257), West Bengal (219) and Uttar Pradesh (117).

The compulsory grading for exports was started in 1942–43. The commodities initially included were tobacco, bristles, sandal wood oil, ghee, grapes, onion, butter and basmati rice. The value of commodities graded for export was only ₹ 29 lakh in 1942–43 and increased to ₹ 903 crore in 1990–91. Later, the function of grading/certification of goods for exports was shifted to other organizations. Obviously, value of graded exports under Agmark did not increase further.

Total value of agricultural commodities graded in India has considerably increased over time. The situation since 1930s to 2015–16 is shown in Table 4.11. Total value of graded products, which was only ₹ 15 lakh in 1938–39, increased to ₹ 1249 crore in 1980–81 and further to ₹ 25112 crore in 2015–16. Considering the total value of agricultural commodities traded, the value of graded products is small. The situation is bound to improve as the electronic trade in primary wholesale markets gets momentum.

Table 4.11: Value of agricultural commodities graded in India

(₹ crore)

Year	Compulsory grading (for export)	Voluntary grading (domestic trade)	Grading at producer's level	Total value of graded products
1938–39	–	*	–	*
1942–43	*	1	–	1
1950–51	14	4	–	18
1960–61	29	40	–	69
1970–71	93	223	121	437
1980–81	334	287	628	1249
1990–91	903	1338	4451	6692
2010–11	181	10644	14072	24917
2015–16	972	17432	6708	25112

* Less than 0.5

Source: Reports of Directorate of Marketing and Inspection, Ministry of Agriculture, Government of India, Faridabad.

Since 2016, E-NAM (electronic National Agricultural Market) has been launched in the country. Already, more than 585 mandies (regulated markets) have been linked to E-NAM portal that is expected to increase inter-state electronic auction of agricultural commodities arriving in mandies. In addition, other 415 mandies are in the process of being linked. Assaying facilities are being established in all these markets. Obviously, grading of agricultural products will increase as and when E-NAM matures (see more in Chapter 7).

OTHER PROVISIONS OF GRADE STANDARDS AND QUALITY SPECIFICATIONS

Apart from the Agricultural Produce (Grading and Marking) Act, 1937, there are quite a few other legal provisions which have been used to lay down grade standards for agricultural products. Some of the important provisions in this category are:

1. For the purpose of minimum price support operations, the central nodal agencies have also laid down quality standards for various commodities. This is necessary as the minimum support price fixed by the government is applicable to the fair average quality (FAQ) of the produce and FAQ needs to be defined properly. The central nodal agencies for this purpose are as follows:
   (a) Food Corporation of India—For cereals.
   (b) NAFED—For pulses and oilseeds including copra.
   (c) Cotton Corporation of India—For raw cotton.
   (d) Jute Corporation of India—For raw jute.
   An important aspect that needs to be noted is that some times when harvested grains are of lower than prescribed FAQ specifications (due to bad weather during grain ripening/harvest/threshing stage), with a view to helping the farmers, the price support purchase agencies (on the direction of the government) lower the quality specifications for FAQ grade. A recent example is of wheat during April–May 2015. The minimum broken and shriveled percentage was hiked from 6 to 9/10 and maximum moisture content was raised from 12% to 14%, for entitlement of full minimum support price.
2. The Central/State Warehousing Corporations lay down standards for agricultural commodities for accepting them for storage.
3. The Bureau of Indian Standards (BIS) under the Indian Standards Institution (Certification Marks) Act, 1952, enables the manufacturers having requisite production and testing facilities to use ISI mark on the products if these conform to the specifications laid down by the BIS. (see more in Chapter 7)
4. In addition, the domestic manufacturers, particularly those who are entering the export trade have obtained international grade certificates under ISO (International Standards Organisation) series.
5. As regards food products, at one time, there used to be 27 laws (Acts) to regulate their quality and standards. These include Prevention of Food Adulteration Act, Food Products Order, Meat Food Products Order, Milk and Milk Products Order, Vegetable Oil Products Order, and several others. The administrative authority, to implement the provisions of these acts, was vested in different Ministries or Departments, which resulted in duplication or confusion in several situations. In order to remove duplicity, multiplicity and confusion in their implementation, a new Food Safety and Standards Act (FSSA) was passed by the Parliament in 2006, which became operational in 2010/11 (see details in Chapter 7).

GRADING FOR EXPORT

The grading for export is done under the Export (Quality Control and Inspection) Act, 1963 and is administered by the Ministry of Commerce. The Act empowers the Government to:

1. Notify commodities which will be subject to quality control and/or inspection prior to export.
2. Establish standards of quality for such notified commodities, and
3. Specify the type of quality control and/or inspection to be applied to such commodities.

After the economic reforms of early 1990s, the operation of compulsory quality control and inspection for export commodities had been simplified by the following steps:

1. State Trading Houses, Trading Houses, Export Houses as well as Industrial units in Export Processing Zones and 100% EOUs have been exempted from the purview of compulsory pre-shipment inspection.
2. The Units approved under the system of In-process Quality Control (IPQC) have been authorized to issue statutory certificates by themselves.
3. Items which were hitherto subjected to compulsory pre-shipment inspection have been exempted from the same, provided the exporter has a firm letter from the oversees buyer stating that the overseas buyer does not require pre-shipment inspection from any official Indian inspection agency.

It may be mentioned that fish and fishery products, honey, egg products, milk products and black pepper for export to USA and basmati rice to EU have been notified for compulsory pre-shipment inspection and certification. Recently DMI has also done certification of grapes from the states of Maharashtra, Andhra Pradesh and Karnataka.

STORAGE AND WAREHOUSING

STORAGE

Meaning and Need

Storage is an important marketing function, which involves holding and preserving goods from the time they are produced until they are needed for consumption. Storage is an exercise of human foresight by means of which commodities are protected from deterioration, and surplus supplies in times of plenty are carried over to the season of scarcity. The storage function, therefore, adds the time utility to products.

Agriculture is characterized by relatively large and irregular seasonal and year-to-year fluctuations in production. The consumption of most farm products, on the other hand, is relatively stable. These conflicting behaviours of demand and supply make it necessary that large quantities of farm produce should be stored for a considerable period of time.

The storage function is as old as man himself, and is performed at all levels in the trade. Producers hold a part of their output on the farm. Traders store it to take price advantage. Processing plants hold a reserve stock of their raw materials to run their plants on a continuous basis. Retailers store various commodities to satisfy the consumers day-to-day needs. Consumers, too, store foodgrains, depending on their financial status.

The storage of agricultural products is necessary for the following reasons:

1. Agricultural products are seasonally produced, but are required for consumption througout the year. The storage of goods, from the time of production to the time of consumption, therefore, ensures a continuous flow of goods in the market;
2. Storage protects the quality of perishable and semi-perishable products from deterioration;
3. Some of the goods, e.g. woollen garments, have a seasonal demand. To cope with this demand, production on a continuous basis and storage become necessary;
4. Storage helps in the stabilization of prices by adjusting demand and supply;
5. Storage is necessary for some period for the performance of other marketing functions. For example, the produce has to be stored till arrangements for its transportation are made, or during the process of buying and selling, or the weighment of the produce after sale, and during its processing by the processor;
6. The storage of some farm commodities is necessary either for their ripening (e.g. banana, mango, etc.) or for improvement in their quality (e.g. rice, pickles, cheese, tobacco, etc.); and
7. Storage provides employment and income through price advantages. For example, traders store foodgrains by purchasing them at low prices in the peak season and sell them in the other seasons when prices are higher.

Traditional Storage Practices in India

About 60 to 70% of foodgrains produced in India remain in the rural sector for varying periods. Various operations carried out before storage are field drying, stacking and threshing. Foodgrains are stored in traditional structures and containers made locally from available local material. The structures differ in size, shape, capacity and type from area to area. The life of the structures depends on construction materials and method of construction. Common construction materials are bamboo, reed, timber, burnt bricks, mud and paddy straw. The structures are mostly cylindrical, elliptical, rectangular or conical in shape. The common storage structures in rural areas are:

1. Gunny bags—made of mesta/jute;
2. Earthen containers—made of burnt clay of capacity ranging from 0.5 to 3 quintals.
3. Bamboo containers—indoor or outdoor structures constructed with split bamboo or locally available plants with walls plastered with mud.
4. Straw structures—constructed with fresh untrampled paddy straw and are usually circular structures.
5. Masonary structures—a room like structure made with wood or brick and lime or brick and cement mortar. The capacity may be of 10 to 30 tonnes.
6. Underground structures—shallow underground pits with depths varying from 1 to 1.5 metres depending upon the water table of the area. These are plastered with clay and straw leaves. After filling the grain, they are packed and compacted with mud to get a dome shape.
7. R.C.C. ring structure—These are made of rings fabricated with cement, concrete and sand.

RISKS IN STORAGE

The storage of agricultural commodities involves three major types of risks. These are:

1. *Quantity Loss:* The risks of loss in quantity may arise during storage as a result of the presence of rodents, insects and pests, theft, fire, etc. Dehydration too, brings

about an unavoidable loss in weight. It has been estimated that about 7% of foodgrains are lost every year because of poor and faulty storage in rural areas.

2. *Quality Deterioration:* The second important risk involved in the storage of farm products is the deterioration in quality, which reduces the value of the stored products. These losses may arise as a result of attack by insects and pests, the presence of excessive moisture and temperature, or as a result of chemical reaction during the period of storage. Dehydration of fruits, vegetables and meat during storage may lower their sale value. Butter, if not properly stored, may become rancid, which reduces its sale value. The loss in the quality of farm products varies with their quality at the time of storage, the method of storage and the period of storage.

3. *Price Risk:* This, too, is an important risk involved in the storage of farm products. Prices do not always rise enough during the storage period to cover the storage costs. At times they fall steeply, involving the owner in a substantial loss. Farmers and traders generally store their products in anticipation of price rise and they suffer when prices fall.

STORAGE STRUCTURES

Storage structures of farm products are of two basic types, depending on whether they are stored underground or above the ground. There is no intrinsic difference between these two provided that basic requirements of safe storage are met. The advantages and disadvantages of these two types of storage structures are as follows:

Underground Storage Structures

Underground storage structures are dug-out structures similar to a well with sides plastered with cowdung. They may also be lined with stones or sand and cement. They may be circular or rectangular in shape. The capacity varies with the size of the structure. These structures are known by different local names in various regions. In Rajasthan the underground storage structures for foodgrains are known as 'khai'. The advantages of underground storage structures are:

1. Foodgrains in an underground storage structure are more free from the seasonal variations in temperature and humidity provided that adequate precautions are taken against the seepage of water in the structures, especially in areas where the water table is high;

2. Underground storage structures are safer from threats from various external sources of damage, such as theft, rain or wind. However, they are not good when the quantity available for storage is small, and there are a number of varieties to store;

3. The underground storage space can temporarily be utilized for some other purposes with minor adjustments; and

4. The underground storage structures are easier to fill up owing to the factor of gravity. However, it becomes cumbersome to take out the grains from these structures. Due to inherent problems, the use of underground structures is limited.

Surface Storage Structures

The advantages of surface storage structures are:

1. They can be maintained in more hygienic conditions by cleaning or whitewashing them;

2. They are more convenient for inspection and the performance of various operations during storage, such as spraying and dusting, fumigation and turning of the grains; and

3. The danger of heating up of grain due to internal heat is less.

However, the cost of storage in terms of the maintenance of storage structure, handling costs, and losses due to external factors are higher in surface structures. Vertical silos on the ground are often constructed for storage; but these have a limited application because of their cost and the requirements of energy for their operation. The losses are low; but the absence of bulk transport facilities, and the use of traditional handling and marketing methods make them unsuitable in larger numbers. Their utility for storage is at the ports for the export/import of foodgrains. The Food Corporation of India, warehousing corporations and private entrepreneuss (under PPP mode) have undertaken the construction of vertical silos (on the ground) at a few places.

Foodgrains in a ground surface structure can be stored in two ways—bag storage and bulk storage.

1 *Bag Storage:* Farm products are stored after placing them in gunny bags made of jute. Storage in bags has following advantages:
   (a) Each bag contains a definite quantity which can be bought, sold or despatched without difficulty;
   (b) Bags are easier to load or unload;
   (c) It is easier to keep separate lots with identification marks on the bags;
   (d) The bags which are identified as infested on inspection can be removed and treated easily; and
   (e) The problem of the sweating of grains does not arise because the surface of the bag is exposed to the atmosphere.

2. *Bulk or Loose Storage:* Farm products are sometimes stored in surface structures in a loose form. The advantages of this method are:
   (a) The exposed peripheral surface area per unit weight of grain is less. Consequently, the danger of damage from external sources is reduced; and
   (b) Pest infestation is less because of almost airtight conditions in the deeper layers.

These two points in favour of bulk storage are significant. The only precaution necessary for bulk storage is to avoid the sweating of grains.

The storage structures used in villages are very faulty, leading to high storage losses. The structures adopted in a village for the storage of farm products in India are made of the raw materials available in the area. Some common storage structures are:

1. Kothi or Mud Pots: These are cylindrical in shape and are made up of unburnt clay mixture with straw and cowdung or cowdung, mud and bricks.
2. Kuthla: These are cylindrical bins of mud-brick mixed with straw and cowdung.
3. Thekka: These are rectangular in shape and are made up of gunny or cotton wound around wooden support.
4. Metal Drums: These are cylindrical in shape and are made up of iron sheets.
5. Gunny Bags: These bags are made up of jute and are used for storing foodgrains and oilseeds.

The Government of India has made efforts to promote improved storage facilities at the farm level. Realising the importance of a scientific storage of grains, the Department of Food at the Centre has launched programmes to impart scientific knowledge to farmers. The first such programme is known as the Save Grain Campaign.

Indian scientists and agriculturists, working together with UNDP/FAO experts, have designed and fabricated improved storage bins for the use of farmers which are moisture resistant and rodent-proof. The drive to improve grain storage facilities resulted in the establishment of the Indian Grain Storage Institute (IGSI) in 1958 at Hapur (Uttar Pradesh). Two Field Stations at Ludhiana (Punjab) and Bapatala (A.P.) were also established to assist the Institute in field research and testing. The institute tries to design and develop suitable structures of all materials, capacities and types for various grains to suit different regions.

Essentials of a Good Foodgrains Storage Structure

Foodgrains are stored at different levels in the food supply chain: the farmer, the miller, the trader, the importer and also the government agencies. Stored grain is a man-made ecosystem undergoing incessant interaction between several abiotic and biotic factors. Abiotic factors are temperature, relative humidity, inter-granular CO_2 levels and moisture content. Biotic factors are insect pests, fungi, mites and rodents. For medium and long term storage, a strong control on these factors is indispensable. A well-managed grain storage system can facilitate strengthening of a stable foodgrain supply chain functioning year round.

Foodgrain storage structures must have the following specific characteristics:

1. There must be enough strength in the walls and floor to support the weight of the grain to be stored;
2. The structure should be inaccessible to insects, birds, rodents, unauthorised persons and moisture to save quality and quantity losses;
3. The structure should be free from excess heat or rapid changes in temperature; and
4. The structure must provide for the easy entry and removal of the grain as well as for the application of insecticides and pesticides, if there is any need for them.

IMPROVED GRAIN STORAGE STRUCTURES

Keeping these points in view, various institutions engaged in research have evolved improved storage structures for farm as well as for large-scale storage. Some of these improved grain storage structures are:

1. For Small-scale Storage

1. PAU bin: This is a galvanized metal iron structure designed by the Punjab Agricultural University, Ludhiana. Its capacity ranges from 1.5 to 15 quintals.
2. Pusa bin: This is a storage structure designed by the Indian Agricultural Research Institute (IARI), New Delhi, and is made of mud or bricks with a polythene film embedded within the walls. The polythene acts as a moisture barrier.
3. Hapur Tekka: This is a storage structure designed by the Indian Grain Storage Institute, Hapur. It is a cylindrical rubberised cloth structure supported by bamboo poles on a metal tube base, and has a small hole in the bottom through which grain can be removed.

2. For Large-scale Storage

1. CAP Storage (Cover and Plinth): This has been developed by the Food Corporation of India. It involves the construction of brick pillars to a height of 14" from the ground, with grooves into which wooden crates are fixed for the stacking of bags

of foodgrains. The whole unit is then covered with a thick polythene sheet. The structure can be fabricated in less than 3 weeks. It is an economical way of storage on a large scale.

2. Warehouse: These are scientific storage structures constructed on a large scale. Warehouse facilities in several areas have been created by the Food Corporation of India, the Central Warehousing Corporation, the State Warehousing Corporations and Cooperative Marketing Organizations. Now, private sector warehouses have also come up in the country.

3. Silos: The Food Corporation of India and others have constructed a few scientific silos for storage of foodgrains in main surplus producing areas like Punjab. In these structures, the grains in bulk are unloaded on the conveyor belts and, through mechanical operations, are carried to the storage structure. The storage capacity of each of these silos is around 25,000 tonnes.

RESEARCH AND TRAINING

In order to develop facilities for applied research and for apex level training in the field of storage and preservation of foodgrains, a Grain Storage Research and Training Centre was established at Hapur in 1958, which was later on expanded into Indian Grain Storage Institute (IGSI) with two field stations at Ludhiana and Bapatala (later shifted to Hyderabad) with the financial assistance from UNDP in 1968. Three field stations at Jabalpur, Jorhat and Udaipur were subsequently established in 1981. In 1996, the IGSI was renamed as Indian Grain Storage Management and Research Institute (IGMRI). While the research, development and training activities are taken by IGMRI, the work of popularizing scientific methods of foodgrain storage developed by IGMRI, among farming community is implemented through the Save Grain Campaign Scheme.

The IGMRI located at Hapur functions under the supervision of Ministry of Consumer Affairs, Food and Public Distribution. Attached to the Institute are five field stations. These field stations are primarily established for conducting intensive studies on the problems of handling and storage of wheat, rice, millets, pulses and oilseeds and are located in different agro-climatic zones of the country.

The main objectives of the IGMRI are:

1. To investigate the nature, extent and degree of losses due to various factors in storage of agricultural commodities under different agro-climatic conditions.

2. To develop code of practices for proper grain storage and handling by recommending cost-effective techniques for control of insects, rodents, birds and microorganisms.

3. To develop improved types of storage structures, grain dryers, grain handling, cleaning and grading equipments, besides improvement in traditional storage structures using locally available and eco-friendly materials.

4. To develop suitable publicity material and semi-technical literature on grain storage and quality control in foodgrains.

5. To train personnel from various organizations viz., FCI, CWC, SWC, Civil supplies Corporations etc., as well as those from other developing countries sponsored by FAO, UNDP and Common Wealth Secretariat on grain storage management practices.

COST AND RETURNS FROM STORAGE—ILLUSTRATION

The gross return on storage may be defined as the increase in the price of the stored product at the time of storage till it is "de-stored" and either sold or consumed. The cost of storage should include the following:

1. The cost of the maintenance of the storage structure, i.e. depreciation, repairs, insurance and interest on sunk capital; or, alternatively, the rent paid for hiring the storage structure;
2. Interest on the value of the stored goods;
3. Value of the quantitative and qualitative loss during storage;
4. Risk premium for a possible price fall and damage during storage;
5. The cost of protective materials; for example, insecticides, pesticides, rodenticides, fumigation, gunny bags, electricity, polythene covers; and
6. Tax payments, payments to labour, etc.

These items of costs may be grouped into fixed or variable costs, depending on whether they vary with the quantity of goods stored or not. For example, for a professional warehouse owner, the maintenance and repair of the storage structure, the salaries of the permanent staff, depreciation on the building, taxes, record keeping, etc., are fixed costs. For a farmer, however, who is trying to decide whether to sell or store the grains for some time for later sale, all the costs are variable.

Whether it pays a farmer to store his farm produce may be worked out with the help of the following formula:

$$NR = GR - C$$

where NR = Net returns to storage; $GR = P_1 - P_0$ (Gross Returns); P_0 = Purchase price or market price at the time of storage; P_1 = Selling or market price at the time of de-storing; C = Cost involved in storage; $NR > 0$, implies positive returns on storage; $NR < 0$, indicates negative returns on storage.

The percentage margin (Ms) from storage may be calculated as:

$$M_s = \frac{P_1 - P_0 - C}{P_0 + C} \times 100$$

where all notations have meanings as defined earlier.

Let us understand this with the help of an example. Let there be a farmer who has the option of selling his wheat immediately after harvest in the month of April or of putting his produce in a warehouse for selling at a later date (say, in the month of December). The market price in the month of April is, say, ₹ 1440 per quintal. Assume that his produce is lying on the auction platform in a regulated market yard, implying that in April, he gets a net price of ₹ 1440 per quintal. The transportation of the produce to the warehouse will cost him (say) ₹ 40 per quintal. The loading and unloading charges prescribed by the Market Committee are (say) ₹ 5.00 per bag or per quintal. The charge of the warehouse is ₹ 30 per quintal per month. There is likely to be a loss of 1% of grains during loading, unloading and storage. The farmer can earn an interest of (say) 10% per annum. Using this information let us work out the minimum price that should be expected in December to make some profit from the storage of wheat. The answer to this question requires estimating the costs of storage as also the return to be foregone. The details for a quintal of wheat are as follows:

Using the notations given earlier, $P_0 = 1440$, $C = 492$, NR will be positive only if P_1 or price of wheat in the month of December is expected to be more than the sum of P_0 and C, i.e. ₹1932.

Items	₹
1. Charges of loading at the auction platform and unloading in the warehouse (twice)	10
2. Transportation charges from the auction platform to the warehouse	40
3. Warehouse charges for 9 months @ ₹ 30 per quintal per month	270
4. Loss during storage and transit @ 1% of grains	14
5. Interest on ₹ 1440 foregone for 9 months @ 10% per year	108
6. Transportation charges from warehouse to the auction platform	40
7. Loading at the warehouse and unloading at the auction platform	10
Total	492

WAREHOUSING

Meaning and Functions

Warehouses are scientific storage structures especially constructed for the protection of the quantity and quality of stored products. Warehousing may be defined as the assumption of responsibility for the storage of goods. It may be called the protector of national wealth, for the produce stored in warehouses is preserved and protected against rodents, insects and pests, and against the ill-effect of moisture and dampness. Warehousing adds time value to the product.

The warehousing scheme in India is an integrated scheme of scientific storage, rural credit, price stabilization and market intelligence and is intended to reduce the post-harvest losses in farm products. The important functions of warehouses are:

1. *Scientific Storage:* Here, a large bulk of agricultural commodities may be stored. The product is protected against quantitative and qualitative losses by the use of such methods of preservation as are necessary.
2. *Financing:* Warehouses meet the financial needs of the person who stores the product. Nationalized banks advance credit on the security of the warehouse receipt issued for the stored products to the extent of 75 to 80% of their value. Now, the warehouse receipt can be bought and sold.
3. *Price Stabilization:* Warehouses help in price stabilization of agricultural commodities by checking the tendency to making post-harvest sales among the farmers. Farmers or traders can store their products during the post-harvest season, when prices are low because of the glut in the market. Warehouse helps in staggering the supplies throughout the year. They thus help in the stabilization of agricultural prices.
4. *Market Intelligence:* Warehouses also offer the facility of market information to persons who hold their produce in them. They inform them about the prices prevailing in the market, and advise them on when to market their products.

This facility helps in preventing distress sales for immediate money needs or because of lack of proper storage facilities. It gives the producer holding power; he can wait for the emergence of favourable market conditions and get the best value for his product. Warehouses help in reducing post harvest losses and in maintaining the quality of the produce and thus enable the farmer to receive higher prices.

TYPES OF WAREHOUSES

Warehouses may be classified on two bases:

1. On the Basis of Ownership

1. *Private warehouses:* These are owned by individuals, large business houses or wholesalers for the storage of their own stocks. They also store the products of others and charge a fee for it.
2. *Public warehouses:* These are the warehouses which are owned by the government and are meant for the storage of goods of any member of the public against a prescribed storage charge. The method of operation and the charges for storage are regulated by the government.
3. *Bonded warehouses:* These warehouses are specially constructed at a seaport or an airport and accept imported goods for storage till the payment of customs duty by the importer of goods. These warehouses are licensed by the government for this purpose. The owner of the warehouse gives an undertaking to the government that customs duty will be collected from the person before he is allowed to remove the goods from the warehouse. In other words, the goods stored in this warehouse are bonded goods. They may be owned by the dock authorities or privately-owned; but they have to work under the close supervision and control of the customs authorities. The following services are rendered by bonded warehouses:

(a) The importer of goods is saved from the botheration of paying customs duty all at one time because he can take delivery of the goods in parts.
(b) The operation necessary for the maintenance of the quality of goods—spraying and dusting, are done regularly.
(c) Entrepot trade (re-export of imported goods) becomes possible. The importer may take delivery of the goods without paying the customs duty if they are to be re-exported. He is thus saved from the botheration of first making the payment of customs duties on imported goods and then getting a refund on re-exported goods.

2. On the Basis of Type of Commodities Stored

1. *General Warehouses:* These are ordinary warehouses used for storage of most of foodgrains, fertilisers etc. In constructing such warehouses no commodity-specific requirement is kept in view.
2. *Special Commodity Warehouses:* These are warehouses which are specially constructed for the storage of specific commodities like cotton, tobacco, wool and petroleum products. They are constructed on the basis of the specific requirements of the commodity.
3. *Refrigerated Warehouses:* These are warehouses in which temperature is maintained as per requirements and are meant for such perishable commodities as vegetables, fruits, fish, eggs and meat. The temperature in these warehouses is maintained below 30° to 50°F or even less, so that the product may not get spoiled by high atmospheric temperature.

COSTS AND RETURNS OF A WAREHOUSING ENTERPRISE

The costs incurred in storage and warehousing can be divided into two groups:

1. *Fixed Costs:* These costs are of permanent nature and remain the same irrespective of the quantity stored in the warehouse. The main components of fixed costs are:
   (a) Depreciation on building and machinery, if any;
   (b) Insurance premium paid to the insurance company;
   (c) Taxes, licence fees etc.;

   (d) Repair and maintenance cost of the warehouse;
   (e) Interest on the investment in construction of the warehouse;
   (f) Salary of the permanent staff;
   (g) Cost of records and book-keeping;
   (h) Fixed part of the electricity charges (meter rent and minimum fixed charges).
2. *Variable Costs:* These costs are of varying nature, i.e. they vary with the quantity stored in the warehouse. The main components of variable costs are:
   (a) Cost of protective material used, viz., insecticides, pesticides, rodenticides, gunny bags, polythene cover, wooden slabs etc.;
   (b) Cost of electric power;
   (c) Wages of temporary labour.

   Consider that a young entrepreneur has constructed a warehouse with a storage capacity of 1000 tonnes at a cost of ₹ 40 lakh. Assume that the owner has to incur following expenditure on the operation of the warehouse:

1. Interest on capital @ 12% per annum (imputed interest);
2. Repair and maintenance cost—₹ 80,000 per annum;
3. Cost of records and book-keeping—₹ 10,000 per annum;
4. Taxes and insurance premium—₹ 30,000 per annum;
5. Wages of the manager and permanent labour—₹ 2,30,000 per annum;
6. Electricity bill—₹ 70,000 per annum;
7. Cost of protective material—₹ 30,000 per annum;
8. Wages of temporary labour—₹ 70,000 per annum.

   Assuming 50 years as the life of the warehouse, the cost structure of the warehouse emerges as follows:

Particulars	Rate	₹ per annum
Fixed costs		
1. Depreciation on the building	2% of ₹ 40 lakh	80,000
2. Interest on the capital	12% of ₹ 40 lakh	4,80,000
3. Repairs and maintenance cost		80,000
4. Cost of account books and records		10,000
5. Taxes and insurance premium		30,000
6. Wages of manager and permanent labour		2,30,000
	Total fixed costs	9,10,000
Variable costs		
7. Eleclricity bill		70,000
8. Cost of protective material		30,000
9. Wages of temporary labour		70,000
	Total variable costs	1,70,000

Total costs of storage per annum = 9,10,000 + 1,70,000 = 10,80,000

   Assuming the capacity utilization of 90%, the cost of storage in the warehouse works out to ₹ 10 per quintal per month (10,80,000 divided by 90% of 10000 × 12). This means, he should fix a charge of ₹ 10 per quintal per month to earn 12% interest on fixed capital investment. In case he aspires to get a remuneration for himself at the rate of (say) ₹ 45,000 per month, he will have to charge at least ₹ 15 (10+5) per quintal per month. This will leave with him 12% annual interest on his capital investment and ₹ 45,000 per month as his own remuneration or profit.

WAREHOUSING IN INDIA

In 1928, the Royal Commission on Agriculture underscored the need for a warehousing system in India. The Central Banking Enquiry Committee, 1931, too, drew attention to this need. The Reserve Bank of India emphasized the need for warehouses as early as in 1944, and proposed that every State Government should enact legislation to regulate the functioning of warehouses. The All-India Rural Credit Survey Committee of the Reserve Bank of India (set up in 1951 and submitted its report in 1954) also made comprehensive recommendations for the development of warehousing as an integrated scheme of rural credit and marketing. As a result of the recommendations of the Committee, the Government of India enacted the Agricultural Produce (Development and Warehousing) Corporations Act, 1956. The Act provided for:

1. The establishment of a National Cooperative Development and Warehousing Board (which was set up on 1st September, 1956);
2. The establishment of the Central Warehousing Corporation (which was established at Delhi on 2nd March, 1957); and
3. The establishment of State Warehousing Corporations in all the States in the country (which were established in various states between July 1957 and August 1958).

In 1962, the Government of India decided to break up the Warehousing Act of 1956 into two separate Acts—the National Cooperative Development Corporation Act, 1962, and the Warehousing Corporations Act, 1962. The Warehousing Corporations Act came into operation on 18th March, 1962. The Act defines the specific functions and the area of operations of Central and State Warehousing Corporations. It enlarged the list of the number of commodities meant for storage.

In 2007, the Government of India enacted another Act 'The Warehousing (Development and Regulation) Act' 2007. The Act envisages bridging the gap that exists between farmers and the credit institutions by trying to address the risk factors like deterioration in quality of agricultural produce stored in warehouses, guarantee of quality and quantity stored in warehouses and repayment capacity of farmers.

Enactment of the new Act is of immense benefit not only to the farmers but also to the industrial and private sector. A farmer by depositing his produce in a registered warehouse can use the Negotiable Warehouse Receipt (NWR) as collateral for obtaining short term loans and thereby avoid the need to sell the produce as soon as it is harvested. A stable NWR system has the potential to induce farmers to produce crops as per the marketable grades and standards for being kept in the warehouses. The new Act is encouraging the private investment in warehousing, which was long over due.

1. National Cooperative Development and Warehousing Board

This board was set up on 1st September 1956 to perform the following functions:

1. To advance loans and grants to State Governments for financing cooperative societies engaged in the marketing, processing or storage of agricultural produce, including contributions to the share capital of these institutions;
2. To provide funds to warehousing corporations and the State Governments for financing cooperative societies for the purchase of agricultural produce on behalf of the Central Government.

3. To subscribe to the share capital of the Central Warehousing Corporation and advance loans to State Warehousing Corporations and the Central Warehousing Corporation;
4. To plan and promote programmes through cooperative societies for the supply of inputs for the development of agriculture; and
5. To administer the National Warehousing Development Fund.

In March 1963, the Board was converted into the National Cooperative Development Corporation (NCDC), and its functions were limited to cooperative development.

2. Central Warehousing Corporation (CWC)

This Corporation was established as a statutory body in New Delhi on 2nd March, 1957. Under the new Act, the Central Warehousing Corporation was formally re-established on March 18, 1963. This Corporation which made a modest start with seven warehouses, with 7,000 tonnes capacity, in December 1957, is now operating 422 warehouses in different places in the country, with a total storage capacity of 101 lakh tonnes. Of this, the present utilization is nearly 83% of the total available capacity. The Central Warehousing Corporation provides safe and reliable storage facilities for about 120 agricultural and industrial commodities. The areas of operations of these central warehouses include centres of all-India and inter-state importance. CWC has 18 regional offices spread throughout the country.

The functions of the Central Warehousing Corporation are:

1. To acquire and build godowns and warehouses at suitable places in India;
2. To run warehouses for the storage of agricultural produce, seeds, fertilizers and notified commodities for individuals, cooperatives and other institutions;
3. To act as an agent of the government for the purchase, sale, storage and distribution of the above commodities;
4. To arrange facilities for the transport of above commodities;
5. To subscribe to the share capital of State Warehousing Corporations; and
6. To carry out such other functions as may be prescribed under the Act.

While foodgrains, sugar and fertilizers occupy 78% of the total utilized storage capacity, in the remaining 22% are stored cement, chemicals and other commodities. Warehouses of the corporation are fairly full all through the year.

Besides the conventional storage godowns, the Central Warehousing Corporation is running air-conditioned godowns at Kolkata, Mumbai and Delhi, and provides cold storage facilities at Hyderabad. Special storage facilities have been provided by the Central Warehousing Corporation for the preservation of hygroscopic and fragile commodities. The Corporation has been able to evolve a technique for a proper and scientific preservation of jaggery during the hot and rainy seasons by selective aeration and controlled conditions. It has set up special warehouses at some centres for the storage of jaggery. The jaggery stored in warehouses fetches a premium price in the market. The Corporation has also evolved techniques for the storage of spices, coffee, seeds and other commodities.

The Central Warehousing Corporation is operating a number of customs bonded warehouses at important centres in Delhi, Amritsar, Ludhiana, Kolkata, Kandla, Ahmedabad, Baroda, Surat, Bhopal, Cochin, Ernakulam and Mumbai to enable exporters/importers to keep their commodities in a good condition, pending their shipment. It has also undertaken the storage and handling of export and import cargo at the international air-port at Palam, New Delhi. At this complex, all the facilities, including inspection and clearance by customs, the payment of duty into the bank, and space for clearing agents, have been provided by the corporation. It has put up a similar air cargo complex at Amritsar for the export/import of goods. It has been expanding its capacity at the port towns to serve the industry and cooperative bodies. It has already established a sizeable capacity at Mumbai, Kolkata, Cochin, Chennai, Mangalore, Paradeep, Kandla, Haldia and Vizag. It also operates container trains on Delhi–JN Port sector.

The CWC also undertakes registration of warehouses under Warehousing Development and Regulation (WDR) Act to enable the warehouses to issue Negotiable Warehouse Receipts (NWRs). Only registered warehouse operators can issue NWRs. Up to March 2019, 139 warehouses were registered. Thousands of NWRs are issued every year. During 2018–19, 11,234 NWRs were issued worth ₹ 989 crore in storage. CWC also helps in constructing mini-godowns for Primary Agricultural Cooperative Societies (PACS) and others. It also offers consultancy services and training for construction of warehouses to different agencies.

The Corporation has introduced a scheme, called the Farmers Extension Service at selected centres to educate farmers in the benefits of a scientific storage and use of public warehouses. It trains 5000 farmers annually at 270 rural based warehouses. The Central Warehousing Corporation also provides a package of services, such as handling and transport, safety and security of goods; insurance, standardization, documentation, and other connected services and facilities. Its annual turn over during 2018–19 was ₹ 1605 crore.

3. State Warehousing Corporations (SWCs)

Separate warehousing corporations were also set up in different States of the Indian Union. The first state warehouse was set up in Bihar in 1956. At the end of March 2019, State Warehousing Corporations were operating 2145 warehouses with a total storage capacity of over 341 lakh tonnes.

The area of operation of the State Warehousing Corporations are centres of district importance. The total share capital of the State Warehousing Corporations is contributed equally by the concerned State Government and the Central Warehousing Corporation. The SWCs are under the dual control of the State Government and the Central Warehousing Corporation.

WORKING OF WAREHOUSES

- *Acts:* The warehouses (CWC, SWCs or licensed private warehouses) work under the respective Warehousing Acts passed by the Central or State Governments. They are licensed under the provisions of the Act.
- *Eligibility:* Any person may store notified commodities in a warehouse on agreeing to pay the specified charges. The person is required to bring his produce to the warehouse for storage. The commodity is inspected, and the quality of the product is determined.
- *Warehouse Receipt (Warrant):* This is a receipt/warrant issued by the warehouse manager/owner to the person storing his produce with them. This receipt mentions the name and location of the warehouse, the date of issue, a description of the commodities, including the grade, weight and approximate value of the produce based on the present price.

- *Preservation of Goods:* The produce accepted at the warehouse is preserved scientifically and protected against rodents, insects and pests and other infestations. Periodical dusting and fumigation are done at the cost of the warehouse in order to preserve the goods.
- *Financing:* The warehouse receipt serves as a collateral security for the purpose of getting credit. Commercial banks advance up to 75% of the value of the produce stored in the warehouse.
- *Delivery of Produce:* The warehouse receipt has to be surrendered to the warehouse owner before the withdrawal of the goods. The holder may take delivery of a part of the total produce stored after paying the storage charges.

The main provisions of the Act governing the grant of a licence to run warehouses were:

1. Any person, including a company, association or corporate body may apply to the State Government for the grant of a licence to carry on the business of warehousing.
2. The government grants the licence after examining the warehouse building and the financial soundness of the party, and after the realization of the prescribed fees.
3. The licence has to be renewed periodically on payment of prescribed fees.
4. The warehouse owner is authorized to receive only notified commodities for storage in his warehouse and issue receipts in a prescribed form.
5. It is the responsibility of the warehouse owner to keep the premises clean, keep different lots of goods separately in the warehouse, and carry on such operations as are necessary to protect the goods against losses from damage and pilferage.

WAREHOUSE RECEIPT (WARRANT)

A warehouse receipt is a receipt/warrant issued by the warehouse manager/owner to the person storing or depositing for storing his produce in the warehouse. This receipt mentions the name and location of the warehouse, the date of issue of receipt, a description of the commodity, including the grade, weight, and approximate value of the produce based on the present price. The warehouse owner of a licensed warehouse is authorized to issue the warrant/receipt, which may be negotiable or non-negotiable.

As per Negotiable Instruments Act, 1981, the negotiable instrument means a promissory note, bill of exchange or cheque payable either to order or to bearer. This negotiable instrument can be transferred by a simple endorsement and delivery. A delivery of part of the goods may be taken through this warrant by the depositor. The essential characteristics of a negotiable instrument are as follows:

1. Easily Transferable: The ownership of the instrument can be transferred simply by delivery of the instrument, in case it is a bearer instrument and by making endorsement in the case of crossed instrument. This saves the bearer from the problem of payment of loan with interest before taking the delivery of the produce from the warehouse.
2. Absolute and Good Title to the Transferee: The negotiable instrument confers absolute and good title to the transferee who takes it in good faith and without the knowledge that the transferer had a defective title.

3. The Transferee becomes holder of the good in due course and the holder in due course is entitled to the full value of the instrument as if he was the rightful owner of the instrument.

Sometimes, the warrant may be non-negotiable. Earlier, the warehouse receipt was not a negotiable instrument as per the Negotiable Instruments Act, 1981. However, section 2(4) of the Sale of Goods Act, 1930 brought the receipt with in the ambit of the definition of a 'document title'. The provisions in the warehouse Acts passed by state Governments provided that, unless otherwise specified in the warehouse receipt, a receipt issued by the warehouse man is transferable by endorsement and entitles to its lawful holder to receive the goods specified in the receipt. However, as per WDR Act, this problem has been solved.

The transfer of warehouse receipt from one person to another means transfer of the right to receive the delivery of the goods covered by that warehouse receipt. The lending institutions viz., banks accept the warehouse receipt as a collateral security specified in it on the same terms and conditions on which the person who originally deposited the goods would have been entitled to receive them. In view of this and the fact that the warehouse receipt is a 'document of title', it enjoys negotiability.

The warehouse receipt commands the same respect as a promissory note in the eyes of lending institution (bankers) due to the fact that the warehouse owner is liable to return the goods of same quality and quantity on surrender of the warehouse receipt by the person who holds it.

Following the Warehousing Development and Regulation (WDR) Act, Warehouse Receipt (WR) financing is being promoted. For example, in 2011, National Bulk Handling Corporation (NBHC), promoted by Financial Technologies Group, along with SBI and MCX, started facilitating bank financing through WR to the farmers. A nation-wide network of 900 locations across 19 states enables banks to facilitate funding against NBHC-issued WRs, ranging in value from as low as ₹ 1000 to ₹ 40 crore. In this way, the banks also got a secure and commercially viable platform to lend to the agriculturists. The facility has been made available for direct funding as well as through group financing, like joint liability groups.

Further, the Government has excluded the items stored in regulated warehouses from the ambit of stock limits imposed by state governments, under ECA, 1955, subject to the condition that the warehouses publish the information on real time basis (the FMC had conveyed to the commodity exchanges in November 2014 about these provisions).

NUMBER AND CAPACITY OF WAREHOUSES

The Government, the Food Corporation of India, Cooperative Marketing Societies and Central and State Warehousing Corporations have taken important measures for the creation of warehousing facilities in the country. As a result, a large number of warehouses/godowns have been built throughout the country in all important rural and urban centres, metropolitan cities, ports and railway stations.

Central and State Warehousing Corporations (CWC and SWCs)

The number and capacity of warehouses of CWC and SWCs in the country at different points of time have been given in Table 4.12.

Considerable efforts were made to increase the storage capacity in the country. The number of warehouses, which had increased from only seven during 1957–58, to 703

Table 4.12: Number and capacity of warehouses in India (including hired)

Year (end)	Number			Capacity in lakh tonnes		
	CWC	SWC	Total	CWC	SWC	Total
1957–58	7	–	7	0.1	–	0.1
1970–71	102	601	703	8.4	18.1	26.5
1980–81	330	1050	1380	37.9	50.0	87.9
1990–91	495	1331	1826	66.5	93.5	160.0
2000–01	466	1639	2105	83.9	149.0	232.9
2006–07	511	1579	2090	102.2	191.9	294.1
2018–19	422	2145	2567	101.0	340.8	441.8

Source:
1. Central Warehousing Corporation of India; quoted in Fertilizer Statistics, Various issues, Fertilizer Association of India, New Delhi, December 1994, p. III-64, Economic Surveys, Various Issues, Ministry of Finance, Government of India, New Delhi, and Annual Report of CWC, 2018–19.
2. Government of India, Annual Report, 1995–96, and Foodgrains — Monthly Bulletin, July 1996, Ministry of Food, New Delhi.

during 1970–71, and 1826 during 1990–91, went up to 2090 during 2006–07 and further to more than 2567 during 2018–19. The total capacity of warehouses which was almost negligible during 1957–58 went up to 442 lakh tonnes at the end of March 2019. Out of the total storage capacity of 442 lakh tonnes, nearly 101 lakh tonnes was with the Central Warehousing Corporation and remaining 341 lakh tonnes with State Warehousing Corporations.

The number of commodities stored in the warehouses has steadily increased. These include foodgrains, fibre crops, fertilizer, cement, rubber, cotton yarn, textiles, paper and leather.

Food Corporation of India

Apart from CWC and SWCs, the Food Corporation of India has also created storage facilities. The Food Corporation of India has total storage capacity of 40.9 million tonnes. It includes owned as well as hired capacity. Most of the capacity is of covered type which include conventional but scientifically designed godowns and silo complexes but a part of the storage capacity is of covered and plinth (CAP) type. The CAP storage capacity consists of cemented floor as the base and tarpaulins or other similar sheets as the cover (Table 4.13).

UTILIZATION OF WAREHOUSING CAPACITY

The utilization of warehousing capacity of the Central Warehousing Corporation was only 42% in 1959–60, which increased over time to 96% in 1970–71. The utilization of the capacity of State Warehousing Corporations increased from 64% in 1960–61 to 75% in 1968–69. At present, about 90% of their storage capacity is being utilized. Of the total storage capacity with CWC, 60% is utilized for foodgrains, 3% for fertilizers and 37% for other purposes. But the available storage capacity, is mostly utilized by traders or public agencies. A study had indicated that only 39% of the warehousing capacity of the Central Warehousing Corporation and 6% of that of State Warehousing Corporations was utilized by farmers or their cooperatives.

Table 4.13: Storage capacity of Food Corporation of India

(lakh tonnes)

At the end of		Covered	Cover and plinth (CAP)	Total
March 1991	Owned	120.0	10.4	130.4
	Hired	76.0	14.7	90.7
	Total	196.0	25.2	221.2
March 2002	Owned	126.1	83.4	209.5
	Hired	141.1	–	141.1
	Total	267.2	83.4	350.6
March 2010	Owned	129.7	25.1	154.8
	Hired	128.9	4.7	133.6
	Total	258.6	29.8	288.4
August 2019	Owned	127.5	26.0	153.5
	Hired	255.1	0.5	255.6
	Total	382.6	26.5	409.1

Source: Food Corporation of India, New Delhi and Ministry of Food, Government of India, New Delhi.

The main reasons for the very low utilization of warehouses by farmers are:

1. Lack of knowledge about the facility of warehousing available for the farmers;
2. Locational disadvantages for warehouses to most of the cultivators located in villages;
3. Complicated and time-consuming procedure of depositing and withdrawing the produce from the warehouses;
4. Non-existence of nationalized bank branches in villages and the problem of arranging finance at the time of taking delivery of the warehouse receipt from the bank; and
5. Small quantity of surplus produce available with most farmers, and the pressing need for finance.

These apart, there are some fundamental factors responsible for lower use of warehouses by the farmers.

1. Most of the Indian farmers are small landowners. Obviously the marketed surplus available with them is small. Often, it is not worthwhile for them to store the produce in a warehouse;
2. Indian agriculture is largely dependent on the monsoon and occasional failures of crops in one or another part of the country are common resulting in lack of regular business for the warehousing;
3. Agricultural products are more bulky and perishable than industrial products;
4. Agricultural commodities are heterogeneous. Their grading is, therefore, essential before placing them in a warehouse. This facility is not available in most of the markets;
5. The warehouses are located in urban centres, near railway stations and big cities. The transport facility from the villages to these centres is not easily available;
6. Till recently, warehouse receipts were treated as papers having no intrinsic value. Often, the private lenders were not interested in lending against this collateral security.

In this context, it must be recognised that storage of the produce is at a cost. It is not only the charges of the warehouse that are to be paid but also the interest on the value of the produce and the premium for risk of a lower price at a later date are to be met. The interest component of the storage cost is no less significant. Moreover, the intra-year price rise may not cover the entire cost of storage every year. The probability of returns from storage being positive is not one. This means that the gains from storage depend on the decision on the timing of the purchase and sale. This necessitates acumen of astute trading, which every farmer does not possess. This apart, farmers in surplus-producing states like Punjab and Haryana sell their produce at the minimum support price to the public agencies and as the minimum support price remains the same till the next harvest season, such farmers do not gain by storing the produce, unless open market prices in the lean season rise to such levels as to cover the cost of storage and still leave a margin as an incentive to store, which rarely happens. The worry on low utilization of warehouses by the farmers should be seen in this light. Even if the facility is utilised by the traders, it indirectly helps the farmers by way of augmenting the demand for commodities stored by the traders.

The situation is however, changing in recent years. The farmers are increasingly realizing the benefits of marketing in groups. The government is encouraging group marketing by way of promoting farmer producer organizations (FPOs) or farmer producer companies (FPCs). The FPOs/FPCs are actively using the warehousing facilities with a view to gaining from storage of the produce of the member-farmers, specially for farm products that are not covered under minimum support price scheme of the government.

WAREHOUSING INDUSTRY

Apart from public sector warehouses, the private sector is also active in developing warehousing facilities. The warehousing capacity is expressed in terms of space rather than volume or business. The available warehousing space in India in 2019 is estimated at around 1440 million square feet. The average size of warehouses has grown from 20000 square feet earlier to around 3 lakh square feet now. In terms of business, it is currently worth ₹ 50000 crore and growing at the rate of 10% per year. The industry is now also using Radio Frequency Identification (RFID) Technology (RFID tags attached to stored goods). For promoting private sector investment, warehousing has been included under priority sector for the purpose of institutional lending.

RURAL GODOWNS

The projected availability of foodgrains and the available storage capacity in India show that there is big gap in storage capacity. This gap has to be bridged as early as possible if advantage is to be taken of the benefits of increased agricultural production. Apart from the warehouses of FCI, CWC and SWCs, there is need for developing rural storage godowns.

Realising the necessity of creation of storage facilities in the rural areas, the Government of India appointed an Expert Committee in March, 1979 under the Chairmanship of Additional Secretary in the then Department of Rural Development for going into the problem of storage of agricultural produce in rural areas. Based on the recommendations of this Committee, a scheme for the establishment of rural godowns, viz., National Grid of Rural Godowns (NGRG) was launched by the Government of India in July, 1979. The scheme was aimed at the creation of a network of rural godowns in the country primarily to take care of storage requirements of agricultural producers particularly of small and marginal farmers in the rural areas. The scheme of rural godowns was intended to achieve the following specific objectives:

1. Prevention of distress sale of foodgrains and other agricultural commodities immediately after harvest;
2. Reduction in quantity and quality losses arising by storage in sub-standard places;
3. Reduction in pressure on transport system in the post-harvest period;
4. Creation of employment opportunities in rural areas;
5. Helping the farmers in getting loans against the stored produce; and
6. Helping in easy procurement of foodgrains by Food Corporation of India.

Rural godowns have been constructed with capacity ranging from 200 to 1000 tonnes depending upon the produce expected for storage in the area. The cost of construction of these godowns have been subsidized to the extent of 50% (shared equally by state and central government) and remaining 50% capital is to be arranged by the implementing agency (Cooperative Marketing Society) in the form of loan from the commercial banks. These rural godowns are constructed as per specifications and designs approved by the state warehousing corporations and are managed by the cooperative marketing societies. The state warehousing corporations provide technical guidance and supervision to the implementing agencies in the maintenance and management of rural godowns. The receipts issued by the managers of rural godowns on the basis of stocks is a negotiable instrument. On the basis of this receipt, farmers can get loan up to 80% of the value of the produce stored from the commercial banks.

The progress of rural godowns and marketing societies godowns constructed in India for creation of storage facilities in rural areas up to 2001 is shown in Table 4.14.

Table 4.14: Cumulative progress of rural godowns in India (up to 2001)			
Year (end of March)	Rural godowns (number)	Marketing societies godowns (number)	Capacity (million tonnes)
1985	1494	NA	NA
1991	50555	8710	12.07
1996	54898	9363	15.00
2001	68876	9414	13.74

Source:

1. Status paper on the Scheme regarding Establishment of Rural Godowns, Department of Rural Development, Government of India, New Delhi, 1986.
2. Status and Potential of Storage in Maharashtra, Pagire, B.V., D.V. Kasar; D.S. Nawadkar and H.R. Shinde, National Seminar on Rural Godowns held at NIAM, Jaipur on April 12–13, 2002.
3. Website of DMI.

To boost up the construction of rural godowns further, the Finance Minister in the budget speech, 2001 announced a Credit Linked Capital Investment Subsidy Scheme for the construction of rural godowns (GBY). The main objectives of this scheme include the creation of scientific storage capacity with allied facilities in rural areas to meet the requirements of the farmers for storing farm produce, processed farm produce, consumers articles and agricultural inputs; promotion of grading, standardization and quality control of agricultural produce to improve their marketability; prevention of distress sale; providing facility of pledge financing to strengthen agricultural marketing infrastructure in the country and to reverse the declining trend of investment in agricultural sector by private and cooperative sector. The main striking features of this new scheme were:

1. Promoters for construction of rural godowns can be individual farmers, groups of farmers, partnership/proprietary firms, self-help groups (SHGs), non-government organizations (NGOs), companies, corporations, cooperatives, Agricultural Produce Market Committees, State Agricultural Marketing Boards and Agro-Processing Corporations.
2. The subsidy component in this scheme was initially 25% of the capital cost as back-ended subsidy to be provided by the central government, 50% is the institutional loan and rest 25% is the owner's contribution.

This scheme of creation of rural godowns in the initial stage was approved for two years (2001–03) for creation of new rural godowns/renovation of existing godowns of the capacity of 2.0 million tonnes. The subsidy under this scheme is being released only for the godowns constructed outside the limits of the municipal corporation area and have a minimum capacity of 100-metric tonnes. The scheme was later extended for the subsequent period.

The rural godown scheme (Grameen Bhandaran Yojana) was liberalized in 2009 to boost the agricultural warehousing capacity. Under the liberalized GBY, the objective was to attract private sector for investment in rural storage/warehousing capacity.

The GBY was, in 2014, subsumed into Agricultural Marketing Infrastructure (AMI) sub scheme, and named as storage infrastructure. All states and union territories are eligible for subsidy for rural storage structures. For special category areas of NE states, Sikkim, UTs of A & N and Lakshadweep islands and hilly areas, the subsidy is 33.33% with a ceiling of ₹ 1333.20 per tonne and maximum of ₹ 4 crore. For FPOs, panchayats, women, SC and ST beneficiaries or their cooperatives/SHGs in other areas, the subsidy rate is the same (1/3) with a ceiling of ₹ 1166.55 per tonne up to 1000 tonnes capacity and ₹ 1000 per tonne for more than 1000 tonnes (up to 30000 tonnes) capacity with a maximum of ₹ 3 crore. For all other categories, the subsidy is 25% with a ceiling of ₹ 875 per tonne up to 1000 tonnes and ₹ 750 per tonne for more than 1000 tonnes capacity, with a maximum of ₹ 2.25 crore. As per DMI, from April 2001 to March 2016, 37795 rural godown projects were sanctioned with a total storage capacity of 61.9 million tonnes (an average of 1640 tonnes per project), involving a subsidy of ₹ 2969 crore.

TOTAL STORAGE CAPACITY

The total storage capacity available in India at the end of 2019 is 161 million tonnes (Table 4.15). According to ASSOCHAM, around 70% of this is owned by the government agencies.

NEW INITIATIVES TO INCREASE STORAGE CAPACITY

The current storage capacity is considerably short of the requirement. It is a common practice of the price support agencies to store precious foodgrains on Cover & Plinth (CAP) formats, resulting quite often in huge wastage. The demand for warehousing services from various sectors of the economy is growing rapidly at the rate of around 9% per annum. For augmenting the warehousing capacity in the country, several new initiatives have been taken.

Table 4.15: Storage capacity available in India		
		(million tonnes)
Agency	**1969**	**2019**
FCI (own and covered)	3.9	40.9
CWC	1.0	10.1
SWCs	0.8	34.0
Cooperative and other godowns	5.3	75.6
Total	11.0	160.6

Source: Government of India, Ministry of Agriculture, Websites of DMI, CWC and FCI.

1. A Warehousing Fund has been created at NABARD to finance agricultural warehousing projects. In 2014–15, a sum of ₹ 5000 crore was provided by the government in this fund. Projects for enhancing warehousing capacity of 49 million tonnes are reported to have been sanctioned.
2. Under Private Entrepreneur Guarantee (PEG) scheme, Food Corporation of India has sanctioned around 20 million tonnes capacity through private entrepreneurs. Under this scheme, the FCI gives guarantee of hiring the warehousing space for a pre-agreed period.
3. The Government has also undertaken the construction of high-tech silos under public private-partnership (PPP) mode. This is based on the experience of a pilot project launched in 2005. Under the pilot project, the FCI entered into a BOO (Build, Own and Operate) agreement for 20 years with Adani Agrologistics for two silos of 50,000 tonnes each at Moga (Punjab) and Kaithal (Haryana). The lease period started in 2008. Since 2012, infrastructure project for the construction of high-tech silos for a total capacity of 2 million tonnes is on-going. The grain storage capacity of each silo is either 50000 tonnes or 25000 tonnes. It needs to be noted that silos need lesser land area than the conventional storage structures. The land required for a 50,000 tonnes capacity silo is 7 acres and for 25000 tonnes capacity silo is 5 acres. The states, number of silos and total storage capacity are as shown in Table 4.16. The main features of the scheme as are as follows:
   (a) The FCI conducts feasibility studies, before initiating the tendering process for each silo.

Table 4.16: Proposed silos with capacity and location		
State	**No. with capacity of each (in thousand tonnes)**	**Total storage capacity (in lakh tonnes)**
Assam	2 (25)	0.5
Bihar	4 (50)	2.0
Gujarat	2 (25)	0.5
Haryana	6 (50)	3.0
Kerala	2 (25)	0.5
MP	7 (50)	3.5
Maharashtra	2 (50)	1.0
Punjab	8 (50)	4.0
UP	6 (50)	3.0
West Bengal	4 (50)	2.0
Total	43	20.0

(b) The FCI provides a guarantee of rentals for 30 years to enable the investor to recover capital investment and maintenance charges.

(c) There is a provision of 20% Viability Gap Finding (VGF) from the government as per infrastructure project norms.

(d) These silos are under Design, Build, Finance, Operate and Transfer (DBFOT) norms, where a private developer is responsible for development of the project. The state government will provide the land for silos.

(e) If private players seek VGF and land from the state government, the rental agreement with FCI is only for 10 years.

(f) The major issue that is coming in the way of success is the compulsion of railway siding and availability of land at such locations.

COLD STORAGES

The term cold storage refers to a refrigerated chamber for the storage of such perishable commodities as fruits, vegetables, fish, eggs, meat, dairy products, etc. In these storage structures, the temperature is controlled and maintained so that the stored perishable products may not deteriorate in quality. In a cold storage, the temperature is maintained in the range of $-1.1°C$ to $10°C$ ($30°$ to $50°F$). The other form of cold storage is the freezer storage, in which the temperature is kept below $1.1°C$ ($30°F$), and the product remains in a frozen state.

In addition to the preservation of the quality of perishable products, the cold storage offers the following advantages:

1. It makes possible the regular supply of perishable commodities in the market. This would not have been possible without the cold storage facility.

2. It helps in the price stabilization of perishable commodities by removing the gluts occurring in the production season.

3. It helps in widening the market for the products, lowering marketing costs, raising the price realised by the producer and lowering the price to consumers, and ensures that products are available throughout the year.

4. Cold storage facilities have made it possible for the consumers to live in greater comfort.

For long, the establishment of cold storage industry remained under regulation. The Central Government issued Cold Storage Order in 1964 and again later in 1980. However, some State Governments like that in West Bengal, Uttar Pradesh, Punjab and Haryana were permitted to promulgate their own orders. The Cold Storage Order was promulgated by the Government of India under Section 3 of the Essential Commodities Act, 1955. It was being administered by the Directorate of Marketing and Inspection to achieve the following objectives:

1. To ensure hygienic and proper refrigeration conditions in the cold storage;

2. To regulate the growth of the cold storage industry in a planned manner;

3. To render technical guidance for scientific preservation of foodstuffs; and

4. To safeguard the interests of farmers and other depositors.

Cold Storage Order, 1964 and also of 1980 was applicable all over the country except in the States of Uttar Pradesh, West Bengal, Punjab and Haryana, where the State Governments have enacted their own Cold Storage Acts. West Bengal and Uttar Pradesh sought permission to enact their own Acts in 1960 and 1975. Punjab and Haryana

Governments were permitted to promulgate their own State orders for regulating the cold storage industry in 1979.

Under the Cold Storage Order, the prospective entrepreneur was required to obtain the permission from the Agricultural Marketing Adviser to the Government of India for construction of a cold storage. With effect from 1st January, 1965, it was obligatory for a cold storage, with a capacity exceeding 8.50 cubic metres to obtain a licence before storing any foodstuff. The Agricultural Marketing Adviser to the Government of India was the authority under the Cold Storage Order, and was empowered to licence the setting up of a cold storage. .

The Cold Storage Order, 1980 was rescinded in May, 1997. The repeal of the cold storage order of 1980 aimed at enabling the government in the removal of licensing, price control and requisitioning of the cold storage space with a view to allowing the functioning of free marketing mechanism for demand based growth of cold storage industry in the country free from all kinds of administrative interference.

Most of the cold storages are in the private sector. The National Commission on Agriculture in 1976 had recommended for adequate measures to be taken by cooperatives and public sector undertakings to provide cold storage facilities in production areas and terminal markets. As a follow-up, the National Cooperative Development Corporation prepared a project for setting up 4.8 lakh tonnes of cold storage capacity in the cooperative sector by 1985 with the World Bank assistance in the states of Uttar Pradesh, Bihar, West Bengal and Madhya Pradesh.

The first cold storage was established in India as early as in 1892 at Kolkata. But noticeable progress in expansion of the cold storage industry was not made until 1947. Even up to 1955, the total cold storage capacity in the country was only 0.83 lakh tonnes. The number of cold storage units and their capacity in India are given in Table 4.17. The rapid strides in the expansion of cold storage capacity were made after 1955. The cold storage capacity increased to 3.06 lakh tonnes in 1960, 16.38 lakh tonnes in 1970, 39.65 lakh tonnes in 1980, 68.15 lakh tonnes in 1990 and further to 349.6 lakh tonnes in 2017. The total number of cold storage units in the country in 1995 has been 3167, which increased to 7645 in 2017. There is a wide inter-state variation in the availability of cold storage facilities in India.

Table 4.17: Growth of cold storage facility in India		
Year	Number of cold storage units in operation	Storage capacity (lakh tonnes)
1947	4	0.03
1955	43	0.83
1960	359	3.06
1970	1218	16.38
1980	2283	39.65
1990	2795	68.15
2001	4199	153.85
2010	5837	283.03
2017	7645	349.58

Source: Economic Survey, Various Issues, Ministry of Finance, Government of India, New Delhi; Directorate of Marketing and Inspection, Ministry of Agriculture and Rural Development, Government of India, Faridabad; NSEL and base line study by MoFPI, Website of MoFPI.

The sector-wise distribution of cold storage facility available in India is shown in Table 4.18. Private sector has played the crucial role in providing cold storage facility. Nearly 90% of total cold storage units are privately owned which account for 95% of the total cold storage facility. The direct involvement of the government is negligible in cold storage sector. Most of the cold storage units are fully used for atleast four to five months of the year.

Table 4.18: Sectorwise distribution of cold storage facilities in India (as on Dec. 31, 2006)

Sector	Number of cold storage units	Capacity of cold storage units (lakh tonnes)
Private	4609 (90.4)	206.5 (95.2)
Cooperative	358 (7.0)	9.5 (4.4)
Public	134 (2.6)	1.0 (0.4)
Total	5101 (100)	217.0 (100)

Note: Figures in parentheses are percentages of the total number and total capacity of storage units.
Source: Directorate of Marketing and Inspection, Government of India; and National Institure of Agricultural Marketing.

Potato is the main product which is stored in the cold storage. Out of the total capacity utilisation, 78% is used for storing potato. Commodity-wise utilization of cold storage units in India can be seen in Table 4.19.

Table 4.19: Commodity-wise percentage distribution of cold storage units in India

Commodity	Number of cold storage units	Capacity (million tonnes)
Potato	2853 (55.9)	16.84 (77.6)
Fruits and vegetables	144 (2.8)	0.08 (0.4)
Milk and milk products	200 (3.9)	0.07 (0.3)
Meat and fish	482 (9.5)	0.18 (0.8)
Multi-purpose	1337 (26.2)	4.27 (19.7)
Others	85 (1.7)	0.25 (1.2)
Total	5101 (100)	21.69 (100)

Figures in brackets are percentages of column total
Source: Agricultural Marketing, NIAM, Jaipur, Statistical Abstract 2006,

The construction of a cold storage requires heavy investment in terms of building and machines. Similarly, to run the cold store, the cost on electricity input is very high and it amounts to 50% of total running or variable cost. As such there is need to develop low cost and energy saving cold storage units.

For promotion of cold storage units in the private sector to meet their increasing needs, a capital investment subsidy scheme for construction/expansion/modernization of cold stores/storages for horticultural produce has been initiated by Government of India. This scheme is implemented by National Horticulture Board. Under this scheme, the promoters of cold storage units are provided 25% back-ended capital investment subsidy, 50% is provided as term loan and 25% is promoter's contribution. The proposals are considered and sanctioned by National Cooperative Development Corporation (NCDC) under cooperative sector. This scheme is implemented in those

States/Union Territories, which do not control rentals for cold storages under any statutory or administrative order.

The available capacity of cold storage is much less than the country's requirements. It is barely sufficient for less than 10% of fruits, vegetables and fish production. The cold storage requirement has further increased in view of the need to promote exports of processed foods.

There is a considerable scope for expansion of the cold storage industry in India. However, due to large capital requirements, lack of proper technical guidance, inadequate and fluctuating power supply and lack of appreciation for stored products, the entrepreneurs are not attracted to establish cold storage units. The lack of cold storage facilities is leading to heavy losses and violent fluctuations in prices of fruits and vegetables. There is a need to encourage cold storage industry in several regions of the producing areas as well as in large urban centres.

In India, the production patterns, dietary habits and economic considerations warrant long period of storage in large quantities of onion and potato. The conditions required for the storage of potato and onions are distinctly different. While the potato requires low temperature and high relative humidity, onions require low temperature and low relative humidity. Most of the cold storages in the country meet the storage requirements of these two vegetables. For other vegetables, temporary storage structures for short period usually not exceeding a week are needed along the route of their movement from producing areas to consuming centres.

The country would also need reefer containers/vans for transport of perishable commodities for domestic and export marketing. Their availability was nil in fifties but increased in recent years to 9000. As of now, at least 53,000 reefer containers/vans are needed to handle the available surplus of perishable agricultural products.

Considering the fact that an average farmer may not need and have access to mechanical refrigerated cold stores; ventilated storages like direct evaporation cooled structures; energy cool chambers; cool homes and forced evaporation cool stores have been developed. These structures provide relatively lower temperature and high humidity as compared to ambient conditions because of natural/forced evaporative cooling. These can be constructed with locally available materials. The zero energy cool home, AADF CIP design cool home and two-tier structures can be afforded by the farmers on their farms. However, other improved structures can be constructed by growers cooperatives or owners of large size farms.

For the benefit of farmers or farmers' group, who are looking for a nearby cold storage, a state-wise list of cold stores along with location and phone numbers is available on the website of NHB and MoFPI.

COLD CHAIN SYSTEM IN MARKETING OF PERISHABLES

Fruits, vegetables and flowers (perishables) are all living products even after harvesting. They still grow and respire. Higher the temperature, higher the respiration rate and shorter the life of horticultural products. Respiration rate slows down on cooling of the produce and thus the product can remain in a good condition for a longer period. The cold chain system is appropriate for prolonging the shelf life of all perishable horticultural and animal food-products.

Cold chain concept is one of the latest developments in marketing of perishable agricultural commodities. This involves cooling of the produce immediately after harvest to the lowest non-damaging temperature and then maintaining the temperature constantly throughout all the post-harvest operations viz., handling, packaging, transportation, storage and marketing up to and including retail sale.

Concept of Cold Chain

The concept of cold chain means setting up of temperature management facilities at production centre, during transportation through the reefer vans to the wholesale markets, equipping the wholesale markets with cold storages and controlled temperature systems supported by pre-cooling and handling centres, further transportation through reefer vans or light commercial vehicles (mobile coolers) to the retailing points; and providing the temperature management system, even in the retailing shops so as to ensure the continuity of cold chain.

Cold chain management is an innovative system that relies on modern technologies to control and monitor the temperature of perishable products to ensure their freshness and quality throughout the entire supply chain. Under this concept, the products, immediately after harvest, should be pre-cooled, then cold stored either in chilled or in frozen conditions, and transported under refrigerated conditions to a main cold store for long term storage. From this main cold store, the products are then distributed to various cities by refrigerated containers so as to reach retail distribution cold stores/ small cold storage devices for ultimate sale to the consumers.

Necessity of Cold Chain

Perishable farm products, by virtue of their constant demand throughout the year, need preservation. Today, food preservation and distribution has become more important as family incomes are rising and urban population, which is increasing at a faster rate, requires a variety of foods which is fresh and of good quality on a daily basis. Most food is produced in rural areas and carried into urban markets by available means of transportation. These food stuffs need to be kept in excellent condition till they reach the consumers.

Most of the food stuffs, by very nature, are perishable and their shelf life is restricted from few hours to few days when left to natural climatic conditions. Hence, a mechanism is needed which can keep food and allied perishables to be transported to consumption areas without deterioration.

It is also necessary that the food which is produced or harvested during a particular season of abundance is to be preserved to help in meeting the demand in subsequent seasons of the year. The cold chain system helps in achieving the above objectives by increasing the shelf life of the perishable farm products. These can reduce the losses by up to 76%.

As mentioned earlier, in a cold chain mechanism, the product is maintained under low temperature and optimum relative humidity. This preserves the shelf life of the product in natural harvested conditions by suppressing and reducing the bacterial, enzymatic and microbial activities.

Post-harvest Losses

Post-harvest losses occur both in quantity and quality of horticultural and other perishables products. The estimated post-harvest losses vary widely as presented in Table 4.20.

Table 4.20: Estimated post-harvest losses of fruits and vegetables in India

Commodity	% loss
Apple	14
Banana	20–80
Lemon	20–95
Orange/Mandarins	20–95
Grapes	27
Papaya	40–100
Cabbage	37
Cauliflower	49
Onion	16–35
Potato	5–40
Tomato	5–50

Source: Indian Horticulture Data Base, 2009

Types of Losses in Food Quality

Loss in product quality can be of many types and depends on the type of products. Some of the common losses in perishable agricultural products are as follows:

1. Melting of frozen products (ice-cream) or refrigerated goods (butter)
2. Increased microbial activity in food products affecting flavour, colour or keeping quality of the product
3. Shrinkage in volumes due to drying (pulpy vegetables and fruits),
4. Loss in texture or texture degradation.
5. Loss in weight resulting in wilting and limpness,
6. Softening of product (ice-cream, butter)
7. Bruising of the product
8. Excessive or unwanted ripening of products (fruits),
9. Change of colour of products.
10. Development of rots and moulds (as in cheese)
11. Loss of properties or effectiveness (e.g. herbs used in preparation of medicines)

Problems Where Cold Chain System Offers Viable Solutions

Cold chain management offers viable solution under the following problem areas:

1. Under concentrated large/surplus supply or production situations such as of vegetables, fruits, flowers and livestock products.
2. Preserving surplus production for sale or use in the lean season to get better prices (e.g. potatoes).
3. Products needing deep freezing or refrigeration for storage (e.g. butter, cheese, chocolates, and other flavored milk products).
4. Increasing shelf life of the products (such as pasteurized milk).
5. Eliminating souring and curdling of products and thus preserving normal good quality of products (e.g. raw milk).
6. Maintaining desired body texture of products like ice-cream, flavored milk, and butter milk.

Logistics of Cold Chains

Cold chains operate at minimum temperature and optimum relative humidity conditions without any heat shocks to the product. The logistics of cold chain system include the following:

1. *Pre-Cooling Facility:* This facility must exist at the point where product is in abundance or at the farm or village level itself. Pre-cooling of the product must be carried out at prescribed conditions of temperature and relative humidity at the needed time.
2. *Refrigerated Containers for Transportation to the Main Warehouse:* Product after pre-cooling should be moved to warehouse through refrigerated transportation in refrigerated containers.
3. *Warehouse/Cold Store:* The facility of washing, grading, and packing of produce should be available before the product is shifted to the cold stores.
4. *Main Cold Stores:* As per the need of the products.
5. *Transportation:* Further distribution of preserved produce through refrigerated transport to distribution cold stores or retail stores.
6. *Display in Retail Stores:* The produce is displayed for sale in refrigerated display counters for buying by the consumers.
7. *Consumer Level:* The consumer buys the produce and either consumes it or carries it to the home for keeping under refrigerated or frozen condition for future use.

Diagrammatically, the cold chain system can be presented as follows:

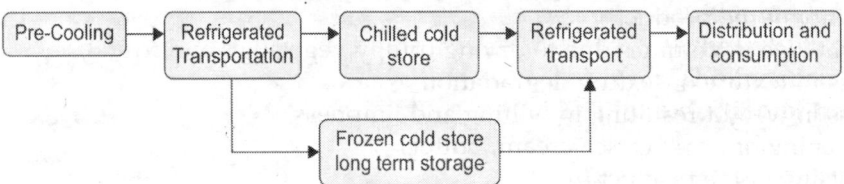

Progress of Cold Chains

So far, adoption of cold chain system has been possible in the marketing of perishable horticultural produce where large integrated organizations are involved, which can control all possible post-harvest aspects and which themselves are the major super market retailers selling large volume of produce. At all other levels, progress of cold chain system is relatively slow. The basic causes for poor progress of cold chain system in India are the following:

1. General lack of marketing and transport infrastructure.
2. Most retail outlets are characterized by small size and low-level of investment.
3. Low capitalization by owners in marketing.
4. Value of many perishable fresh farm produce is low to warrant the investment in the cold chain system.

Nevertheless four important developments need to be noted.

1. During the 12th Five Year Plan, the National Horticulture Mission had planned to launch special trains to carry fruits and vegetables between major producing and consuming areas. As a first step, a banana train was launched in September 2012 to transport banana from Raver and Jalgaon in Maharashtra to Azadpur market in Delhi. This horti-train runs once or twice a week and carries specially designed 80

insulated and ventilated containers. The capacity of each container is 12 tonnes, thus the train carries about 1000 tonnes of banana in each round. It is a joint ventue of National Horticulture Board and Container Corporation of India (CONCOR). The travel time is 26 hours (one way) including 8 hours for loading and unloading. Each rack can be owned by one or a group of farmers. It has reduced the cost of transportation as well as wastage in transit. The cost of transportation has come down by 50%. It may be noted that around 12000 to 15000 wagon loads of fruits and vegetables are transported every year from Maharashtra to the markets in UP, AP, Karnataka and Rajasthan. A special Kisan train has also been launched.

2. India, currently, has around 9000 actively refrigerated trucks handling nearly 4.0 million tonnes of perishable items. To help the transportation system, Government of India has launched a toll free help line number to provide assistance to refrigerated transporters stuck in the middle of the highways. The calls are handled by Mumbai-based Reefer Vehicle Call-in Centre (RVC), set up jointly by Government's National Centre for Cold-Chain Development (NCCD) and Mahindra Logistics.

3. There is an encouraging response from the private sector logistic companies also. Some companies (like Snowman Logistics Ltd) are providing 'source to stores' services. The operations comprise ambient to chilled and frozen (–25°C to +25°C) warehousing, primary and secondary transportation, and value added services, including kitting, labeling, sorting and bulk breaking. The Snowman has 23 temperature controlled warehouses at 14 locations and operates 370 reefer vehicles.

4. According to one estimate, during the last 10 years (since 2004–05), cool chain facility, (including cold storages) for handling 32 million tones of perishable goods has been added in the country.

5. The gap in terms of need for pack houses, reefer vans and ripening units is very wide that needs to be attended.

Government Schemes for Investment in Cold Chains

Modern cold chain system involves a number of equipments such as pre-cooling facilities, cold storages, humidity controlled atmospheric storage facilities, reefer containers, ripening rooms, mobile coolers (walk coolers) and integrated handling and storage systems. All these need huge capital investment. To meet this increased financial requirement, APEDA, National Horticulture Board (NHB) and Ministry of Food Processing Industries (MoFPI) of Government of India have introduced various schemes. Under these schemes, a back-ended-subsidy @ 20 to 25% of fixed capital investment is provided to prospective entrepreneurs. The MoFPI is currently supporting implementation of 299 cold chain projects, of which 141 are reported to have been completed and others are under implementation.

▌ PROCESSING AND VALUE ADDITION

Processing is an important marketing function in the present-day marketing of agricultural commodities. A little more than 100 years ago, it was a relatively unimportant function in marketing. A large proportion of farm products was sold in an unprocessed form, and a great deal of the processing was done by the consumers themselves. At present, consumers are dependent upon processing for most of their requirements. Many technological changes have occurred in the recent past, such as the introduction of refrigeration, modern methods of milling and baking foodgrains, new processing methods for dairy products, and modern methods of packing and preservation. These technological changes have had a significant impact on the standard of living of the consumers, on the economic and social organizations of society, and on the growth of trade in the country. Value addition in food in India continues to be low at around 15% as against 23% in China, 45% in Philippines and 188% in U.K. However, the situation is now changing. For example, the gross value added from food processing industries went up from ₹ 1.47 lakh crore in 2011–12 to ₹ 2.49 lakh crore in 2017–18, registering a compound growth rate of 9.50% per annum (NAS, quoted by MoFPI, website).

Meaning

The processing activity involves a change in the form of the commodity. This function includes all of those manufacturing activities which change the basic form of the product. Processing converts the raw material and brings the products nearer to human consumption. It is concerned with the addition of value to the product by changing its form.

Value addition, in general, is the process of changing or transformation of a product from its original state to a more valuable state. Many raw commodities have intrinsic value in their original state but the act of processing and marketing adds value to it. The examples are processing of gram (chick pea) in to gram flour (besan), mustard seeds in to oil and cake and sugarcane into gur or sugar.

Value addition term is frequently used while discussing the ways of increasing the profitability of agricultural commodities. However, the popularity of the word 'value addition' rose substantially during 1990's and has become one of today's buzz words. Value addition in agricultural products can be brought about by (a) processing of the product; (b) packaging, branding and grading of the product; (c) developing innovative uses of the product; and (d) developing fast-food technologies.

There are three possible levels of value addition or processing of farm products (NAAS, Policy Paper No. 16, 2002):

- Level 1: Post-harvest or Primary Processing: proper cleaning, grading and packaging (fruits, vegetables, potato)
- Level 2: Secondary Processing: basic processing, packaging and branding (packed flour, suji, rice, fruit pulp)
- Level 3: Higher End Processing: Supply chain management, modern processing technology, packaging of processed foods, branding and marketing (potato chips, breakfast foods, noodles)

The processing of agricultural commodities can take any of the following forms:

1. Food Processing

1. Milling of grains: Wheat flour milling, rice milling and pulse milling;
2. Manufacture of edible oils and oilcakes from oilseeds;
3. Manufacture of hydrogenated and vanaspati oils;
4. Manufacture of sugar, gur and khandsari from sugarcane;
5. Manufacture of dairy products like ghee, butter, khoa, milk powder and cream from liquid milk;
6. Canning and preservation of fruits and vegetables;

7. Manufacture of juices and other products from fruits and vegetables;
8. Slaughtering of animals and preparation of meat;
9. Canning and preservation of fish;
10. Manufacture of bakery and confectionary products;
11. Tea processing;
12. Coffee curing and roasting;
13. Cashewnut processing;
14. Manufacture of starch;
15. Manufacture of prepared animal feed.
16. Preparation of Ready-to-Eat (RTE) foods.
17. Manufacture of soft drinks and bottling of water.

2. Agro-related Processing (Other than Food Processing)

1. Processing of tobacco leaves and manufacture of tobacco products;
2. Brewing and manufacture of beverages;
3. Cotton ginning and making bales;
4. Manufacture of yarn, cloth and made ups (cotton and jute);
5. Leather processing and manufacture of leather and fur products;
6. Manufacture of wood products, furniture and fixtures;
7. Manufacture of paper and paper products.

There are several methods of processing agricultural commodities. But they continually undergo a change because of—

1. A change in technology;
2. A change in capital-labour price ratio;
3. A change in managerial skill;
4. A change in the demand for product quality in the market;
5. A change in the volume to be processed; and
6. A change in habits and life styles of people.

ADVANTAGES

The processing of agricultural products is essential because very few farm products—milk, eggs, fruits and vegetables—are consumed directly in the form in which they are obtained by the producer-farmer. All other products have to be processed into a consumable form. Even milk, eggs, fruits and vegetables need some degree of processing and value addtion before consumption. Processing is important, both for the producer-sellers and for consumers. This increases the total revenue of the producer by regulating the supply against the prevailing demand. It makes it possible for the consumers to have articles in the form liked by them. The specific advantages of the processing function are:

1. It changes raw food and other farm products into edible, usable and palatable forms. The value added by processing to the total value produced at the farm level varies from product to product. It is nearly 7% for rice and wheat, about 79% for cotton and 86% for tea. It is generally higher for commercial crops than for food crops. Examples of the products in this group are: the processing of sugarcane to make sugar, gur, khandsari; oilseeds processing to make oil; grinding of foodgrains to make flour; processing of paddy into rice; and conversion of raw mango into pickles.

2. The processing function makes it possible to store perishable and semi-perishable agricultural commodities which otherwise would be wasted and facilitates the use of the surplus produce of one season in another season or year. Examples of the processing of the products in this group are: drying, canning and pickling of fruits and vegetables, frozen foods, conversion of milk into butter, ghee and cheese and curing of meat with salting/smoking.

3. The processing activity generates employment. The baking industry, the canning industry, the brewing and distilling industry, the confectionary industry, the sugar industry, oil mills and rice mills provide employment to a large section of society.

4. Processing satisfies the needs of consumers at a lower cost. If it is done at the door of the consumer, it is more costly than if it is done by a firm on a large scale. Processing saves the time of the consumers and relieves them of the difficulties and botherations experienced in processing.

5. Processing serves as an adjunct to other marketing functions, such as transportation, storage and merchandising.

6. Processing widens the market. Processed products can be taken to distant and overseas markets at a lower cost.

VALUE ADDITION AND PROCESSING OF AGRICULTURAL COMMODITIES

A large number of units are engaged in processing and value addition. Majority of these are in unorganized sector. There are 13.9 million agribusiness units in India of which 5.11 million are food processing units in the unoganized sector. Agricultural products are processed by employing different types of machinery and technology. The type of processing ranges from simple drying, parboiling, husking, polishing, and grinding to the complex form of producing an altogether new product. Hydrothermal treatment is one of the most common steps in the processing of foodgrains. Hydrothermal treatment of grains refers to the addition of moisture and heat to the grains for improving the quality and quantity of their product or to facilitate different milling operations for the desired products. This treatment is also called conditioning of grains and is a pre-milling treatment. This treatment is used for such purposes as:

1. Improving shelling efficiency;
2. Improving nutritional quality;
3. Improving milling quality in paddy;
4. Facilitating dehulling of corn and wheat;
5. Facilitating de-husking and splitting of kernels during milling of pulses;
6. Reducing toxic effect by soaking into hot water, as in the case of Kesari dal;
7. Removing disagreeable odour as in case of soyabean.

The methods used in processing of some important commodities have been given in the paragraphs that follow.

PROCESSING OF WHEAT

In India, about 90% of the wheat is consumed in the form of chapatis and 10% as bread, biscuits, buns and other bakery products. Irrespective of the form of consumption, wheat grains are required to be milled to convert them into flour or other forms of brokens (like dalia). Wheat grain consists of 85% endosperm, 12% bran and 3% germ.

Milling

Wheat milling involves grinding the kernel into a whole wheat flour and separating the bran from the white flour (endosperm). The milling of wheat in India is done in the following ways:

1. *Stone grinding by hand:* This method is used to grind wheat in some rural households. A housewife, by using stone chakkis which are operated by hand, mills 0.5 to 1 kg of wheat per hour.
2. *Chakkis:* This is a low capacity power-operated grinding device used in small villages. The cost of miling is high; but because of their convenience, they are widely prevalent in Indian villages. In recent years, several companies have come out with small electric-operated milling sets which are being used by several middle and high income families to make flour for their domestic use.
3. *Roller mills:* Most of the wheat flour in India is produced by roller flour mills. The steps involved in roller flour milling are:
   (a) *Cleaning:* Wheat is first cleaned of stones, dirt, weeds and foreign matter by separators, aspirators, scourers, magnets and washers.
   (b) *Tempering:* The cleaned wheat is moistened and held in tempering bins for 8 to 24 hours to toughen the outer coat and mellow the endosperm.
   (c) *Blending:* Wheat grains of various protein content are mixed to produce flour of the desired quality.
   (d) *Grinding and separating:* Wheat grain is first broken by a pair of corrugated rollers. The whole wheat flour is repeatedly rolled, sifted and purified till a complete separation of the bran has been achieved.

The first roller flour mill in India was set up in 1880. The number increased to 40 in 1940 and 111 in 1958. By 1970, the number of roller flour mills went up to 206. Since then, the trend in the number of roller mills is shown in Table 4.21. By 1980, the number of roller flour mills increased only marginally to 232. But there has been a sharp increase in their number during the later period. The number increased to 454 in 1985, 820 in 2008 and 900 in recent years. At present, around 15 million tonnes of wheat is converted to various wheat products by these roller flour mills.

Table 4.21: Number of rice mills and roller flour mills in India (as on January 1)

Year	Rice					Wheat roller flour mills
	Hullers	Shellers	Huller cum shellers	Modern rice mills	Total	
1970	51888	2302	4832	–	59022	206
1975	80077	3676	7240	340	91333	232
1980	73306	4283	8065	5071	90725	232
1985	79197	4484	6654	17826	108161	454
1990	86007	4447	7859	29614	127927	NA
1995	90091	4237	8362	33557	136247	812
2008	91287	4538	8385	35088	139298	820
2014	90000	NA	NA	35000	NA	900

Source: Ministry of Food Processing, Quoted in Bulletin on Food Statistics 1987–89, Directorate of Economics and Statistics, Ministry of Agriculture, Government of India, New Delhi, pp. 99–101 and Annual Reports, and MoFPI website.

The state-wise distribution of roller flour mills shows wide variation. The leading flour-producing states, as reflected from the number of roller flour mills, are Uttar Pradesh, Maharashtra, Karnataka, Tamil Nadu, Bihar, Andhra Pradesh, West Bengal, Haryana and Punjab.

In India till 1991, the roller flour mills were regulated by the government through licensing. However, since July 1991, licensing has been abolished for setting up of new roller flour mills or for expansion in the capacity of existing units. No licence is required for the manufacture of wheat products. There are no controls on price and distribution of wheat products either. The mills are free to obtain their requirement of wheat from any source, including that from the Food Corporation of India.

Costs and Margins in Processing of Wheat

For illustration, the structure of costs and margins of flour mills in the processing of wheat into flour as reported by Malik et al.[10], can be seen in Table 4.22. The cost of processing of wheat in late eighties was estimated as ₹ 17 to ₹ 18.80 per quintal depending on the size of the mill. The unit cost of processing is less and the net return per unit is more in the case of a mill with higher capacity as compared to a mill with lower capacity.

Table 4.22: Average cost of processing of wheat into wheat flour (1990 prices)

Particulars	20 HP flour mill		10 HP flour mill	
	₹	Percentage	₹	Percentage
1. **Fixed cost**				
(a) Depreciation	2032	8.3	1372	6.8
(b) Interest on fixed capital @ 14% per annum	2624	10.7	1740	8.6
Total fixed cost	4656	19.0	3112	15.4
Quantity of wheat processed (quintals)	1440	–	1080	–
Fixed cost per quintal	3.20	–	2.90	–
2. **Variable cost**				
(a) Labour charges	8400	34.2	8300	40.9
(b) Electricity charges	5040	20.6	3780	18.6
(c) Repairs and maintenance	1252	5.1	1210	6.0
(d) Rent of the building	2250	9.2	1700	8.3
(e) Loss in processing	2880	11.7	2160	10.6
(f) Interest on working capital @ 14% per annum for one month	46	0.2	44	0.2
Total variable cost	19868	81.0	17194	84.6
Variable cost per quintal	13.80	–	15.90	–
3. Total cost (1 + 2)	24524	100.0	20306	100.0
4. Cost of processing per quintal of wheat	17.0	–	18.80	–
5. Charges of the processor per quintal of wheat	20.00	–	20.00	–
6. Net return per quintal of wheat	3.00	–	1.20	–

Source: Malik, H.S., Sriniwas and A.C. Gangwar, Comparative Efficiency of Processing Units and Marketing Channels of Wheat Flour in Hisar Market of Haryana, Indian Journal of Agricultural Marketing, IV (2), July–December, 1990, p. 211.

Grain-based Processed Foods

India has a large baking industry, which is engaged in the manufacture of biscuits, bread, bun, cakes and pastries. There are more than 60,000 bakeries in India. This industry is growing rapidly because of the increase in urbanization, in population, in the per capita income of the masses, and the changes in tastes. In urban areas, a large number of bakeries, both small and large, are functioning.

The production of bread and biscuits, both in the organised and unorganised sectors is estimated to be 37 lakh tonnes and 20 lakh tonnes respectively. Out of the total biscuits manufactured in India, nearly one-third is in the organised sector and remaining two-thirds in the small-scale and unorganised sector.

In addition to bread and biscuits, grain-based snacks, pasta products and branded flour (atta) are other grain based processed products. The number of registered food processing units, manufacturing grain mill products, was 18,899 in 2016–17 (MoFPI and ASI).

PROCESSING OF PADDY

Paddy rice milling is one of India's largest industries, for the output of this industry exceeds the total of all the other foodgrain processing industries. Paddy consists of about 20% husk, 6% bran, 2% germ and 72% endosperm (rice). There are six major steps in the processing of paddy, depending upon the method used for processing.

1. *Drying:* Drying refers to the reduction of the moisture content in paddy to about 14%. At the time of harvesting, paddy contains 16 to 18% moisture. Drying can be done either in the sun or by means of a mechanical drier (forcing heated or unheated air through the paddy in a bin or a thin moving stream). Mechanical drying was introduced in India in 1965.
2. *Cleaning:* Cleaning is done to remove the foreign matter present in the paddy.
3. *Parboiling:* Parboiling involves soaking and steaming paddy to impart a desired flavour to it and to increase the out-turn. It reduces breakage in milling, improves storage life, and helps in the preservation of vitamins and protein in the rice grains.
4. *Husking:* Husking refers to the removal of husk from the rice grain. Rice milling is undertaken to remove the husk and a specified percent of bran from the seeds and endosperm.

Four principal rice milling methods are employed in India. These are:

1. *Hand Pounding:* This method involves the pounding of paddy with poles or a pestle and mortar.
2. *Huller Mills:* The heart of a huller rice mill is a fluted metal cylinder rotating with 500 to 600 rpm within a hollow stationary cylinder. Husk and bran are removed in one operation by abrasive action. The rice is polished by a second and third pass through the machine.
3. *Sheller Mills:* An under-run disc sheller consists of two stones or composition wheels, each 18" to 56" in diameter, and laid on top of one another. Between these two stones, paddy is husked by the rotation of the adjustable lower wheel. The bran is removed by polishing cans or rice hullers. The husk, bran and rice are separated mechanically.
4. *Rubber Roller Mills:* Each mill consists of a pair of rotating rubber rollers between which paddy is poured at one to four tonnes per hour, depending on the design of the mill. Shelling results from the abrasion created by the two rollers turning at slightly different speeds. Soft rollers minimize breakage.

A comparison of the four rice milling methods is given in Table 4.23. In 1970, there were 59020 mechanized rice mills of all types in India. This number increased to 139298 in 2008. The increase in the number of different types of rice mills in

Table 4.23: Comparison of different rice milling methods (as in 1970)

| Particulars | Hand pounding | Rice milling methods | | |
		Huller mill	Sheller mill	Rubber mill
When introduced in India	Oldest	1920	Quite popular	1965
Quantity of paddy milled in India through the method	Over 40%	31% of paddy crop and over 50% of the paddy not hand pounded	25% of paddy crop and 45% of that not hand pounded	About 4% (balance)
Capital cost (₹)	75	3,750 to 4,000	37,500 to 45,000	65,000* (₹ 25 lakh for complete unit)
Paddy intake per day of 8 hours	60 kg	2 tonnes	8–16 tonnes	8–16 tonnes
Yield of processed rice from paddy (remaining is bran, brokens and husk)	60–65%	65%	65–68%	70–72%
Cost of milling per tonne of rice (₹)	37.50 to 45.75	9.25 to 15.50	6.25 to 12.50	NA

* For milling equipment only. This does not include mechanical drying, parboiling and silo storage.

Source: Moore, J.R., S.S. Johl and A.M. Khusro, Indian Foodgrain Marketing, Prentice-Hall of India Private Limited, New Delhi, 1973, p. 148.

India between 1970 and 2014 is shown in Table 4.21. The relatively much sharper increase in the modern rice mills is due to their much superior milling efficiency at 70 to 72% as compared to a meagre 60 to 65% for hand pounding, 61 to 65% for hullers, and 62 to 63% for sheller and huller-cum-sheller mills.

5. *Polishing:* Polishing is the removal of bran and germ from the rice grain.
6. *Separating:* This means separation of the parts of broken grain from whole grain.
7. *Grading:* Grading is separation of rice by size. Head rice are the grains that are 3/4th of a whole grain and larger brokens are smaller grains.

PROCESSING OF PULSES

Pulses are rich in protein (lysine) and constitute 10 to 15% of India's foodgrain diet. Pulses are main source of protein in vegetarian diet. They are mainly consumed in the form of de-husked split pulses. The important pulses are gram (Bengal gram), arhar or tur (red gram), moong (green gram) and urad (black gram). It has been estimated that more than 75% of the pulse crops are consumed as dal (split grain). The other uses of pulse crops are parched gram, besan (dal flour) and cattle feed. Pulse processing is generally known as milling. There are nearly 10,000 pulse milling units in the country. Pulse milling is the third largest foodgrain industry (after rice and wheat) in India.

Milling of pulses means removal of the outer husk and splitting of the grain into two equal halves. Generally, the husk is more tightly held by the kernel of some pulses than most cereals. Therefore, dehusking of some pulses, poses a problem. The method of alternate wetting and drying is used to facilitate dehusking and splitting of pulses. In India, the dehusked split pulses are produced by traditional method of milling. In traditional pulse milling methods, the loosening of husk by conditioning is insufficient and as such a large abrasive force is applied for the complete dehusking of the grains. This results in high losses in the form of brokens and powder. The yield of split pulses in traditional milling method is only 65 to 70% in comparison to 82 to 85% potential yield.

There is no common processing method for all types of pulses. Dal is made in a series of steps, both at home and in the mills. The important steps involved in dal-making are:

1. *Cleaning:* Removing foreign matter from the main pulse grains.
2. *Dampening:* Soaking the grains in water for the desired period of time.
3. *Tempering:* Keeping the soaked grains after removal from water for drying in the sun.
4. *Splitting:* Grinding the grain to make dal.
5. *Husking:* Removing the husk from the dal.

Dal may be further processed by grinding it into flour (besan). Dal processing plants are located in the main pulse-producing and trading States of India. They range in size from cottage industries to multi-storeyed plants, using pneumatic conveyors.

PROCESSING OF OILSEEDS

Oilseeds are a group of farm products which are mainly consumed only after they have been processed. They contain oil and cake. The following methods are generally used in processing oilseeds. The method depends on the type of oilseed to be processed and the availability of power.

1. *Bullock-driven Ghani Method:* Ghani is a traditional method of oilseed processing. A ghani consists of a wooden vessel in which oil is collected and which is attached to

a wooden log rotated by a single bullock. The quantity of oilseeds to be processed at a time is about 10 kg, and the process takes about 2 hours. The oil and cake extracted by this method are of better quality than those extracted by other methods.

2. *Electric-driven Ghani Method:* This is similar to the traditional bullock-driven ghani except that, here electric, instead of bullock, power is used to rotate the ghani. The process takes less time. The quantity of oilseed processed per day is higher than that processed by the bullock-driven method.

3. *Expeller Method:* This method has become very popular and is generally used nowadays. Expellers of different sizes are available. The capacity and time taken by each expeller vary with its size. The oil recovery percentage is higher by this method than by the ghani method.

4. *Solvent Extraction Method:* By the above methods, 5 to 12% oil is left in the cakes. To remove this extra quantity of oil, the solvent extraction method is used. The oilcake is treated with hexine to dissolve the oil present in it. The solution containing the oil and the solvent is then passed through condensers, where oil is distilled. The distilled oil is non-edible and is used mainly in the soap industry. The solvent-extracted meal is used as a feed for dairy cattle because of its high protein content. This methods is also used for extracting oil from soyabean seeds.

The number of oilseeds processing units and their installed capacity in India are given in Table 4.24.

Table 4.24: Number and Capacity of Oilseed Processing Units in India		
Type of processing plant	**Number of units**	**Installed capacity (million tonnes)**
Ghanis (or Kolus)	1,30,000	42.5
Small scale oil expellers	15,000	
Oil mills	15,500	2.0
Solvent extraction plants	633	27.4
Refineries (oil)	1000	10.2
Vanaspati units (hydrogenation)	235	4.4

Source: MoFPI—Rabo Bank Report.

Most of the oil mills have processing capacities of 1 to 5 tonnes per day. Nearly 10% of the oil mills in the country have the capacity of more than 25 tonnes per day. Number of oil mills with a capacity of more than 50 tonnes per day is only around 1% of the total number of oil mills.

The returns and costs of an oilseeds processing unit can be worked out easily. Suppose an entrepreneur has installed a unit consisting of two oil expellers for processing of groundnut in a major groundnut growing district. He has borrowed a sum of ₹ 4 lakh from a commercial bank at an interest rate of 15% per annum. Out of this loan he invested ₹ 2 lakh each on the construction of the building and on the purchase of expellers and other necessary accessories. The capacity of each expeller is 7.50 qtls. of groundnut per day. The expeller remained in use for 200 days during the year. He purchased groundnut at price of ₹ 2000 per quintal and could sell oil and cakes at average prices of ₹ 66 and ₹ 10 per kg, respectively. The recovery of oil was 28% of the groundnut-in-shell. He paid ₹ 80,000 per annum as salary of the permanent workers. His electricity bill was ₹ 48,000 and repairs and maintenance cost during the year was ₹ 40,000. Other petty expenses and bill for casual labour amounted to ₹ 40,000 and ₹ 72,000 per year respectively. Based on this information, the cost of processing and the profit of the entrepreneur are as follows:

Particulars	Rate	₹ per annum
Fixed cost		
1. Depreciation on building	@ 2% per annum on ₹ 200,000	4,000
2. Depreciation on expellers	10% per annum on ₹ 200,000	20,000
3. Interest on the capital	@ 15% on ₹ 4 lakh	60,000
4. Salary of permanent workers		80,000
	Sub total	1,64,000
Variable costs		
1. Electricity charges		48,000
2. Repairs and maintenance		40,000
3. Other petty expenses		40,000
4. Wages of casual labourers		72,000
	Sub total	2,00,000
	Total costs (Fixed + Variable)	3,64,000
Purchase value of groundnut-in-shell (15 × 200) = 3,000 quintals	@ ₹ 2,000	60,00,000
Returns		
Sale of oil (840 quintals)	@ ₹ 6600 per quintal	55,44,000
Sale of oilcakes (42% of groundnut-in-shell, i.e. 1260 quintals)	@ ₹ 1000 per quintal)	12,60,000
	Total returns	68,04,000
Net profit (68,04,000 − 60,00,000 − 3,64,000)		4,40,000

PROCESSING OF FRUITS AND VEGETABLES

Although fresh fruits and vegetables are still a delicacy, but with the increase in surpluses and the need to carry the surpluses from producing areas to the consumption centres, commercial processing of fruits and vegetables has gained importance in the recent years. The fruits are converted into such items as jam, jellies, squash, syrups and canned fruits through simple processing methods. Fresh vegetables are converted into pickles, sauces, dehydrated vegetables and frozen foods. The shelf life of such processed foods is more than that of fresh fruits and vegetables. Though the share of the farmers in the price paid by the consumer of such processed foods continues to be low, the processing industry has helped in providing market clearance to the growers of fruits and vegetables. Earlier, some processing of these was a common practice at the household level.

The growth of fruits and vegetables processing industries in India up to 2006 is shown in Table 4.25. Most of these are in the unorganized sector. The number of factories/units registered with MoFPI up to 2016–17 is 1254.

Half of the installed capacity of fruits and vegetables processing units in the country remains under utilized as majority of these processing units (72%) are in the cottage sector which are constrained by the technological backwardness. According to one estimate, only 2% of total fruits and vegetables are processed in the country and 98% is consumed in raw form. There is a wide inter-state variation in the spread of fruits and vegetables processing units. About 20% of the processing units are located in Maharashtra followed by Uttar Pradesh (10%), Tamil Nadu (9%) and Kerala (8%).

Table 4.25: Growth of fruits and vegetables processing industries in India

Year	No. of units (licensed)	Installed capacity ('000 tonnes)	Capacity utilization ('000 tonnes)	Percentage capacity utilization
1980	2026	275	69.6	25.3
1985	3100	405	179.2	40.8
1990	3846	894	260.0	29.0
1995	4368	1760	850.0	48.3
2000	5293	2100	990.0	47.1
2006	5413	2100	NA	NA

Source:
1. Indian Journal of Agricultural Marketing, Vol.14(3), Conference Special, September–December, 2000, p. 78.; and Vol. 17 (2), July–Dec 2004, p. 191.
2. Economic Survey, 1999–00, Ministry of Finance, Government of India, New Delhi.
3. Website of MoFPI.

Further the distribution of fruits and vegetable units are not directly correlated with the availability of raw material in the state. For example, the State of Bihar, which contributes 13% of the total production of fruits and vegetables, accounts for only 1.28% of total processing units in the country.

The size-wise growth of fruits and vegetable processing units up to mid-1990s is presented in Table 4.26. More than one-third units are in the home sector (processing less than 10 tonnes fruits and vegetables per annum) followed by cottage sector. The low capacity utilization of the processing units is on account of their existence in home, cottage and labeller (72%) sector, which are constrained by technological backwardness. One general observation for processing sector units is that there exists vast untapped potential for fruits and vegetables processing in India, which should be explored to provide employment to the educated youth and save losses in the marketing chain. The packaging segment of fresh fruits and vegetables is also growing. Around 5 million tonnes of fresh fruits and vegetables are sold as packaged products.

Table 4.26: Scale-wise distribution of fruit and vegetable processing units in India

Scales	Size (tonnes per annum)	1980–81	1990–91	1994–95
Large	>250	218 (10.8)	442 (11.5)	497 (11.4)
Medium	100–250	263 (11.6)	331 (8.6)	343 (7.9)
Small	50–100	163 (8.0)	323 (8.4)	371 (8.5)
Cottage	10–50	398 (19.6)	768 (20.0)	854 (19.6)
Home	1–10	763 (37.7)	1303 (33.9)	1520 (34.8)
Labeller	<1	248 (12.3)	679 (17.6)	783 (17.8)
Total	–	2026 (100.0)	3846 (100.0)	4368 (100.0)

Note: Figures in parentheses are percentage of the column totals.

PROCESSING OF MILK

Liquid milk is not only highly perishable, but is also difficult to carry long distances. Therefore, markets for milk were normally local markets. It is the introduction of the function of processing in the milk marketing chain which has lead to the evolution of milk markets from purely localised ones to the national market. The milk which is surplus in the villages is collected at specified centres, tested for quality, transported to chilling plants, pasteurised, chilled, stored under low temperature, packed as pasteurised milk in suitable containers, converted into such milk products as packaged milk, butter, ghee, milk powder and cheese and sold to the consumers located far away from the milk-producing areas. Milk processing activity enables the consumers to get liquid milk and milk products at a time and place desired by them on the one hand and makes it possible for the milk producers to get a reasonable price for their produce on the other.

Dairy processing received major impetus in the country during the Operation Flood Programme. By the end of the first phase of the operation flood programme, raw milk processing rose to 31.90 lakh litres per day from only 6.55 lakh litres per day in 1970, i.e. before the start of the operation flood programme. The milk processing was further strengthened during the second phase of operation flood programme. With a view to inter alia providing milk producers greater access to the organized milk processing sector, National Development Plan, Phase I (NDPI) is being implemented (by NDDB) in the country from 2011–12, in 18 states that account for 90% of India's milk production. Recently, Government of India has launched a Dairy Processing and Infrastructure 'Development Fund (DIDF) with an outlay of more than ₹ 10000 crore for the period 2017–18 to 2028–29, with the aim of increasing dairy processing infrastructure. Out of total marketed surplus of milk in the country, nearly 50% is processed; 26% in the organized sector and 24% in the unorganized sector.

The number of liquid milk plants (dairy) has gone up to over 500 with per day milk handling capacity of 56 million litres. There are around 1.86 lakh dairy cooperative societies, which is an essential part of milk handling and processing network. Out of total milk processing capacity, 39% is in cooperative sector, 44% in private sector, and 17% in other segments.

PROCESSING OF MEAT, POULTRY AND MARINE PRODUCTS

Total meat production in India is about 8.1 million tonnes, of which about 20% is processed into meat products, compared to about 80% in Finland and 50% in Russia. Livestock slaughter in India is very low (1.5 %) compared to the developed countries. According to Ministry of Food Processing Industry, there are 181 meat processing units in the organized sector. In addition, there are 1228 fish processing units registered with MPEDA, which include 388 freezing plants, 98 IQF plants, 13 canning plants, 156 ice plants, 12 fish meal plants and 561 pre-processing plants.

So far, little attention has been paid in the country on development of processing technology for meat, poultry and marine products. Age-old slaughtering practices are still followed in majority of cities without any quality control and hygiene. Very few mechanized abattoirs are functioning in India. Extension of shelf life of meat and marine products is mostly through freezing. Canning of these is practiced only at a small scale. For addressing the related issues, the National Meat and Poultry Processing Board was set up at New Delhi in 2009.

PROCESSING OF CASHEWNUTS

Cashewnuts are converted into finished kernels by a simple process. Processing of cashewnuts can be conveniently divided into following eight stages:

1. *Drying of nuts:* Procured raw nuts are spread on the floor for sun-drying.

2. *Roasting of nuts:* Roasting of cashewnuts is done by drum roasting method. A specific temperature is provided by burning cashewnut shells. Roasting is done for one to two minutes.
3. *Shelling:* Roasted nuts are cooled for sometime and are broken by beating them with wooden mallets. It is a time-consuming process and each labourer can shell 10 kg of nuts per day of eight hours.
4. *Drying of shelled kernels:* The shelled kernels are dried to loosen the kernel coat which facilitates easy peeling. A tray drier is used for this purpose.
5. *Peeling:* Kernel coat is peeled off by hand or by bamboo sticks.
6. *Grading:* The kernels are sorted out according to their size and nature. Prevailing grades in cashewnuts are special Jumboo, Jumboo, American, standard, splits, pieces and other wastes.
7. *Conditioning:* Conditioning is done only for export purposes. The kernels packed in tins are given carbon dioxide treatment to avoid breakage. Cashew kernels are heaped on mesh bottom plates on the ground, sand is spread out and water is sprinkled on it. Fans are run overhead which evaporate the moisture in the sand and it is absorbed by kernels.
8. *Packing:* The conditioned kernels are packed in 11 kg tins.

India imports unprocessed cashewnuts and exports cashew kernels after processing. The processing cost of 80 kg of raw cashewnuts (which yield 22 kg of kernels) in early 1990s was estimated as ₹ 125. Out of the total cost of processing, nearly 60% is the labour cost. The cashewnut processing is thus, a labour-intensive activity (Srinivas and Raju).[11]

SUGARCANE PROCESSING

Very small percentage of sugarcane is used in unprocessed form, and that too mainly for planting material. Some small part is used for consumption as juice, which also requires minimum processing. A major part of sugarcane (about 85%) is processed for conversion into sugar. Remaining 12% is used for production of jaggery and khandsari. For processing of sugarcane, there are 741 sugar mills in the country, with a total capacity of producing 26 million tonnes of sugar. About half of the sugar mills are in the cooperative sector, and remaining are mostly private sugar mills. Most of the sugar produced in the country (more than 60%) is consumed by small and medium business establishments and high income households.

FOOD PROCESSING INDUSTRY

A strong and dynamic food processing sector plays a vital role in diversification and commercialization of agriculture. As already mentioned, food processing enhances shelf-life, ensures value addition, enhances farmers' incomes, and expands markets for agro foods. It is, therefore, considered a sunrise industry. There are 5.11 million food processing units in the unorganized sector. These include 2.22 million MSMEs (registered).

In order to give impetus to the development of food processing segment of the Indian economy, the Ministry of Food Processing Industries was set up in 1988. There is a National Food Processing Development Council (NFPDC), which has now 29 members. The key segments of food processing sector in India are the following:

1. Apart from a large number of unorganized food processing units, there are many registered food processing units (factories). Their number has increased from 26219

in March 2008 to 39,740 in March 2017. These include 1254 fruits and vegetable processing units, 18,899 grain milling units and 2039 dairy processing units. The details are shown in Table 4.27.

2. According to the latest information available from the Annual Report of MoFPI for 2016–17, the value of output of registered food processing industries in India was ₹ 10.95 trillion (₹ 10.95 lakh crore).

3. The growth rate of food product industry has accelerated from 2.49% per annum during 1997–2002 to 4.05% per annum during 2002–07 and further to 11.4% during 2011–18. However, the level of processing across segments varies considerably. For example, the level of food processing varies from 2.2% in fruits and vegetables to 35% in milk (Table 4.28).

Table 4.27: India: Number of registered food processing units (industries) (at the end of March)

Commodities processed	2008	2011	2017
Fruits and vegetables	735	1052	1254
Grain mill products	12807	17792	18899
Dairy products	1096	1493	2039
Vegetables and animal oils and fats	2515	3307	3112
Sugar	778	895	741
Fish, meat and related products	77	115	716
Bakery products	955	1450	1767
Others	7256	9734	11212
Total	26219	35838	39740

Source: Annual Reports, MoFPI (website).

Table 4.28: India: Level of food processing across segments (%)

Segment	India	Developed countries
Fruits and vegetables	2.2	65 to 78
Milk	35.0	60 to 75
Poultry	6.0	60 to 70
Meat	20.0	60 to 70
Marine	26.0	60 to 70

Source: Britt-Louise Anderson, SIWI, quoted in Financial Express, July 5, 2010

During the last ten years, processing of food in India has gone up from 6% in 2004 to 11% in 2010 and further to 15% in 2014. This has helped in reducing the wastage of food by 16%. Overall value addition in food products in India is 8%, which is projected to go up to 35% by 2025. The share of food processing in total manufacturing is around 9.8%.

Recognizing that most of the food processing units are micro, small and medium enterprises (MSMEs), the government has declared several SME clusters for their development. There are 59 SME clusters related to food processing. These include 10 oil mill clusters, 26 rice mill clusters, 11 food product based clusters, 7 dal mill clusters and 5 other clusters.

Union Finance Ministry has set-up a Venture-Capital Fund (VCF) to solve the funding woes of SME sector. NABARD has been provided funds to refinance banks as and when they lend to the food processing sector. Further, a special fund has been set up in NABARD to make available affordable credit to food processing units in designated clusters. The excise duty on machinery for food processing has also been reduced. These apart, the Mega Food Parks are being set up to give boost to the food processing industry. Now, state governments or their agencies have also been made eligible to set up food parks.

The Ministry of Food Processing Industry has also launched a Scheme of Financial Assistance for Implementation of (HACCP/ISO 22000, ISO 14000/GHP/GMP) Quality Management Systems in food processing. The objectives of the scheme include (a) to prepare the FPIs to face global competition; (b) to enable adherence to stringent quality and hygiene norms; and (c) to keep Indian Food Industry technologically abreast of international best practices. Under this scheme, all implementing agencies, i.e. central/state government organizations, IITs, universities and private sector will be eligible for reimbursement of 50% of the cost of consultant fee, fee charged by certification agency, plant and machinery, technical civil works, and other expenditure towards implementation of TQM in general areas subject to the maximum limit of ₹ 15 lakh (for difficult areas, it is 75% subject to a maximum of ₹ 20 lakh).

The Ministry of Food Processing Industries has set up two important national institutes:

1. Indian Institute of Crop Processing Technology at Thanjavur (Tamil Nadu) for promoting education, research and skill development in food processing, food science and food technology.
2. National Institute of Food Technology, Entrepreneurship and Management (NIFTEM) at Kundli in Sonepat district of Haryana. The mandate of NIFTEM is to work as a one stop solution provider and sector promotion or business promotion organization. The objectives inter alia include providing business incubation services; conduct frontier research and foster innovations for development of food processing sector; to produce world class managerial talent in food science and technology; to work for up-gradation of SME food processing clusters; and to function as knowledge repository.

FOOD PARKS AS INSTRUMENTS OF PROMOTING FOOD PROCESSING

Food parks provide basic facilities necessary for setting up of food processing industry. The food parks encourage processing entrepreneurs as these provide external economies of scale to the processing units set up in the food parks. The rationale for setting up of food parks is that small and medium entrepreneurs face problems in investment of huge amount of capital necessary for creation of facilities such as construction of cold storage, warehouses, quality control laboratories and effluent treatment plants. Provision of assistance for development of these facilities, as common facilities, make the cluster of food processing units in food parks cost competitive and as a consequence help in promoting food processing and developing the food processing industry of the country.

The Ministry of Food Processing Industries (MFPI), Government of India is pro-actively pursuing the task of setting up of food parks in different parts of the country as a part of strategy to develop food processing infrastructure. The Ministry is implementing a number of schemes and programmes to take this industry forward. Till recently 49 food parks have been sanctioned by the Ministry of Food Processing Industries. Public sector units, joint/private/assisted sector, NGOs, and cooperatives are eligible to get assistance up to ₹ 4 crore for development of common facilities viz., un-interrupted power supply, water supply, cold storage, ice plants, warehousing facilities, treatment plants, quality control and analytical laboratories and major processing facilities (fruits concentrate/pulp making units) as a part of the food park development.

The MFPI is also pursuing the setting up of mega food parks. Each mega food park may consist of at least 30 food processing units, with an investment of about ₹ 370 crore per park. This will include ₹ 120 crore investment from Special Purpose Vehicle (SPV), of which ₹ 50 crore will be from Government of India. It is expected that such a mega food park will generate considerable employment opportunities for rural youth. The gestation period of such food parks is about two years. Mega food parks will improve post-harvest management, reduce wastage and increase farmers' incomes. These will encourage clustered and market-driven farming. These are basically joint food parks with collaboration of investors from other countries. Foreign investors will bring new processing technology, packaging and quality assurance. Quality testing will be done in India (with foreign labs) before consignments leave for exports.

The overall objective of setting up mega food parks is to facilitate state-of-the-art infrastructure with efficient supply chain management from farm gate to the retail; help in value addition; reduce wastage; create employment; and increase income of the farmers. Each mega food park is likely to benefit 6000 farmers directly and around 25000 farmers indirectly.

The scheme of mega food parks was launched in 2009 and by the middle of 2012, 15 mega food parks were sanctioned. Each mega food park requires around 30 acres area. The target for the 12th Five Year Plan period was to set up 15 more mega food parks. In March 2015, the Government allotted/sanctioned 17 mega food parks. Out of these 10 are allotted to private players and 7 to the state governments. Total investment expected is ₹ 6000 crore. There will be 40 to 50 food processing units in each mega food park. A special fund of ₹ 2000 crore made available to NABARD will help in attracting private investment in these parks. The maga food parks are expected to offer following facilities to the food processing entrepreneurs: (a) availability of developed plots; (b) assured availability of power and water; (c) plug and play facilities for micro and small enterprises; (d) cold storage and dry storage facilities; (e) multi-commodity automated washing, sorting and grading units; (f) handling and processing of fruits and vegetables; (g) milk chilling and processing facilities; and (h) common effluent treatment plant.

There are currently 42 Mega Food Parks in India, at different stages of development. These are spread over 25 states (one to three in each state). Four of these are reported to have been completed, 11 are operational and 22 are under implementation.

Food processing sector in India is contributing significantly to India's economic growth, and also attracting the interest of foreign investors. Foreign Direct Investment (FDI) in this sector, which was around $ 500 million per year, is now approaching a level of $ 1 billion per year. This has happened after 100% FDI was permitted in this sector. This sector has a potential of creating employment of around one million person years.

❚ BUYING AND SELLING

MEANING

Buying and selling is the most important activity in the marketing process. At every stage, buyers and sellers come together, goods are transferred from seller to buyer, and the possession utility is added to the commodities.

The number of times the selling-and-buying activity is performed depends on the length of the marketing channel. In the shortest channel where no middleman is involved, this activity takes place only once, i.e. the producer or farmer sells and the consumer purchases. But, usually, in the case of farm commodities, selling/buying activities are undertaken each time when the produce moves from the farmer to the primary wholesaler, from the wholesaler to the retailer, and from the retailer to the consumer.

The buying activity involves the purchase of the right goods at the right place, at the right time, in the right quantities and at the right price. It involves the problems of what to buy, when to buy, from where to buy, how to buy and how to settle the price and the terms of purchase.

The buying function seems to be a very simple function. But it involves the following subsidiary functions before the actual buying takes place:

1. *Planning the Purchase of Goods:* Deciding the quantity of each good to be purchased.
2. *Contractual Function:* Determining the sources of supply and establishing contacts with them.
3. *Negotiation of Price and Terms and Conditions of Buying.*
4. *Final Agreement and Transfer of Goods.*

The selling activity involves personal or impersonal assistance to or persuasion of, a prospective buyer to buy a commodity. The objective of selling is to dispose of the goods at a satisfactory price. The prices of products, particularly of agricultural commodities vary from place to place, from time to time, and with the quantity to be sold. Selling, therefore, involves the problems of when to sell, where to sell, through whom to sell, and whether to sell in one lot or in parts. The selling function thus includes the following sub-functions, the performance of which enables one to get a good price for the produce:

1. *Product Planning and Development:* This sub-function includes the activities of determination of the variety/quality of the product to be produced, grading it, and deciding about the trade or brand names, if any, to be adopted for the product.
2. *Contractual Function:* This involves the determination of potential buyers of the product and of entering into contracts with them (pre-harvest contractors or mandi traders).
3. *Demand Creation:* This includes the activities which are designed to stimulate an already existing desire for the satisfaction of the want of a given product. In other words, it means selling the products with which potential consumers are not familiar (for example organic wheat, baby corn or new variety of brinjal).
4. *Negotiating the Prices and Settling the Terms and Conditions of Sale with the Buyers:* At the time of buying and selling, the following terms and conditions must be settled to avoid future problems:
    (a) Whether the weight of packing material (gunny bag) is included in the weight of the commodity;

(b) Whether the empty gunny bags will be returned to the seller;

(c) The total quantity negotiated;

(d) The terms of payment—whether it will be in cash or after a grace period;

(e) The delivery of the produce whether it is "on spot" or "on arrival", or "forward" delivery;

(f) The final agreement and transfer of ownership of the product.

(g) Whether the price settled is on 'as is where is' basis or some other term.

METHODS OF BUYING AND SELLING

The following methods of buying and selling of farm products are prevalent in Indian primary wholesale markets where farmers take their produce for sale:

1. Under Cover of a Cloth (Hatha System)

In this method, the prices of the produce are settled by the buyer and the commission agent of the seller by pressing/twisting the fingers of each other under cover of a piece of cloth. Code symbols are associated with the twisting of the fingers, and traders are familiar with these. The negotiations in this manner continue till a final price is settled. When all the buyers have given their offers, the name and offer price of the highest bidder is announced to the seller by the commission agent.

This system provides opportunities for cheating the seller, for the seller is not aware of the price that has been offered by other buyers; the commission agent may not communicate various prices to the seller; and may strike a deal in favour of one who offers a somewhat lower price. This method has been banned by the government because of the possibility of cheating, though it continues to be used in some markets.

2. Private Negotiations

In this method, prices are fixed by mutual agreement. This method is common in unregulated markets or village markets. Under this method, the individual buyer comes to the shops of commission agents at a time convenient to the latter and offer prices for the produce, which, they think, are appropriate, after the inspection of the sample. If the price is accepted, the commission agent conveys the decision to the seller, and the produce is given, after it has been weighed, to the buyer.

In villages too, private negotiations take place directly between buyer and sellers. The sellers take the sample to the buyer and asks him to quote the price. If it is acceptable to the buyer, a verbal contract is executed. This however is a slow and time-consuming process and is not suitable when either large quantities have to be sold or a large number of buyers exist in the market. The advantage of this method is that the seller gets a good price, for buyers are not aware of the price offered by other buyers. Each buyer, therefore, tries to bid the highest possible price to get the produce.

3. Quotations on Samples Taken by Commission Agent

In this method the commission agent takes the sample of the produce to the shops of the buyer instead of the buyer going to the shop of the commission agent. The price is offered, based on the sample, by the prospective buyers. The commission agent makes a number of rounds of prospective buyers until none is ready to bid a price higher than the one offered by a particular buyer. The produce is given to the one whose bid has been the highest.

4. Dara Sale Method

In this method, the produce in different lots is mixed and then sold as one lot. The advantage of this method is that, within a short time, a large number of lots are sold off. The disadvantage is that the produce of a good quality and one of a poor quality fetch the same price. There is, therefore, a loss of incentive to the farmer to cultivate or bring good quality products. This method is common for such crops as zeera (cumin) in many markets of the country.

5. Moghum Sale Method

In this method, the sale of produce is effected on the basis of a verbal understanding between buyers and sellers without any pre-settlement of price, but on the distinct understanding that the price of the produce to be paid by the buyer to the seller will be the one as prevailing in the market on that day, or at the rate at which other sellers of the village sold the produce. This method is common in villages, for farmers are indebted to the local moneylenders. Often the buyer pays less than the prevailing market rate on the plea of the poor quality of the produce.

6. Open Auction Method

In this method, the prospective buyers gather at the shop of the commission agent around the heap of the produce, examine it and offer bids loudly. The produce is given to the highest bidder after taking the consent of the seller-farmer. This method is preferred to any other method because it ensures fair dealing to all parties, and because the farmers with a superior quality of produce receive a higher price. In most regulated markets, the sale of the produce is permissible only by the open auction method.

The following are the merits of the open auction method:

1. A sale by this method inspires confidence among the buyers and sellers. The seller is able to follow the bidding easily.
2. The auction serves as a meeting place for the supply of, and demand for, goods.
3. It disposes of the market supply promptly.
4. A wide variety of goods are available to buyers for selection.
5. The auction method reduces the number of salesmen needed in the process.
6. The buyers of small lots are not put to a disadvantage against the buyers of large lots.
7. All the sections of the society interested in the sale and purchase are well informed about the prevailing prices and can take judicious decisions about the sale and purchase of agricultural commodities.
8. The payment of the price of the goods is made immediately after the sale if an auction has been completed.

The disadvantages of the open auction method are:

1. The auction method requires more time on the part of both the buyer and the seller, for they have to wait for the day and time of the auction. An open auction is a very time-consuming process because of the variation in the quality of the various lots.
2. In big market centres, specially in the peak marketing season, the time allotted for auction is short. Both the buyers and the sellers are in a hurry. As a result, sellers may receive a low price.

3. In an open auction, buyers sometimes join hands. Active participation in it is then reduced.

4. The auction leads to a "buyers' market", for buyers have full information about the supply of, and demand for, the product.

Some of the problems arising out of the open auction method may be overcome if the grading of agricultural produce is adopted by the cultivators. This will reduce the time involved in inspection and bidding for each lot separately, and will result in increasing the overall efficiency of the marketing system.

Three types of open auctions are prevalent in different markets. These are:

1. *Phar System of Open Auction:* By this method, one bid is given for all the lots in a particular shop and all the lots are sold at that price. One extreme case of this method is when one bid is given for the product in the whole market.

2. *Random Bid System of Open Auction:* By this method, the commission agent invites a few buyers when the produce is brought to his shop for sale. All the prospective buyers are not informed. As a result, the competition is poor. Sometimes, the commission agent informs only those buyers who are either his relatives or whom he wants to oblige. Bidding may continue simultaneously at a number of places to reduce competition.

3. *Roster Bid System of Open Auction:* This is a systematic method of open auction. Bidding starts from a point in the market at a notified time about which the prospective buyers are given information in advance. This overcomes the defects existing in the previous two methods of open auction. The bidding party, after the auction of the produce at one shop, moves to the next in a clockwise or anti-clockwise direction till the auction of the produce at all shops is over, or the scheduled auction time expires. On the following day, the auction starts from the next point, and so on. This method is in vogue in most of the regulated markets. The auction is supervised by the auction clerk or the person nominated by the market committee.

7. Close Tender System

This method is similar to the open auction method, except that bids are invited in the form of a close tender rather than by open announcement. The produce displayed at the shop of the commission agent is allotted lot numbers. The prospective buyers visit the shops, inspect the lots, offer a price for the lot which they want to purchase on a slip of paper, and deposit the slip in a sealed box lying at the commission agent's shop. When the auction time is over, the slips are arranged according to the lot number, and the highest bidder is informed by the commission agent that his bid has been accepted and that he should take delivery of the produce.

Some of the regulated markets have adopted this method of sale, which is time-saving and involves the minimum physical labour. There is no possibility of collusion among the buyers because each has quoted the price on the basis of his individual assessment of profit margins, taking into consideration the price prevailing in terminal and other secondary markets. The smooth functioning of this method depends on the efficiency of, and the supervision exercised by, the market committee officials.

The methods employed for the sale of agricultural commodities in Indian markets differ from market to market and also from commodity to commodity. However, in regulated markets, either the open auction or the close tender system is prevalent. In

Tamil Nadu, the buyers have adopted the close tender system which, it is claimed, is quicker and tends to give a higher price to the farmer than in the open auction system.

8. E-Mandi or E-Auction

E-auction brings transparency in the auction system and creates a level playing field for all the participants in the market. The process starts when the farmer reaches the mandi gate. The APMC issues a unique number which captures details of farmer's name, village and quantity of the produce. Technical experts grade the produce and confirm the quantity. Farmers are also offered warehousing facilities where the farmers store their produce along with the unique number tags given to them. Only ½ kg sample is taken for grading. The buyers inspect the samples and based on Unique Identification Number (UIN) quote their bids on computer. The highest price is treated as successful bidder. The announcement is made on PAS and farmer is informed on his mobile number. E-mandi is a win-win situation for all the stakeholders viz:, farmer, buyer and APMC.

NCDEX Spot Exchange (NSPOT) launched this as a pilot project in Gulbarga APMC of Karnataka for tur (arhar) and was very successful. Taking into account the success in Gulbarga, it was extended to more market places in Karnataka as well as other states.

9. Online Platform

Extended on-line buying/selling of agricultural products has started taking shape in India. Karnataka state had taken lead in this matter. Initially 155 APMC markets (mandies) in the state were operating under unified online market where traders across the state and outside are using online platform for buying 92 agricultural commodities from the farmers. Within two years, the state linked all the markets to the common online market place. Rashtriya e-Market Services Private limited (ReMS) is a special purpose vehicle created by Karnataka Government and NCDEX spot exchange, which has been handling the integration of mandies (markets) through single online platform. In this method, through a single registration, the traders are allowed to participate in the auction of various mandies across the state. The farmers have the choice of taking their produce for sale in any regulated or private market in the state and receive timely online payment. The model provides increased competition, easy and fast trading, and better price discovery mechanism. The platform developed by ReMS provides value-added services, including assaying and grading. The selling of fruits and vegetables is not included in the online marketing platform. For meeting the expenses required for online platform, ReMS charges 0.2% of the gross sale amount from the buyers and sellers. Some other states also studied the Karnataka Model. The Government of India also decided to launch a National Agriculture Market (NAM) on this pattern.

Government of India in July 2015 launched a central sector scheme to promote E-NAM linking 585 selected markets of the country. In April 2016, electronic trading platform (E-NAM) was formally launched. The aim was to electronically link the 585 APMCs, to give farmers large choice to sell their produce and also to traders to buy their trade requirements from markets across the country. It is in a sense spatial integration of markets, which can improve price discovery, give price signals and reduce transaction costs. So far 585 APMCs in 16 states and two UTs are already on the E-NAM portal. Other 415 markets are in the pipe line. However, the needed infrastructure is still at evolving stage (more details in Chapter 7).

10. Price Support Purchases by Government Agencies

All above methods are in vogue where buyers are private traders. These apart, considerable quantities of farm products are also purchased by government agencies at pre-announced support prices. These operations are carried out usually during post-harvest months. There is no compulsion on farmers to sell to these agencies. The need for such purchases arises only when market prices tend to dip below the announced support prices (see details in Chapter 7).

DEMAND CREATION

Demand creation is a sub-function of the selling function in the marketing of products. It includes all the special efforts that are made to stimulate the desire for the goods of prospective buyers, with the ultimate objective of sale at a profit. The seller uses various techniques to arouse the desire for the product by dwelling on its beneficial qualities, of which the prospective buyers are not aware.

Demand creation is an important function in the present day marketing of processed agricultural products. The demand for most of the agricultural products arises automatically, for these commodities are basic necessities. With the increasing modernization of the processing techniques for various agricultural products, the importance of demand creation has increased in the agricultural sector.

Some of the common methods, which may be used to create a demand for the products, are:

1. *Personal Solicitation by Salesman:* By this method, the salesman personally demonstrates the good points of the product to the buyers and answers their queries by giving them the information sought for. In this way, sales are effected.
2. *Advertisement:* By this method, a message about the product is conveyed to the buyers by the seller who uses various media (non-personal method) for the purpose—such as written or printed material; pictures and diagrams, speeches and announcements. These may generate a demand and result in immediate sale or sale at some time in the future. The media used for these advertisements are hand-bills, posters, calendars, circular letters, newspapers, magazines, radio, television, and slides.
3. *Other Methods:* The other methods employed for the creation of demand are:
   (a) Display of goods in stores, at railway stations and other important public places;
   (b) Arrangement of trade fairs and exhibitions and presentation of the goods;
   (c) Distribution of free samples among the prospective buyers; and
   (d) Offers of various types of facilities, such as an extensive credit facility, the home delivery facility, the facility of guarantee against a price fall in the future, combination offers (giving one article free on purchase of another article), free deals (giving 13 pieces if a dozen are purchased), price reductions, a trade-in allowance (taking old used articles and giving concessions on the new purchase), etc.

The practices used in the creation of demand add to the cost of marketing. Whether the advertisement pays should be judged by the *added cost-added return* principle. If an increase in the sales revenue is higher than the advertisement cost, it pays to advertise the sale of goods. The converse, too, holds true.

RETAILING

As mentioned earlier, buying and selling of agricultural commodities or processed products is an essential function along the entire marketing channel (details of marketing channels in Chapter 6). Retailing is the last point of marketing channel where the seller is the retailer and buyer is the consumer. The actual interaction of consumer demand and supply (availability with the retailer at that time) takes place at this stage of marketing. In the case of processed food products, this buying/selling function becomes quite critical. The processors or processing firms use all possible methods of advertising or demand creation to attract consumers towards their products. It is in this context that grocery and food retail sector has become an important sector of agriculture and food economy of India. The grocery and food retail sector that was worth ₹ 18.1 trillion in 2015 is estimated to have grown to ₹ 61 trillion or ₹ 61 lakh crore in 2019.

RETAIL SALES OF FOODGRAINS THROUGH FPS

Sales of cereals (staple food grains), as a part of public distribution system, is a special selling function that covers 67% of the India's population. Rice, wheat and nutri grains are sold to ration card holders through government licensed outlets called fair price shops (FPSs). Fixed quantities, as per entitlements, are sold at pre-announced prices. (The current prices are ₹ 3, ₹ 2 and ₹ 1 per kg respectively of rice, wheat and nutri cereals). These shops are part of the public distribution system (details in Chapter 7). There are around 240 million eligible households who buy the grains from these shops, every month at fixed prices.

▌PRICE DISCOVERY AND PRICE DETERMINATION

Under the marketing system, there is the responsibility of realising the value of the goods delivered to the final consumers and distributing it to various marketing agencies and farmers. This process is accomplished by a system of pricing the products at each stage as they move through the marketing channels. The system of pricing is important, for prices perform the following functions:

1. They guide the allocation of resources in the marketing activities over time, space and form;
2. Prices guide the allocation of goods and services among prospective buyers; and
3. The level of prices forms the basis of the distribution of incomes or gains among producers, middlemen and consumers.

Middlemen do not determine the prices, for they do not determine the consumer's demand or quantum of market supplies. They merely discover the prices based on their evaluation of the supply (quantities available); and the prospects of what the buyers are likely and willing to pay for these quantities at each stage of marketing.

Prices are discovered in an individual market place by the traders whereas price determination takes place in the total market and not in an individual market place. Prices are determined by the forces of demand and supply in the total market. (as illustrated in Chapter 2).

CHARACTERISTICS OF PRICES DISCOVERED

1. The price discovered should clear the available supply from the market and help in the distribution of the product among the persons who need it;

2. The price discovered should act as an incentive for the producers; and

3. The price discovered should be such as to ensure the continuance of the marketing agencies in the business, i.e. it should provide a reasonable profit to the market functionaries.

PROCESS OF PRICE DISCOVERY

In the days of, primitive trading, large markets and sources of price information were not in existence. Buyers and sellers were forced to make price discovery on the spot. Buyers offered as low a price as possible and sellers demanded as high a price as possible. The final price was settled by negotiation. Even today, for some commodities, this method of price discovery is in operation. However, this is a time-consuming method. The growth of primary and secondary wholesale markets has brought buyers and sellers together; and they can now observe price-affecting conditions in a better way. Market news helps in this process. The bargaining between buyer and seller over a particular lot takes place, with a full knowledge of the level of prices prevailing not only in the nearby primary or secondary wholesale market, but also in the terminal markets. Sometimes, the factors operating in international markets also figure in their calculations. One way of price discovery in local markets is that dealers take a round of the market in the early morning and collect information on the total supplies for the day, the quantity demanded, and the views of the competitors or fellow-traders. They keep track of the prices in other markets and make their offers. As the day progresses, when the sellers find that their stocks are not moving satisfactorily, they lower their sale price to clear the available supply. Price discovery is a continuous process, for information about the market conditions continues to flow regularly.

The process of price discovery has two phases:

1. Evaluating the conditions of demand and supply, and determining the general level of prices for the commodity; and

2. Determining the price of the specific lot of the commodity being exchanged relative to the general price level.

MARKET INFORMATION

Market information is an important marketing function which ensures the smooth and efficient operation of the marketing system. Accurate, adequate and timely availability of market information facilitates decision about when and where to market the products. Market information creates a competitive market process and checks the growth of monopoly or profiteering by individuals. It is the lifeblood of a market.

Everyone engaged in production, and in the buying and selling of products is continually in need of market information. This is more true for agricultural products because their prices fluctuate more widely than those of the products of other sectors. Market information is also essential for the government, for creating a policy environment for a smooth conduct of the marketing business, and for the protection of all the groups of persons associated with this. Market information is essential at all the stages of marketing, from the sale of the produce at the farm until the goods reach the last consumer.

MEANING

Market information may be broadly defined as a communication or reception of knowledge or intelligence. It includes all the facts, estimates, opinions and other information which affect the marketing of goods and services.[12]

IMPORTANCE

Market information is useful for all sections of society which are concerned with marketing. Its importance may be judged from the point of view of individual groups. These groups are:

1. *Farmer–producers:* Market information helps in improving the decision-making power of the farmer. A farmer is required to decide when, where and through whom he should sell his produce and buy his inputs. Price information helps him to take these decisions.

2. *Market middlemen:* Market middlemen need market information to plan the purchase, storage and sale of goods. On the basis of market information, they are able to know the pulse of the market, i.e. whether the market is active or sluggish, the temperature of the market (whether prices are rising or falling), and market pressure (whether supply is adequate, scarce or abundant). On the basis of these data, they project their estimates and take decisions about whether to sell immediately or to stock goods for some time, whether to sell into the local market or to go in for import or export, whether to sell in their original form or process them and then sell, and so on. The failure of a business may partly be attributed to either the non-availability of market information or its inadequate availability and interpretation. Cooperative marketing societies operating as commission agents make use of market information for advising their members so that they may take decisions about when to sell their product. Processors make use of market information to plan their purchases of raw material so that they may run their plant continuously and profitably.

3. *General economy:* Market information is also beneficial for the economy as a whole. In a developed economy, there is need for a competitive market process for a commodity, which regulates the prices of the product. The competitive process contributes to the operational efficiency of the industry. However, a perfectly competitive system is difficult to obtain, but the availability of market information leads towards the competitive situation. In the absence of this system, different prices will prevail, leading to the profiteering by specialized agencies. The business of forward trading is based on the availability of market information.

4. *Government:* Market information is essential for the government in framing its agricultural policy relating to the regulation of markets, buffer stocking, import-export, and administered prices. Market information also helps the government in taking decisions related to fiscal and monetary measures, support prices, PDS supplies and open market operations.

TYPES OF MARKET INFORMATION

Market information is of three types:

1. *Market Intelligence:* This includes information relating to such facts as the prices that prevailed in the past and market arrivals over time. These are essentially a record of what has happened in the past. Market intelligence is, therefore, of historical nature. However, an analysis of the past helps us to take decision about the future. Thus, market intelligence helps in building outlook for current and future decisions.

2. *Market News:* The term 'market news' refers to current information about prices, arrivals and changes in market conditions. This information helps the farmers in taking decisions about when and where to sell his produce. The availability of market news in time and with speed is of the utmost value. Sometimes, a person who gets the first market news gains a substantial advantage over his fellow-traders who receive it late. Market news quickly becomes obsolete and requires frequent updating. Nevertheless, market news at different points of times improves market intelligence and outlook.

3. *Market Outlook:* Market outlook refers to the market situation that is likely to prevail in the near future. This is of considerable use to farmers, traders and processors. For example, any information on the prices that are likely to prevail during the forthcoming post harvest season helps farmers in planning their area allocation to alternative crops. During the last ten years, two comprehensive five year network projects (first led by Tamil Nadu Agricultural University–NAIP project) and the second in continuation led by ICAR National Institute of Agricultural Economics and Policy Research) were implemented to evolve and test the statistical models for generating the forecasts on agricultural prices. In both cases, the price forecast accuracy was more than 90%.

CRITERIA FOR GOOD MARKET INFORMATION

Good market information must meet the following criteria so that it may be of maximum advantage to the users:

1. *Comprehensive:* Market information must be complete and comprehensive. It must cover all the agricultural commodities and their varieties, and all the geographical regions. It must cover prices, production, supply movements, stocks and demand conditions.

2. *Accuracy:* The accuracy of market information is essential. The collection of accurate market information is a tedious and expensive task under changing market situations. There must be honesty in the collection of the information. Constant efforts should be made to improve its accuracy. The information reporter must be thoroughly acquainted with the market and the product so that he may collect accurate information about them.

3. *Relevance:* Market information must be relevant in the sense that it must be collected, arranged and disseminated, keeping in view the user's interest. Generally, a lot of information that is collected is not used; the time and energy spent on its collection, therefore, become a colossal waste. It is not enough to simply collect a mass of data and report them through various media; the data must be presented in a form that is relevant for the users.

4. *Confidentiality:* There must be a sense of confidentiality among the firms for whom the information has been collected. The information revealed under this situation of confidentiality will be more correct and may assist in drawing policy implications. The names of firms or respondents should not be leaked out.

5. *Trustworthiness:* Trustworthiness is another criterion of good market information. The agency that collects it must create faith, and the users must trust the organisation which is making this information available to them.

6. *Equal and Easy Accessibility:* Every person engaged in marketing, whether big or small, farmer, wholesaler, retailer, government or a private agency, must have equal and easy access to the available information. There should not be any sort of restriction on individuals in the use of this information.

7. *Timeliness:* Market information must be made available in time. For this purpose, a speedy transmission is necessary. Late dissemination of market information is of no use. Often, this information becomes stale, particularly when it is disseminated too late to be of any use. A system for speedy dissemination of information should be devised.

COLLECTION AND DISSEMINATION OF MARKET INFORMATION

There are three major steps in the collection and dissemination of market information:

1. *Collection of Market Information:* Both official and non-official agencies collect market information. They may be public or privately owned agencies. Some private companies publish their findings in addition to sending information to clients. Dealers in inputs and trade associations collect market information. However, public agencies play a major role in India in the collection of this information. The main agencies collecting market information are the State Agricultural Marketing Departments, the State Agricultural Marketing Boards, the Food Department, and the Directorates of Economics and Statistics of the State and Central Governments.

2. *Dissemination of Market Information:* The collected information has no meaning until it reaches the persons who need it. The sources through which market information is disseminated are:

   (a) *Personal Contacts:* This is the most important source of dissemination of market information. Information is given orally, i.e. by one businessman to another businessman, by a businessman to a farmer, or by one farmer who has sold the produce to another farmer.

   (b) *Post and Telephones:* Businessmen get information from other markets on the telephones. These days, mobile phones are used but prior to that commission agents used to convey the information on the prices of different commodities to their client-farmers in postcards. They filled the prices on these postcards and posted them daily or at some time intervals.

   Telecommunication is becoming increasingly important in dissemination of market information. There has been a phenomenal growth in the last two decades in the country. Both public and private sector companies have taken up the function of providing telephone services. Private sector companies have entered recently in this area and are providing information services in urban as well as in rural areas. The teledensity (number of telephone per 100 population), which was 2.86 in March 2000 jumped to 5.72 in August 2003 and currently is around 90. However, there is a considerable difference in tele-density between urban and rural areas. In rural areas, it is 57.50% as against 159.7% in urban areas. Nevertheless, it has experienced tremendous growth both in rural and urban areas. Total number of telephone connections (subscription) in India in March 2019 was 118.34 crore, of which 98.17% (116.17 crore) were wireless (mobile) connections. The land line connections were only 1.83% (2.17 crore).

   The internet use has also increased phenomenally. The mobile industry has witnessed exponential growth over the last few years owing to affordable tariffs, wider availability, roll out of mobile number portability, expansion in 3G and 4G coverage, changing life styles and conducive policy and regulatory environment.

(c) *Newspapers:* The newspapers in English, Hindi and regional languages publish the wholesale prices of important agricultural crops in the selected markets of the country/State. In addition, the Economic Times and the Financial Express contain a lot of information on the various aspects of marketing and on prices. Almost all dailies carry a special section on business, trade and prices.

(d) *Magazines:* Magazines, such as the Economic and Political Weekly and Business Today are important weekly trade journals, which collect and disseminate a lot of information connected with trade.

(e) *Government Agencies Reports:* The regulated markets, the Agriculture Marketing Department in the States, the Directorates of Economics and Statistics in the States, the Central Directorate of Marketing and Inspection, Reserve Bank of India, and the Directorate of Economics and Statistics, Ministry of Agriculture, Government of India are some of the government agencies which disseminate the collected market information through their regular publications, broadcasts on All India Radio or telecast on D.D. channels. The Agmarknet displays the real time price information of various agricultural commodities in more than 7000 wholesale markets of the country.

(f) *Price Bulletins:* These are issued daily, weekly, or every month. The important bulletins through which price information is disseminated are: Bulletin of Agricultural Prices (Weekly), Agricultural Situation in India (Monthly), Agricultural Prices in India (Annual) and Bulletin on Food Statistics (Annual). The monthly situation and outlook reports are published by the Directorate of Marketing and Inspection, Government of India.

(g) *Radio and Television:* Information on prices and related aspects is regularly relayed/telecast on various channels. Almost all channels have now reserved a slot on trade and business in their programmes. Several state governments and National Informatics Centre (NIC) of the Government of India have taken initiatives to inter-link the markets with NIC-NET with a view to ensuring quick flow and accessibility of information on prices and arrivals. The information on prices and trade is also available on Internet at several websites. Recently, Door Darshan (India's official TV channel) has launched a 24-hour Kisan Channel (Farmers Channel). It telecasts, apart from the outlook information, farm product prices prevailing in different/ important markets of the country. Though, it is named as Farmers' Channel, it is useful for all the stakeholders, including village traders, wholesalers, processors, retailers, agricultural entrepreneurs and rural families. The viewership of this channel is very high.

(h) *Mobile Services:* The price information, along with other agricultural related information is also being provided by various government and private organizations through SMS on mobile phones. As the incoming is usually free of charge, it is becoming increasingly popular among farmers. An example is of Green Phablet launched by ICRISAT in 444 villages in Karnataka, Andhra Pradesh and Telangana. The 3G device uses a Green SIM, a special SIM card, which can work with any mobile phone. The extension workers are using it as push-based and pull-based advisory services. Several Krishi Vigyan Kendras, IFFCO and ICT have also launched such mobile messaging services.

3. *Interpretation of Market Information:* The utility of market information varies with the user. One user, who is good enough in his interpretation of the available market

information, gets the benefit from it; the other man, who is poor in the interpretation of the available information, is not able to get the same advantage. Educated businessmen, who are well-equipped with modern means of communication and have experts to guide them, take a better advantage of market information.

CRITICISM OF MARKET INFORMATION

The market information system is criticised by the users on the following grounds:

1. Market information provided to the farmers cannot be evaluated by them because of their illiteracy and poor communication ability.
2. Market reports are incomplete in many respects. Often, there is no mention of quality when the price is quoted, and the price quoted is other than a modal price. In many cases, the reported prices vary considerably from the actuals because of inaccurate field reporting, sampling errors and other factors.
3. There is manipulation in the collection of information meant for market reports, specially if the collector is biased and has pre-conceived notions.
4. Most of the time, the news reported is so late that it is of no use. There is a considerable time lag in the publication of the magazines and reports carrying such information. The information contained in the magazines, therefore, becomes obsolete from the practical utility point.
5. Most of the information pertains to wholesale marketing. Very little information on retailing is available.
6. The market information that is made available is of greater use to the buyers of farm products than to the farmers selling their products.

There is, therefore, a strong case for effecting improvements in the existing market information system.

SUGGESTIONS FOR IMPROVEMENT IN MARKET INFORMATION

There have been improvements in the collection and dissemination of market information during the past few decades. Some of the suggestions for improvement in the existing market information system, for agricultural commodities for making it orderly and efficient, are:

1. There should be a standardized system of quoting the prices of the different varieties of the commodity and units of quotations, so that the prices may be compared over time and space.
2. The price announced should be in the local language and should cover more local markets of the area rather than secondary and terminal markets located far from the area. The frequency of, and time allotted for the announcement should be increased so that the farmer may profit from the information. The broadcast or telecast time should be such that the farmers can listen to the announcement and take decisions about taking the produce to the market.
3. Arrangements for the display of prices on notice boards at important public places like market yards or panchayatghars, etc., should be made.
4. The staff posted for the collection of market news should be knowledgeable and trained. There must be thorough and frequent check to ensure that guesswork, manipulation and bias are excluded.
5. Market news should have no place for rumours. Rumours are harmful for the farmers as well as consumers.

6. Market news should be provided as fresh as possible so that it may create confidence and trust among the users.

7. The information on the arrivals of the commodities in the market, on demand, market tone, etc., should be announced along with the information on prices.

8. A correct and intelligent interpretation of market information should be made and announced at the same time as market information. This is very important, for farmers may not be able to interpret it correctly.

9. There must be proper co-ordination between market intelligence and policy-making departments so that the latter may better understand the problem and can make such adjustments in the information as may be called for.

10. In many markets, a price range is reported instead of a single price. This is so because of the variation in quality and the large number of transactions taking place in the market. The range is very wide, and may not serve the purpose of the users. Therefore, the price range should not be very wide or ambiguous.

11. There must be an educational programme for the users of market information so that they may evaluate it and take the best advantage of it.

12. Apart from the information on current prices, the most important information for farmers is the forecast of prices, i.e. prices likely to prevail during the coming harvest season. These price forecasts should be made available to the farmers well before the sowing season. As mentioned earlier, the human resource capacity has been created and perfection of methodology has been done within Indian Council of Agricultural Research. There is an urgent need for nationwide permanent institutional structure, to regularly generate price forecasts at the national and regional levels, which will help the farming community as well as the policy makers and implementing agencies.

GENESIS OF MARKET INTELLIGENCE SYSTEM IN INDIA

Market intelligence is an essential function for the formulation of a sound price and trade policy and its successful implementation. The formulation of a sound price policy requires an analysis of long-term trends in the data on prices, arrivals, demand, supply, and other information. A market intelligence system provides the necessary data for such an analysis and for an understanding of the behaviour of relevant factors; and helps in the evolution of a proper price policy and generating outlook information.

The role of market intelligence can be judged from its following objectives:

1. To provide better understanding of the forces that are operating in a particular situation as well as anticipating the situation that is likely to develop.

2. To provide regular and continuous appraisal of market behaviour and of various factors that influence the market behaviour.

3. To offer a clue to the probable behaviour of the market and the forces that are likely to influence it in the near future.

4. To undertake evaluation of the functioning of the marketing organizations/ institutions with a view to ensuring efficient and effective implementation of agricultural marketing and price policy.

5. To offer advice on the measures needed for influencing the decisions of traders and market players (a) for ensuring remunerative prices to the farmers; (b) for assuring the supply of the products to the consumers at reasonable prices; and (c) for maintaining stability in the market prices.

Market intelligence is required by the government organizations, traders and their organizations, farmers, consumers and researchers as well. Government organizations need market intelligence for formulation of plans and policies and also for their effective application. Traders and their organizations require market intelligence for competitive sales, purchases and efficient commercial operations in order to carry out the functions of distribution efficiently. Farmers need market intelligence for proper adjustments in cropping pattern and to decide when, where and how much to sell. Consumer need market intelligence to understand market forces for making purchases in a rational manner. Researchers require market intelligence in order to assess the efficiency of the marketing system, identify the bottlenecks in marketing programme/projects and for suggesting future remedial steps and strategies.

The different committees which have been appointed from time to time—the Foodgrains Policy Committee, 1943, and the Prices Sub-Committee of the Policy Committee on Agriculture, Forestry and Fisheries, 1948—had emphasized the importance of data on market intelligence and of a market intelligence authority. Prior to 1953, a large mass of data on agricultural prices was collected by a variety of agencies at state and district levels. But there was no uniformity in the concepts and definitions of the terms, in the centres for which data were collected and in the coverage of data at various levels. To bring about uniformity in the collection of information necessary for market intelligence and also to remove difficulties experienced in the functioning of the then system, the Ministry of Food and Agriculture. Government of India appointed a committee in November, 1953 under the chairmanship of Shri P.N. Thapar. In pursuance of the recommendations of this Committee, an integrated scheme for Market Intelligence was prepared by the Directorate of Economics and Statistics, Government of India and launched during the Second Five Year Plan. This Integrated Scheme on Market Intelligence was implemented in all the states and union territories except Nagaland, Arunachal Pradesh, Andaman & Nicobar Islands, Dadra & Nagar Haveli, Mizoram and Lakshadweep.

It may be mentioned here that the National Commission on Agriculture, 1976 also made several recommendations for improving the market intelligence system in the country. Some of these are:

1. Efforts be made to issue periodical reports on outlook for future;
2. The scope of market intelligence should be enlarged to include pulses, oilseeds, important fruits and vegetables and spices crops;
3. The studies on costs and margins should also form a part of market intelligence;
4. The market intelligence should be extended to more centres and crops;
5. All regulated markets should be made reporting centres for the purpose of market intelligence;
6. Whole-time technical reporting agencies should be set up in all the important wholesale markets;
7. The scope of the market news service should be extended; and
8. Foreign market intelligence should also be made an essential part of the market intelligence in the country.

MARKET INTELLIGENCE SCHEME IN INDIA

The market intelligence scheme in India provides for the collection of data on prices, arrivals, despatches and stocks of important agricultural commodities for the selected market centres of the country. These data are reported by technical persons such as market intelligence inspectors, market reporters, agriculture assistants, price inspectors, statistical investigators, market secretaries and/or auction clerks posted in the selected markets of the country. Several changes have been brought about in the scope and coverage of the scheme from time to time. However, there is a considerable gap in the ideal system of market intelligence and the existing coverage.

The market intelligence scheme covered 137 agricultural commodities and 1300 markets of the country. The market intelligence centres set up under the scheme work under the supervision of Directorate of Agricultural Marketing or Directorate of Economics and Statistics of the states. In Rajasthan,there are 58 market intelligence centres, which collect the daily market price quotations from 58 selected markets and supply to the Directorate of Agricultural Marketing, Rajasthan and Directorate of Economics & Statistics, Government of India. The weekly review reports on retail and wholesale prices, arrivals, sale and stock position and fluctuations in prices of the selected commodities are also prepared and made available for wider use.

The importance of market intelligence scheme can be realized from the wide spread concerns expressed by the people at large about the behaviour of prices from time to time. The rate of rise in prices (inflation) of different groups of commodities (food, fuel, power, manufactured goods etc.), influences the monetary and fiscal policy, investors' behaviour and also the consumers. The rate of price rise also affects the salaries of employees in the government and organized sector as dearness allowances are linked to the rate of inflation. Therefore, government has put in place an elaborate system of collecting the price quotations of various commodities from a large number of markets and computing the price index numbers. The index number of prices, by very definition, has a well defined base period (usually a financial year or a triennium). The rate of price rise (or inflation) needs to be monitored both at wholesale and retail levels. Corresponding to these, wholesale price index (WPI) and consumer price index (CPI) numbers are worked out and published for wider use. The base year, number of commodities and the number of markets/quotations are changed or updated after five or ten years as and when it is felt that some structural changes in the economy have taken place.

The WPI series in India, started in 1942, with week ended 19th August 1939 as the base which had only 23 items and one price quotation for each. In 1947, it was expanded to include 78 items and 215 price quotations, with base remaining the same. Since then, the series was revised many times. The current series has 697 items, 8331 price quotations and 2011–12 as the base year. The details of eight series of WPI since 1947 are shown in Table 4.29.

As regards measurement of inflation at consumer/retail level, Consumer Price Index (CPI) is computed and brought out for different sections. Currently, the index number series are available for Industrial Workers (CPI-IW) with base 2001, for Agricultural Labourers (CPI-AL) with base 1986–87, and for Rural Labourers (CPI-RURAL) with base 2012. These are compiled and released by Labour Bureau, Government of India. Looking at the implications of Consumer Price Index Numbers for different sections of the society, the Government has introduced in 2011 a new series of combined CPI (CPI-C) with base 2012. This is compiled and released by central statistical organization (CSO). The weights for important agricultural/food commodities in the current series of WPI and CPI-C can be seen in Table 4.30.

The importance of flow of market information has considerably increased in recent years. This is particularly so in the case of fruits and vegetables where fluctuations

Table 4.29: Salient features of WPI series in India

Year of introduction	Base period	No. of items	No. of quotations	Averaging
1947	August 13–19, 1939	78	215	Weighted GM
1952	1952–53	112	555	Weighted AM
July 1969	1961–62	139	774	Weighted AM
January 1977	1970–71	360	1295	Weighted AM
July 1989	1981–82	447	2371	Weighted AM
April 2000	1993–94	435	1918	Weighted AM
September 2010	2004–05	676	5482	Weighted AM
May 2017	2011–12	697	8331	GM/Weighted AM

GM = geometric mean, AM = arithmetic mean
Source: Economic Advisor, DPIIT, Government of India.

Table 4.30: Weightage for agricultural products in WPI and CPI-C

WPI (2011–12 = 100)		CPI-C (2012 = 100)	
Commodity	Weight	Commodity	Weight
All commodities	100.00	General index	100.00
Primary articles	22.62	Food and beverages	45.86
Fuel and power	13.15		
Manufactured products	64.23	Cereals and cereals products	9.67
Total	100.00	Pulses and pulse products	2.38
Food grains	3.46	Fruits and vegetables	8.93
Food articles	15.56	Milk and milk products	6.61
Oilseeds	1.12	Sugar and confectionary	1.36
Veg. and animal oils/fats	2.64	Spices	2.50
Fruits and vegetables	3.48	Meat, fish and eggs/oils and fats	4.04/3.56
Condiments and spices	0.53	Non-alcoholic beverages	1.26
Fibres	0.71	Prepared meals, snacks and sweets	5.55

Source: Economic Adviser, DPIIT for WPI, and CSO, MOSPI for CPI-C.

and inter-spatial price differences are considerably more. It is in this connection that the Market Planning and Design Centre (MPDC) in the Directorate of Marketing and Inspection has developed a model of market information service for fruits and vegetables in India. In this system the market is the basic information unit, responsible for collection and transmission of information relating to its own activities with a two-tier dissemination system—for distant as well as for local markets. This system has been established in Azadpur Wholesale Market, Delhi for apple, citrus, mango, potatoes and onion.

The National Horticulture Board set up under the Ministry of Agriculture, Government of India has also given emphasis on information service for fruits and vegetables. The Board has provided computers in all the important fruits and vegetable markets. These markets are interlinked through computerized machines connected to a computer in Central Data Processing Unit at Delhi. The unit analyses the data received by it and the information regarding the prices and arrivals of different commodities in a particular market centre is flashed to other market centres. The National Horticulture

Board is also disseminating the market information through publications in newspapers, relay from Radio Stations, telecasts on T.V. channels and real time display on its website.

It needs to be noted that Indian Council of Agricultural Research (ICAR) under National Agricultural Innovation Project (NAIP), implemented a research sub-project titled "Establishing and Networking of Agricultural Market Intelligence Centres in India" from June 2009 to September 2013. The Leader was Tamil Nadu Agricultural University and there were 10 collaborating State Agricultural Universities. A model methodology of price forecasts was developed, which was used to forecast pre-harvest and post-harvest prices for a large number of agricultural crop products, regularly for eight crop seasons. The validity of forecasts was estimated as 90 to 99%. The forecasts were distributed (before the sowing season) through voice SMS, Text SMS, thousands of hard copies, publications in national, regional and local newspapers and TV/radio broadcasts.

The price forecast sub-project was continued (after the NAIP was over) as a network project of ICAR under the leadership of National Institute of Agricultural Economics and Policy Research, New Delhi. The methodology was further perfected and forecast accuracy was almost the same. It is hoped that this will soon become a regular institutional framework at the national level (like weather forecasts by IMD).

INFORMATION TECHNOLOGY (IT) APPLICATIONS IN AGRICULTURAL MARKETING

Agricultural produce marketing requires connectivity between the markets, the growers/exporters/traders, industry and consumers through a wide network of national and international linkages so as to provide day to day information with regard to the commodity arrivals and prices; to provide links for on-line international market information; to provide export related documentation; to inform about the latest research in agricultural marketing, packaging and storage; and to provide connectivity with the World Trade Centers (WTC), National Horticulture Board (NHB), National Institute of Agricultural Marketing (NIAM), Agricultural and Processed Export Development Authority (APEDA), State Agricultural Marketing Boards (SAMBs), Universities and other such organizations.

Information Technology (IT) is being regarded as the fifth factor of production along with land, labour, capital and management. It has integrated the world by the use of Internet. Information technology is basically concerned with e-commerce, i.e. on-line information facilitating transactions, future planning for purchases and selling of agricultural products and inputs; and various other aspects which World Wide Web provides. E-commerce has revolutionized trade in developed economies and is picking up in developing countries.

IT in agricultural marketing comprises of the following tasks:

1. Linking and networking of agricultural markets;
2. Computer aided auction displays and trading;
3. Marketing information system;
4. Commodity information system;
5. Highway Automation system.

There are at least following four ways for making use of IT in improving agricultural marketing:

1. Holistic and Integrated Information Management

With a view to fully utilizing the information technology, agricultural marketing information service system has been set up at the national level. This service should be an integrated service incorporating the farmers advisory service with Decision Support System (DSS). Such a system will help the farmers in taking intelligent decisions related to storage, pricing and marketing. One of the major problems in designing of agricultural marketing information service is that the information needs of the individual target groups are diversified. For evolving information system, assessment of the information needs of the diversified target groups is very important, so that the information management is holistic and integrated.

The information system should be broad based to cover information related to aspects such as storage, transport, weather forecasts and export potential. At the state level, marketing Boards/Directorate should provide consolidated information to all the market users. The coordinating agency should take the stock of information available and its generation process and finally distribute it to the need based target groups. Attempts are ongoing in this regard.

2. Electronic Auctioning System (EAS)

The Electronic Auctioning System (EAS) is the system to perform the bidding process electronically. The EAS can help in larger markets where number of lots for auction is large and the time allotted for biding is limited to 3 to 4 hours. During the peak season the biding work is so hectic that the biding for three to four lots has to be completed each minute. Hence the introduction of EAS is necessary. The main advantage of EAS are:

1. Minimizes the paper work
2. Automates the billing and revert system
3. Improves the efficiency of the system
4. Increases the transparency
5. Provides a disciplined structure which matches with typical functioning to manage the bidding system
6. Provides easy and efficient summarized information, and
7. Ensures security of the data/information.

The basic objective of the system is to promote trade transparency and help farmers by checking malpractice and ensure recording of each auction. This system of bidding was first implemented in the auction of fruits and vegetables in the biggest fruit and vegetable market of Azadpur (Delhi). It was later introduced in Karnataka state and now through E-NAM, it is likely to become a nationwide practice.

3. E-Catalogue for Commodity Profiles

In the context of need to increase export competitiveness, each and every product needs to be publicized highlighting its characteristics on nutrition values, chemistry, quality standards, seasonality, quantity available for supply and prices. A brief commercial profile of the commodity on Internet can help the buyer in making comparative analysis of costs and margins. If for each commodity, commercial profiles giving these details are prepared and transmitted to international markets through 'web-pages', it can greatly widen markets for such products. APEDA and MEA are rigorously working on these lines.

4. National Atlas of Markets

Mapping of the agricultural markets of the country is a pre-requisite for carrying any planning/developmental activity. All the regulated markets along with their classification on the national maps can give a synoptic view of the distribution of the markets. The infrastructural facilities, the volumes transacted, the area and population served, and the outflow and inflow of the commodities are the various aspects, which should be mapped out. Mapping of country's markets would also be useful for research, planning and policy formulation. The National Atlas of Agricultural Markets should be based on the application of GIS tools. Such mapping activity can also be put on the Internet for its greater usages. National Institute of Agricultural Marketing (NIAM) has taken initiatives in this regard to bring out a National Atlas of Indian Agricultural Markets. It needs to be regularly updated and made openly accessible to all stakeholders.

▌ FINANCING FOR MARKETING

There is a long interval between the time of production and consumption. Between these two points, the ownership of commodities shifts many times—a fact which necessitates financial arrangements. Middlemen need finance not only for the purchase of stocks, but for the performance of various marketing functions, such as processing, storage, packaging, transport and grading. The financing function of marketing involves the use of capital to meet the financial requirements of the agencies engaged in various marketing activities.

No business is possible nowadays without the financial support of other agencies because the owned funds available with the producers and market middlemen (such as wholesalers, retailers and processors) are not sufficient. The financial requirements increase with the increase in the price of the produce and the cost of performing various marketing services. In the words of Pyle: "Money or credit is the lubricant that facilitates the marketing machine." The government has taken several measures to provide financial support to market functionries.

FACTORS AFFECTING CAPITAL REQUIREMENTS FOR AGRICULTURAL MARKETING

The capital requirement of a marketing agency for its marketing business varies with the following factors:

1. *Nature and volume of business:* Financial requirements for trading in high value crops like cumin, chillies, cotton and oilseeds are higher than for trading in foodgrains. For the wholesale business too, financial requirements are higher than for retail business.
2. *Necessity of carrying large stocks:* It is essential to carry over large stocks throughout the year, of goods which are seasonally produced and marketed on a wholesale basis throughout the year. Financial requirement is higher for trade in such goods.
3. *Continuity of business during various seasons:* If business is continuous throughout the year, the financial requirements will be greater than if business is conducted only during a particular season or for few months.
4. *Time required between production and sale:* Some goods are sold immediately after production—perishables, for example—while others are disposed off after a certain time—rice and cheese, for example. Financial requirements in the marketing of the latter goods are, therefore, higher.

5. *Terms of payment for purchase and sale:* The terms of transactions—whether payment will be in cash, on credit or by instalments—affect the financial requirements of the marketing middlemen.

6. *Fluctuations in prices:* Financial requirements are higher for goods which suffer frequent price fluctuations than for goods that are subject to less frequent price fluctuations.

7. *Risk-taking capacity:* The financial needs of the market middlemen vary with their risk-taking capacity. A middleman with a low risk-taking capacity often resorts to hedging, and needs less finance than a middleman who takes risks.

8. *General conditions in the economy:* During the period of price fall or recession, the financial requirements increase. The marketing agency has to hold stocks for a longer period in anticipation of a price rise. Moreover, the recovery of old bills tends to be slow. Whenever, a new product is introduced, the dealer needs more finance temporarily till the demand for it picks up in the economy.

TYPES OF MARKETING FINANCE

The marketing finance required by the marketing middlemen is of two types—fixed capital for land, buildings (shops and godowns); equipment and machinery (weighbridge, grading equipment, etc.), and working capital which is required to meet the marketing costs, purchase value, and salaries of the employees. In the trade of agricultural products, the proportion of working capital is higher than that of fixed capital. It is also necessary to make arrangements for financing the farmers during the period between the production and sale of their produce. This is necessary to improve their holding capacity and to avoid the post-harvest sale of the produce when prices are low in the market. Because of their acute financial needs, many farmers market their standing crops—of fruits, for example–or borrow money in advance from local traders/commission agents against their crops, and bind themselves to sell the crop through the trader/commission agent. This checks their freedom to sell the produce in the open market.

To improve the financial position of the farmers and to strengthen their holding capacity, the following, steps have been taken by the government:

1. Since July, 1969, with the nationalization, commercial banks have started financing the agricultural sector in a big way and meeting the increasing needs of the farmers for production purposes.

2. The cooperatives, too, have developed and entered the field of agricultural financing. An integrated scheme of credit and marketing has been introduced. Under this scheme, cooperative credit societies can realize their credit, together with the interest due on it, by the sale proceeds of the produce directly by intimation to Cooperative Marketing Societies. These may make the payments for the produce to the farmer after deducting their dues. A rapid progress has been made in this area.

3. With the development of warehousing facilities in the country, farmers can now get credit equivalent to around 80% of the value of the produce kept in the warehouses. Banks extend the financing facility to farmers against the mortgage of the warehouse receipt. This scheme has lessened the financial problems of the farmers and of market middlemen. As a result, the tendency to sell the produce immediately after the harvest has been checked to some extent. However, it has met with only limited success. So long as the interest rate continues to be more than the intra-year rise in prices, storage cannot be a profitable proposition.

NABARD IN AGRICULTURAL MARKETING FINANCE

The National Bank for Agriculture and Rural Development (NABARD) was set up as an apex organization in the sphere of rural finance on 12th July, 1982. It is a national institution for providing refinance, regulating credit, and extending other related facilities to agriculture, small scale industries, cottage industries, handicrafts and rural development sectors. It provides refinancing support to financial institutions for financing wide range of activities pertaining to agriculture and rural development. NABARD in addition to providing refinance facilities for agricultural production, also provides refinance facilities to financial institutions for development of infrastructure, cooperative marketing, construction of warehouses and cold storages, creation of transportation facilities, construction of market yards and processing of farm products.

The refinance facilities available from NABARD in the sphere of agricultural marketing are:

1. Marketing of Produce: NABARD provides refinance support to the state cooperative banks by way of short-term credit limits for assisting the cooperative marketing societies to help the members in marketing of their produce at remunerative prices and also to enable them to repay their dues to primary cooperative credit societies.
2. Construction of Godowns and Storage Facilities: NABARD provides refinance support for financing of construction of godowns and cold storages.
3. Construction of Market Yards: NABARD provides refinance support for construction of market yards to cooperative and other banks. This also includes construction of various amenities in the market yards like construction of shops, platform, rest houses, canteen, bank and post office premises.
4. For Means of Transportation: NABARD makes available the refinance facilities for financing the purchase of trolleys and other means for transportation of marketable surplus from fields to the markets.
5. Establishing of Processing Units: NABARD makes available the refinance facilities for financing of agro-based processing units such as rice mills, flour mills, oil crushing, canning of fruits and vegetables, gur and khandsari units, cold stores and cool chains.

Union Ministry of Food Processing Industries and National Horticulture Board have launched several schemes of subsidised finance for food processing activities. A recent development has been the emphasis on subsidy-cum-credit support for marketing of horticultural commodities, as per the details shown in Table 4.31, along with the estimated cost of various items. The scale and rate of subsidy continue to change over time. The current details are available on the websites of MoFPI and NHB.

INSTITUTIONAL CREDIT AND FINANCIAL INCLUSION

Owing to the refinance facilities available (from NABARD/NCDC) the commercial banks, cooperative banks and Regional Rural banks are extending credit facilities for several agriculture activities, including input marketing, storage of agricultural commodities, agro-processing and value addition, cold storage and cold chains etc. Ministry of Food Processing Industries, National Horticulture Board and Union Ministry of Agriculture and Farmers Welfare have launched several schemes (discussed elsewhere) that entail both subsidy and loan component. Some major recent financial initiatives are (a) viability gap funding for silos (FCI), (b) Warehousing finance (NABARD), (c) Rural godown scheme (DMI), (d) Mega food parks (MoFPI), (e) cold

Table 4.31: Estimated cost and subsidy for horticultural marketing activities

Items	Cost (₹ in lakh)	Subsidy
Pack house	3.0	50%
Pre-cooling unit	15.0	40%
Cold storage (5000 tonnes capacity)	300.0	40%
Mobile/primary processing unit	25.0	40%
Ripening chamber (5000 tonnes capacity)	15.0	40%
Low-cost processing unit	2.0	50%
Low cost onion storage unit	1.0	50%
Controlled environment retail outlet	10.0	40%
Mobile/static vending cart + platform + cool chamber	–	50%

value chains (MoFPI). (f) NWR system of financing; (g) NDDB's scheme for milk marketing and dairy processing, (h) Fisheries Development Boards efforts for fish marketing; and (i) Agricultural Marketing Infrastructure (AMI) initiative of Ministry of Agriculture (DMI).

Under the Financial Inclusion Plan, that aims to increase the access of institutional credit, by the end of March 2019, there were 5, 97,155 banking outlets in villages. These include 52489 bank branches and 5,41,129 bank's business correspondents (providing all kinds of banking services to rural people). Nearly 4.9 crore Kisan Credit Cards have been issued. During 2018–19, total agricultural credit advanced in India was ₹ 12,54,800 crore. During 2020–21, the target for credit advancement for agricultural sector is ₹ 15 lakh crore.

The Integrated Scheme for Agricultural Marketing (ISAM), a sub-scheme of Agricultural Marketing Infrastructure (AMI), includes up-gradation of more than 20000 grameen haats. Subsidies and credit facilities are being extended to rural institutions to create facilities in these haats and link these to the E-NAM portal, so as to provide access of farmers to sell their produce in the markets spread throughout the country.

RISK-TAKING/RISK BEARING

MEANING AND IMPORTANCE OF RISK

Hardy[13] has defined risk as uncertainty about cost, loss or damage. Risk is inherent in all marketing transactions. There is the risk of the destruction of the produce by fire, rodents or other elements, quality deterioration, price fall, change in tastes, habits or fashion, and the risk of placing the commodity in the wrong hands or area.

There is a time lag between the production and consumption of farm products. The longer the time lag, the greater the risk. The risk associated with marketing cannot be dispensed with for this risk contributes to profit. Someone has to bear the risk in marketing process. But most of the risk is taken by market middlemen, for they have the capacity to bear it.

Whenever risks are greater and varied, the margin taken by the risk-bearers is higher, and vice versa. One who holds the commodity in the process is the bearer of the risk, because of which he may be better off or worse off.

TYPES OF RISKS IN MARKETING

The risks associated with the marketing process are of three basic types:

1. *Physical Risk:* This includes a loss in the quantity and quality of the product during the marketing process. It may be due to fire, flood, earthquake, rodents, insects, pests, fungus, excessive moisture or temperature, careless handling and unscientific storage, improper package, looting or arson. These together account for a large part of the loss of the produce at the individual as well as at the macro level. Such losses are a loss to society, too, and must be averted to the extent possible.

2. *Price Risk:* The prices of agricultural products fluctuate not only from year to year, but during the year from month to month, day to day and even on the same day. The changes in prices may be upward or downward. Price variation cannot be ruled out, for the factors affecting the demand for, and the supply of, agricultural products are continually changing. A price fall may cause a loss to the trader or farmer who stocks the produce. Sometimes, the risks are so great that they may result in a total failure of the business, and the person who owns it may become bankrupt. To illustrate the kind of price risk faced by the farmer, let us take a real case. There was a farmer sitting on the heap of gram (weighing about two tonnes) on the auction floor in a mandi of Bikaner district. He had tears in his eyes. On enquiry, the farmer said that he brought his produce yesterday and as per mandi's practice, his gram was auctioned. The highest price offered to him was (say) ₹ 'X' per quintal. This price was below his expectation, so the farmer refused to strike a deal and decided to wait for the next day (as per mandi rules, a farmer has the right to refuse the sale if he so decides even at the highest price offered to him in open auction). Next day, again his lot was put for auction. As the market arrivals were very large on this day, the highest price offered to him this day was around ₹ 200 per quintal lower than that of previous day. The upshot of the case is that the farmer, in the hope of getting a higher price, decided (took risk) to wait for one more day and in the process, lost ₹ 4000 (200 × 20).

3. *Institutional Risks:* These risks include the risks arising out of a change in the government's policy, in tariffs and tax laws, in the movement restrictions, statutory price controls, imposition of levies, ban on exports, etc. A recent case of such a risk faced by the exporters of onion is the sudden imposition of a ban on exports (due to sharp rise in onion prices in the domestic markets, resulting from crop failure/ shorter crop as a result of heavy rains in the kharif harvest season, 2019). The exporter, who aspired to gain by exporting and had accumulated some stocks for this purpose, felt hurt by this government decision (ban on exports). But his actual loss may not turn out to be high or may be even zero, as domestic prices had skyrocketed.

MINIMIZATION OF RISKS

The agencies engaged in marketing activities worry about the risks associated at every stage; and they continually try to minimize the effects of these risks. A risk cannot be eliminated because it also carries profit. The agencies which do not take risks hardly earn profit. The risk management by the adoption of some of the measures listed below may minimize the risks:

1. Reduction in Physical Loss

The physical loss of a product (quantity and quality both) may be reduced by the adoption of the following measures:

1. Use of fire-proof materials in the storage structures to prevent accidents due to fire;
2. Use of improved storage structures and giving necessary pre-storage treatment to the product to prevent losses in quality and quantity arising out of excessive moisture, temperature, attacks by insects and pests, fungus and rodents;
3. Use of better and quicker transportation methods and proper handling during transit;
4. Use of proper packaging material; and
5. Use of cool chain for perishable commodities.

2. Transfer of Risks to Insurance Companies

The burden of physical risk may be minimized by shifting it to insurance companies. There are specialized professional agencies to bear such risks. They collect some premium and provide full compensation to the party in case of loss due to the reasons for which the products are insured. In this way, the company insures a number of farmers against losses.

The insurance is available for the goods transported from one place to another, for the goods stored in registered storage godowns or warehouses, agricultural or food processing industries, cold stores, cool chains and even retail stores. The premiums depend on the risk involved and the sum insured. General insurance companies, both public and private, provide insurance cover.

As regards insurance for farmers, the Government of India in the 12th Five Year Plan launched Integrated Scheme for Farmers' Income Security (ISFIS) by integrating various on-going insurance schemes (since earlier five year plan periods). There were five schemes under ISFIS:

1. *National Agricultural Insurance Scheme (NAIS):* Started in 1999–2000 and implemented by Agricultural Insurance Company of India (AIC) in 24 states and two Union Territories.
2. *Pilot Modified NAIS (MNAIS):* Implemented on pilot basis in 50 districts since 2010–11. The unit area has been reduced to panchayats. The private insurance companies have also been permitted to implement.
3. *Pilot Weather Based Crop Insurance Scheme (WBCIS):* It was launched in Kharif 2007 in 20 states. In addition to AIC, private companies were also implementing.
4. *Pilot Coconut Palm Insurance Scheme (CPIS):* AIC was implementing since 2009–10 and being overseen by the Coconut Development Board.
5. *Livestock Insurance Scheme:* Being implemented since 2005–06 in 100 districts and regularized from 2008–09 in additional 100 districts.

However, neither the MNAIS nor the WBCIS could attract farmers in large numbers. There were several reasons for the partial success. Hence, in April 2016, the GoI launched the Pradhan Mantri Fasal Bima Yojana (PMFBY) to plug the gaps in the existing schemes and bring more farmers under crop insurance cover. The responsibility of implementation was given to selected insurance companies under the guidance of Departments of Agriculture of union and state governments. The insured cropped area in 2016–17 was 57 million hectare, which was 30% of GCA, as against 23% in 2015–16. However, in 2017–18, the insured area decreased slightly due to several issues. The number of farmers covered under PMFBY was 5.81 crore in 2016–17, 5.27 crore in 2017–18 and 5.64 crore in 2018–19.

3. Minimization of Price Risk

The risk associated with the variations in the prices may be minimized by the adoption of the following measures:

1. Fixation of minimum and maximum prices of commodities by the government and allowing movements in prices only within the specified range. In India, there is a system of announcing minimum support prices (MSPs) for 23 crop products since mid-1960s. For effective implementation of MSPs several efforts are being made, including public purchases and price deficiency payments. As regards consumer prices, staple cereals (rice, wheat and nutri cereal) are assured to 67% of the eligible households at very low prices (see details in Chapter 7). In addition, exports and imports are allowed or kept under check to influence domestic prices in the interest of farmers and consumers, as the case may be (see details in Chapter 11).

2. Making arrangements for the dissemination of accurate and scientific price information to all sections of society over space and time. This should include information on market demand, acreage under a particular crop, estimates of market supply and of the import and export of commodities.

3. An effective system of advertising may reduce price uncertainty and create a favourable atmosphere for commodity.

4. Operation of speculation and hedging. The price risk associated with the commodities for which the facility of forward trading is available may be transferred to professionals through the operation of hedging. A detailed exposition of speculation, hedging and futures trading follows.

SPECULATION AND HEDGING

Speculation and hedging are important ways of minimizing price risk in business. In the former, risk is taken by the person specializing in the business without much consideration of business trends, while in the second, a calculated risk is taken.

SPECULATION

The fundamental idea underlying speculation is the purchase or sale of a commodity at the present price with the object of sale or purchase at some future date at a favourable price. The speculator is normally concerned with profit-making from price movements. He purchases when prices are low. He is, therefore, not a normal or regular trader. The difference in the prices prevailing at two times constitutes his profit. Speculator may lose in this process. The essentials of a speculator are:

1. He enters the trade at current prices;
2. The transactions of speculators are completed on some future date;
3. The speculators enter the trade with the sole object of making profit from price movements. Sometimes, they indulge in hoarding as well;
4. Except in a few cases, the physical delivery of produce is neither taken nor given. Only the difference in the prices is paid or taken; and
5. Speculators are not regular buyers and sellers in the market. They do not conduct any regular business apart from speculative business.

Based on the legalities involved, speculation is of two types:

1. Speculation Proper

Speculation proper refers to speculation on the part of a person who makes it his profession. Such professional speculators devote their whole time and energy to the collection of information about the future course of price movements. The decisions of the speculator are not hunch decisions. These are intelligent forecasts based on predicted trends. This type of speculation is beneficial for the economy as a whole and is usually accepted by the society.

2. Illegitimate Speculation

This is a gamble in business. The speculators adopt such manipulative practices as creating conditions of artificial scarcity in the market and leading to a rise in prices. The main aim of the speculator is to earn a big profit. This type of speculation is not based on any rationale, though it influences the prices of products. Such speculation is prohibited by the government in the best interest of the economy.

Economic Benefits of Speculation

1. Speculation Dampens Price Fluctuations: Speculators buy at current prices in anticipation of a rise in prices in the future which results in pushing up the current prices. This encourages production and discourages consumption. Other speculators, who sell in the present period in the expectation of a fall in future prices, bring about a fall in the current prices, which encourages consumption and discourages production. The sum total of the effects of these speculative activities results in dampening price fluctuations.
2. The price differentials in different markets are bridged to some extent.
3. Speculation helps in the adjustment of the supply of, and demand for, commodities at normal prices.

Related Terms

1. *Spot/cash transactions:* A transaction in which payment is made on the spot or within a prescribed short period, and delivery of the product is taken on the same day or within a specific period are known as spot or cash transactions. Three things are essential in cash transactions:
   (a) The purchaser has to take the delivery of the produce immediately after sale;
   (b) The seller has to deliver the goods immediately; and
   (c) Payment for the produce has to be made immediately.
2. *Futures Transactions:* This is a transaction in which prices of commodities are settled in cash but the commodities are delivered on some future date as agreed. Generally, in futures transactions, the loss or profit is paid or received on the expiry of the time instead of the physical handing over of the commodity.

   In futures transactions, two groups of persons are involved, i.e. the bulls and the bears. Persons who expect that prices will go up in future are bulls; but those who expect that prices will go down in future are bears. The futures transactions take place as a result of action on the part of these two groups of persons.
3. *Contract:* A contract is a promise to deliver or accept delivery of specific grade of a commodity at a specified time in future.

HEDGING

Meaning

Hedging is a trading technique of transferring the price risk. It protects traders from extreme crash in prices. Hedging has been defined as follows:

- Shepherd: "Hedging is executing opposite sales or purchases in the futures market to offset the purchases or sales of physical products made in the cash market".[14]
- Hoffman: "Hedging is the practice of buying or selling futures to offset an equal and opposite position in the cash market and thus avoid the risk of uncertain changes in prices."

Hedging refers to the purchase or sale of a commodity in a futures market accompanied by a sale or a purchase in the cash market. In this approach, each sale is entered into with an equivalent, purchase of the commodity. It is assumed that prices in the two markets move exactly parallel, and that the losses arising in one market are offset by profit in another market. Hedging is based on two assumptions:

1. The future and cash commodity prices move up and down together, i.e. the basis of price changes remains unchanged.
2. The mechanics of hedging includes the making of simultaneous transactions, but of opposite nature, in the futures and cash markets.

Benefits of Hedging

The benefits of hedging are:

1. It protects the hedger from sustaining loss and enables him to earn his normal trade profit;
2. Hedging enables him to keep the trade margins at a lower level because there is no risk; and
3. Hedging facilitates the financing of inventories of stored commodities to the maximum possible extent.

Illustration of Hedging

The procedure for hedging may be illustrated by the example given in Table 4.32.

Table 4.32: Example of hedging transactions	
Cash transactions	**Future transactions**
December 12 buys 100 bags of wheat @ ₹ 2200/qtl.	Decemer 12 sells 100 bags of wheat of May future @ ₹ 1800/qtl.
December 19 sells 100 bags of wheat @ ₹ 2000/qtl.	December 19 buys 100 bags of wheat for May future @ ₹ 2000/qtl.
Loss in cash ₹ 200 per qtl.	Gain in future ₹ 200 per qtl.

This example shows that the gain or loss in the spot (cash) market is compensated by a loss or gain in the futures market transactions. The assumption is that cash and future prices move up and down together, i.e. by the same amount. However, in practice, the spread between cash and future prices may be more or less—by some small amount or sometimes even more. This is the margin earned by hedgers.

Hedging is employed by many traders to protect themselves against losses due to market price fluctuations by executing cash purchases and sales practically simultaneously with future transactions in the opposite side. It is the performance of mainly the two contracts of an opposite, though corresponding, nature at the same time, one in the spot market where the commodity physically is handled, and the other in the futures market; where the commodity exchange takes place. In short, there are two opposite responsibilities balancing each other.

One other example should make the operation and logic of hedging clear. Suppose, a cotton trader contracts a deal with some overseas firm in February 2019 to supply 1000 quintals of cotton lint at a price of ₹ 8400 per quintal to be shipped in May 2019. In order to protect himself from a possible loss, he buys cotton futures at a ruling futures price of say ₹ 8420 per quintal. Now in the month of May 2019, he discovers that the ruling spot price of cotton is ₹ 8500 per quintal. As he had contracted to ship 1000 quintals at a price of ₹ 8400, he loses ₹ 100 per quintal on this deal. But the future prices also have moved up (say) to ₹ 8520 per quintal, in sympathy with the spot or ready or cash prices. Hence, he sells cotton future at ₹ 8520 per quintal (which he purchased at ₹ 8420 per quintal) and gains ₹ 100 per quintal. This way, his loss on the spot or ready or cash market is compensated by the gain in futures market.

Difference between Speculation and Hedging

The basic differences between speculation and hedging are:

Speculation	Hedging
1. Purchases and sales in the cash as well as in futures markets are made with the objective of making profit.	The purchases and sales in the cash and futures markets are made to protect oneself against excessive price fluctuations.
2. The activities of buying and selling are not necessarily opposed to each other.	The activities of buyers and sellers are always opposed to each other.
3. It is not necessary that the two types of transactions should be of equal quantity.	It is obligatory to buy and sell the goods in equal quantities in the two markets.
4. Under speculation, the speculator purchases goods and sells them when prices rise as per his expectations.	The commodities are not stored by traders. Only the difference in the price is given or taken on the due date.

▌ COMMODITY FUTURES TRADING

MEANING

Futures trading is a device for protection against the price fluctuations which normally arise in the course of the marketing of commodities. Stockists, processors or manufacturers utilize the futures contracts to transfer the price risk faced by them.

Futures trading includes both hedging and speculation. But since hedging is its raison detre, it is also known as hedge-trading. Futures markets are, therefore, known as "hedge" markets.

Conceptually, 'futures trading is an agreement between a buyer and seller obligating the seller to deliver a specified asset of specified quality and quantity to the buyer on a specified date at a specified place and the buyer in turn is obligating to pay to the seller a pre-negotiated price in exchange of the delivery'. Futures trading, thus, performs two important functions viz., price discovery and hedging of price risk in a commodity.

Widely divergent views exist on the effects of futures trading. A few are convinced that commodity futures trading tends to stabilize prices and reduce price variations. Others not only disagree with this view but vigorously allege that, more often than not, futures trading aggravates the price trends and increases both the magnitude and frequency of price variations. A third group denies that futures trading has any influence, either favourable or adverse, on commodity prices.

It is useful to understand the technical difference between forward and futures contracts. Both of these are commitment contracts, which impose a performance obligation in both long and short terms. While forward contracts are customized or 'over-the-counter' products (OTC), futures contracts are standardized or exchange traded products.

In forward contracts, all aspects such as the lot size or number of units of the underlying asset to be delivered per contract; the grade of the underlying asset, for commodities which are characterized by multiple grades; the location for delivery; and the delivery date are decided by bilateral discussions between the two parties to the contract. Consequently the contract may be structured in any way that the two parties deem fit.

In futures contracts, there is an Exchange (MCX, NCDEX etc) which will specify the lot size; one or more deliverable grades and delivery locations; as well as the delivery period. While the exchange, in practice, does give an element of choice to two counter parties, they cannot design a contract beyond the boundaries specified by the Exchange.

NATURE OF COMMODITIES FOR FUTURES TRADING

The commodities permissible under futures trading must satisfy the following conditions:

1. Commodities should be in plentiful supply. If a commodity is in short supply, a few traders may corner the whole supply and charge any price they like to the buyers.
2. The commodity must have a minimum degree of perishability, i.e. it must be storable for futures delivery.
3. The commodity should be homogeneous and capable of being graded so that its future deliveries may be made without problems regarding quality.
4. The commodity should have a large demand from a number of independent consumers so that a single buyer may not be in a position to impose his terms for purchases.
5. The supply of the commodity should not be controlled by a few large firms. It should be available with a large number of suppliers.
6. The price of the commodity should be liable to fluctuations over a wide range, and
7. There should be free flow of the commodity to and from the market without any outside interference/control.

SERVICES RENDERED BY A FORWARD MARKET

The forward market renders the following services to the economic system:

1. It enables the merchants, stockists and processors to protect themselves against the risk of adverse fluctuations in the prices of the commodity. It reduces price fluctuations so that the margin of profit may be small;

2. The highly competitive character of the market smoothens out price fluctuations and ensures an even flow of goods from the seller to the buyer, avoiding gluts in the peak season and shortages in the slack seasons;

3. It brings about an integration of the price structure of commodities at different points of time in the same way as transportation and communications bring about an integration of prices in different parts of the market;

4. It facilitates large purchases and sales of the commodity at short notice in advance of delivery and in the absence of production; and

5. It brings about a co-ordination of the current and future expectations by a continual revaluation of stocks of goods in the light of the changing supply and demand conditions.

DANGERS OF FORWARD MARKET

The dangers arising out of the forward market are:

1. The forward market opens out the way for a large number of persons with insufficient means, inadequate experience and information to enter into commitments which may be beyond their means. In such conditions, market gets demoralized.

2. It enables unscrupulous speculators, with little interest in the actual supply of, and demand for, a particular commodity, to corner the supplies and organize bear raids and bull raids on the market in the hope of making easy money for themselves. This results in violent fluctuations in prices.

ROLE OF COMMODITY FUTURES EXCHANGES

Commodity futures exchanges provide an electronic trading platform, that helps create a transparent price discovery mechanism. The price discovery happens when the seller enters the quantity and the price at which he is willing to sell, and the buyer enters the quantity and the price at which he wishes to buy. When a sell and buy order match, a trade gets generated. On the electronic platform, the buyer and seller remain anonymous, with no scope for manipulation. Trading in commodity futures is attracting investors because of (a) transparency in price mechanism, (b) low margins, (c) risk management, (d) benefits to farmers by way of price clarity, and (e) an organized market place. Also, investment in commodity futures (a) is less volatile compared with equities and bonds, (b) is high liquid asset class, and (c) provides opportunity to gain from price movements in the commodity space. Small traders find low margins and leverage to invest in large number of lots as per their convenience.

For risk management, exchanges have well structured settlement procedures and prudent risk management practices. The clearing houses stand as a legal counter party between the buyer and the seller, thus the clearing house becomes buyer to every seller and seller to every buyer. The farmers can also sell their anticipated produce in advance on these platforms and thus reduce the price/market risk.

FORWARD MARKET COMMISSION

The Forward Market Commission (FMC) was established in 1953 under Section 3 of the Forward Contracts (Regulation) Act, 1952 and has executive as well as advisory functions. The functions initially assigned to the commission were:

1. To advise the government in respect of recognition or withdrawal of recognition of associations conducting forward trading.

2. To keep forward markets under observation.
3. To draw the attention of the government to the various developments that are taking place in the different forward markets with suitable recommendations.
4. To collect and publish information as regards trading conditions in respect of markets falling under its jurisdiction.
5. To submit periodical reports to government on the operation of the Act and on the working of the forward markets, and
6. To inspect accounts of recognized associations generally with a view to improving the organization and working of forward markets.

The FMC was merged with SEBI in October 2015.

PROGRESS OF COMMODITY FUTURES TRADING IN INDIA

Commodity futures trade in India has been in existence since late 18th century but their growth has been stunted mainly due to restrictive policies implemented till the early 1990s. They could not develop fully as an efficient mechanism of risk management and price discovery. With the initiation of the economic liberalization policy and signing of the agreement on agriculture of the World Trade Organization (WTO), interest in these markets has been revived for their risk management and price discovery roles as agricultural prices are expected to be determined mainly by the domestic and international forces.

Futures trading in various groups of commodities was established at the end of 19th century. In cotton, futures trading was started in Bombay. The Europeans took a hand in founding the Bombay Cotton Traders Association in 1875 for the regulation of cotton trade, which was the first step in the evolution of an organized futures market. The futures markets were established for oilseeds at Bombay in 1900, for wheat at Hapur in 1913, for raw jute and jute goods at Kolkata in 1912, and for bullion at Mumbai in 1920. Subsequently, similar markets for these commodities were established at other places also. To provide against unhealthy speculation, forward trading in agricultural commodities was regulated under the Forward Contracts (Regulation) Act, 1952. The Act was enacted with a view to regulating forward contracts prohibiting options in goods and dealing with certain other related matters. The government has regulated or banned forward trading in several commodities in order to check unhealthy speculation. The Act was amended from time to time to plug the loopholes.

The Forward Markets Review Committee, set up by the Government of India under the chairmanship of Prof. M.L. Dantwala, recognised the need for futures trading even in conditions of short supply, and upheld the view that speculations in futures markets should be recognised as a necessary factor for their proper working.

A major boost to revitalization/setting up of futures exchanges came after the submission of Kabra Committee Report in 1994. The Committee recommended futures trading in 17 commodities and establishment of international futures exchanges for trading in castor and pepper. The Kabra Committee also recommended strengthening of existing commodity exchanges and Forward Markets Commission. The World Bank and United Nations Conference on Trade and Development (UNCTAD) in the report for 1996 had also recommended the revival of the commodity futures markets in the country. The commodity futures received bigger and much needed thrust after the launch of a World Bank aided project in September, 1998 as this project was aimed to support the strengthening of commodity futures exchanges and the Forward Markets Commission in India.

Initially futures trading was allowed for nine commodities, i.e. pepper, castorseed, potato, gur, turmeric, hessian, sacking, coffee and cotton in 20 exchanges. The performance of the Indian commodity futures market overtime varied across the commodity exchanges. Among all the commodity future markets, Ahmedabad castorseed futures market was an efficient and unbiased market. Hessian and turmeric futures markets were inefficient. A major reason for the poor performance of Indian futures market could be the lack of inadequate participation of hedgers in these markets.

The Government later lifted the ban on futures trading in 54 commodities. With this, it was made possible to carry out futures trading in these commodities through organized commodity exchanges, subject to the regulations of the Forward Markets Commission. The number of commodities for futures trading went up to 148. The move behind lifting of the ban was to bring about stability in prices of these commodities and prevent wide fluctuations. Among the commodities in which future trading was allowed include wheat, gram, jowar, bajra, maize, urad, tur, moong, masur, peas, barley, rice or paddy, khandsari, linseed, cotton, cotton yarn, cotton cloth, art silk, yarn, raw jute, methi, pepper, beetelnut, chilies, cloves, ginger and nutmeg.

The scenario of futures trading in agricultural commodities at the end of 2006 was as follows:

1. Negative list for futures trading abolished in April 2003.
2. Prohibition and regulation of non-transferable contracts abolished.
3. 22 exchanges recognized for futures trading.
4. Four new national multi-commodity exchanges approved.

In January and February 2007, futures trading in rice, wheat, pigeon pea and black matpe was banned or suspended. Commodities, which were traded in the commodity futures market during 2007 included a variety of agricultural commodities, spices, metals, bullion and crude oil.

The Forward Markets Commission (FMC), under the Department of Consumer Affairs, which was the regulator for commodity futures trading under the provision of Forward Contracts (Regulation) Act, 1952, continued with its proactive approach towards regulation and development of the markets. FMC initiated various steps to attract larger participation of all the stake holders in the supply chain so as to make the price discovery process more efficient. FMC was later made an independent regulatory body by an Ordinance issued on January, 31, 2008. FMC, in association with various state governments, agricultural universities, academic institutions, NGOs, commercial banks and others, had been organizing the awareness/training programmes exclusively targeted at farmers. The awareness programmes were designed to make farmers aware of the benefits of the futures market and to utilize the price signals emanating from the futures exchanges in taking various decisions.

In 2007, Government had set up an expert committee headed by Dr. Abhijit Sen, Member Planning Commission, to study the extent of impact of futures trading on wholesale and retail prices of agricultural commodities. The Committee, in its report submitted to the Government, recommended that ban on futures trading in foodgrains should continue. However, later in September 2010, the Union Cabinet approved the trading of future options in several commodities. The changes in the policy regime for futures trade since 2003 can be summarized as follows:

April 2003　Futures trading in commodities allowed

Jan 2007　　Futures trading in pigeon pea and black matpe banned

Feb 2007　　Trading in rice and wheat futures suspended

May 2008　 Futures in soya oil, rubber, chickpea and potato suspended

Dec 2008　　Futures trading restored in soya oil, rubber, chickpea and potato

May 2009　　FMC banned launch of new sugar futures

Sept 2010　 Cabinet approves futures options in commodities

July 2015　　Futures trading is continuing in chana, castor, mustard, soyaoil, guarseed, wheat, coriander, sugar, cumin and turmeric.

Oct 2015　　FMC was merged with SEBI.

April 2017　SEBI allowed options trading.

There are three national commodity exchanges, apart from several regional exchanges. These are NCDEX (National Commodity Derivative Exchange), MCX (Multi Commodity Exchange) and ICEX (Indian Commodity Exchange). Agricultural commodities are traded mainly in NCDEX and some commodities also in MCX. There are several outlets or locations for trading. NCDEX is the largest farm commodity exchange in India. It has more than 13000 terminals for trade. It has nine service providers and a total of 249 warehouses at 85 locations to facilitate trading. The farmer producer organizations on board for trading are 249 (of 13 states) with a membership of more than five lakh farmers, who have mobile/SMS linkages. However, actual quantity traded by farmers is still not substantial. There are 36 agricultural commodities traded at NCDEX which include cereals, pulses, oilseeds, oilcakes, crude palm oil, guar seed and gum, sugar and gur and spices.

A snap shot of number of contracts and value of agricultural futures trade in agricultural commodities in India is shown in Table 4.33. As can be seen, the trade volume has come down over the years since 2013.

Table 4.33: India: Volume and value of agricultural futures trade

Year	No. of contracts (million)	Value of trade (₹ lakh crore)
2004	7	3
2005	42	8
2006	51	9
2007	40	7
2008	25	8
2009	32	9
2010	41	12
2011	45	15
2012	50	17
2013	39	13
2014	35	11
2015	32	11
2016	24	8
2017	12	4

These are approximate rounded figures. Source: NCDEX and MCX (quoted by Gulati et al., 2017).

NORMS FOR ACCREDITION OF WAREHOUSES

For strengthening the warehousing facilities in the commodity futures markets, certain norms for the Warehouse Service Providers (WSPs) were notified. The notified norms are:

1. A WSP should have a minimum networth of ₹ 5 crore for providing services to the single commodity and ₹ 10 crore for multiple commodities.
2. The promoter of WSP should have been in the business for at least 3 years.
3. WSP should be of repute with credible standing in the market. It should preferably be a corporate body with professional management team.
4. It should furnish security deposit of 2% of the value up to ₹ 250 crore of goods stored, 3% of the value between ₹ 250 and 500 crore and 5% of the value above ₹ 500 crore of goods stored.
5. The ownership (promoter) and management should be separate.
6. WSP should set up a number of panels to oversee the operations. There should also be an audit committee and customer grievance cell.
7. It should have sound infrastructure to store the commodities.
8. It should abide by the rules of the Exchange, besides of the centre and state governments.

FORWARD/FUTURES TRADING: AN ILLUSTRATION

How forward/futures trade can become a hedging instrument can be easily understood with the help of a simple example. Consider that there are two participants A and B. A is a buyer and B is a seller. In the month of May 2019, the two participants, come together for trade in maize futures. The quantity of maize is (say) 10 tonnes. They enter into a contract to buy/sell this quantity of maize in October 2019 for a price of (say) ₹ 1500 per quintal.

For a regulator (Exchange), the usual practice is to collect from both, an amount equal to (say) 10% of the total value (which in this case is ₹ 15000). Both parties, separately deposit ₹ 15000 each to the Exchange/Regulator and thus have entered or signed the futures contract in May 2019.

Note that this contract has been signed by both parties because B expects market price of maize in October to be lower than the contracted price (₹ 1500) whereas A expects that market price will be higher than the contracted price.

In October 2019, there are three possible outcomes.

• First Scenario: The market price of maize is the same as contracted price, i.e. ₹ 1500 per quintal. None of them loses or gains. Both may buy and sell as in the open market at the agreed price. Both get their deposits from the regulator back (of course after adjusting the specified charge/fee).
• Second scenario: In October 2019, the market price is higher at (say) ₹ 1580 per quintal. In this case, owing to the contract, A gains ₹ 8000 (80 × 100) and B loses ₹ 8000. Obviously as the market price is higher by ₹ 80/q, B may not agree to sell maize to A @ ₹ 1500/quintal. The regulator will debit ₹ 8000 (80 × 100) from the account of B and credit to A's account. The B will sell in the market @ 1580, but lose ₹ 80 per quintal in cash from its deposit. The A will buy from the market @ ₹ 1580 but will gain ₹ 80/q by way of forward contract in the form of credit to its account.
• Third scenario: In October 2019, the market price is lower at (say) ₹ 1450 per quintal. In this case, A refuses to buy as per agreement at ₹ 1500 per quintal. The regulator

debits ₹ 5000 (50 × 100) from the account of A and credits to the account of B. In turn, B sells the produce in the market @ ₹ 1450/q and gets ₹ 50/q (cash credit) from the futures trade agreement.

FORWARD/FUTURES TRADING AND FARMERS

The forward/futures markets provide an alternate market place for farmers. They also help farmers address price risks by helping to lock in a price. Farmers associations can hedge on futures platform in two ways.

1. By hedging price on futures platform and deliver goods through exchange approved warehouses.
2. By hedging price on futures platform and later squaring off the position and selling goods on the spot markets.

However, futures markets and exchanges need the following for their efficient functioning and growth:

1. High volumes and enough market depth
2. Frequent trading and effective participation of trading members.
3. Well developed spot market in the vicinity of futures market.
4. Adequate physical delivery in all commodities
5. Well developed grading and harmonized standards

OPTIONS AND OPTIONS TRADING

An option is a financial derivative that represents a contract sold by one party (the option writer) to another party (the option holder). The contract offers the buyer the right, but not the obligation, to buy (call) or sell (put) a security or other financial asset at an agreed-upon price (the strike price) during a certain period of time or on a specific date (exercise date)

As mentioned earlier, in 2017, SEBI allowed options trading in commodity derivatives, which is expected to help farmers or FPOs. The options carry lower trading costs as compared to futures trading. Further as the options carry limited downside risk, it would encourage/help the farmers to hedge their commodity price exposure. However, the design of the derivative product and awareness creation of derivative options, will determine as to how far farmers can benefit.

With commodity options being available soon, farmers will be getting another instrument to hedge their risk which is cheaper and more efficient. Options would give the farmers benefit of price protection in case the price falls below (say) cost of production, as well as the benefit of any rise in price. This will be a better instrument for farmers than futures.

It must be noted that MSP also works as options contract. If prices fall below the MSP, the government has the obligation to buy from the farmers at MSP. But the farmer is under no obligation to sell to the government and is free to sell in the open market.

As regard options, there are two types viz., "call" option and "put" option. The buyer of "put" option has the right but not the obligation to sell or make a delivery at a predetermined price and date. Therefore, "put" option could be used by farmers/ FPOs as it empowers them with a right to sell at predetermined prices without any obligation to sell. The farmers can lock in their price (by put option) by paying a premium. If the government decides to subsidise the premium, it can be of great help to the farmers in options trading and minimizing their price risk.

MARKETING INFRASTRUCTURE

The structure, conduct and performance of the marketing system depend, apart from the regulatory measures, on the status of infrastructural facilities. Infrastructure consists of a combination of public and private assets, which sustain the addition of place, time and form utilities to the products and services. These include, apart from the institutions and organizations, roads, railways, warehouses, cold stores, processing units, research and training institutions, means of communication and transportation and market yards and sub-yards.

IMPORTANCE

Marketing infrastructure serves as the wheels for carrying economic activities. Market infrastructure is important not only for the performance of marketing functions and the expansion of the size of the market but also for transfer of appropriate price signals leading to improved marketing efficiency. Infrastructural facilities lead to reduction in marketing costs which is crucial for increasing the realization of farmers and reducing the costs to the consumers.

The basic rationale of any infrastructure is the sustenance it provides to production activity, income generation and social service supplies. It has also positive effect on income distribution because the low per capita infrastructure limits the access of small and marginal farmers to the market. The relationship between agricultural development and investment in infrastructure has been long recognized. A study conducted by Ahmed[15] while attempting to quantify the impact of investment in rural infrastructure concluded that improved infrastructure is a primary driving force under every condition for commercialization. The benefits of commercialization and specialization to a great extent depend upon infrastructure and both have push and pull relationship. The availability of infrastructure not only affects the choice of technology, reduces transportation costs and produces powerful impetus to production but also affects income distribution in favour of small and marginal farmers by increasing their access to the market. The expansion of different infrastructural facilities has been instrumental in increasing the integration of spatially separated markets. Studies have shown that market infrastructural facilities (transport and communication) have significantly increased horizontal and vertical integration of agricultural produce markets, which improved the process of price discovery and transmitting the price signals from deficit to surplus areas (Acharya[16], 2003).

The role of adequate infrastructure for accelerated growth of the agricultural sector and in turn of the entire economy has assumed great importance in recent years due to several developments viz.,

1. Growth of agricultural production depends almost entirely on the growth of productivity of land and availability of modern technologies. Infrastructure development is necessary for transfer of technologies, supply of modern inputs and facilities for market clearance.
2. The creation of adequate infrastructural facilities in a liberalized and market driven economic environment is necessary particularly in rural area for minimizing economic disparities between rural and urban areas.
3. Creation of infrastructure in rural areas is justified for reducing the migration of people from rural to urban centers; and
4. Development of infrastructural facilities is also necessary to reduce the marketing costs for increasing the realization of farmers.

TYPES OF MARKETING INFRASTRUCTURAL FACILITIES

The infrastructural facilities for marketing can be classified in various ways. One of the ways to classify marketing infrastructural facilities is into physical and institutional.

1. Physical Marketing infrastructure includes roads, railways, transport vehicles, electrification, storage structures, cold stores and cold chains, telecommunication, grading, packing and processing units. Creation of physical infrastructures is a capital-intensive activity with a long payback period (PBP).
2. Institutional Marketing infrastructure can be grouped into the following:
   (a) Public sector organizations – Food Corporation of India, Cotton Corporation of India, Jute Corporation of India, Commodity Boards for tea, coffee, tobacco, spices, rubber, cardamom, coir, silk etc; National Horticulture Board; National Dairy Development Board; Commodity Export Councils; State Trading Corporation; Directorate of Marketing and Inspection; Commission for Agricultural Costs and Prices; Agricultural Produce Market Committees; State Agricultural Marketing Boards and Council of State Agricultural Marketing Boards are some of the marketing institutions which have been created in the country during the last 70 years.
   (b) Cooperative Sector Organizations – Primary, Central and State level marketing societies/unions/Federations; Special Commodities marketing societies viz., for sugarcane, cotton and milk; Processing Societies viz., for cotton, oilseeds, milk, sugarcane, fruits and vegetables; National Agricultural Cooperative Marketing Federation (NAFED); and Tribal Cooperative Marketing Federation (TRIFED) are some of the marketing institutions created in the country in the cooperative sector.

The other way to classify the marketing infrastructure is on the basis of capital requirements.

1. Capital Intensive Marketing Infrastructure—Most of the physical infrastructure viz., roads, storage structures and processing plants require large initial capital investment and are included under capital intensive marketing infrastructure.
2. Capital Extensive Marketing Infrastructure—The institutional infrastructure falls in this category. They require limited initial capital investment but their operational and maintenance cost is quite substantial.

The difference between capital intensive and capital extensive marketing infrastructure is of degree rather than of kind. More details of physical and institutional marketing infrastructure are given in Chapters 4, 7 and 8.

ROLE OF PUBLIC VERSUS PRIVATE SECTOR IN MARKETING INFRASTRUCTURE

Till about the late eighties, marketing infrastructure was created mainly in the public sector. Electricity, railways, roads, telecommunication, postal services and ports were among the marketing infrastructure, which remained reserved for the public sector. However, after 1991, virtually all sectors of infrastructure have been opened for private sector investment. A large number of infrastructure projects have come up under public-private partnership mode. However, the public sector even today continues to play an important role in creation of infrastructure in backward, remote and difficult desert and hilly areas because of their low utilization and poor returns to investment. The corporate sector, under the mandatory requirement of CSR is also contributing to creation of infrastructure, particularly in the social sector.

INTER-STATE VARIATION IN THE STATUS OF MARKETING INFRASTRUCTURE

There is a considerable regional variation in the availability of marketing infrastructure in the country. National Productivity Council (1992) as well as the Centre for Monitoring Indian Economy (CMIE) have constructed indices for the infrastructural facilities in the states of India at different points of time. According to National Productivity Council, in 1992, the state of Punjab ranked first and Gujarat second in terms of general infrastructural facilities. The indices of infrastructural facilities in the states of India as constructed by CMIE at two points of time (1980–81 and 1993–94) and by the XI finance commission for 2004–05 are given in Table 4.34.

Table 4.34: Infrastructural development index in States of India

(All India = 100)

States	1980–81	1993–94	2004–05
1. Andhra Pradesh	98.1	96.1	103.3
2. Assam	77.7	78.9	77.8
3. Bihar	83.5	81.1	81.3
4. Gujarat	123.5	122.4	124.3
5. Haryana	145.5	141.3	137.5
6. Himachal Pradesh	83.5	98.8	NA
7. Jammu & Kashmir	88.7	84.0	NA
8. Karnataka	94.8	96.9	104.9
9. Kerala	158.1	157.1	178.7
10. Madhya Pradesh	62.1	75.3	76.8
11. Maharashtra	120.1	107.0	112.8
12. Orissa	81.5	97.0	81.0
13. Punjab	207.3	191.4	187.5
14. Rajasthan	74.4	83.0	75.9
15. Tamil Nadu	158.6	144.0	149.1
16. Uttar Pradesh	97.7	103.3	101.2
17. West Bengal	110.6	94.2	111.2
INDIA	100	100	100

Source:
1. Centre for Monitoring Indian Economy (CMIE), March 1997.
2. XI Finance Commision Report, quoted in Planning Commision, Government of India, XI Five Year Plan 2007–12, p. 129 for 2004–05.

There appears to be no change in the relative position of the states in terms of availability of infrastructural facilities overtime. Bhatia (1999)[17] also constructed state wise index numbers of rural infrastructure based on 14 indicators for the period 1994–95. Both the studies reveal nearly the same ranking of the states. The ranking of states in 2004–05 is almost the same.

Marketing infrastructure is well developed in the states of Punjab, Kerala, Tamil Nadu, Haryana and Gujarat but continue to be weak in the eastern Uttar Pradesh, Bihar, Rajasthan, Orissa, Assam and Madhya Pradesh. Farmers in the states with poorly developed marketing infrastructural facilities do not get adequate price signals for adoption of new technology which may be a reason for the lower economic status of farmers in these states.

INFRASTRUCTURE INVESTMENT

At one time, the availability of marketing infrastructure in the country was quite inadequate. The requirements of marketing infrastructure are likely to increase with the increased agricultural production and expansion in demand for marketing services. Government of India has estimated the investment requirements for marketing infrastructure, keeping the projected supply of agricultural commodities in the country. It was estimated that an investment of ₹ 2,68,742 crore is required up to 2012 to create adequate marketing infrastructure as per details given in Table 4.35.

Table 4.35: Investment requirement for development of marketing infrastructure and potential sources of investment up to 2012

(₹ crore)

Infrastructure items	Investment requirements	Potential sources		
		State sector	Private sector	Central sector
1. Rural roads	74000	37000	–	37000
2. Developing market yards	6026	6026	–	–
3. Development of F&V markets	970	485	–	485
4. Rural periodical markets development	2146	–	–	2146
5. Storage structures	5400	1350	2700	1350
6. Cold storage	27000	–	20250	6750
7. Cleaning and grading	2000	900	100	1000
8. Reefer vans	600	–	480	120
9. EOAZs	600	50	400	150
10. Value addition/processing	150000	–	112500	37500
Total	268742	45811	136430	86501

Source: Government of India (2001), Report of the Expert Committee on Strengthening and Developing of Agricultural Marketing, Ministry of Agriculture, New Delhi, pp. 1–52.

Realizing the huge investment requirement for creation of adequate marketing infrastructure in the economy the Expert Committee on Strengthening and Developing of Agricultural Marketing suggested that out of the total investment requirement, 50% is to be tapped from private sector. For this, the Government has created enabling and investment-friendly atmosphere for private sector investment. This includes reforms through different legislative measures. Different measures for promotion of private investment in the agricultural marketing sector were identified by the Government of India and several schemes were launched and are on going to create infrastructural facilities for improvement of the agricultural marketing system and increase the efficiency. Under these schemes, subsidy as well as institutional loans is made available to the cooperatives, private entrepreneurs, farmer producer organizations, state governments or their agencies and others (as specified in each scheme). Some of these are as follows:

1. Schemes of Ministry of Food Processing Industries, (GoI)
2. Schemes of Department of Animal Husbandry, Dairying and Fisheries (GoI)
3. Mission for Integrated Development of Horticulture, (MIDH-DAC and FW), GoI
4. Rashtriya Krishi Vikas Yojana (DAC and FW), GoI
5. Integrated Scheme for Agricultural Marketing, ISAM (DAC and FW), GoI.

6. Programs supported by FCI, DFPD (GoI).
7. Programs supported by APEDA, Ministry of Commerce and Industries (GoI).
8. Schemes of NCDC, DAC and FW (GoI).
9. Schemes for Dairying Sector supported by NDDB.

The schemes cover almost all kinds of incentives for creation of desired marketing infrastructure. These include expansion of storage infrastructure, development of infrastructure in more than 20000 grameen (rural primary) markets and basic facilities for E-NAM. However, the emphasis is also on marketing and processing of perishables, including cold chain infrastructure. According to some estimates, the gaps in cold chain infrastructure and investment needs are as shown in Table 4.36.

Table 4.36: India: Cold chain infrastructure: Gaps and investment needs

Component	Required (No.)	Available (No.)	Gap (No.)	Unit cost (₹ in lakh)	Investment required (₹ crore)
Integrated pack house	70080	249	69831 (99.6)	95	66340
Reefer transport	61826	9000	52826 (85.0)	30	15848
Cold stores (bulk)	NA	NA	650	400	2600
Cold stores (hub)	NA	NA	360	350	1260
Ripening units	9132	812	8320 (91)	40	3328
				Total	89376

Figures in parentheses are percentages.
NA = break-up not available.
Source: Report on DFI, MOAFW, Vol. III, p. 109 (assessed by NCDC in 2015).

The upshot of the discussion in this section is that the institutional infrastructure is quite strong. The legal framework is also in place and is being continuously updated or reformed. The capital intensive infrastructure is being expanded. The road length is 59 lakh km and railway rout length is 68400 km. Systems for grading and quality standards are in place. For non-perishables (food grains and oilseeds), the storage capacity is 161 million tonnes and efforts are on going to expand it further. However, the infrastructure in terms of processing and cold value chains continue to be less than what ought to be. Hence, most of the current ongoing initiatives are focusing on these kinds of infrastructure. The Union government is targeting to invest ₹ 103 lakh crore during the next five years (2019–24) on various kinds of infrastructure.

▌REFERENCES

1. Thomsen, F.L., Agricultural Marketing, McGraw-Hill Book Company, New York.
2. Kohls, R.L. and J.N. Uhl., Marketing of Agricultural Products, Macmillan Publishing Co. Inc., New York, 1980, p. 23.
3. Converse, P.D., H.W. Huegy and R.V. Mitchel, Elements of Marketing, Seventh Edition, Englewood Cliffs, N.J., Prentice-Hall Inc., 1965.
4. Sundaresan, R., K.N., Selvaraj, C. Ramaswamy and Anil Kuruvila, Agricultural Marketing Technology and Increasing Market Orientation—Issues and Future Strategies; Indian Journal of Agricultural Marketing, 14(3), September–December, 2000, p. 54.
5. Acharya, S.S., Agricultural Production, Marketing and Price Policy in India, Mittal Publications, New Delhi, 1988, p. 265.

6. Naqvi, S.A.I. and H.P. Singh, "A Review of Grading at Producer's Level", Agricultural Marketing, Vol. XIX, No. 4, January, 1977, p. 2.
7. Singh, Balwinder and Prakash Mehta, "Effective Marketing of Apples", Agricultural Marketing, Vol. XII, No. 2, June, 1969, pp. 8–10.
8. Singh, Balwinder and D.S. Sidhu, "Marketing of Mangoes in Punjab State", Agricultural Marketing, Vol. XIX, No. 3, October, 1976, p. 58.
9. National Commission on Agriculture, Report No. XII, Ministry of Agriculture, Government of India, New Delhi, 1976.
10. Malik, H.S., Sriniwas and A.C. Gangwar, Comparative Efficienry of Processing Units and Marketing Channels of Wheat Flour in Hisar District of Haryana, Ind. Journal of. Agri. Marketing, Vol. IV (2), July–Dec. 1990, p. 211.
11. Srinivas, T. and V.T. Raju, "Economics of Processing of Cashewnut", Bihar Journal of Agricultural Marketing, Vol. III, No. 3, July–Sept. 1995, pp. 284–88.
12. Tousley, R.D. and Others, Principles of Marketing. Macmillan Publishing Company, New York, 1968, p. 496.
13. Hardy, C.O., Risk and Risk-Bearing, University of Chicago Press, 1932, p. 1.
14. Shepherd, G.S., Marketing Farm Products—Economic Analysis, The Iowa State University Press, Ames, Iowa, Fourth Edition, 1965, pp. 153–54.
15. Ahmed, R., Infrastructure, Occasional Paper, quoted in the Annual Report of 1992, International Food Policy Research Institute, Washington.
16. Acharya, S.S., Agricultural Marketing in India, Millennium Study of Indian Farmers, Ministry of Agriculture, Government of India, New Delhi, 2004.
17. Bhatia, M.S., Rural Infrastructure and Growth of Agriculture, Economic and Political Weekly, 34(13), March 27, 1999, pp. A-43–48.
18. Gulati, A., T. Chatterjee and S. Hussain, "Agricultural Commodity Futures: Searching for Potential Winncrs", ICRIER, New Delhi, Dec. 2017.
19. Pal P., "Impediments to the Spread of Crop Insurance in India", Economic and Polirical Weekly, Vol. LII. No. 35, Sept. 2, 2017, p. 16–19.

5

Marketing Agencies,
Institutions and Channels

In this chapter, we discuss marketing agencies, marketing institutions and marketing channels through which farm products move from producers to consumers. A very small proportion of farm produce moves directly from farmers to consumers. Most of the farm products move to consumers through several agencies/institutions and channels. The role played by marketing agencies and institutions in the marketing system is quite indispensable as these perform important marketing functions. They also help in expanding the markets for farm products and add value to the products.

The production of any commodity is complete only when it reaches the hands of those who need it—the consumers. All the commodities cannot be produced in all the areas because of variations in agro-climatic conditions. Hence, there is a need for their movement from producers to consumers.

There are two main routes through which agricultural commodities reach the consumers:

1. *Direct Route:* Sometimes, agricultural commodities directly move from producers to consumers. There is a complete absence of middlemen or intermediaries. But now only a very small proportion of the agricultural commodities move directly from producers to consumers.
2. *Indirect Route:* Agricultural commodities generally move from producers to consumers through intermediaries or middlemen, which are basically market functionaries performing important marketing functions. The number of intermediaries may vary from one to many. In the modern era of specialized production, both the horizontal and vertical distance between the producer and the consumer has increased, resulting in a reduction of direct sales. The role of market middlemen has increased in the recent past because a substantial part of the produce moves through them.

MARKETING AGENCIES

In the marketing of agricultural commodities, the following agencies are involved:

1. FARMERS OR PRODUCERS

Most farmers or producers, perform one or more marketing functions. They sell the surplus either in the village or in the market. Some farmers, especially the large ones, assemble the produce of small farmers, transport it to the nearby market, sell it there and make a profit. This activity helps these farmers to supplement their incomes. Frequent visits to markets and constant touch with market functionaries, bring home to them a fair knowledge of market practices. They have, thus, an access to market information, and are able to perform, to same extent, the functions of market middlemen.

2. MIDDLEMEN

Middlemen are those individuals or business concerns which specialize in performing various marketing functions and rendering such services as are involved in the marketing of goods. They do this at different stages in the marketing process. The middlemen in foodgrain marketing may, therefore, be classified into five groups as follows:

(a) Merchant Middlemen

Merchant middlemen are those individuals who take title to the goods they handle. They buy and sell on their own and gain or lose, depending on the difference in the sale and purchase prices. They may, suffer loss with a fall in the price of the product. Merchant middlemen are of following four types:

Wholesalers

Wholesalers are those merchant middlemen who buy and sell goods in large quantities. They may buy either directly from farmers or from other wholesalers. They sell goods either in the same market or in other markets. They sell to retailers, other wholesalers and processors. They do not sell significant quantities to ultimate consumers. They own godowns for the storage of the produce. Those wholesalers who buy from the farmers are called primary wholesalers and those who buy from primary wholesalers are called secondary wholesalers.

The wholesalers perform the following functions in marketing:

1. They assemble the goods from various localities and areas to meet the demands of buyers;
2. They sort out the goods in different lots according to their quality and prepare them for the market;
3. They equalize the flow of goods by storing them in the peak arrival season and releasing them in the off-season;
4. They regulate the flow of goods by trading with buyers and sellers in various markets;
5. They finance the farmers so that the latter may meet their requirements of production inputs; and
6. They assess the demand of prospective buyers and processors from time to time, and plan the purchase and movement of the goods over space and time. They play a critical role in discovery of prices.

Retailers

Retailers buy goods from wholesalers and sell them to the consumers in small quantities. Sometimes, they also buy directly from farmers. They are producers' personal representatives to consumers. Retailers are the closest to consumers in the marketing channel.

Itinerant Traders and Village Merchants

Itinerant traders are petty merchants who move from village to village, and directly purchase the produce from the cultivators. They transport it to the nearby primary or secondary market and sell it there. Village merchants may have their small establishments in villages. They purchase the produce of those farmers who have either taken finance from them or those who are not able to go to the market. Village merchants also supply essential consumption goods to the farmers. They act as financers of poor farmers. They often visit nearby markets and keep in touch with the prevailing prices. They either sell the collected produce in the nearby market or retain it for sale at a later date in the village itself.

Mashakhores

This is a local term used for big retailers or small wholesalers dealing in fruits and vegetables. Earlier, the mashakhores used to deal only in one or two fruits and vegetables, purchasing from the commission agents or wholesalers in substantial quantities usually three to four quintals of vegetables like potato, onion, carrot, okra, tomato, and spinach. They usually sell to the bulk consumers like hotelwalas, para-miliary units or small retailers/vendors in lots of around 5 kg to 10 kg each. However, in recent years, mashakhores have started retailing to all types of customers without the condition of a minimum quantity. In other words, the mashakhores are now working more like ordinary retailers.

(b) Agent Middlemen

Agent middlemen act as representatives of their clients. They do not take title to the produce and, therefore, do not own it. They merely negotiate the purchase and/or sale. They sell services to their clients (buyers or sellers) and not the goods or commodities. They receive income in the form of commission or brokerage. They serve as buyers or sellers in effective bargaining. Agent middlemen are of two types:

Commission Agents or Arhatias

A commission agent is a person operating in the wholesale market who acts as the representative of either a seller or a buyer. He is usually granted broad powers by those who consign goods or who order the purchase. A commission agent normally takes over the physical handling of the produce, arranges for its sale, collects the price from the buyer, deducts his expenses and commission, and remits the balance to the seller. All these facilities are extended to buyer-firms as well, if asked for.

Commission agents or arhatias in unregulated markets are of two types, Kaccha arhatias and Pacca arhatias: Kaccha arhatias primarily act for the sellers, including farmers. They sometimes provide advance money to farmers and itinerant traders on the condition that the produce will be disposed of through them. Kaccha arhatias charge arhat or commission in addition to the normal rate of interest on the money they advance. A Pacca arhatia acts on behalf of the traders in the consuming market. The processors (rice millers, oil millers and cotton or jute dealers) and big wholesalers in the consuming markets employ Pacca arhatias as their agents for the purchase of a specified quantity of goods within a given price range.

In regulated markets, only one category of commission agent exists under the name of 'A' class trader. The commission agent keeps an establishment—a shop, a godown and a rest house for his clients. He renders all facilities to his clients. He is, therefore, preferred by the farmers to the cooperative marketing society for the purpose of the sale of their produce. Commission agents extend the following facilities to their clients:

1. They advance 40 to 50% of the expected value of the crop as a loan to farmers to enable them to meet their production expenses;
2. They act as bankers of the farmers. They retain the sale proceeds, and pay to the farmers as and when the latter require the money;
3. They offer advice to farmers for purchase of inputs and sale of products;
4. They provide empty bags to enable the farmers to bring their produce to the market;
5. They provide food and accommodation to the farmers and their animals when the latter come to the market for the sale of their produce;
6. They provide storage facility and advance loans against the stored product up to 75% of its value;
7. They arrange, if required by the farmer, for the transportation of the produce from the village to the market; and
8. They help the farmers in times of personal difficulties.

Brokers

Brokers render personal services to their clients in the market; but, unlike the commission agents, they do not have physical control of the product. The main function of a broker is to bring together buyers and sellers on the same platform for negotiations. Their charge is called brokerage. They may claim brokerage from the buyer, the seller or both, depending on the market situation and the service rendered. They render valuable service to the prospective buyers and sellers, for they have complete knowledge of the market—of the quantity available and the prevailing prices.

Brokers usually have no establishment in the market. They simply wander about in the market and render services to clients. There is no risk to them. They do not render any other service except to bring the buyers and sellers on the same platform. In most regulated markets, brokers do not play any role because goods are sold by open auction. Their number in foodgrain marketing trade is decreasing over time. But they still play a valuable role in the marketing of other agricultural commodities, such as gur, sugar, edible oil, cottonseed and chillies.

(c) Speculative Middlemen

Those middlemen who take title to the product with a view to making a profit on it are called speculative middlemen. They are not regular buyers or sellers of produce. They specialize in risk-taking. They buy at low prices when arrivals are substantial and sell in the off-season when prices are high. They do the minimum handling of goods. They make profit from short-run as well as long-run price fluctuations.

(d) Processors

Processors carry on their business either on their own or on custom basis. Some processors employ agents to buy for them in the producing areas, store the produce and process it throughout the year on continuous basis. They also engage in advertising activity to create a demand for their processed products. They add form utility to agricultural commodities.

(e) Facilitative Middlemen

Some middlemen do not buy and sell directly but assist in the marketing process. Marketing can take place even if they are not active. But the efficiency of the system increases when they engage in business. These middlemen receive their income in the form of fees or service charges from those who use their services. The important facilitative middlemen are:

- *Hamals or Labourers:* They physically move the goods in marketplace. They do unloading from and the loading on to bullock carts or trucks. They assist in weighing the bags. They perform cleaning, sieving, and refilling jobs and stitch the bags. Hamals are the hub of the marketing wheel. Without their active co-operation, the marketing system would not function smoothly.
- *Weighmen:* They facilitate the correct weighment of the produce. They use a pan balance when quantity is small. Generally, the scalebeam balance is used. They get payment for their services through the commission agent. The weighbridge and electronic balance system of weighing also exist in big markets.
- *Graders:* These middlemen sort out the product into different grades, based on some defined characteristics, and arrange them for sale. They facilitate the process of price settlement between the buyer and the seller.
- *Transporters:* They assist in the movement of the produce from one market to another. The main transport means are the railways and trucks. Bullock carts or camel carts or tractor-trolleys are also used in villages for the transportation of foodgrains, oilseeds, cotton and vegetables. They add place utility to goods.
- *Communication Agency:* It helps in the communication of the information about the prices prevailing, and quantity available, in the market. Sometimes, the transactions take place on the telephone. The post and telegraph, telephones, newspapers, the radio and informal links are the main communication channels in agricultural marketing. All these agencies, providing these services, play an important role in the marketing system.
- *Advertising Agency:* It enables prospective buyers to know the quality of the product and decide about the purchase of commodities. Newspapers, the radio, television and cinema slides are the main media for advertisements.
- *Auctioners:* They help in exchange function by putting the produce for auction and bidding by the buyers.
- *Storage and Warehouse Operators:* These add 'time utility' to the farm products. Relevance of storage and warehouse operators is considerably more for agricultural commodities as the production is seasonal.
- *Financers:* They perform important function of financing the other middlemen who do not have their own finances. They can be individuals as well as institutional financing agencies.

MARKETING INSTITUTIONS

Marketing institutions are business organizations which have come up to operate the marketing machinery, including formulation and implementations of rules of the game. In addition to individuals, corporate, cooperative and government institutions are operating in the field of agricultural marketing. These apart, associations of market functionaries play an important role.

They perform one or more of the marketing functions. They assume the role of one or more marketing agencies, described earlier in this section. Some important institutions in the field of agricultural marketing are:

(a) Public Sector Institutions

1. Directorate of Marketing and Inspection (DMI)
2. Commission for Agricultural Costs and Prices (CACP)
3. Food Corporation of India (FCI)
4. Cotton Corporation of India (CCI)
5. Jute Corporation of India (JCI)
6. Specialized Commodity Boards
   - Rubber Board
   - Tea Board
   - Coffee Board
   - Spices Board
   - Coconut Development Board
   - Tobacco Board
   - Cardamom Board
   - Coir Board
   - Silk Board
   - National Horticulture Board (NHB)
   - National Dairy Development Board (NDDB)
   - Fisheries Development Board
7. Others
   - Central Warehousing Corporation (CWC)
   - State Warehousing Corporations (SWCs)
   - State Trading Corporation (STC)
   - Agricultural and Processed Food Export Development Authority (APEDA)
   - Export Inspection Council
   - Marine Products Export Development Authority (MPEDA)
   - Silk Export Promotion Council (SEPC)
   - The Cashewnuts Export Promotion Council of India (CEPCI)
   - Agricultural Produce Market Committees (APMCs)
   - State Agricultural Marketing Boards (SAMBs)
   - Council of State Agricultural Marketing Boards (COSAMB)
   - State Directorates of Agricultural Marketing
   - Research Institutions and Agricultural Universities

(b) Cooperative Sector Institutions

1. National Cooperative Development Corporation (NCDC)
2. National Agricultural Cooperative Marketing Federation (NAFED)
3. National Cooperative Tobacco Growers Federation (NTGF)
4. National Consumers Cooperative Federation (NCCF)
5. Tribal Cooperative Marketing Federation (TRIFED)
6. Special Commodity Cooperative Marketing Organizations (Sugarcane, Cotton, Milk)
7. State Cooperative Marketing Federations.
8. Primary Agricultural Cooperative Marketing Societies (PACS)

(c) Others

- Traders Associations (like Grain Traders Associations and Fruit Merchants Associations)

- Processors Associations (like Soybean Processors Association—SOPA and Oilseed Processors Association)
- 'Hamal' or Market Labour Associations.
- Organized Retailers, Retail Chains and E-Tailers
- Farmer Producer Companies or Groups.

The role, functions and other details of some of these institutions have been discussed at different places in relevant chapters.

MARKETING CHANNELS

Marketing channels are routes through which agricultural products move from producers to consumers. The length of the channel varies from commodity to commodity, depending on the quantity to be moved, the form of consumer demand and degree of regional specialization in production.

DEFINITION

A marketing channel may be defined in different ways. According to Moore et. al.,[1] the chain of intermediaries through whom the various foodgrains pass from producers to consumers constitutes their marketing channels. Kohls and Uhl[2] have defined marketing channel as alternative routes of product flows from producers to consumers. At every stage of the marketing channel, one or other form of value (or utility) is added to the product. Hence, these are also called value chains.

FACTORS AFFECTING LENGTH OF MARKETING CHANNELS

Marketing channels for agricultural products vary from product to product, country to country, lot to lot and time to time. For example, the marketing channels for fruits are different from those for foodgrains. Packagers play a crucial role in the marketing of fruits. The level of the development of a society or country determines the final form in which consumers demand the product. For example, consumers in developed countries demand more processed foods in a packed form. Wheat has to be supplied in the form of bread. Most eatables have to be cooked and packed properly before they reach the consumers. Processors play a dominant role in such societies. In developing countries like India, however, most foodgrains are purchased by consumers in the raw form and processing is done at the consumer's level. Again, the lots originating at small farms follow different route or channels from the one originating in large farms. For example, small farms usually sell their produce to village traders; it may or may not enter the main market. But large farms usually sell their produce in the main market, where it goes into the hands of wholesalers. The produce sold immediately after the harvest usualy follows longer channel than the one sold in later months.

With the expansion in transportation and communication network, changes in the structure of demand and the development of markets, marketing channels for farm products in India have undergone a considerable change, both in terms of length and quality.

MARKETING CHANNELS FOR CEREALS

Marketing channels for various cereals in India are more or less similar, except the channel for paddy (or rice) where rice millers come into the picture. For pulse crops,

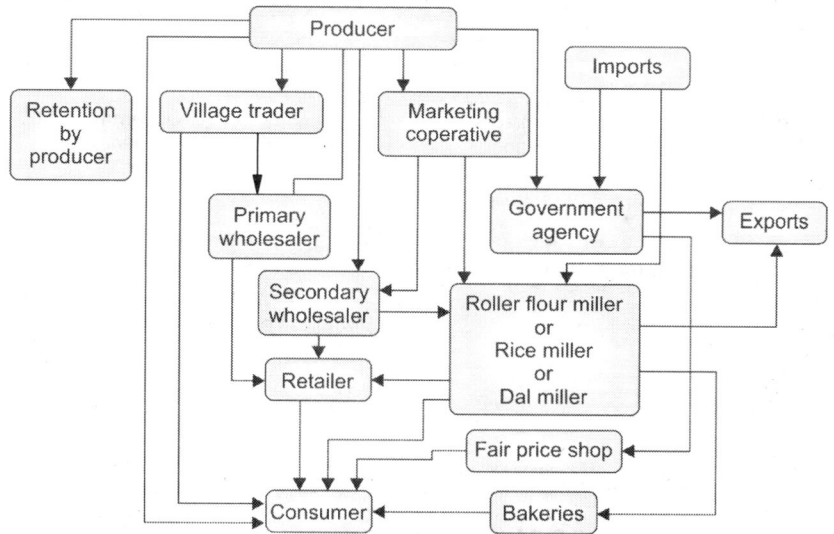

Fig. 5.1: Marketing channels for foodgrains

dal mills appear prominently in the channel. The flowchart in Fig. 5.1 enables us to know the marketing channels for general foodgrains in India.

Some common marketing channels for wheat have been identified as follows:

1. Farmer to consumers;
2. Farmer to retailer or village trader to consumer;
3. Farmer to wholesaler to retailer to consumer;
4. Farmer to village trader to wholesaler to retailer to consumer;
5. Farmer to cooperative marketing society to retailer to consumer;
6. Farmer to a government agency (FCI, etc.) to fair price shop-owner to consumer;
7. Farmer to wholesaler to flour miller to retailer to consumer.

The channels for paddy-rice and pulses are broadly the same, except that the rice millers or dal millers come into the picture before the produce reaches retailers or consumers.

MARKETING CHANNELS FOR OILSEEDS

Marketing channels for oilseeds are different from those for foodgrains, mainly because the extraction of oil from oilseeds is an important marketing function of oilseeds. The flowchart in Fig. 5.2 reveals the movement of oilseeds from producers to consumers in India.

The most common marketing channels for oilseeds in India are:

1. Producer to consumer (who either directly consumes the oilseeds or gets it processed on custom basis);
2. Producer to village trader to processor to oil retailer to consumer;
3. Producer to oilseed wholesaler to processor to oil wholesaler to oil retailer to oil consumer;
4. Producer to village trader to processor to oil consumer;
5. Producer to government agency to processor to oil wholesaler to oil retailer to oil · consumer.

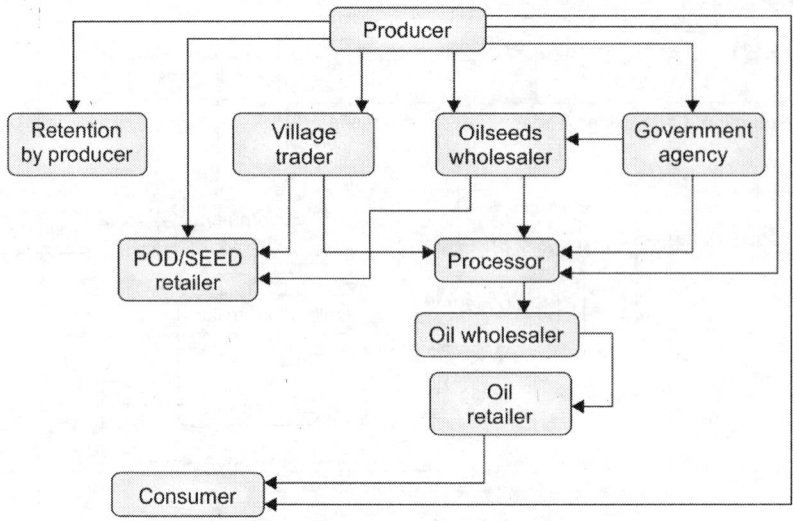

Fig. 5.2: Marketing channels for oilseeds

MARKETING CHANNELS FOR FRUITS AND VEGETABLES

Marketing channels for fruits and vegetables vary from commodity to commodity and from producer to producer. In rural areas and small towns, many producers perform the function of retail sellers. Large producers directly sell their produce to the wholesalers or processing firms. Some of the common marketing channels for vegetables and fruits are:

1. Producer to consumer;
2. Producer to primary wholesalers to retailers or hawkers to consumer;
3. Producer to processors (for conversion into juices, preserves, etc.);
4. Producers to primary wholesalers to processors;
5. Producers to primary wholesalers to secondary wholesalers to retailers or hawkers to consumers;
6. Producers to local assemblers to primary wholesalers to retailers or hawkers to consumers.

An important feature of marketing channels for fruits and vegetables is that these commodities just move to some selected large cities/centres and subsequently are distributed to urban population and other medium size urban market centres. The wholesale markets of these urban centres work as transit points and thus play an important role in the entire marketing channel for fruits and vegetables. Large wholesale markets for fruits and vegetables are concentrated in 10 major cities viz., Delhi, Kolkata, Bangalore, Chennai, Mumbai, Jaipur, Nagpur, Vijayavada, Lucknow and Varanasi. These cities account for 75% of vegetables marketed in major urban areas in India. Further, the transit trade takes place through the cities with more than 20 lakh population which account for 68% of the fruits and vegetables grown in the respective regions. There are 65 urban wholesale markets for fruits and 81 for vegetables. Each market, on an average, serves a population of about 7 lakh.

With the development of cold stores and cool chains, the marketing channels for fruits and vegetables can be understood from Fig. 5.3.

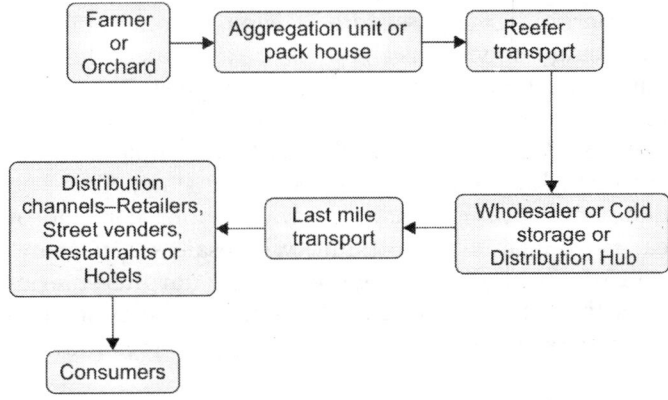

Fig. 5.3: Marketing channels for fruits and vegetables

MARKETING CHANNELS FOR EGGS

The prevalent marketing channels for eggs are:

1. Producer to consumer;
2. Producer to retailer to consumer;
3. Producer to wholesaler to retailer to consumer;
4. Producer to cooperative marketing society to wholesalers to retailers to consumers;
5. Producers to egg powder factory.

Sometimes, the wholesaling and retailing functions are performed by a single firm in the channel.

MARKETING CHANNELS FOR LIVE POULTRY

A study conducted in the Punjab[3] has identified the movement path for live poultry as shown in Fig. 5.4.

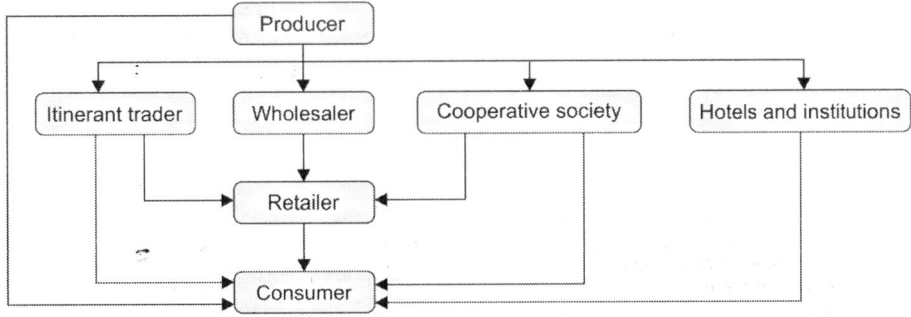

Fig. 5.4: Marketing channels for live poultry

MARKETING CHANNELS FOR PULSES

Most of the studies on the identification of marketing channels for agricultural commodities have concentrated on a concept of marketing channel which defines the flow of the produce from the producer (farmer) to the consumer. But as the commercialization (market orientation) of agriculture is increasing and as the farmers and consumers are located in different states or different countries, the marketing channels that are emerging go across state or even national boundaries. This apart, unless quantities flowing into various channels are estimated, the relative importance of alternative channels cannot be assessed. Such an analysis was done by Acharya[4] for gram grains in Rajasthan. According to this study, there are three points of entry of gram grain in the marketing channel, viz., farmer level, wholesaler level (from outside the state) and processor level (also from outside the state). There are 28 marketing channels, village traders appear in 8 channels, grain wholesalers appear in 18 channels, processors appear in 15 channels, dal (split) wholesalers appear in 5 channels and retailers appear in 15 channels. Assuming the farmers' surplus entering the marketing channel as 100 units, the entry from outside the state at wholesaler and processor level was 4.24% of the farmers surplus. The percentage quantities moving in 28 channels are given in Table 5.1.

MARKETING CHANNELS FOR CORIANDER

Coriander is an important spice crop. The marketing channels for coriander, along with the marketed surplus moving in each of the channels, as revealed from a study in Rajasthan, are shown in Table 5.2. The marketed surplus of coriander was estimated as 95% of total production.

SUPPLY (EXPORT) CHAIN FOR BANANA

A supply chain for banana exports from India is shown in Fig. 5.5. It is a case of tissue culture (TC) banana, where the exporter contracts the farmers for producing TC banana for export.

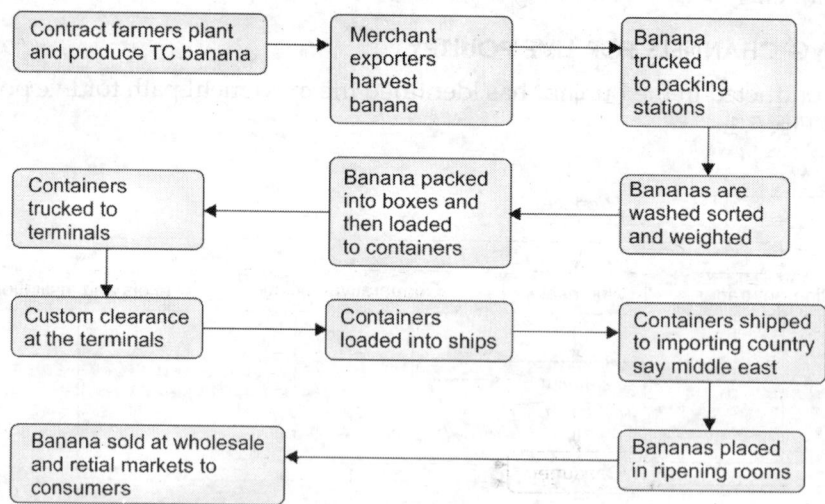

Fig. 5.5: Banana supply (export) chain

MARKETING CHANNELS FOR MILK

Milk is very perishable liquid product. The consumers demand is both for liquid milk as well as for milk products. In India, milk is mainly produced by marginal and small land holders. Therefore, cooperatives and private organized sector are also active in milk marketing. The main milk marketing channels in India are as follows.

Table 5.1: Quantity of marketed surplus of gram moving in various marketing channels

Channel No.	Agencies involved							Quantity (%)
1.	F	–	–	–	–	–	C	0.17
2.	F	–	–	–	–	R	C	0.76
3.	F	V	–	–	–	–	C	0.91
4.	F	V	–	–	–	R	C	0.17
5.	F	V	W	–	–	R	C	0.65
6.	F	V	W	–	–	–	G	0.13
7.	F	V	W	P	–	R	C	0.02
8.	F	V	W	P	S	R	C	0.70
9.	F	V	W	P	–	–	O	1.68
10.	F	V	W	–	–	–	O	3.30
11.	F	V	W	–	–	R	C	8.80
12.	F	–	W	–	–	–	G	1.76
13.	F	–	W	P	–	R	C	0.32
14.	F	–	W	P	S	R	C	9.44
15.	F	–	W	P	–	–	O	22.80
16.	F	–	W	–	–	–	O	44.88
17.	F	–	–	P	–	R	C	0.04
18.	F	–	–	P	S	R	C	1.02
19.	F	–	–	P	–	–	O	2.45
						Sub total		100.00
20.	O	–	–	P	–	–	O	1.45
21.	O	–	–	P	–	–	C	0.02
22.	O	–	–	P	S	R	C	0.60
						Sub total		2.07
23.	O	–	W	–	–	R	C	0.22
24.	O	–	W	–	–	–	G	0.04
25.	O	–	W	–	–	–	O	1.11
26.	O	–	W	P	–	–	O	0.56
27.	O	–	W	P	S	R	C	0.23
28.	O	–	W	P	–	R	C	0.01
						Sub total		2.17
						Grand total		104.24

F = Farmer, C = Consumer, R = Retailer, V = Village trader, W = Wholesaler, G = Government agency, P = Processor, S = Dal wholesaler, O = Outside Rajasthan

Source: Acharya, S.S., Agricultural Production, Marketing and Price Policy in India, Mittal Publication, New Delhi, 1998, pp. 308–12.

1. Farmer–Consumer
2. Farmer–Milk Trader/Vender–Consumer
3. Farmer–Milk Trader–Processor–Consumer
4. Farmer–Milk Cooperative network–Booth/Outlet–Consumer
5. Farmer–Milk Collection or Pooling Point–Bulk Chilling Unit–Milk Tanker–Milk Processing Unit–Milk Booth or Retailer–Consumer

It may be mentioned that farmers sell their surplus milk to different agencies as shown in Table 5.3. Nearly 20% is sold directly to consumers, 38% to local traders, 5% to commission agents, 34% to cooperatives and 2% to processors.

Channel No.	**Agencies involved**	**% of marketed surplus**
	Table 5.2: Marketing channels for coriander	
I	F-CA-W-R-C	80.0
II	F-CAW-R-C	18.6
III	F-R-C	1.3
IV	F-C	0.1
	Total	100.0

F = Farmer, CA = Commission agent, W = Wholesaler, R = Retailer, C = Consumer, CAW = Commission agent-cum-wholesaler.
Source: Ajjan, N., S.S. Burark, D.C. Pant, GL Meena and N. Raveendran, "Commodity Report—Coriander", ICAR-NAIP Project, MPUAT, Udaipur, March 2012.

Table 5.3: Milk marketing outlets by size groups of farmers

(% of total milk sold)

Outlets	Marginal	Small	Medium	Large	All
Consumers	24.7	14.8	15.9	16.4	20.4
Local traders	38.7	38.6	34.8	42.2	38.4
Commission agents	3.2	4.8	4.0	11.8	4.8
Cooperatives	30.3	40.3	42.1	26.0	33.6
Processors	2.1	1.2	2.7	2.2	2.0
Others	1.0	0.3	0.5	1.4	0.8
Total	100.0	100.0	100.0	100.0	100.0

Source: Kumar A. et al. (2018).

SOME INNOVATIVE MARKETING CHANNELS

Innovative marketing channels being encouraged in India are the following:

1. Farmer's Markets
2. Direct Sourcing from Farmers
3. Milk Marketing through Cooperatives
4. Unified Market Platform
5. E-NAM
6. Group Marketing by FPOs/FPCs
7. Contract Farming/Marketing

FARMER'S MARKETS

It has been realized that the marketing channel for farm products which are highly perishable (fruits, vegetables and flowers) should be as short as possible. Perishable farm produce should move quickly from farmers to consumers. If farmers directly sell their produce to the consumers, it will not only save losses but also increase farmer's share in the price paid by the consumers. Therefore, direct marketing by the farmers is being encouraged as an alternative channel. Some examples of these channels are given below:

1. Apni Mandi/Kisan Mandi

An innovative concept of 'Apni Mandi' has been introduced in some states. Apni Mandi is also called 'Kisan Mandi', as it is different from the traditional mandi or market yard, where the produce moves to the buyer through either a commission agent or trader. In Apni Mandi there is a direct contact between the farmer producer and the buyer who is generally the consumer. This system does away with the middlemen. In Apni Mandi, farmers sell their produce directly to the consumers without involvement of the middlemen. The price spread in Apni Mandi is considerable low. These are working satisfactorily in the case of fruits and vegetables. These, 'Apni Mandis' are similar to the Saturday markets of United Kingdom and United States of America.

The main objectives of popularising the concept of Apni Mandi are:

1. better marketing of agricultural produce especially of fruits and vegetables;
2. ensuring direct contact of the producer-farmers and the consumers and thereby enhancing the distributional efficiency of the marketing system;
3. increasing the profitability of agricultural crops for the producers by minimization of marketing costs and the margin of the middlemen;
4. ensuring the availability of fresh fruits and vegetables and other farm produce at reasonable prices to the consumers;
5. removing social inhibitions among the farmers for retail sale of their produce;
6. encouraging additional employment to the producers and thereby enhancing their incomes;
7. promoting national integration by inviting the farmers of other states to sell the produce grown by them directly to the consumers in Apni Mandis of other states; and
8. providing business techniques to the farmers so that in the long-run they may adopt this practice for other crops and enterprises too.

The first Apni Mandi was started in Punjab state by the Punjab Mandi Board at Chandigarh in February, 1987. Punjab Mandi Board took the initiative with a view to providing small farmers around cities a direct access to consumers. Similarly, in Haryana, the first Apni Mandi was started at Karnal in 1988. In Rajasthan also, this scheme has been introduced in several district towns. The initiative is worth emulating.

The market committee of the area where Apni Mandi is located provides space, water, sheds, counters, balances and other facilities to the farmers in Apni Mandis. The Market Committee Staff need to work hard with dedication for the success of Apni Mandis. The State Agricultural Marketing Boards provide financial assistance to the Market Committees for these services rendered by them to the Apni Mandi. This scheme is being implemented with certain resistance from middlemen. Some farmers also have reservations about the success of the scheme as it assumes adequate skills of retailing on the part of farmers. However, farmers as well as consumers would benefit from the Apni Mandi Scheme and its popularity is picking up over time.

2. Hadaspar Vegetable Market

Hadaspar vegetable market is a model market for direct marketing of vegetables in Pune city. This sub-market yard is situated nine kms away from Pune city. This belongs to the Pune Municipal Corporation and the fee for using the space in the market is collected by the municipal corporation from the farmers. This is one of the ideal markets in the country for marketing of vegetables. In this market there are no commission agents/middlemen. The market has modern weighing machines for weighing the produce. Buyers purchase vegetables in lots of 100 kgs or 100 numbers. The produce is weighed in the presence of licensed weighmen of the market committee and sale bill is prepared. The purchasers make payment of the value of produce directly to the farmer. The purchaser is allowed to leave the market place along with the produce after showing the sale bill at the gate of the market. Disputes, if any, arising between buyers and sellers are settled by the supervisor of the market committee after calling the concerned parties. The market committee collects 1% sale proceeds as market fee for the services and facilities provided by the committee to the farmers and buyers.

3. Rythu Bazars

Rythu Bazars have been established in the major cities of Andhra Pradesh state with the prime objective to provide direct link between farmers and consumers in the marketing activity of fruits, vegetables and other essential food items. Both producers and consumers are benefited from Rythu Bazars as producer's share in the consumers rupee is more by 15 to 40% and consumer's get fresh vegetables, fruits and food items at 20 to 35% less prices than the prevailing prices in nearby markets. Further, marketing costs are at the minimum level as middlemen are completely eliminated from the marketing activities in Rythu Bazars. The maintenance expenditure of Rythu Bazars is being met from financial sources of Agricultural Produce Market Committee (APMC) nearer to the Rythu Bazars.

Rythu Bazars started functioning in the Andhra Pradesh State from January 20, 1999. Within five years 96 Rythu bazars were operating in all the 23 districts of the state. There is no government involvement in price fixation. This function is left to farmers who organise themselves into committees and these committee are fixing sale prices daily after taking into consideration the wholesale and retail prices prevailing in the nearby towns. Generally, in the Rythu Bazar, prices are fixed 20% over the wholesale prices and 15 to 20% less than local market prices. Prices fixed are displayed at several places all over the Rythu Bazar for the benefit of the consumers.

The major highlights of Rythu Bazars are:

- District collectors are making the land available for the Rythu Bazars.
- Permanent infrastructure with all support system are being constructed in the Rythu Bazars by the concerned Agricultural Produce Market Committee.
- The vegetable cultivators in the identified villages are provided the photoidentity cards and only these cultivators are permitted to sell vegetables in these bazars.
- State Government arranges special buses on most routes for transport of vegetables.
- Temporary storage facilities are also available.
- Coordination exists between revenue, marketing and horticulture departments for smooth functioning of these markets.
- A distinct and common identity of such markets across the state is being established.
- Other essential commodities like pulses and edible oils are also sold in these markets at reasonable prices.
- Vegetable production programme in the area is also undertaken by the horticulture department of the state to ensure regular supplies of vegetables to the consumers.

Rythu Bazars have generated a great deal of enthusiasm both among farmers and consumers as farmers get better prices for their produce due to curtailment of commission and overhead costs on account of the non-existence of middlemen, and consumers get vegetables at low prices compared to the prices in other markets.

4. Uzhavar Sandies

Uzhavar Sandies (Farmers' Market) were established in selected municipal and panchayat areas of the Tamil Nadu in 1999 by the state government. In these markets, farmers enjoy better marketing infrastructure free of cost and also receive considerably higher prices for the products than what they use to receive from middlemen at village or primary markets of towns. Farmers are additionally benefited in the form of interaction with other farmers and with departmental personnel. Farmers also get good quality seeds and other inputs in the market yard itself. The consumers in these markets are benefited by getting fresh vegetables at relatively lower prices.

The objectives of establishing uzhavar sandies/santhai are:

1. Establishing direct contact between vegetable growers-cum-sellers and final consumers;
2. To assure fair price of the produce to the farmers;
3. To provide vegetables to the consumers at fair prices;
4. To promote honest trading; and
5. To ensure timely availability of fresh fruits and vegetables to the consumers.

5. Shetkari Bazar

On the lines of farmers' markets in other states, the Shetkari Bazars were established in the state of Maharashtra for the marketing of fruits and vegetables. The Shetkari Bazar, by eliminating intermediaries, links producers direct to the consumers, reduces price-spread (marketing margin of intermediaries) and enhances producer's share in consumer's rupee. Thus, these markets increase the farm income, well being of the farmers and bring stability in prices of horticultural and plantation crops[5].

6. Krushak Bazars

On the lines of Rythu Bazars in Andhra Pradesh and Uzhavar Sandies in Tamil Nadu, Government of Orissa has taken up a programme of establishing Krushak Bazars in the state of Orissa in the year 2000–01 with the purpose to empower farmer-producer to compete effectively in the open market to get a remunerative price for his produce and to ensure products at affordable prices to the consumers[6].

The government provides following incentives for opening of the Krushak Bazars in the state:

1. Provides 1 to 2 acres of land at suitable place, free of cost, for establishing the bazar.
2. A cluster/group of villages within the proximity of market area and farmers growing vegetable are identified having the surplus produce for sale.
3. The identified farmers are allowed to use marketing facilities so that there is no intervention of middlemen and farmers get better prices for their produce.
4. Public utility facilities viz., drinking water, electricity, toilet, canteen and rest house are provided to farmers by the Krushak Bazars.
5. Identified farmers are provided inputs like seeds and fertilizer at the reasonable prices in the Krushak Bazars, and
6. Storage facilities in the market area are also provided to the farmers in Krushak Bazars.

7. Mother Dairy Booths

Mother Dairy, basically handling milk in Delhi, decided to try its hand in retail vegetable marketing by direct purchasing vegetables from the farmers, moving them in specially built vehicles, storing them in air conditioned godowns and distributing them to the consumers through its retail outlets in 1989. Mother Dairy management has opened retail outlets in almost all important colonies of Delhi for providing vegetables to the consumers at reasonable prices.

8. Safal Market

Safal market is a move by National Dairy Development Board (NDDB) to introduce a transparent and efficient platform for marketing of fruits and vegetables by connecting farmers to the market through growers association. The NDDB started a unit of SAFAL at Delhi as a fruit and vegetables retail chain. Later, NDDB established an alternative system of wholesale market for fruits and vegetables at Bangalore.

9. Vegetable and Fruit Promotion Council, Keralam (VFPCK)

Vegetable and fruit farmers of Kerala state were under the mercy of inefficient marketing system as there is no APMC Act in the state, until a revolutionary change was brought about by Vegetable and Fruit Promotion Council, Keralam (VFPCK) under the aegis of Kerala Horticulture Development Programme[7].

The VFPCK initiative is a unique alternative marketing model for marketing the vegetables and fruits in Kerala. The markets are organized and run by the farmers. The VFPCK has organized/established three hundred plus Swasraya Karshaka Sangam (SKS), each of which has 15 to 20 SHG (Self Help Groups) of farmers under them.

Farmers bring their produce to the SKS markets, where it is auctioned and produce is given to the highest bidder. The aim of the model is better price realization for the farmer. This VFPCK-SKS-SHG model has helped in mobilizing more than one lakh fruits and vegetable farmers in Kerala state under its fold.

The Council (VFPCK) has organized SHGs as the basic local institutional units for introducing innovative interventions in horticulture. Now, with nearly 5800 SHGs and about 1,11,000 farmers in its fold, the Council enjoys increased social interaction, collective bargaining, quality input sourcing, advanced production technology, micro finance and development of farmer markets. Production centre oriented and farmer participatory group marketing concept followed by the Council has helped in increasing farmer's share in consumer rupee. Under this system, a group of 7 to 15 neighbouring SHGs constitute a Field Centre (FC), where in the SHG farmers bring their produce to a common place for marketing. Traders are coming to the Field Centres for collection and this increases the bargaining power of the farmers. To give additional support for bargaining, the VFPCK's Market Information Centre (MIC) makes available the daily market prices of vegetables collected from different markets in Kerala state and outside.

DIRECT SOURCING FROM FARMERS

Apart from farmers markets, this is another format of direct marketing. In this format, bulk buyers like processors, retail chains or exporters are authorized to directly source the farm produced goods from the farmers/farmers' fields. This helps the farmers by way of avoiding the need for transportation of goods to distant places. The market committees of the area issue such licenses.

MILK MARKETING THROUGH COOPERATIVES

Most innovative marketing system for agricultural commodities ever evolved in India has been for milk, which is the most perishable and a liquid farm product. It was spear-headed by National Dairy Development Board, (NDDB), with head quarters at Anand in Gujarat. The NDDB conceptualized and implemented a series of programmes during the last five decades like Operation Flood, Perspective Plan and National Dairy Development Plan (still ongoing) to increase milk production and arrange its marketing (including processing) through a network of milk cooperatives. About 1.9 lakh dairy cooperatives are engaged in collecting milk and helping 1.7 crore milk producers in the country. Milk producers are provided market access and input services. Milk processing capacity in the cooperative sector is 81.5 million litres per day (more in Chapter 8).

UNIFIED MARKET PLATFORM (UMP)

In Karnataka, UMP was launched in 2013, well before the launch of E-NAM at the national level. Under UMP, 157 mandies in Karnataka are using e-trading, e-permits, e-payments and scientific grading and assaying services. There are reports to show that farmers price realization has increased owing to the facility of UMP in Karnataka (see details in Chapter 7).

ELECTRONIC NATIONAL AGRICULTURAL MARKET (E-NAM)

E-NAM portal was announced in 2015 and was formally launched in April 2016. This portal provides the facility of bidding by a trader for a lot of farm commodity put for sale in any physical market of the country. This has the advantage of providing fair price to the farmers by improving the process of price discovery. So far, 585 mandies have already been linked and other 415 are in the process. For a perfect E-NAM to emerge, there are several facilities that are to be developed in all the markets of the country. E-NAM will be an effective alternative to the APMC yards for buying and selling of the agricultural commodities. As and when, a perfect E-NAM takes shape, it will be a boon to the farmers as well as the entire agricultural marketing system (see more details in Chapter 7).

GROUP MARKETING (FPOs and FPCs)

The idea of group marketing was operationalised in the case of milk by late Verghese Kurien in 1970s. Later in 2001, he extended the idea and conceptualized Farmer Producer Companies and also proposed a legislation to regulate FPCs and help farmers become the only share holders and make these self-governed. In 2002, a Producer Companies Act was enacted. Later in 2013, policy and process guidelines for FPOs were framed by the Union Ministry of Agriculture and Farmers welfare (MOAFW) to enable cooperatives to convert themselves to companies, while ensuring the unique element of cooperative business intact. However, the real push from the government came in the 12th five year plan (2012–17), when SFAC was mandated for promotion of FPCs and FPOs.

While cooperative model did not seem effective in all the areas and commodities (except milk/dairy and sugarcane in Gujarat, Maharashtra and North Karnataka), FPCs have shown promise for small landholders who have low marketed surplus. These help reduce their marketing cost and increase bargaining power through aggregation. FPCs are also attracting private investment in agriculture. These have also triggered an increase in contract farming. FPOs/FPCs are also helping farmers by way of futures/forward trading for reducing price risks. Women farmers, fishers and small farmers of drought prone areas are also using FPOs/FPCs to market their produce at better prices. The number of FPOs/FPCs in the country has gone up to around 5000.

CONTRACT FARMING/CONTRACT MARKETING

MEANING

Contract farming or contract marketing essentially is an arrangement between the farmer-producers and the agri-business firms to produce certain pre-agreed quantity and quality of the produce at a particular price and time. It can only be a pure procurement transaction or can extend to the supply of inputs or even beyond.

Contract farming is emerging as an important mode of procurement of raw materials by agri-business firms in India due to the developments in the field of agricultural marketing, changes in food habits and in agricultural technology in the new economic environment. This is an important initiative for reducing transaction costs by establishing farmer-processor linkages in addition to the already existing methods of linking the farmers to the consumers.

The distinction between 'sales' and 'contract to sell' needs to be understood clearly. In the case of sale, the title or ownership of goods is transferred at once whereas in the 'contract to sell', the goods are transferred at a later date. A contract to sell is not in the true sense of the word a sale, rather it is merely an arrangement to sell. A contract is an agreement but an agreement is not necessarily a contract.

In contract farming, companies or organizations engaged in processing and marketing of agricultural products are entering into contracts with the farmers. They provide inputs to the farmers and buy back the product at a rate specified in advance. Following type of inputs and services are normally provided by the company to the farmers:

- Seeds of the variety they need for processing/marketing
- Guide lines to grow the crops
- Pesticides which do not result in residual toxicity
- Extension services
- Fertilizers/harmones required for the crop
- Other material if not locally available.

The contract may be entered into by parties anytime from the start of the sowing or planting to the harvesting, processing, packaging and marketing stage of the crop. Normally, the contract is entered before the start of the sowing or planting because the buyer can then stipulate the conditions of cultivation, use of the seed variety needed by them, use of pesticides and insecticides, and requirement of onfarm grading, sorting, packaging and processing. The buyer of the product generally keeps the right to monitor the crop at every stage of its growth.

Following documents are obtained/given to selected farmers by the companies:

- Application/Registration form
- Contract farming agreement
- Issue of pass book
- Issue of ID Card

ADVANTAGES OF CONTRACT FARMING

Contract farming/marketing is beneficial both for the producer-farmers as well as to the processing or contracting company in several ways:

To the farmer, contract farming

1. Reduces the risk of price/production
2. Ensures the price as market is assured
3. Increases the quality consciousness
4. Ensures higher production because of better quality seeds and pesticides
5. Reduces marketing costs
6. Provides financial support in cash or kind
7. Ensures efficient/timely technical guidance almost free of cost.

To the processing or contracting company, contract farming—

1. Ensures supply of quality agricultural produce at right time and at lesser cost to the company
2. Canalizes direct private investment in agricultural activities.
3. Ensures that the toxicity level is reduced as per requirement for export or for domestic market.

Government is increasingly looking towards the corporate sector to augment the rural incomes and employment through agro-processing. In this context, policy makers see the contract farming/marketing as an important avenue to ensure greater private sector participation in agriculture.

FLIP SIDE OF CONTRACT FARMING

The important weaknesses of contract farming are:

1. Contract farming is involved mostly in cash crops which may lead to shift in area from food crops which, beyond a limit may endanger food security, biodiversity and agricultural crops cycle of the country.
2. Contract farming may create the danger of imposition of undesirable seeds.
3. The temptation of getting commercial profits from cultivation of a variety of the crop may cause permanent damage to the land.
4. Market making outside the country may cause market breaking inside the country.

However, contract farming is a welcome development. But the contract should be made under high scrutiny possibly because of exploitation of the farmers. The terms of the contract should be spelt out in advance and a consent letter is obtained both from the farmer and the company. The government should ensure a monitoring mechanism and a dispute settlement body to ensure that both parties adhere to the terms of contract.

TYPES OF CONTRACT FARMING

Contract farming can be of the following five broad models (formats) depending on the product, the resources of the sponsorer (contractor) and the intensity of the relationship between the farmers and the sponsorer[8].

1. Centralized Model

This model involves a centralized processor and/or packer buying from a large number of farmers. It is used for tree crops, annual crops, poultry, and dairy products and for commodities requiring a high degree of processing. It is vertically coordinated with quota allocation and tight quality control. The sponsorer's involvement in production varies from minimal input provision to the opposite extreme where the sponsorer takes control of most of the production aspects.

2. Nucleus Estate Model

It is a variation of the centralized model where the sponsorer also manages a large farm or estate. The nucleus estate is usually used to guarantee the raw product to the processing plant but is sometimes used only for research or breeding purposes. The model involves a significant provision of material and management inputs.

3. Multipartite Model

This model involves a variety of organizations frequently including statutory bodies. It can develop from the centralized or nucleus estate model, e.g. through the organization of farmers in to cooperatives or the involvement of a financial or input supply institution.

4. Intermediary Model

This is another version where the sponsorer is involved in sub-contracting the linkages with farmers. In this model, there is a danger of the sponsorer losing the control over production and quality as well as the prices paid to or received by the farmers/ producers.

5. Informal Model

The informal model is characterized by individual entrepreneurs or small companies. It involves informal production contracts usually on a seasonal basis. The model often requires government support services such as research and extension. It involves greater risk of extra-contractual marketing.

METHODS OF PRICE FIXATION IN CONTRACT FARMING

There are several forms of price fixation methods under the contract farming[9]. One is when price fixation is done by the negotiation between the farmers and the company. The farmer enters into contract production with an assured price. The terms and conditions of contract are decided informally and are often liable to be distorted by the company. There may be frequent breaches of contracts from both sides as there is no arbitration mechanism. It is considered as an exploitative model in contract farming. The second method is where some government agency is involved. In this model, the processing/procuring firms make contract with the farmer and the price fixation is done by a government agency in consultation with both the parties. None of the parties is expected to breach the contract. In the event of breach of the contract, there can be a penalty for non-compliance of the terms and conditions of the contract. The basic point is that the contract is enforced by the government agency. The third model is operating in Tamil Nadu where both farmers and company are free to settle the price mutually, but once the price is settled, it is submitted to the Enforcement Officer of the state government. In the case of dispute, the Enforcement Officer does arbitration, which is a method of settling the differences between the parties. Both parties are liable to comply with the award of arbitration.

EXPERIENCE IN CONTRACT FARMING

Contract farming in India owes its origin to the entry of Pepsi Foods in tomato processing in Punjab state. Following the recommendations of the Johl committee to diversity the cropping pattern in Punjab, the state government adopted contract farming as one of the ways for diversification. Several companies followed PepsiCo in Punjab and other parts of the country for contract farming in agricultural and horticultural crops.

Initially the following companies started tie-ups with farmers in India for contract farming for the products specified:

Tomato pulp	Pepsi Company and farmers of Punjab and Rajasthan for tomato growing.
Poultry	Contract farming of broilers between the Coimbatore hatchery with farmers.
Pulpwood	ITC/WIMCO/JK Papers and farmers in Andhra Pradesh, Orissa, Punjab and Uttar Pradesh.
Organic dyes	Marigold farmers and extraction units in Coimbatore.
Dairy processing	Chitale of Pune and small farmers in Maharashtra and Gujarat.
Exotic vegetables	Trikaya Foods/VST and small farmers of Maharashtra and Andhra Pradesh.
Mushrooms	NAFED and Sonepat (Haryana) farmers.
Gherkins	Exporters with farmers of Bangalore.
Edible oils	ITC Agro-Tech and sunflower cultivators in Andhra Pradesh and Karnataka.

Other products where farmer processor linkages (contract farming) are being practised in India are:

- Baby corn cultivation
- Tomatoes for manufacture of sauce and ketchup
- Chillies for manufacture of chilly paste
- Garlic and onion for manufacture of paste, powder and dehydrated products
- Special varieties of banana
- Potato for making chips and wafers
- Barley in making of bears
- Onions and mandarin oranges
- Durum wheat
- Basmati rice
- Pulses and oilseeds
- Medicinal plants
- Mint
- Groundnut
- Flowers

Several corporate houses such as Hindustan Unilever, United Breweries, McCain, Daburs, Thapars, Mahindras, Escorts, and Tatas (Rallis) have adopted contract farming, besides several other small companies to produce their needed raw material. Till the end of 2007, around 4.3 lakh hectares of area was under contract farming. Contract farming is a versatile institutional innovation which can be adopted for different situations, crops and for various purposes and can benefit both the farmers and processing firms, if properly modeled and implemented. Contract farming is fast evolving as a mechanism of alternative marketing in the country. Contract farming has been successful in effecting crop diversification in many states. Punjab, Karnataka, Maharashtra, Madhya Pradesh and Tamil Nadu have proved to be the front ranking states in this regard.

Various studies conducted in different states for various crops have shown positive impact of contract farming, i.e. farmers have been benefited in terms of increased returns. The future of contract farming is quite promising. This is more so due to the increasing consciousness about food safety among the rising middle class population with increasing disposable income and the food safety requirements of export markets of the developed countries.

RISKS IN CONTRACT FARMING

There are several risks in contract farming both for farmers and sponsorers (processing units). These are of the following nature.

Risk for Farmers

1. Farmers face the risk of both market failure and production, particularly while adopting new and unfamiliar crops and their varieties.
2. All contracted production may not be purchased due to inefficient management and marketing problems of the contracting company.
3. Sponsorer may be tempted to manipulate quality standards to reduce purchases due to inefficient management and over production of the crop output.
4. Sponsoring companies may exploit their monopsony position.
5. The staff of the sponsoring organization may adopt corrupt practices especially in allocation of quota.
6. Farmers may become indebted in the long run due to low production, poor technical advice, significant changes in market conditions, company's failure to honor contracts or excessive amount of advances given by the company.

Risks for Sponsorers/Companies

1. Contracted farmers may face land constraints due to lack of security of tenure and thus jeopardizing sustainable long term operations.
2. Social and cultural constraints may affect farmer's ability to produce according to company's specifications.
3. Farmer discontent may arise due to poor management and lack of consultation.
4. Farmer may break the contract and sell the produce in alternative markets as he may be encouraged by rival companies or due to ruling of higher open market prices.
5. Farmers may divert the use of inputs (supplied by the company) to other crops and as a consequence expected production may not be realized.

INCENTIVES FOR PROMOTING CONTRACT FARMING

Contract farming is a means of allocating/distribution of risk between processor and the farmers. It will succeed if both the parties share the risks and rewards.

A few years before, the Ministry of Food Processing Industries of Government of India has launched a scheme entitled 'Grant Under Backward Linkages' to promote contract farming. Under this scheme, a grant of 10% of value of raw material purchased from the contract farmers (subject to a maximum of ₹ 10 lakh per annum) was provided to food processing units up to three years. The Ministry had also prescribed a model agreement form. The criteria for the grant were:

1. The processing unit should provide seed, insecticides, fertilizers and extension services to contract farmers at reasonable charges;
2. The number of contract farmers should be atleast 25;
3. There should be an agreement prior to the period of contract farming for a maximum period of one year;
4. The processing unit should give advance intimation about its contract with farmers to the Ministry as well as State Nodal Authority (one month before the contract comes into operation).
5. The claim for reimbursement should be recommended by the State Nodal Authority.

Contract Farming included in APMR Act (2003)

In 2003, a new model APMR Act was circulated to states that included permission and encouragement of contract farming by APMCs. As a follow up, 21 states included the provision of contract farming in their respective APMR Acts. In addition, the Punjab state formulated its own contract farming Act. It provided for registration of contracts with concerned APMCs.

New Model Contract Farming Act, 2018

In 2018, the Union Ministry of Agriculture circulated a Model "Agricultural Produce and Livestock Contract Farming and Services (Promotion and Facilitation) Act and appealed to the state governments to enact such a law. The Act is promotional and facilitative. It is not regulatory in character. The emphasis is on protecting the interests of the parties entering into contract. The contract farming, according to this Act, will be outside the ambit of APMR Act.

The new Model Act provides for the following:

1. Setting up of an unbiased state level agency called "Contract Farming and Services (Promotion and Facilitation) Board".
2. Establishing a Registering and Agreement Recording Committee or Officer at district, block, and/or taluka level.
3. No rights, title, ownership or possession of land to be transferred or alienated or vested in the contract farming sponsor.
4. A dispute settlement mechanism at the lowest level.
5. It will include agronomic and horticultural crops, livestock, dairy, poultry and fishery.
6. Apart from marketing, it will include services relating to pre-production, production and post-production, that will be provided by the company to the farmers.

The contract farming will help transfer the risk of post-harvest market unpredictability from farmer to the sponsor. While the market risk cover constitutes the fulcrum of contract farming, it also enjoys the latitude of greater partnership between the two parties, whereby the sponsor agrees to professionally manage inputs, technology, extension education, and post-harvest infrastructure and services as per mutually agreed terms. The new model Act is intended to facilitate (rather than regulate) contract farming.

The contract farming is the future of Indian agriculture. The farmers, specially small farmers, can benefit more from contract farming if they organize themselves into groups, clubs, FPOs, cooperatives or any other form.

FOOD AND GROCERY RETAILING

As mentioned earlier, retailing is the last stage in the marketing channel/chain. Generally, food and grocery retail outlets are combined with general merchandise outlets. The history and evolution of retailing in India can be summarized as follows:

Till 1947	Haats, fairs and weekly bazaars
1970s	Organized retail in textiles started
1980s	Start of branded retail show rooms
1991	Doors for organized retail trade opened
1991–95	Beginning of multi-product sales at one place
1995–99	Modern malls started
2000–05	Beginning of high street shopping centers
2006	Government permitted 51% FDI in single brand retail (SBR)
2006–10	Increase in SBR stores and modern malls in metros and large cities.
2010–15	100% FDI in SBR permitted.
2016–17	100% FDI in marketing of food items, produced and manufactured in India, permitted.

The food, including grocery, retail industry in India comprises of both organized and unorganized sectors. The unorganized sector is highly fragmented and dominated by conventional local markets, and small scale single outlet businesses whereas the organized sector has super markets, hyper markets and retail chains and privately owned large retail business centres.

UNORGANIZED FOOD RETAIL SECTOR

The unorganized sector in food retail constitutes tiny grocery stores (56%), small kiosks (17%), general provision stores run by single traders (14%), and others (13%). They have a floor space of around 100 to 500 sq. ft. and maintain stock-keeping units (SKUs) of 500 to 1000. In this sector retailers have low operating costs, minimum overheads, low margins and are in close proximity to customers. Generally, they keep their shops open for long hours, maintain good personal relationship with the neighbourhood customers and offer facilities of home delivery, and credit. The supply chain of the unorganized food retailing is inefficient due to lack of refrigeration, transportation and warehousing facilities. Most of the retailers in this sector buy their supplies from local distributors or wholesalers. There are around 12 million kirana stores that account for 90% of total retail business.

ORGANIZED FOOD RETAIL SECTOR

The organized food retail sector includes supermarkets, hyper markets, discount chains, convenience stores and cooperatives. The hyper markets like Metro and Big Bazar, located in metropolitan cities occupy a floor area of about 25000 to 100,000 sq. ft. and offer increased value to price sensitive consumers with SKUs handling around 20,000 to 40,000 products. Super markets like Food Bazar and World Food have a floor area of 3000 to 6000 sq. ft. in (and around) major cities with a self service approach. They maintain around 6000 SKUs with margins of 14 to 16%. The investment required in the organized food retail is high because of high costs involved in building the infrastructure and establishing a reliable supplier base.

Two of the major players in the supermarket sector in the country are Reliance Industries and Bharti-Walmart tie-up. Other key players include ITC, Food world, Spencer, Godrej, Pantaloon, Subhiksha, and Aditya Birla Group. A list of some new generation retail chains, covering horticultural commodities, is given in Table 5.4. Several new players have entered in recent years.

Table 5.4: Some new generation organized retail chains in India

Private retailer markets	Year entered	Ownership (group)	No. of stores	Location
1. Nilgiris	1971	Local		Major cities in south India
2. Trinethra	1986	Trinethra Group (since 2007, Aditya Birla Group)	170	Major cities in south India
3. Margin Free	1994	Local cooperatives		Major cities in south India
4. Spencers	1996	RPG Group	97	Major cities in south India
5. Subhiksha	1997	Subhiksha Trading Services Pvt. Ltd.	1000	Major cities in AP, TN, Pondicherry and Delhi region
6. Food World	1999	Dairy Farm International	90+	Mumbai and south India
7. Tru Mart	2001	NA	NA	Maharashtra, Gujarat, Bangalore, Chennai, Hyderabad
8. Food Bazar	2002	Futures Group	60+	National level-major cities
9. Metro Cash and Carry	2003	NA	NA	Bangalore, Hyderabad, Mumbai, Kolkata, Chennai
10. My Dollar Store	2004	NA	NA	Nationwide
11. Shoprite	2004	NA	NA	Mumbai
12. Star-India Bazar	2004	NA	NA	Nationwide
13. Reliance Fresh/ Retail	2006	Reliance Industries Ltd.	40	Nationwide
14. Spinach	2006	NA	NA	Nationwide
15. Max Hyper Markets	2007	NA	NA	Nationwide
16. Bharatil	2007	NA	NA	Nationwide
17. Bharti-Walmart	2007	Joint	NA	Nationwide
18. More	2007	Birlas	NA	Nationwide

COMPARISON OF ORGANIZED RETAIL CHAINS AND FARMERS' MARKETS

There are several apprehensions in some circles about the impact of organized retail chains verses the farmers' markets. In this context, a study undertaken in Hyderabad city of Andhra Pradesh by Dastagiri et. al.[10] provides some interesting results. The study compared some important features of Rythu Bazars, and retail chains of Reliance and Subhiksha groups.

It was found that in private retail markets (Reliance and Subhiksha) the produce is procured directly from the farmers at the farm gate. The farmers get higher share in the consumer rupee because as the channel size is reduced, the associated cost of transportation, incidental costs as well as the margin of intermediaries are reduced. The middlemen are completely eliminated which results in greater efficiency of the retail markets in comparison to farmers' markets (Rythu Bazars). It was reported that in public sponsored farmers' markets, intermediaries still exist indirectly and they take away some margins. Another obvious finding is that public markets were operating at no profit-no loss basis whereas companies owning the private markets derive certain amount of profit from retailing, but it does not adversely affect the farmers. The study concluded that the marketing models of private companies, particularly of Reliance Fresh and Subhiksha, were more efficient than that of Rythu Bazars due to their low cost on marketing. Some salient features of three markets are given in Table 5.5.

Table 5.5: Some features of marketing system for fruits and vegetables in public (Rythu) and private markets

Particulars	Rythu Bazars (public sector)	Reliance (private)	Subhiksha (private)
1. Marketing model	Govt. acts as facilitator between farmer and consumer	Direct purchase from the farmer	Direct purchase from the farmer
2. Farmer's share in consumer rupee	Farmer gets good share but less than under private retail chain (Reliance and Subhiksha)	Higher share than Rythu Bazar as transport and marketing costs are saved	Higher share than Rythu Bazar as transport and marketing costs are saved
3. Role of middlemen	Partially and indirectly existing	Complete elimination	Complete elimination
4. Consumer status	Middle income group	High income group	High income group
5. Organization	Government	Private	Private
6. Goal	No profit/loss	Commercial/profit	Commercial/profit
7. Grading	No grading	Grading	Grading
8. Prices	Low	Low/High	Low/High
9. Sales participants	Farmers	Company	Company
10. Marketing channel	Farmer–Consumer	Farmer–Company–Consumer	Farmer–Company–Consumer

Source: Dastagiri et al. (2009).

SUPPLY CHAIN MANAGEMENT (SCM)

Supply chain management concept is being used by organized retail chains as well as exporters. Supply chain management (SCM) refers to the management of the entire set of production, distribution and marketing processes by which a consumer is supplied with a desired product[11]. The supply chain activities start from selection of varieties for demand driven marketing and end with the supply of the commodity to the consumers with higher satisfaction at lower prices.

As mentioned earlier, the organized retail chains for agricultural commodities are increasing at an exponential growth rate in India. After success stories in large cities, retail investors are entering the state capitals and tier II cities (population exceeding

5 lakh). The supply of vegetables by farmers to modern retail outlets brings a new form of organizing production based on consumer demand. All the activities from plough to plate is organized and managed by trained staff under SCM[12].

Supply chain management is important for perishable horticultural products (fruits, vegetables and flowers) both for domestic and international marketing as large quantity of these is lost (quantitatively and qualitatively) at the farm and during transit because of the poor and inefficient system for their handling and marketing.

India being a vast country, with highly diverse agro-climatic and socio-economic conditions, there is a need to have an efficient supply chain system for perishable horticultural products. Although, the development and maintenance, on continuous basis, of an efficient SCM is a complex phenomenon, an efficient SCM is necessary for the growers for increasing production and for the economy to increase value addition and exports.

India is bestowed with a vast network of markets and infrastructural facilities but they lack proper linkages. In the absence of this, producers of horticultural products get very low share in the price paid in the domestic markets and consumers have to pay high prices. In the world market, the credit for Indian products is poor due to loss in quality in transit because of poor SCM of exports.

The SCM system is highly efficient in the markets established by the organized corporate sector companies due to better linkages and efficient management. Their business is expanding at a faster rate. The SCM, therefore, poses a threat to the domestic small and petty traders of horticultural products. However, several companies have involved petty local traders in their SCM models.

The post-harvest loss is estimated at 25 to 30% in fruits and vegetables at the distribution stage of supply chain. In terms of value, it is estimated as over ₹ 30,000 crore per annum. The post harvest losses in horticultural products are due to (a) mishandling of produce, (b) poor and improper storage facilities, (c) poor and improper means of distribution, and (d) seasonal over-production. Hence, the supply chain management of horticultural products needs improvement both in terms of technology infusion and management logistics. One solution for creating an ideal and efficient supply chain system for horticultural products is through integrating the production centres with a network of marketing places supported by modern amenities needed for increasing the efficiency of the supply chain system both in rural and urban areas. In this context, for export marketing, the centre for perishable cargo and walk-in type cold rooms at airports and seaports are vital platforms.

With incentives from the government, several start-ups have emerged in India that are linking farmers to retailers by developing new technologies and helping in improving the farm-to-fork supply chain. They procure from farmers and supply to retailers. Farmers are updated on demand, quality and price of the product through apps and SMS. Retailers (or even bulk buyers) place orders through apps and are offered better prices, scheduled deliveries and quality assurance. The deliveries are traceable on apps. Payments are made directly to farmers' bank accounts. The tech-enabled supply chain is farmer to start-up to retailer/bulk buyer.

▌WHY THE NEED FOR ORGANIZED RETAIL CHAINS

There are several factors for encouraging organized retailing in farm products (food and grocery). (a) The marketed surplus of farm products is increasing continuously. (b) The demand for value added services is expanding due to rising per capita incomes, changes in life styles, increase in urbanization and consequently increased demand for safe and packaged food and their year-round availability. (c) The avoidable wastage of food in the supply chain continues to be very large, estimated to be nearly ₹ 1 lakh crore per annum. (d) The domestic organized trade sector did not invest enough to improve the situation. (e) Direct marketing through farmers markets have been promoted but their share in handling total available marketed surplus is very low and will continue to be so. (f) Indian consumer's preference for fresh produce (F & V) continues to be high at around 75%, which requires cool chains at a large scale. The organized retail sector can, to some extent, help in overcoming these emerging challenges. However, there has been an intense debate in the country when FDI in multi brand retail (MBR) was first permitted by the Government (it was held back, again allowed and the uncertainty continued for many years). Now, 100% FDI in marketing of food items produced and manufactured in India has been allowed. Keeping aside the question of FDI, let us look at the arguments in favour and against organized food and grocery retail (Acharya, 2012, 2013, 2014).

ADVANTAGES OF ORGANIZED RETAIL CHAINS (ARGUMENTS IN FAVOUR)

1. They can and do spend on back-end infrastructure (from farm to store shelves) and thus minimize wastage and losses in the supply chain.
2. They directly procure from the farmers/farmers' groups and thus reduce cost and wastage of farm products.
3. They bring new technology in marketing of farm products.
4. They introduce expertise and modern management practices in marketing.
5. The farmers gain in the form of sales convenience, assured prices, less marketing risks and may even get higher prices of their graded produce.
6. They reduce gross marketing margins, part of which may be transferred to farmers as well as consumers.
7. The consumers gain in terms of more choices and shopping comfort.
8. These maintain better food safety standards.
9. Better tax compliance is expected from them, which implies higher tax revenue for the government at lower cost.
10. Quite often, they serve as source of procurement for small retailers.
11. Investment by organized retailers raises property and land values in nearby locations, and increases demand for skilled youth, thus leading to a step-up in other economic activities around their locations (brick and mortar stores).
12. As organized retailers procure from micro and small enterprises (MSEs), these experience growth of their outputs without own outlets.
13. Organized retailers increase competition in retail and thus compel other retailers to keep their costs/prices low, which partly benefits the consumers.

FEARS FROM ORGANIZED RETAIL CHAINS

There are basically three fears being expressed against the entry of organized retailers. These fears mainly are in the context of FDI in MB retail.

1. The first fear is that the farmers may eventually face a few buyers (oligopsony). In this regard, based on the international experience, it is being pointed out that, owing to several peculiar characteristics of Indian's purchase habits, the share of organized retailers in the farm products (food and grocery retail) is not likely to exceed 20 or 25% of the total. For the remaining 75 to 80% of the marketed surplus,

there shall be many buyers. Further, as the farmers are increasingly organizing themselves into marketing groups, they will have added bargaining power.

2. The second fear is that organized retailers will throw out traditional kirana stores and hawkers out of business. In this context, it must be realized that as in other developing countries, kirana stores and hawkers will still exist, but they will face competition which will compel them to improve their efficiency.

3. The third fear is that organized retailers will maximize their profits. It is true, but they will save avoidable losses and share part of these with the farmers and consumers to attract business for them. They will face stiff competition from other organized retailers and traditional unorganized retail formats.

RESPONSE OF KIRANA STORES

There are two important responses/developments of traditional kirana stores in the aftermath of more than15 years of organized sector entry in food and grocery retail.

1. The kirana stores are resorting to modern retail practices to compete with organized retail chains. These include (a) constant upgrading of categories, (b) systematic staking of products, (c) expansion of floor space, (d) better visual appearance, (e) improved billing system, (f) self service format, (h) home delivery facility, and (i) sale on credit.

2. Larger retail chains have started wooing the local kirana stores for partnership. This is more so for e-commerce retail companies. In fact, if organized retailers make kirana stores as their franchisee, it may benefit both the producers and consumers.

OVERALL ANALYSIS

A dispassionate analysis is, therefore, needed. The arguments on both sides, perhaps, err towards going to extremes. The entry of organized retailers/super markets is neither an unmitigable disaster nor a magic wand to solve all the marketing problems of Indian agriculture. The fears of higher competition are real, but competition is necessary to improve marketing efficiency. Efficiency improvement due to increased competition in airlines, road transport, telecom, automobiles and farm machinery is sufficient evidence. The international food giants, when allowed entry in the country, were painted as anti-people and it was feared that Haldirams, Narula's, Nathu's and Bikanerwalas will vanish. But the reality is that these have also upgraded themselves and doing roaring business both within the country and overseas, and even forced global food giants to Indianise their menus. Not only this, just by the side of international/air-conditioned restaurants, ₹ 150—ordinary lunch, ₹ 70—dhaba lunch and ₹ 25—thela lunch are co-existing because they cater to different sets of consumers. It will not be fair to underestimate the acumen/ingenuity of our corner stores, hawkers and traditional retailers. The kind of service/comfort these provide to varied customers can not be provided by super markets.

As regards FDI in multi-brand retail, the arguments in favour and fears against are more or less the same as for domestic organized retailers/ retail chains. The fears should be seen in the light of conditions imposed on Foreign Direct Investors, before allowing them to start multi-brand retail. These conditions are as follows.

1. A minimum investment of $ 100 million will be necessary.

2. At least half of the investment should be in back-end infra (like cold chain, warehousing, packaging, design improvement, quality control and distribution).

3. Retail stores will be permitted only in cities of 1 million-plus population.
4. At least 30% of merchandise items should be sourced from Indian micro, small or medium enterprises (those enterprises which have plant and machinery investment of less than ₹ 5 crore).
5. Farm products should not be retailed under any brand name.
6. Government will have the first right of procurement of farm products.
7. Self-certification will be complied by the foreign investor.

E-COMMERCE OR E-BUSINESS

A new alternative system of marketing, trade, commerce or business has emerged in recent years. With the advent of computers and internet, the trading networks have become the common words in business. These are revolutionizing the business world by opening a new horizon of trade, commerce or business on a global scale. There is no universally accepted definition of e-commerce. The World Trade Organization (WTO) has defined e-business as production, distribution, marketing, sales and delivery of goods and services by electronic means. E-commerce starts with setting up of virtual shop in virtual location exhibiting the products and services one wants to sell.

The commercial transaction is divided in three stages: (a) Advertising and searching stage, (b) Ordering and payment stage, and (c) Delivery stage. Any one of or all of these stages of business may be carried out electronically.

Benefits of E-Business

The benefits of e-business are:

1. More affordable than traditional one.
2. More business partners can be reached.
3. Caters to more geographically dispersed customer base.
4. Procurement cost is lower.
5. Purchase costs, sales and marketing costs are lower.
6. Reduction in inventories.
7. Better customer service.
8. Lower cycle times.

Benefits to Consumers

1. Increased choice of vendors and products.
2. Convenience of shopping at home or office.
3. Greater information access on demand.
4. More competitive prices and increased price comparison capabilities.
5. Greater customization in the delivery of services.

Benefits to Business Community

1. Exchange of a larger quantity of information.
2. Global visibility.
3. Rapid planning cycles and strategies.
4. Increased market share.
5. Avoids communication gaps.
6. Access to new consumer groups.
7. Zero inventories.
8. Better return on investment.

9. Reach to more persons at a lower cost.
10. Relationship building is easy.
11. Continuous process improvement possible.
12. Better decision making.
13. Human and intelligent service.

Broad Activities under E-Business

1. Business to Business (B2B)
2. Business to Consumer (B2C)
3. Business to Government (B2G)
4. Intra-organizational business activities.

Of these, business to business segment is currently the biggest one but B2C and B2G are also catching up. The growth of e-commerce is mostly due to the rapid popularity of the World Wide Web system, and rapid growth in internet users even in remote rural areas.

Challenges of E-Business

E-Business is the catalyst for fast growing economies but it faces a number of challenges. These are as follows:

1. **Limited Consumer Exposure and Buying:** Web users are doing more surfing than buying. Hardly 20% of surfers actually use web for shopping or to obtain commercial services. The main customers in E-business are businessmen rather than consumers.
2. **Skewed User Demographics:** The general population is not able to operate the internet. They are used mostly by technically oriented and upscale persons.
3. **Chaos and Clutter:** As the internet offers millions of websites and a staggering volume of information, navigating the web can be frustrating. Many sites go unnoticed and every visited site needs to be upgraded depending upon the growing population needs and ever increasing demand for internet.
4. **Security:** Consumers worry that unscrupulous interlopers may intercept their credit card numbers. The internet is becoming more aware of security. At present there is a continuous race between the pace of new security measures and new code breaking measures.
5. **Ethical Concerns:** Consumers worry about privacy and fear that the companies might make unauthorized use of their information. There are also chances of unfairness, deception and frauds.

However, the day is not far off when general population will be more knowledgeable about internet and the security measures will be tightened. It is likely that very soon e-marketing would become a significant feature of business world.

E-business or E-commerce is catching up in India. (a) The e-business communication has been going on for many decades. (b) In the last one and half decades, Electronic Spot Exchanges have extended their activities. (c) In many of the regulated markets, e-auction has started in agricultural commodities. (d) State-wide online trading (e-auction included) has commenced in Karnataka. (e) Government of India has launched a common National Agricultural Market, which intends to e-link all agricultural produce markets on a national electronic platform for trading. (f) These apart, E-tailing ((electronic retailing) has also commenced, which includes grocery and food retailing and fast-food retailing. According to some estimates (CII), the business size of E-tail in India is projected to go up to $ 32 billion by 2020, registering a growth rate of 55% per annum. As regards the share of food and grocery in total E-tail, it is about 1% which is projected to grow to 4% by 2020.

E-RETAILERS (FOOD AND GROCERY)

There are two kinds of retail businesses. One is brick and mortar (fixed shops) and two is e-tailers or e-retailers. The present share of e-retailers in total food and grocery retail is around 2.5%. It is projected to grow to 4% by 2025. As already mentioned, these are increasingly tying up with local kirana stores for quick delivery. For illustration, some of the e-retailers operating in India are Bigbasket (since 2011), Zopnow (since 2011), Local Banya (since 2012), and Grofers (since 2013). Their delivery time varies from 90 minutes to 3 hours and in few cases 'the same day'.

Quite a few new e-shops or startups have recently emerged in cities like Bengaluru, Hyderabad, Gurugram, Delhi and Chennai that are changing the way wet markets operate. These are offering fresh fish, poultry and meat at the door steps of consumers. Consumers are also using their services. The startups have obtained quality certificates from FSSAI and these assure quality fresh products and quicker delivery to the consumers at their door steps.

Recent decision of the government to permit 100% FDI in food retail will help in compressing the food value chain, reducing the huge wastage and, in turn, benefitting both the farmers and consumers. This is so owing to the conditions prescribed for foreign retailers like (a) 15% of the investment should be in back-end infrastructure, and (b) food products marketed should be manufactured in India, entirely from local inputs. It may also help the local small retailers like kirana/corner shops as the foreign retailers are likely to tie-up their business with these. A new concept of Phygital in the retail sector may emerge. Phygital is a situation of harmonization of physical (brick and mortar) retailers with digital (online) retailers.

FAST FOOD CHAINS AND RESTAURANTS

One other important aspect of food retail (apart from grocery retailing) is the retail of ready-to-eat foods. The organized segment comprises of fast food chains and restaurants. Fast food is the term used to describe restaurants and outlets dishing out food that is quick to prepare and serve. Fast food chains or restaurants are also termed as Quick Service Restaurant (QSR) chains. These are increasing in popularity as the pace of life hastens and leisure hours become shorter. India's fast food market, which has seen a robust run in metro and tier-I cities, is now taking off in tier-II and tier-III cities and big towns. The market size is growing at the rate of around 15% per annum.

There are several Indian chains successfully operating since many decades. Out of the World's top 10 fast food chains, many have their presence in India. These are McDonald's, KFC, Subway, Pizza Hut, Starbucks, Domino's, Dunkin Donuts and Papa Jonn's. Each one of these have many outlets in India.

In addition to brick and mortar outlets (restaurants), there are now a few online food ordering companies that have partnerships with thousands of restaurants.

Online RTE food delivery market (including tiffin startups) is currently around ₹ 28000 crore. The main operators are Zomato, Swiggy, Food panda and Uber Eats. This business has entered in tier II cities also. Zomato is now the largest food aggregator in the world and works in 550 cities in 24 countries, including India. Every month, nearly

45 million Indian employees are reportedly collecting food from restaurants and supplying to individuals. The company also serves meals to lakh of poor persons, as a part of its CSR activities. Swiggy is the other big player. In India, there are around 60 million online food orders per month with an average value of ₹ 250. Their businesses are growing exponentially over the years. These provide employment to a large number of delivery agents and delivery persons.

REFERENCES

1. Moore, J.R., S.S. Johl and A.M. Khusro, Indian Foodgrain Marketing, Prentice-Hall of India Private Limited, New Delhi, 1973, p. 35.
2. Kohls, R.L. and J.N. Uhl., Marketing of Agricultural Products, Macmillan Publishing Co., Inc., New York, 1980, p. 595.
3. Randhawa, B.S., A.S. Kahlon and J.S. Sahota, "Costs and Margins in Marketing of Live poultry in Gurdaspur District of punjab", Agricultural Marketing, Vol. X, No. 1. April, 1967, p. 11.
4. Acharya, S.S., Agricultural Production Marketing and Price Policy in India, Mittal Publications, New Delhi, 1988.
5. Pawar, P.P., K.R. Waykar, B.K. Mali and S.S. Bhosale, Need of Shetkari Bazar for Marketing of Fruits and Vegetables in Maharashtra, Ind. Journal of Agri. Marketing, 15(3), Sept.–Dec., 2002, pp. 53–54.
6. Atibudhi, H.N. and Binodini Sethi, Krushak Bazar: An Ideal Bazar: A Case Study in Orissa, Indian J. of Agri. Marketing, 15(3), Sept.–Dec. 2001, pp. 35–40.
7. Nair, Gireesh S. and V. Sreejith, "A Novel Marketing System for Fruits and Vegetables in Kerala: A Case of VFPCK", Indian Journal of Agricultural Marketing, Vol. 21 (3), Sept–Dec 2007, p. 72–73.
8. Bhardwaj S.P. and A.K. Vasishtha, "Policy Support for Contract Farming in Lac Cultivation—An Alternative Marketing Model", Indian Journal of Agricultural Marketing, Vol. 21 (3), Sept–Dec 2008, p. 47–48.
9. Kumar Shiv and Puran Chand, "Prevailing Practices and Dimensions of Contract Wheat Seed Farming in Haryana State", Agricultural Economics Research Review, Vol. 17 (2), July–December, 2005, p. 152–3.
10. Dastagiri M.B., B. Ganesh Kumar and S. Diana, "Innovative Models in Horticultural Marketing in India", Indian Journal of Agricultural Marketing, Vol. 23 (3), 2009, p. 90–91.
11. Malaisamy, A, M. Chandra Sekaran and R. Parima Larangan, "An Economic Analysis of Supply Chain Management and Marketing Efficiency of Mango in Tamil Nadu, India," Indian Journal of Agricultural Marketing, Vol. 21 (3), September–December 2007, p. 125.
12. Lokanathan, K, "Supply Chain Management Analysis of Tomato: Farm to Modern Retail Outlets", Indian Journal of Agricultural Marketing, Vol. 21 (3), September–December, 2007, p. 65–71.
13. Acharya S.S., "Organized Retailing in Agri-Business in India: Some Facts and Emerging Issues", Keynote Address, International Seminar on Organized Retailing vis-a-vis Farm Economy in India, CESS, ISAM and IPE, Hyderabad, Sept. 21–22, 2012.
14. Acharya S.S., "FDI in Multi-Brand Retail: Will it Benefit Smallholder Farmers", Commemorative Volume, MIP Program, ICRISAT, February 2013.
15. Acharya S.S., "Reforms in Retailing of Farm Products: Why and How", Text of special Lecture, 28th conference on Agricultural Marketing, ISAM and AERC, SPU, Vallabh Vidya Nagar (Gujarat), December 5, 2014.
16. Kumar Anjani, Mishra A.K., Siraj P. and Jha G.K., "Farmers' Choice of Milk Marketing Channels in India", Economic and Political Weekly, December 29, 2018, p. 58–67.

6

Marketing of Farm Inputs

Output marketing is an aspect of agricultural marketing which has been dealt with so far. Another equally important aspect is the marketing of farm inputs. A timely and adequate supply at fair prices of farm inputs—seeds, chemical fertilizers, plant protection chemicals, farm equipment and machinery, labour, electricity, diesel oil and credit—are of great importance in the production of farm output. The importance of purchased farm inputs has significantly increased in the recent past with the technological breakthrough in Indian agriculture. The value of purchased inputs, at current prices, has gone up from ₹ 10511 crore in 1950–51 to ₹ 3.24 lakh crore in 2016–17.

The timely supply of modern farm inputs to the farmers of all categories at reasonable prices depends on the existence of an efficient marketing system for them. The importance of an efficient marketing system for farm inputs may be judged by the following:

1. Farm products are produced in the countryside. The effect of change in technology can, therefore, be realised only if the farm inputs reach the farmers in time at the least cost.
2. The use of modern inputs by farmers largely depends upon the spread of information about them. The marketing system has to perform this function. Marketing agencies have to adopt persuasive methods to induce even the most tradition-bound farmer to use modern farm inputs. Dynamic and efficient channels for marketing of farm inputs are, therefore, essential for the popularization of knowledge about modern inputs among the farmers.
3. An efficient marketing system for farm inputs is essential for the development of the inputs manufacturing and supplying industries in the country. These industries provide employment opportunities in manufacturing, selling and handling of farm inputs.

THEORY OF DETERMINATION OF INPUT PRICES

From the point of view of economic theory, the demand for an input is a derived demand, i.e. derived from the demand for the product in whose production the input is used. But more than one inputs are usually required to produce a given agricultural product. To understand this, let the production of a commodity (Q) be expressed as a function of three inputs, x_1, x_2, x_3, as follows:

$$Q = f(x_1, x_2, x_3)$$

The first-order conditions for profit maximization are:

$$\frac{\partial Q}{\partial x_1} = \frac{P_1}{P_q}$$

$$\frac{\partial Q}{\partial x_2} = \frac{P_2}{P_q}$$

$$\frac{\partial Q}{\partial x_3} = \frac{P_3}{P_q}$$

where, P_1, P_2, P_3 and P_q are prices of x_1, x_2, x_3, and Q, respectively.

Simultaneous solution of these first-order conditions would give normative demand for all the inputs and this level of demand would depend on all the prices. In this sense, the demand for an input can be expressed as function of its own price; prices of other complementary and substitute inputs and price of the product. For a given level of prices of other inputs and product prices, the demand function for an input as a function of its own price can be obtained.

Another situation most commonly encountered on individual farms is that some inputs such as fertiliser and insecticides remain variable while others such as land and water are fixed. In such situations, the production (Q) can be expressed as a function of (say) three variable inputs (x_1, x_2 and x_3) and (say) three fixed inputs (Z_1, Z_2 and Z_3) as follows:

$$Q = f(x_1, x_2, x_3 \,/\, Z_1, Z_2, Z_3)$$

In such situations, while the three first-order conditions remain the same as stated above, the simultaneous solution of these first-order conditions would give normative demand for three variable inputs which depends on the prices of variable inputs and that of output and on the levels of the three fixed resources.

Since it is a derived demand, the quantity of input demanded by an individual producer will depend on its marginal product. For a profit maximizing firm, the downward sloping portion of the marginal value product curve, is the relevant demand curve for the input. Similarly, for a profit maximizing input manufacturing firm, the rising portion of the marginal cost curve is the supply curve for the input. The intersection of these two curves indicates the market clearing price and quantity of inputs.

But the situation of supply curve for inputs like labour is different. The supply curve for labour is backward bending, i.e. at some level of wages, the individual may decrease his supply. The exact point, of course depends on the relative weights attached by the individual to money income and leisure. For other inputs like irrigation water from public-owned canals, the supply curve is a straight line, i.e. required quantity at administratively fixed price.

The supply, demand, distribution, marketing pattern and pricing of important farm inputs have been discussed in the following paragraphs.

CHEMICAL FERTILIZERS

Fertilizer is decidedly the most important among all the inputs purchased by the farmer for use in present-day agriculture with a view to accelerating agricultural production. It now accounts for around 20% of the value of total purchased inputs by the farmers. It has been estimated that 53% of the incremental foodgrain production in India during the seventies was due to fertilizer use, and its contribution is expected to have increased since then. The demand for chemical fertilizers has increased with the evolution of new hybrid and dwarf variety seeds, which are more responsive to chemical fertilizers.

The use of fertilizers increases land productivity, due to the increase in yields and easing the nutrient constraints on multiple cropping and land development programmes. Fertilizers relax the land constraint. Since the yield increase is proportionately more than the corresponding incremental labour applied, fertilizer use increases labour productivity. The production, distribution and consumption of fertilizers create additional employment opportunities—a fact which is extremely important in labour-surplus countries.

SUPPLY

The sources of supply of fertilizers in India are indigenous production and imports. The chemical fertilizers are produced in all the three sectors—public, cooperative and private. Although even before independence, Tata Iron and Steel Company had a plant for manufacturing ammonium sulphate, but it was only after Independence that fertilizer production received serious attention. The licensed manufacturing capacity in 1950 was only 16700 tonnes of N and 20500 tonnes of P.

India does not manufacture potassic (K) fertilizers and all its requirements are met through imports. Even in the case of phosphatic fertilizers, the import dependence is substantial. Up to the middle of 1970s, almost half of the country's requirements were met by imports. The domestic production of N and P fertilizers got real boost after the first world oil shock of 1973–75.

Due to the oil shock of mid-seventies, the fertilizer prices in the world market spurted and within a few weeks, the prices of imported fertilizers had almost doubled. Obviously, this led to a sharp increase in the import bill creating pressure on the already scarce foreign exchange reserves. The increase in the landed cost of imported fertilizers also had serious implications for the sale price of fertilizers, which was a critical complementary input, with high yielding and fertilizer-responsive seeds, for enhancing the production of foodgrains in the country. It was, therefore, recognized that a high degree of self-reliance in fertilizers is necessary for achieving the objective of food security for the nation. As a follow-up, a strategic decision was taken to increase the domestic production of fertilizers. Efforts were directed at expanding the capacity of existing plants/units as well as at establishment of new fertilizer manufacturing plants. As a result, the fertilizer manufacturing capacity increased to 8.2 million tonnes of N and 2.7 million tonnes of P by 1991 and further to 13 million tonnes of N and 6.2 million tonnes of P by 2009.

The trend in production, imports and consumption of N, P and K fertilizers at selected points of time is shown in Table 6.1. The domestic production of N increased from only 0.1 million tonnes in 1960–61 to 7 million tonnes in 1990–91 and further to 13.4 million tonnes in 2017–18. In the case of P fertilizers, it went up from 0.1 million tonnes in 1960–61 to 2.1 million tonnes in 1990–91 and further to 4.7 million tonnes in

Table 6.1: Production, imports and consumption of fertilizers in India

(million tonnes of nutrients)

Particulars	1960–61	1980–81	1990–91	2000–01	2010–11	2017–18
Production						
N	0.1	2.2	7.0	11.0	12.2	13.4
P	0.1	0.8	2.1	3.7	4.2	4.7
K	–	–	–	–	–	–
Total	0.2	3.0	9.1	14.7	16.4	18.1
Imports						
N	0.4	1.5	0.4	0.2	4.5	3.6
P	–	0.5	1.3	0.4	3.8	2.1
K	*	0.8	1.3	1.5	4.1	2.8
Total	0.4	2.8	3.0	2.1	12.4	8.5
Consumption						
N	0.2	3.7	8.0	10.9	16.6	17.0
P	0.1	1.2	3.2	4.2	8.1	6.8
K	*	0.6	1.3	1.6	3.5	2.8
Total	0.3	5.5	12.5	16.7	28.2	26.6

* Less than 0.05 million

Source: Economic Survey 2009–10, Ministry of Finance, Government of India, and Agricultural Statistics at a Glance, 2014 and 2018.

2017–18. Out of the total production of fertilizers, 51% is in private sector, 27% in the cooperative sector and 22% is in the public sector. The imports of all the three nutrients (NPK) also went up during this period. However, the imports have come down after 2010–11.

DEMAND (CONSUMPTION)

The use of chemical fertilizers started in India at the beginning of last century in 1906. Initially the major consumers of chemical fertilizers were the growers of plantation crops and gradually their consumption spread to other crops. Consumption of chemical fertilizers was low up to 1950s. The growth in consumption of chemical fertilizers commenced with the advent of planned economic growth after independence, and specially since the mid-1960s, when the new agricultural development strategy for food security (green revolution) was launched in the country.

The demand for fertilizers is a derived demand. It depends on the prices of fertilizers, of other complementary inputs and the levels of availability of complementary inputs such as irrigation and output prices. Most of the fertilizers are consumed in assured irrigated areas of the country.

The fertilizer use by farmers in India has considerably increased during the last six decade. As shown in Table 6.1, fertilizer corisumption was barely 0.3 million tonnes in 1960–61 which increased to 26.6 million tonnes of nutrients in 2017–18. During 2017–18, out of the total fertilizer use, N accounted for 64%, P accounted for 26% and K accounted for 10% with N: P: K ratio at 6.4: 2.6: 1.

For assessing the emerging scenario of demand, apart from the aggregate consumption of fertilizers, it is necessary to look at the trends in fertilizer use per

Table 6.2: All India consumption of fertilizers (plant nutrients) per hectare of gross cropped area

(kilogrammes)

Year	N	P	K	Total
1951–52	0.4	0.1	0.1	0.6
1960–61	1.4	0.4	0.2	2.0
1970–71	8.9	3.3	1.4	13.6
1980–81	21.3	7.0	3.6	31.9
1990–91	43.0	17.3	7.1	67.4
2000–01	56.7	21.9	8.1	86.7
2008–09	77.9	33.6	17.1	128.6
2017–18	81.7	33.0	13.4	128.1

Source: Fertilizer Statistics, Fertilizer Association of India, New Delhi, and Agricultural statistics at a glance, 2009, 2014 and 2018, Government of India.

hectare of gross cropped area. The consumption of different fertilizers per hectare of gross cropped area in India is shown in Table 6.2.

The fertilizer use per hectare of gross cropped area which was even less than one kilogram up to 1955–56 increased to 13.6 kg in 1970–71, 67.4 kg in 1990–91 and further to 128.1 kg in 2017–18. However, the level is quite low as compared to that in other countries. In terms of total fertilizer use although India ranks fourth in the world after U.S.A., erstwhile U.S.S.R. and China but the average fertilizer use is much lower than the global average. Even compared to the Asian countries, the consumption of fertilizers on unit area basis in India is about two-thirds. Notwithstanding the low consumption of fertilizer use in India vis-a-vis other countries, there are some states where fertilizer use is quite high.

For example, per hectare fertilizer use is 265 kg in Telangana, 213 kg in Punjab, Haryana and Bihar, 171 kg in Andhra Pradesh, 168 kg in Uttar Pradesh, 165 kg in Karnataka and 163 kg in West Bengal.

IMPORT DEPENDENCE

The gap between demand and domestic production is met through imports. As mentioned earlier, in the case of potassic fertilizers, India is entirely dependent on imports. As shown in Table 6.1, with the increase in requirement/demand for potassic fertilizers, the imports were bound to go up. From only 0.8 million tonnes of imports in 1980–81, these have gone up to 2.8 million tonnes in 2017–18. India's import dependence in 2017–18 was 21% for N fertilizers and 30% for P fertilizers. If imported raw material for domestically produced P fertilizers is taken into account, the import dependence for P fertilizers is around 90%.

Fertilizer being a critical input and keeping in view its role in accelerating the growth of agricultural production, it would be in the overall interest of the country to achieve high degree of self-sufficiency in fertilizers. This is all the more necessary in the case of a country of the size of India as its demand for fertilizers is a considerable part of total fertilizer produced in the world. However, Indian farmers should not be penalised by protecting inefficient domestic fertilizer producing units. Several Indian Manufacturers, are setting up joint venture plants overseas to produce fertilizers and supply to India.

PRICES

In India, fertilizers are manufactured by various units which use a variety of basic feedstocks and technologies. The feedstocks are fuel oil, naphtha, coal and gas. It is, therefore, natural that cost of production per unit of fertilizer varies from plant to plant. This apart, the economic cost of imported fertilizer is different than the domestically produced fertilizer. One of the elements of fertilizer pricing policy followed in the country till 1991 was to have a single price for a specific grade of fertilizer throughout the country irrespective of the plant in which they are produced. Therefore, the selling prices of different grades of fertilizers were statutorily fixed by the Government of India. The margins retained by various agencies engaged in the sale of fertilizers were also statutorily fixed by the government. The prices of some important fertilizers fixed by the government from time to time till July 1991 are given in Table 6.3.

Table 6.3: Prices of some fertilizers in India (up to July 1991)			
			(₹ per tonne)
Fertilizer	11-7-81 to 28-6-83	29-6-83 to 30-1-86	31-1-86 to July 1991	w.e.f. August 1991
Urea (46% N)	2350	2150	2350	An average increase of 30% for all fertilizers
Di-ammonium phosphate (18:46:0)	3600	3350	3600	
Muriate of potash (60% K)	1300	1200	1300	-do-
Sulphate of potash (50% K)	2100	1950	2100	-do-
Mixed fertilizers				
15:15:15	2100	1950	2100	-do-
10:26:26	2950	2750	2950	-do-
12:32:16	3250	3000	3250	-do-

Source: Government of India, Agriculture in Brief, 19th and 23rd Editions, Ministry of Agriculture, New Delhi.

In order to reconcile the uniform sale price and varying cost of production across manufacturing plants, a Retention Price Scheme (RPS) was introduced in late seventies in the country. Under this scheme, if a manufacturer's net realization from the sale of fertilizers at the government controlled rate (the same farm gate price throughout the country) falls short of the retention price, the government pays the difference as subsidy to the fertilizer manufacturers. The retention price was fixed after taking into account the type of feedstock or raw material being used, cost of other inputs and maintenance such that under the assumption of 80% capacity utilization, the plant is able to earn 12% post tax return on its net worth. The basic rationale of adopting the RPS was that it reduced uncertainty of returns on investment, which encouraged fertilizer manufacturers to increase capacity utilization and also attracted investment from both new and existing entrepreneurs. One other element of fertilizer price policy adopted in India has been that the prices of fertilizer were kept low so that farmers have the incentive to use more fertilizer, which has been the very important purchased input for increasing the productivity of crops. In an endeavour to keep fertilizer prices at levels lower than the economic cost of domestic production or landed price of imports, the gap between the two widened over time. This gap is what is usually termed as fertilizer subsidy.

Since July 1991 (the period when a programme of economic reforms was launched in the country), there have been considerable changes in the fertilizer pricing policy. The low analysis nitrogenous fertilizer (like Ammonium Sulphate and Calcium Ammonium Nitrate) were decontrolled with effect from July 25, 1991. These were again brought under price control with effect from August 25, 1992 but were finally decontrolled with effect from June 10, 1994.

Phosphatic, Potassic, NP and NPK fertilizers were decontrolled with effect from August 25, 1992. Thus beginning from June 10, 1994, all fertilizers, except urea, have been decontrolled. The import of MoP was decanalized in June 2003. Though the objective of decontrol was to reduce the burden of subsidies on the government exchequer, the actual out go on this account has not decreased. Owing to the decontrol of phosphatic and potassic fertilizers, their prices shot up and the government had to announce adhoc subsidy to reduce the burden on the farmers. The prices of some important fertilizers since July 1991 are shown in Table 6.4.

Table 6.4: Prices of fertilizers in India (1991 to 2015)

(₹/tonne)

Year	Urea	DAP	MOP
Controlled prices			
July 25, 1991	3300	5040	1820
Aug 14, 1991	3060	4680	1700
Aug 25, 1992	2760	Decontrolled	Decontrolled
June 10, 1994	3320	Decontrolled	Decontrolled
Prices paid by farmer			.
1994–95 (K)	3320	7500	3800
1995–96 (K)	3320	9800	4450
1995–96 (R)	3320	10,000	4600
1996–97 (K)	3320	11000*	4800*
1996–97 (R)	3320	9000**	4300**
1997–98	3660	8300	3700
1998–99	3660	8300	3700
1999–2000	4000	8300	3700
Since Feb 2000	4600	8900	4255
Since Feb 2002	4830	9350	4455
Since Feb 2003	5070	9350	4655
Since March 2003	4830	9350	4455
March 2015	5360	24000	16700

* Up to 6th July, 1996, K = Kharif season, R = Rabi season
** With enhanced price concession on DAP/MOP
DAP and MOP prices after decontrol are net of subsidy (except March 2015).
Source: Economic Survey, Ministry of Finance, Government of India, New Delhi and Fertilizer Statistics, Fertilizer Association of India, New Delhi. (various issues)

The prices of urea continue to remain under the control of the government (administered prices) and there continues a provision of subsidy on decontrolled P and K fertilizers. The index of fertilizer prices since 2000–01 has increased only marginally owing to the increase in prices of decontrolled P and K fertilizers. The official index of fertilizer prices shows only a marginal annual increase up to 2013–14 as shown in Table 6.5. The fertilizer price index went up mainly due to the increase in prices of decontrolled phosphatic and and potassic fertilizers. The controlled urea is sold at statutorily notified uniform sale price and decontrolled P and K fertilizers are sold at indicated MRPs. Since November 2012, price of urea is ₹ 268 per bag of 50 kg. Obviously, the fertilizer price index has not shown much increase during the last five years.

Table 6.5: Index numbers of prices of fertilizers		
Month/Year (April–March)	2004–05 = 100	2011–12 = 100
2005–06	102	–
2009–10	108	–
2010–11	117	–
2011–12	133	100
2012–13	149	114
2013–14	152	117
2014–15	NA	120
2015–16	NA	121
2016–17	NA	118
2017–18	NA	118

Source: Economic Survey, Reports of CACP, Government of India (various issues).

FERTILIZER SUBSIDY

The subsidy on fertilizers as defined in India is the difference between net realization by the domestic fertilizer manufacturers (farmer's price minus distribution margins) and the ex-factory retention price (inclusive of equated freight) fixed by the government. In the case of imported fertilizers, the subsidy is the difference between the C.I.F. (cost, insurance and freight) price of imported fertilizer plus the pool handling charges and the farmer's price (excluding dealer's margins and sales tax). After the decontrol of, P and K fertilizers, there is direct subsidy on these nutrients. The subsidy on urea is based on each plant's cost of urea production and it is the difference between MRP fixed by the government and company's delivered cost at the retail level (at present, the fertilizer subsidy is calculated on the basis of delivered cost at factory gate and the transport cost up to retail outlet is paid as freight subsidy). The Government of India, on an average, pays ₹ 500 per bag of 50 kg as subsidy, which is around 186% of the current MRP of urea. In the case of P and K fertilizers, the government pays a fixed subsidy to companies, which decide their own retail prices.

Although it is a fact that fertilizer subsidy exists because fertilizer is sold to the farmers at a price lower than its economic cost but it is not true that the benefits of this subsidy accrue only to the farmers who use this fertilizer. In this connection, it is important to recognise that the fertilizer subsidy although induces farmers to use more fertilizer which increases the production of agricultural commodities but if as a part of the overall price policy, the prices of farm products are also kept low, the benefits of fertilizer subsidy also accrue to the consumers of farm products[1]. The consumers are able to get food products at affordable prices and the industries which use raw material from the agricultural sector are able to keep the cost of production of such manufactured goods low.

In this context, it is relevant to look at the prices of fertilizers in relation to the prices of farm products. In 1980, an Indian farmer had to forego 4.58 kg of paddy to buy one kg of N and 5.55 kg of paddy to buy one kg of P_2O_5. During the same period, a farmer in Japan had to sell only 0.68 kg and 0.55 kg of paddy to buy one kg of N and P_2O_5 respectively.

In 1993, to buy a kg of N, a farmer in India had to sell 1.94 kg of paddy, whereas the quantity of paddy required to be sold by the farmer was 1.87 kg in Bangladesh, 1.79 kg in China, 0.32 kg in Japan and 0.40 kg in Korea. For buying a kg of P, the quantity of paddy required was 4.60 kg in India as against 3.24 kg in Pakistan, 2.26 kg in China and 0.72 kg in Japan.

In 2012–13, the quantity of paddy required to buy one kg of N was 0.93 kg. It was 3.81 kg for buying one kg of P (DAP) and 3.27 kg for one kg of potash (K_2O). Similarly, the quantity of wheat was 0.86 kg for N, 3.53 kg for P, and 2.10 kg for one kg of potash. An overtime comparison clearly shows that prices of different nutrients, in relation to the prices of paddy and wheat, have continuously declined. In recent years, this continues to be so for N fertilizers, which is the major fertilizer used by the farmers.

Over the years, although the cost of domestic and of imported fertilizers has increased but the sale prices of fertilizers were not raised commensurate with the increase in the cost. This apart, the consumption of fertilizer has expanded rapidly. Therefore, subsidy bill has continuously increased over the years. The amount of subsidy outgo on fertilizers in India during the 50 years is given in Tables 6.6 and 6.7.

Table 6.6: Subsidy on fertilizers in India (1971 to 2001)

(₹ in crore)

Years	Indigenous fertilizers	Imported fertilizers	Decontrolted fertilizers	Total
1971–72	–	–20	–	–20
1974–75	–	371	–	371
1975–76	–	242	–	242
1976–77	60	52	–	112
1980–81	170	335	–	505
1985–86	1600	324	–	1924
1990–91	3730	659	–	4389
1995–96	4300	1935	–	6235
2000–01	9480	1	4319	13800

Sources:
1. Fertilizer Statistics Various Issues, The Fertilizer Association of India, New Delhi.
2. Economic Survey, Various Issues, Government of India, New Delhi.

There was no subsidy on indigenous fertilizers up to 1975–76. It was only the imported fertilizer that was subsidised. Following the introduction of fertilizer retention price and subsidy scheme with effect from November, 1977 and due to the increasing trend in production/consumption, subsidy on fertilizers increased from 112 crore in 1976–77 to as much as ₹ 6235 crore in 1995–96. It went up to ₹ 13800 crore in 2000–01.

Between 1980–81 and 1990–91, there has been no increase in the fertilizer prices. In 1991–92, the government raised the prices of fertilizers by 30% to minimise the drain on budgetary resources. However, the small and marginal farmers were exempted

Table 6.7: Subsidy on fertilizers (2002–03 to 2019–20)

(₹ in crore)

Year	Urea	P&K	Total
2002–03	7791	3225	11016
2004–05	10986	5142	16128
2007–08	26385	16934	43319
2010–11	24337	41500	65837
2011–12	37683	36108	73791
2012–13	40016	30576	70592
2013–14	41853	29427	71280
2014–15	54400	20667	75067
2015–16	50478	21938	72416
2016–17	47470	18843	66313
2017–18	43762	22836	66598
2018–19	45307	25291	70598
2019–20*	53629	26367	79996

* Budget estimates
Source: Department of Fertilizers, Reports of CACP, Government of India.

from this fertilizer price increase. This policy amounted to a dual price framework in which the small and marginal farmers were to be given fertilizer at low prices while other farmers had to pay a higher price. As it was not easy to implement a dual price policy for such a farm input, it was given up later.

Subsidy on fertilizers is common in most of the developing countries, though the rate of subsidy varies across countries and types of fertilizers. Though the direct effect of fertilizer subsidy is to provide fertilizer to the farmers at a price lower than its economic price, the ultimate benefits of the subsidy accrue either to the farmers or to the consumers or both depending on the package of other policies including product price policies. Differential rate of subsidy on nitrogenous, phosphatic and potassic fertilizer was used to correct the imbalance in the use of nutrients by the farmers. However, after the decontrol of P and K fertilizers, their prices rose sharply leading to the distortion in the use ratio of N, P and K.

In order to correct the imbalance, the government started subsidy on decontrolled fertilizers in 1996–97. Since then, there has been considerable increase in subsidy on P and K fertilizers. The distribution of fertilizer subsidy according to nutrients can be seen in Table 6.7. The figures of subsidy shown in Table 6.7 are actual releases by the government. However there are considerable unpaid dues of fertilizer companies. According to Fertilizer Association of India, the unpaid dues have piled up over the years and cumulatively these, at the end of December 2019, were around ₹ 1.80 lakh crore. Given this, actual fertilizer subsidy is more than the figures shown in Table 6.7.

MARKETING OF FERTILIZERS

Fertilizers are produced only at selected locations and imported fertilizers arrive at seaports. The marketing system has to carry out the functions of storage, transportation and selling to the farmers spread throughout the country. Over time the marketing system for fertilizers has undergone rapid change both in terms of its capacity and mode of operation. Its evolution has been mainly guided by the public policy. Since fertilizer was a new input for the farmers, the spread of know-how and incentives had to accompany the marketing of fertilizers. In fact, initially the demand for fertilizer had to be created. But the objective of demand creation was not to sell more fertilizers and earn profit but was to increase agricultural production. Up to the end of the First Five Year Plan (1951–56), the sale of chemical fertilizers was the sole responsibility of cooperative societies and State Agriculture Departments. During the Second Five Year Plan (1956–61), the sale of fertilizers became almost the monopoly of cooperative societies. This step aimed at popularising the cooperative movement and achieving a higher efficiency of the distribution system. The village panchayats were also given this responsibility wherever the cooperatives did not exist. In 1965, the Government of India constituted a committee, under the chairmanship of B. Shivraman, to suggest changes in the policy relating to production, distribution, and promotion of fertilizer consumption. Based on the recommendations of Shivraman committee, several changes were brought about to bring marketing of fertilizers into the mainstream of agricultural development strategy. Later, the Government allowed the fertilizer production units, which had been licensed before December 31, 1967, to sell 70% of their produce through their own agencies for a period of seven years from the date of commencement of production. The remaining 30% of the production was required to be sold through public or cooperative agencies.

At the end of March, 1970, the number of sale points of fertilizers were 71652, out of which 53% were in the private sector and 47% were in the cooperative or public sector (Table 6.8).

Table 6.8: Number of fertilizer sale points in India					
Year (as on 31st March)	Number			% to total	
	Cooperatives or other institutions	Private	Total	Cooperatives or other institutions	Private
1969	36505	30071	66576	55	45
1970	33418	38234	71652	47	53
1980	51560	64862	116422	44	56
1990	80040	151130	231170	35	65
2000	73933	205360	279293	26	74
2009	59288	208832	268120	22	78
2014	71373	232962	304335	23	77

Source: Fertilizer Association of India, Fertilizer Statistics.

During the early seventies, the proportion of private sector sale points increased at a faster rate but slowed down in the later half of seventies. However, subsequently, the private sector fertilizer outlets have expanded at a rate higher than that of the cooperative or the public sector. At the end of March, 2014, there were 3.04 lakh sale points of fertilizers in the country, out of which 77% were in the private sector and remaining 23% were operated by either cooperative societies or other public sector institutions like State Agro Industries Corporations. In order to make available the fertilizers to farmers, the temporary sale points are also provided in some areas during a part of the year.

The flow of indigenous fertilizers from manufacturing plants and imported fertilizers from seaports is depicted in Fig. 6.1.

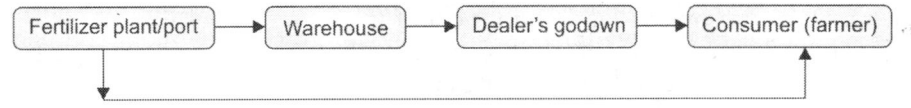

Fig. 6.1: Flow of fertilizers in India

Though the manufacturers prefer to move the stocks of fertilizer from plant to the dealer godowns on a regular basis but this rarely happens because dealers want fertilizers only in the selling season. The dealers are reluctant to purchase and store the fertilizers in the off season, as they are not sure of the quantity of fertilizers that is expected to be demanded by the farmers in the ensuing sowing season.

Normally, the fertilizer is transported to each district headquarter and stocked in warehouses to arrange for their dispatch to the retail outlets in the sowing season. There is a reasonably adequate and widespread network of warehouses or godowns at all the places in the country, managed by both public and private sectors.

Agricultural credit plays an important role in fertilizer marketing. Farmers invariably look for credit for purchase of fertilizers. Institutional credit is made available to the farmers by either the cooperatives or nationalized banks for the purchase of fertilizers. Fertilizer dealers also extend credit to their prime customers.

Promotion of correct use of fertilizers to different crops to get higher yields is also an important activity carried out by fertilizer manufacturers and market dealers. The fertilizer manufacturers appoint field officers for sales promotion. Manufacturers also have their own Farmer Advisory Services through which farmers' meetings, training programmes and demonstrations are conducted to guide them to use the recommended dosages of fertilizers in different crops grown by them.

Legal Aspects of Fertilizer Marketing

1. The Essential Commodities Act, 1955

This Act was passed by the Parliament in 1955 in the interest of general public with the object of controlling production, supply and distribution of trade and commerce in certain commodities. Fertilizers was declared as an essential commodity and as such provisions of this act are applicable to fertilizer production, supply and distribution too.

2. The Fertilizer (Control) Order, 1957

The Government of India declared fertilizers as an essential commodity on March 29, 1957 and issued the Fertilizer (Control) Order under section 3 of the Essential Commodities Act, 1955. The basic objective of the Fertilizer (Control) Order, 1957 is to regulate the manufacture, sale, distribution, price and quality of fertilizers. Under the provisions of this Order, it is a statutory requirement that any person carrying on the business of selling fertilizers will have to obtain a license from a competent Registering Authority appointed by the State Government. Till recently the prices of various fertilizers were also controlled by the Government of India. Urea still continues to be under price control.

The Order, provides that (a) if a dealer is found selling at a higher price than what is fixed by the Government, he is liable for punishment; and (b) if the fertilizer sold is found of sub-standard quality, the trader selling such a substandard or adulterated fertilizer can be prosecuted.

The Order also provides for appointment of Inspectors of Fertilizers under clause 19 of the order. These persons are authorised to collect samples during inspection and send them for analysis to fertilizer testing laboratories. If the samples are found to be of sub-standard quality, the Fertilizer Inspectors are required to take immediate action against the dealer in the form of seizing or detaining the stocks, books or accounts and launch prosecution.

The contravention of any of the clauses of the Fertilizer (Control) Order, 1957 is punishable, including imprisonment. Every offence punishable under this Act is a cognizable offence. The offence can be summarily tried by special courts appointed for this purpose.

3. The Fertilizer (Movement Control) Order, 1973

The Central Government issued the Fertilizer (Movement Control) Order, 1973 in order to ensure equitable distribution of fertilizers in the states of India. No person was allowed to export or attempt to export any fertilizer from any state without a valid authority issued by the Government of India, Ministry of Agriculture or the Director of Agriculture of a State Government or any other officer appointed by the State Government. Under this Act, Fertilizer Inspectors were given the powers of entry, search and seizure of any vehicle/ conveyance used for the purpose of moving fertilizers from one state to the other.

4. The Fertilizer Control Order (FCO), 1985

In order to ensure adequate availability of fertilizers of standard quality to farmers and to regulate trade, quality and distribution in the country, fertilizers have been declared as essential commodity. The FCO, 1985 was promulgated under section 3 of the Essential Commodities Act 1955.

Defects in Fertilizer Marketing

Notwithstanding the fast expansion of sale points of fertilizers, the defects in the marketing system of fertilizers are identified as follows:

1. The number of sale points are still inadequate. The farmers in hilly and desert areas have to travel long distance to buy the fertilizers.
2. Quite often, the supplies of the fertilizers at many sale points are not sufficient to meet the demand for fertilizers in the area.
3. At many sale points, the fertilizers are not stocked at a time when farmers want to purchase. For example, if the supplies to the sale point do not reach before the sowing of crops, the farmers are not able to buy the fertilizer which they wish to use as basal dose.
4. Quite often, the makes and grades of the fertilizers which the farmers wish to buy are not available at the nearest sale point.
5. Fertilizers are prone to adulteration and several cases of adulteration have been reported.
6. Sometimes, the quantity of fertilizers in the bags is less than the specified one. Although, this happens because of mishandling but it is deliberate also.
7. When the supply is less than the demand for fertilizers in an area, during a specified season, the dealers charge a price higher than the statutory or normal price.
8. Sometimes, the farmers are forced to buy another kind of fertilizer along with the kind desired by them. For example, at some sale points farmers are forced to

purchase some phosphatic fertilizer along with nitrogenous fertilizer. Technically, this may be a right practice, but farmer as a buyer feels this practice as undesirable compulsion.

9. Farmers in many areas do not have cash to pay for the fertilizers. Short-term loan or crop loan from the banks is meant to meet this requirement. But if credit proposals are not processed in time to enable the farmers to buy the fertilizers on credit, some sale of fertilizers gets a set-back in such areas.

10. In many areas, the salesman do not possess the requisite know-how on the use of fertilizers which farmers wish to seek from them.

Suggestions for Better Fertilizer Marketing

Suggestions for improving the fertilizer marketing system are as follows:

1. There is a need to increase the number of sale points specially in hilly, tribal and desert areas so that the farmers have not to travel much distance to buy the fertilizer. This will save time and also minimize the travel cost.

2. There is also a need to develop proper distribution arrangements involving a combination of cooperatives, government and private agencies, depending on the potential of the area. Restriction on the entry of marketing firms should be relaxed by making the fertilizer licensing policy liberal so as to increase competition and efficiency in the fertilizer trade. Wherever, cooperative institutions have not been successful, private dealers should be encouraged to supplement the sales efforts. The basic objectives of the policy should be to make fertilizer available to all the farmers at the time of need at reasonable prices.

3. Sales points should be developed into good agro-service or agro-business centres. Such centres should provide advice to the farmers on different aspects of fertilizer application in addition to making the fertilizer, other inputs and services available to them.

4. Packing material and technology for fertilizers should be improved to minimise the chances of loss during transit and storage as also of pilferage from the bags. High density poly ethylene (HDPE) bags are used in packging of fertilizers to prevent losses in transit and storage, and also from pilferage.

5. The procedure of linking credit with fertilizer supply should be simplified.

6. Fertilizer should also be made available in smaller packets of 5 to 10 kg.

7. There is need to check adulteration and underweighment of bags. This can be done by strengthening the quality control organization (drawing of samples at different stages of marketing and laboratory testing) in addition to the use of good packing material.

8. There is also a need to minimise the number of brand names to avoid confusion among the farmers specially those who are illiterate or have poor educational level.

9. The ratio of prices of three nutrients (NPK) should be maintained at levels consistent with the normative use under different cropping patterns and soil conditions.

10. Soil testing fecilities should be up scaled to make these facilities available to all the farmers within easy reach, which would help in optimum use of different nutrients by the farmers.

11. Micronutrients and sulphur should be recognized as critical inputs at par with NPK and subsidy should be extended to secondary and micronutrients.

Costs and Margins in Fertilizer Marketing

The gross marketing margin is the difference between the import or ex-factory price and the retail price of the fertilizer. It includes the commission of wholesalers, agents and retailers, transportation costs, storage costs, interest and other overhead costs. In general, the dealer's commission accounts for 30 to 35%, transportation cost 20%, handling cost 10%, storage cost 10% and miscellaneous items account for remaining 25 to 30% of the gross marketing margin. Marketing margins for fertilizers vary across countries. According to one study, they vary between 5.2% in Taiwan and 62.2% in Thailand. In India, the marketing cost for fertilizers has been estimated at 10% of the retail price.

When the selling price is fixed by the government, as it has been happening in India and when the difference in the economic cost and sale price is borne by the government in the form of what is called the fertilizer subsidy, the whole market structure becomes over-shadowed by the public intervention. This is more so when even the prices of raw material used by the fertilizer manufacturers and the wages are also administratively determined. While fertilizer manufacturers claim that their cost of production is not fully met, others throw the responsibility of higher prices of fertilizer and higher budgetary subsidies on the inefficiency of fertilizer manufacturing units. These issues over-shadow the analysis of marketing costs and margins in the context of overall efficiency of production and supply of fertilizers.

RECENT STEPS TO IMPROVE FERTILIZER MARKETING

1. **Nutrient-based Subsidy:** A nutrient-based subsidy (NBS) scheme was introduced in April 2010 to encourage balanced use of fertilizers. Initially, it was implemented for P and K fertilizers and made applicable for DAP, MOP, TSP and other grades of complex fertilizers. The NBS is the same for both domestic and imported fertilizers. Both primary nutrients (N, P, K) and secondary nutrients are covered under NBS. Manufacturers of customized and mixed fertilizers are eligible to source subsidy. Subsidy is given on the basis of per kg of N, P, K and S in the fertilizers. At present, 21 grades of P and K (DAP, MAP, TSP, MOP, AS, SSP) and 15 grades of NPKS fertilizers are covered under NBS scheme. There is an additional subsidy for boron (B) and zinc (Zn) fortified fertilizers to encourage use of micro-nutrients.

2. In addition to NBS, the government also pays freight subsidy to fertilizer manufactures (factory-gate to retail outlets) with a view to keeping the sale price of fertilizers low for the farmers.

3. The fertilizer companies are mandated to print both MRP and subsidy on the packets.

4. Special efforts have been made to expand the network of soil testing laboratories (STLs), mobile soil testing labs and strengthening these for micro-nutrients testing. By the end of 2014, there were 1206 STLs, consisting of 1018 static and 188 mobile labs. In addition, there are many fertilizer quality testing labs.

5. Soil Health Card (SHC) scheme was launched by the government in January 2015, which aims at providing a SHC to every farmer. The card carries crop-wise recommendations on nutrients/fertilizers required for the farm. The scheme was implemented in two rounds/cycles of two years each. Combined for both cycles (2015–16 to 2018–19), 23.4 crore Soil Health Cards were distributed to farmers.

6. The procedure for incorporation of new products was liberalized to encourage production of soil-specific, crop-specific and fortified fertilizers.

7. Organic fertilizers (compost and vermi-compost) and bio-fertilizers (rhizobium, azotobacter and azospirilium) were incorporated in FCO, 1985.

8. Neem-coating of urea was initiated and in August 2015, the target of 100% of urea with neem coating was achieved. This helped in curbing the diversion of urea for non-agricultural uses and also helped in checking leaching of urea to sub-soils.

9. **Direct Benefit Transfer (DBT):** Direct Benefit Transfer Scheme was introduced in 2017. Under this, the subsidy is released to fertilizer companies based on the actual delivery of fertilizer to the farmers by retailers. It is reported that by March 2018, pan-India roll out of DBT was completed. Under this scheme 2 lakh fertilizer retailers were trained and PoS (point of sale) devices provided to them. There are some hiccups being reported that are being tackled. However, DBT scheme is improving the transparency and is reported to have resulted in saving in fertilizer subsidy. The advantages of DBT are as follows:

   (a) Ready and timely availability at retail point.
   (b) Ensures sale at subsidized rate through PoS device.
   (c) Receipt is generated with sale price.
   (d) No over-pricing or over charging.
   (e) Sale receipt also shows subsidy born by the government.

▌SEEDS

The process of modernizing agriculture primarily involves intensive use of non-conventional inputs such as quality seeds, chemical fertilizers, pesticides, weedicides, irrigation, farm machinery and a network of research and extension infrastructure. The seed is a trigger point which sets in motion the process of technological change. The returns to investment depends significantly on the quality of seed that is used in the production of crops. The need of a suitable seed having desired characteristics such as high yield, better grain quality and resistance to pests and diseases, is well recognised for increasing the crop yields in any agro-climatic region. Although seed accounts for only a small part of the total cultivation expenses, yet without good seed, the investment on fertilizers, water, pesticides and other inputs does not pay the dividend.

Under the traditional system of farming, most of the farmers retained a part of their produce for use as a seed during the next crop season. No distinction was made in the marketing channel between the ordinary produce and the seed except that, at the end of the marketing channel, the farmer selected a better quality lot if purchase was made for use as seed. Advances in farm-technological research resulting in the evolution of high-yielding dwarf and disease-resistant varieties have added a new dimension to the marketing aspect of seed input. Every year new seeds of hybrid varieties have to be procured and used which necessitates a very close linkage between the production and marketing of seeds of these varieties.

The seeds used by the farmers should be genetically true to type and should also possess other desired qualities viz., vigour, stout germination potential, free from exogenous mixtures and undesirable weeds. When the farmer does not get seeds possessing genetic and physical qualities up to the desired standards, the yield of the crops is adversely affected, which, in turn entails multiple losses, both to the farmer and the society at large.

Recognising the importance of quality seeds in agriculture, the National Seeds Corporation (NSC) was set up in 1963 for production and distribution of quality seeds.

Later, State Farms Corporation of India (SFCI) was set up in 1969 to develop modern farms, mainly for the production of quality seeds. In addition, 16 State Seeds Corporations (SSCs) have been set up in the states of the country.

SEED PRODUCTION AND SUPPLY

Seed is available to the farmers in India through three major sources:

1. Seed retained from the previous year's crop;
2. Seed obtained from fellow farmers or grain traders; and
3. Seed purchased from formal seed industry.

For many crops, more than 70% of the seed used by the farmers in India is obtained from the first two sources. The share of organised seed sector is meager on account of high prices of certified seeds and their non-availability in terms of proper place and time. The importance of use of self-retained seed for self pollinated crops like cereals can be hardly emphasised.

There are basically two types of seeds. One are the varieties, which pollinate by themselves and can be reused but with diminishing returns. The others are hybrids, which are developed through controlled pollination and cannot be reused. Except wheat and soyabean, crops have both types of seeds.

Seed production and its marketing involve a high level of technology and a high standard of proficiency. The seed business is a business of trust. It is one of the most difficult areas of management in agricultural development. The nonavailability of improved seeds of high quality deprives the farmers of the advantages of modern technology. With the introduction of the high-yielding varieties and hybrids in mid-sixties, production and distribution of improved seeds gained importance in the national programme of agriculture development. Following steps are involved in the production and marketing of quality seeds:

(a) Production of Quality Seeds

The quality seed for ultimate use by the farmers is produced in three stages viz., breeder seed, foundation seed and certified seed. Breeder seed is the primary stage of the seed production cycle. The production of breeder seed is organised by the Indian Council of Agricultural Research through the concerned breeders and scientists attached to its Institutes and State Agricultural Universities as per the requirements indicated by various states. The production of breeder seeds of crops was 526 tonnes in 1980–81 which has increased to 6870 tonnes in 2005–06 and further to 10508 tonnes in 2017–18.

The breeder seed is multiplied into the foundation seed by NSC, SFCI and SSCs. Now the private seed producers have also entered into the production of breeder and foundation seeds. While the state's local requirements of foundation seeds are met by the SSCs, the NSC and SFCI take care of the requirement of national varieties. The total production of foundation seeds which was 10915 tonnes in 1980–81 has increased to around 47600 tonnes in 1995–96 and further to 1.95 lakh tonnes in 2017–18.

The foundation seed is supplied to the selected farmers for multiplication into what is called the certified seed. The farmers who undertake the work of production of certified seeds have to meet certain standards like isolation of the fields, the cultivation practices and removal of plants of undesirable varieties from the plots. The certification agencies keep close supervision over such plots. The seed is purchased by the agencies, tested in the laboratories for purity and germination, graded and packed with certification mark before releasing for sale in the market The production and marketing of certified seeds have expanded very fast in the country. While in 1980–81, 2.18 lakh tonnes of certified seeds were produced and supplied to the farmers, in 2010–11, it increased to around 27.73 lakh tonnes. During 2017–18, it went up further to 35.2 lakh tonnes. Production of breeder seeds, foundation seeds and certified seeds during 1980–81 to 2017–18 is shown in Table 6.9. The crop-groupwise data on production of certified seeds can be seen in Table 6.10.

Table 6.9: Production and distribution of quality seeds

Year	Breeder seed (tonnes)	Foundation seed (thousand tonnes)	Certified quality seeds (lakh tonnes)
1980–81	526	10.9	2.18
1995–96	4265	47.6	6.99
2000–01	4269	59.1	8.63
2005–06	6870	79.6	15.50
2010–11	11885	180.6	27.73
2017–18	10508	195.4	35.20

Source: Ministry of Agriculture, quoted in Economic Survey 2002–03, and 2009–10, Ministry of Finance, Government of India, New Delhi, Ag. Statistics at a Glance, 2018.

Table 6.10: Production of certified seeds in India

(lakh tonnes)

Crop groups	1980–81	1990–91	2000–01	2010–11	2016–17
Cereals	1.87	3.60	5.95	18.26	25.97
Pulses	0.10	0.40	0.39	2.08	3.13
Oilseeds	0.01	1.00	1.25	5.06	5.91
Fibres	0.07	0.20	0.29	0.26	0.27
Potato	0.12	0.80	0.72	2.01	2.71
Others	0.01	–	0.03	0.06	0.03
Total	2.18	6.00	8.63	27.73	38.02

Source: Government of India; Indian Agriculture in Brief, 21st Edition and Annual Report 1990–91; Agricultural Statistics at a Glance 2014, 2018, Ministry of Agriculture and Fertilizer Statistics, Fertilizer Association of India, New Delhi.

Until recently seed production and multiplication programmes have been formulated and implemented under the aegis of government initiative and agencies. For successful implementation of these, Government of India and the state governments have been taking necessary promotional programmes. Private sector has also recently taken up seed research on a large scale. Main interests of the public sector agencies in seed research is in the evolution of hybrid seeds. Admittedly, the overhead expenditures on seed research and distribution being high in public sector seed agencies, the cost of seed has been high. Over the years, public sector seed agencies incurred heavy losses and most of them are financially not very sound. Nevertheless, their presence in the seed business is essential to safeguard against monopoly or oligopoly powers, which the private seed companies may acquire in their absence.

(b) Seed Processing, Packing and Storage

After the purchase of the seeds from registered seed growers, the seeds are appropriately processed. This is an integral part of improved seed technology. Seed processing primarily consists of five major steps viz.: (a) Extraction, (b) Drying, (c) Purification, (d) Treatment, and (e) Packing. Packing is important for the prevention of adulteration with the inferior seeds. A label is put on each tested seed packet by the certification agency (NSC or SSC). The certified seed packets are stored up to the next crop season for sale to the farmers. It is necessary to prevent deterioration in the quality and the loss in the germination percentage during the period of storage.

(c) Quality Control and Certification

To retain the confidence of the farmers and provide them protection against unscrupulous practices, maintenance of quality and its certification is very important. For this purpose, the Indian Seeds Act was passed in 1966, which came into force from October 1, 1969 in all the states and union territories of the country. Later, Seed Control Order, 1983 came into existence.

The Indian Seeds Act (1966) aims at regulating the quality of seeds sold to the farmers. The Act envisages a two-pronged approach namely, compulsory labelling and voluntary certification. Under compulsory labelling any one selling seed of a notified variety should ensure that the seed conforms to the prescribed limits of germination and purity and its container is labelled in the prescribed manner. Under voluntary certification, any one desirous of producing certified seeds may apply to a certification agency for the grant of a certificate to the applicant after satisfying that the seed has been produced according to the prescribed norms and that it conforms to the prescribed standards. The Act stipulates that any person offering to sell seeds of notified varieties should label them and give information about germination, purity and moisture percentage—all conforming to the minimum standards. Any discrepancy between the label and the contents is a criminal offence.

The main provisions of the Indian Seeds Act and the Rules thereof are:

1. Standard of seeds: The Central Seed Committee of the Government of India through its sub-committees—Central Sub-Committee on Crop Standards and Notification and Central Sub-Committee on Release of Varieties — has prescribed the standards for the seeds of notified varieties so that the uniformity is maintained throughout the country. In addition, the Central Seed Certification Board also goes into the standards of certified seeds.

2. Certification of seed: This is voluntary in nature and is done by 19 seed certification agencies set up in various states under the provisions of the Indian Seeds Act. Certification of seed includes field inspection, proper grading, representative sampling and careful laboratory testing. The certificate is affixed securely to each sealed bag or container to show that the seed is certified.

3. Seed Testing: Seeds are tested for purity and other qualities in seed testing laboratories. There are many state seed testing laboratories in the country.

4. Enforcement of Seeds Act: Enforcement of the Seeds Act has been entrusted to the State Governments. The seed inspectors of the government visit the premises of distribution agencies, check up the quality and take samples in case of doubt for testing in laboratories

5. The Central Seed Committee is the apex body to administer the Seeds Act in the country and to offer the technical advice to the State Governments.

SEED MARKETING AND DISTRIBUTION

Seed marketing is more complicated and specialised process as compared to marketing of other inputs and of agricultural products. Production of good quality seed is of no value if it does not reach the farmer in time. Seed is a biological entity. In most cases, seeds are produced far away from the consumption centres. Further, seed produced in one season is supplied to the farmers in the following season. Hence, it requires proper storing. Moreover, it has to be taken for sale to the farmers during the sowing period. Any delay in the supply by a few days may mean accumulation of unsold seed stocks. There are chances of loss in the germination percentage if it is to be stored for another year. As the marketing of seed involves procurement, distribution, sales promotion and linking credit with sales, effective coordination of all the related agencies is necessary to achieve the objective of making available good quality seeds to the farmers in time.

The seed marketing involves taking of bags of certified seeds to the needy farmers through the network of sales outlets of government, cooperative societies and private agencies. The sale of certified seeds of cereals, pulses and oilseeds is handled by the private sector as well as government and cooperative organisations. The National Seeds Corporation as also the State Seeds Corporations have their own sale points for seed marketing. In some states, Department of Agriculture also sell seeds through their field staff. The sales of seeds of vegetables, flowers and other crops are mostly handled by private traders.

The Agro-Industries Corporations of the states encourage private entrepreneurs to establish agro-service centres in rural areas by providing them with training and arranging supplies of farm inputs for subsequent sales to the farmers. Over the years, private trade has come up in the seed marketing activity in a big way.

Seed companies market most of their seed through a network of private dealers. There are hundreds of private seed companies operating in the country but they differ significantly in terms of type and quantity of seed sold. A majority of them are small local companies who do not have their own breeding programs and sell only seeds of popular varieties or hybrids. Larger seed companies produce and sell seeds of proprietary hybrids. They also produce seeds of popular varieties and hybrids. Transnational seed companies mainly develop, produce and sell seeds of proprietary hybrids.

There are basically two seed distribution channels in vogue in the country.

1. The first is an organised /formal seed industry comprising public seed agencies (NSC, SSCs), private seed companies, and agricultural research organisations. Under this channel seed is distributed through a network of distributors, dealers, and retail outlets. The public seed agencies take the help of state departments of agriculture and cooperatives in the distribution of seeds.
2. The second channel of seed delivery system (SDS) is informal farmer driven localized SDS, where the farmers obtain, produce, conserve, improve and distribute seed in an unorganized open market, without any regulation. The key characteristics of informal channel are (a) operates at local level, (b) wide range of exchange mechanisms; (c) addresses immediate farmers' needs; and operates on mutual trust. Nearly 70% of small/marginal landholders obtain seed from this channel or use their own seed.

EXPORT AND IMPORT OF SEEDS

Though Indian seeds are popular in many countries but domestic requirements are given priority over demands from foreign countries. Certified seeds of all cereals had remained on the Restrictive List under the export policy. These seeds cannot be exported without prior permission of the Ministry of Agriculture. However, in view of great demand for Indian seeds in other countries and opportunity for earning foreign exchange, the National Seeds Corporation has been exporting certified seeds to countries like Bangladesh, Burma, Ethiopia, Nigeria, Vietnam, Maldives, Egypt, Kampuchea and Denmark. Export of breeder and foundation seeds is generally not permitted.

A new policy on seed development introduced from October 1, 1988 aims at securing for the farmers high quality seeds available anywhere in the world. As a result, there has been significant increase in the import of high quality seeds, particularly those of oilseeds, pulses, coarse grains, vegetables and flowers.

NATIONAL SEED PROJECT (NSP)

The thrust of National Seed Policy has been to develop the infrastructure required for production and distribution of improved seeds. As such the Government of India in collaboration with the World Bank launched a National Seed Project in October, 1976. The main components of the project were:

1. Development of farms for production of foundation and breeder seeds;
2. Setting up of seed processing plants for foundation seeds;
3. Construction of specialised seed stores;   •
4. Building up of buffer stock of good quality seed;
5. Strengthening of seed certification agencies and seed testing laboratories;
6. Strengthening of universities farms for production of breeder seeds;
7. Establishment of State Seed Corporations; and
8. Creation of training facilities for seed technology and marketing.

The NSP was implemented in phases with details as follows:

- **Phase I (NSP I):** This phase of the NSP was aimed at expanding the production and supply of high quality seeds and seed processing capacity in the States of Andhra Pradesh, Maharashtra, Punjab and Haryana. The period of this phase was from October, 1976 to December, 1984.
- **Phase II (NSP II):** This phase had the same objectives as of phase I and covered the states of Bihar, Orissa, Karnataka, Rajasthan and Uttar Pradesh. The period was December, 1976 to December, 1985.
- **Phase III (NSP III):** The third phase was launched in March 1990. This phase sought to provide facilities for the growth of private sector seed industry through adequate institutional financing; to improve the working of national and state level public sector seed corporations and to ensure timely and adequate availability of qualify seeds at reasonable prices.

SCHEME FOR DEVELOPMENT AND STRENGTHENING OF INFRASTRUCTURE FACILITIES FOR PRODUCTION AND DISTRIBUTION OF QUALITY SEEDS

This is a central sector scheme under the Union Ministry of Agriculture with the aim of making quality seeds available to farmers in time at affordable prices. The scheme is being implemented on all India basis from the year 2005–06. The seed component of

Prime Minister's Relief Package for 31 suicide-affected districts is also being implemented under this scheme. The thrust of the scheme is improving quality of farm-saved seeds through seed village programme to enhance seed replacement rate, boost seed production in the private sector and help public sector seed companies to contribute to enhancing seed production. More than 25,000 seed villages have been organized, which help in increasing certified quality seed production. In addition, several seed infrastructure development projects have been implemented.

STRUCTURE OF SEED INDUSTRY

The Indian seed industry (value of seed business) in 2010 was around ₹ 8000 crore, which is expected to have grown to ₹ 15000 crore. Out of total size, hybrid seeds accounted for around 62%. Across crops, the highest share in seed business is of cotton (23%) followed by paddy (18%), maize (11%), vegetables (11%), wheat (9%) and sunflower (4%). Hybrids and high-yielding varieties of cotton and vegetables (which have low volume seeds) are sold at very high prices in the market. The private seed companies, therefore, take more interest in these high-value and low-volume seeds.

The seed industry consists of public sector seed corporations and private sector seed companies, along with registered seed growers.

Public Sector

The public sector consists of the:

(a) National Seeds Cooporation (NSC);
(b) State Farms Corporation of India (SFCI); and
(c) State Seeds Coorporations (SSCs).

These Corporations multiply and market the varieties bred by the public sector institutions, i.e. the research institutions financed by ICAR and the State Agricultural Universities. The details of these Corporations are given below:

(a) National Seeds Corporation (NSC)

The importance of improved seeds was realised in India long ago. But systematic efforts for the development of improved seeds began only during the Second Five Year Plan period. In 1957, the Government of India in collaboration with the Rockefeller Foundation of the USA, initiated a coordinated maize improvement programme, which was an important milestone in the development of an improved seed industry in the country. Later, the year 1961 was celebrated as the World Seed Year by the Food and Agriculture Organisation (F.A.O.) of the UNO. Coordinated crop improvement programmes were initiated for wheat, jowar, maize, bajra and rice. The seeds of hybrid or high yielding varieties were made available to Indian farmers during the mid-sixties. At this time, the necessity of an organization at the central level was realised for the multiplication and development of improved seeds. As a result, the National Seeds Corporation, a central organization was established in March, 1963. This is a Government of India undertaking set up under the administrative control of the Ministry of Agriculture. The main functions assigned to the Corporation are:

1. To establish a strong seed production industry in the country;
2. To produce foundation seeds based on the breeder seeds evolved at research stations;
3. To establish seed processing plants in the country;

4. To impart technical training in seed technology and to arrange for extension education of the farmers;

5. To enter into contracts for the distribution and selling of seeds; and

6. To undertake, by inspection and other means, quality control measures in all phases of the seed business carried on by or in cooperation with the State Seeds Corporations.

The National Seeds Corporation produces and markets the certified seeds of wheat, rice, maize, jowar, bajra, jute, fodder crops and vegetable crops. It has also created facilities for the production, processing and storage of certified seeds and for the production of foundation seeds. The Corporation selects seed certification agencies for the States in consultation with Central and State Governments under the Indian Seeds Act, 1966. The National Seeds Corporation has established quality control and seed testing laboratories at many places. It also exports seeds to other countries.

The National Seeds Corporation arranges for the sale of seeds through government agencies as well as through private selected traders with a view to making improved seeds available to the farmers in time. The desisting of State Departments of Agriculture from stocking and distribution of seeds in 1966–67 made it necessary for the NSC to develop an organised seed marketing system in the country. The extended marketing activities of the National Seeds Corporation include:

1. Strengthening of the marketing department at headquarters and regional offices;

2. Introduction of the dealer system;

3. Development of the export market in seed;

4. Enlarging other ancillary services like processing facilities, storage and movements; and

5. Intensive publicity to focus on the benefits of certified seeds.

With the launching of the National Seed Project, the National Seeds Corporation assumed the role of a leader to develop the seed industry on sound lines. Specifically the National Seeds Corporation is contributing towards the States Seeds Corporation's share capital, coordinates the certified seed production programme of several States Seeds Corporations, assesses the demand for seeds, looks after the inter-state marketing of certified seeds, plans and organizes the production of foundation seeds, plans the production of breeder seeds in consultation with ICAR, provides market research and sales promotion efforts, provides training facilities to the staff participating in the seed industry, maintains a reserve stock of seeds, provides certification services to those States which do not have independent seeds certification agencies and produces vegetable seeds for local and export market. It makes available seeds of more than 600 varieties of crops.

The Corporation has 10 regional offices and 77 area or sub-offices. It has around 12500 registered seed growers in its fold. The dealer-system of NSC includes more than 2800 dealers, which account for around two-thirds of its total seed sales. It is also producing and selling TC banana plantlets and seedlings/ saplings of selected fruits. It is a nodal agency for the central scheme on Assistance for Boosting Seed Production in the Private Sector.

(b) State Farms Corporation of India (SFCI)

The SFCI was established in 1969 under the Companies Act, 1956 to set up and run agricultural farms, primarily for the production of seeds of foodgrains, fibre crops, oilseeds, plantation crops, fruits and vegetables in various parts of the country. The Corporation operates large-scale farms in all the States where State Seeds Corporations have been set up (except Maharashtra). It participates in each State Seeds Corporation as a share holder; prepares development plans for those of its farms involved in National Seeds Project regarding production of foundation and certified seeds on behalf of National Seeds Corporation and State Seeds Corporations and acts as a consultant for farm development plan of the State Agricultural Universities and other institutions.

Seeds are produced on 12 central government farms under the control of SFCI. Some of these are Suratgarh and Jetsar farms in Rajasthan, Jhasugarha farm in Orissa, Jalandhar farm in Punjab; Hisar farm in Haryana, Raichur farm in Karnataka, Mizo Hills farm in Assam and Cannanore farm in Kerala. The total area under the farms of State Farms Corporation of India is 36141 hectares.

(c) State Seeds Corporations (SSCs)

State Seeds Corporations have been established in 16 States to widen the network of production and distribution channels for certified seeds in the country on the lines of the Tarai Development Corporation. In Rajasthan, the State Seeds Corporation was established on March 28, 1978. The main functions of State Seeds Corporation are: (i) production, (ii) processing, (iii) storage, and (iv) marketing of certified seeds. They work in coordination with NSC, agricultural universities and state departments of agriculture for executing the state seed plans.

Private Sector

There are many private seed firms existing in the country. Private seed firms are heterogeneous with respect to size, research capacity and product segments. There are more than 500 private seed companies in the seed sector. Of these, top 10 seed companies account for more than two-thirds of the domestic seed market. Major domestic companies are Nuziveedu, Kaveri, Rasi, Ajeet, Ankur and Rallis. Global players active in India are Bayer, Monsanto, Syngenta and Dow DuPont. The share of private sector in total seed business is little over 50%. The striking difference between private and public sector seed firms is

1. In public sector, research is separate from seed production and marketing whereas these functions are integrated in the private firms.
2. Product type: Private sector largely focuses on hybrid seeds especially of vegetables, oilseeds, cereals and cotton whereas public sector focuses mainly on major crops of the area.
3. Ownership: Private firms are closely held and not listed in the stock-exchanges although some of the large firms have sold equity to foreign seeds companies.

The seed industry in India, indeed had its beginning in the early 1960s with the establishment of the public sector National Seeds Corporation (NSC). The NSC provided foundation seed, training and technical assistance to state governments and private companies. This was followed in 1969 by the establishment of Tarai Seed Development Corporation (TSDC) that became the model for the State Seeds Corporations (SSCs) established in 1970s and 1980s. The primary purpose of these and related public sector organisations as state seed certifying agencies was to produce, certify and distribute high quality seeds that were the product of public research. They also stimulated private sector actively both directly or indirectly.

As the imports of commercial seeds was prohibited and foreign direct investment was not permitted, private sector actively depended on home grown firms. Thus, private sector grew in incremental steps focussing first on vegetables and later moving on to sorghum and pearl millet. In recent years private seed industry has grown and in many crops it has a sizeable presence. In the last decade, regulatory reforms have relaxed the restrictions on the entry of large and foreign owned private firms in seed industry.

DEMAND FOR SEEDS

Every plot of land meant for sowing of crops requires seed. In that sense, there is a demand for seed in the country for about 200 million hectares of gross cropped area. For field crops which are mostly seasonal, the demand for seed is a recurring demand in the sense that every year new seed is to be used and this seed has to be necessarily produced in the previous season and stocked for use in the next season. Another important characteristic of the demand for seed is that aggregate demand is relatively price inelastic because the seed is an indispensable input for crop production. For individual crops or crop groups also, the demand is price inelastic but the degree of inelasticity is less than that for aggregate of all crops.

From the point of view of the market demand, the portion of the demand for seed that is met by the farmers from their own retained stocks has limited relevance. Those who enter the market to buy their seed requirements are essentially of two categories viz., those looking for quality seeds of certified category and new varieties and those who, look for quality seed but are indifferent to the certification or the kind of variety. The discussion on the production and marketing of quality seeds of improved and high-yielding varieties presented in the earlier section should now be related to these categories in the sense that in our effort to increase agricultural production through technological upgradation, the move is to eliminate the kind of indifference to the choice of varieties from the second category of farmers. To the extent that our seed marketing and extension education systems are not able to reach all the farmers, the residual group of farmers will continue to meet its demand for seed from normal commodity trade channels.

The organised seed industry mainly caters to the demand of those farmers who are conscious of the quality seeds. The consciousness is being created as a part of agricultural production programme and development efforts. One way of looking at the effective demand for quality/certified seeds is through the quantities of certified seeds of various crops that were produced and sold to the farmers in various years. This, we have seen in the previous sections. The other way is to look at the area covered by high-yielding variety seeds and how has the gap been reduced over the years. Seeds of high-yielding varieties were introduced in India in mid-sixties. The main foodgrain crops covered under high-yielding varieties programme are paddy, wheat, maize, jowar and bajra.

The area under high-yielding variety seeds of foodgrain crops has increased over the years. However, in the case of coarse cereals, the coverage is lower as compared to wheat and paddy. As already indicated earlier, the demand for seed is an annually recurring phenomenon and as the coverage under high-yielding variety seeds is still not complete, the demand for such seeds will expand in future, the rate of expansion, of course will depend upon the intensification of extension education efforts and expansion of infrastructural facilities specially irrigation to new areas. In 2017–18, India produced 2.64 lakh tonnes of hybrid seeds against the requirement of 2.35 lakh tonnes.

In the case of other crops like pulses, oilseeds, fibres, vegetables, fruits and flowers, demand for quality seeds is more than what the organised seed industry is able to meet, specially the demand for quality seeds suitable for diverse agro-climatic conditions. In order to fulfil this demand-supply gap, a New Programme of Seed Development was launched with effect from October 1, 1988. In this programme, special emphasis was given to the import of high quality seed, strengthening and modernization of quarantine facilities and the development of domestic seed industry.

PRICING OF SEEDS

The prices of ordinary qualities of seeds transacted between farmers—usually from large farmers to small farmers or from traders to the farmers are derived from the prevailing grain prices. Prices of such seeds, therefore, tend to be higher than the grain prices. However, the prices of improved varieties of seeds and seeds of high-yielding varieties tend to rule much higher than the grain prices.

Prices of certified seeds marketed by the public agencies are announced by the government at the commencement of every sowing season. The government fixes these prices keeping in view the cost of production. The cost of production of certified seeds is higher than the cost of production of grain by 15 to 20% in the case of self-pollinated crops and by about 40% in the case of hybrid seeds. As the cost of production of seeds of hybrid varieties is generally two to three times of cost of production of ordinary grains, the prices of seeds of these varieties are many times more than the prices of ordinary grains.

The National Seeds Corporation is the leading agency in seed marketing. State agencies and the private traders set their prices in relation to prices fixed by the National Seeds Corporation. The prices of seeds marketed by National Seeds Corporation are determined by the government on a cost-plus basis. Apart from the procurement price of seed, cost of processing, certification cost, interest on capital, profit margin, transportation cost, handing costs and dealer's commission are also considered in arriving at the prices of seeds. The National Seeds Corporation revises the prices of seeds after taking market forces into account in addition to covering their costs to prevent diversion of seed into consumption.

Despite official attempts at containing prices of certified seeds, prices have increased sharply over the years, reflecting the increase in foodgrain prices in general and the increase in the cost of seed production in particular.

SEEDS POLICY AND REGULATION

Government regulates the seed industry and trade in seed in various respects. The main components of seed regulation are

(a) The Central Seeds Act of 1966;
(b) The Seeds Control Order of 1983;
(c) The Seeds Policy of 1988 and 2002; and
(d) The Protection of Plant Varieties and Farmers Rights (PPV & FR) Act, 2001.

The Central Seeds Act of 1966 and the Seeds Control Order of 1983

With the promulgation of the Central Seeds Act in 1966, the legal foundation was laid for India's present day seed industry, which provides legislative framework for regulation of quality seeds produced and distributed in the country. In addition, the Central Seeds Act prescribed certification standards of seeds and assigned responsibility for their enforcement to the State Governments. The Central Seeds Committee (CSC) and its various sub committees and the Central Seed Certification Board (CSCB) are apex agencies set up under the Act to deal with all the matters pertaining to quality seed production and distribution. These provide statutory backing to the system of variety release, seed certification, and seed testing. Varieties are released after evaluation at multi-location trials for a minimum of 3 years. Varieties approved are 'notified' which is a pre-requisite for certification. While all public sector varieties go through this process, it is not mandatory for private sector released varieties.

Certification is a process that certifies that seed is of a specific variety and is of acceptable genetic purity. Usually seeds are also tested for physical characteristics such as germination capacity, analytical purity and pathogen levels. Certification requires that the certifying agencies have access to the parent lines of the variety. The uncertified seeds released by the private sector are though not certified but the seeds are required to be truthfully labeled (TL) listing quality attributes on the label.

The Seed Control Order brings seeds within the scope of the Essential Commodities Act that regulates the marketing of essential items. All seeds sales outlets have to be licensed and must observe certain marketing practices such as public display of stocks and prices.

Seeds Policy

Prior to 1991, the seed industry was also subjected to the policies on industrial licensing and foreign direct investment that applied generally. The seed sector was reserved for the small scale sector and the entry of the foreign firms was tightly regulated. These controls have fallen by the wayside as a consequence of the economy-wide reforms of 1991.

There had been two later developments which need to be noted.

1. In September 2001, the Plant Variety Protection and Farmers' Right Act came into being; and
2. In June 2002, Government announced new seeds policy that significantly alters the framework of regulation.

Variety registration (i.e. notification) is now mandatory for all varieties, new and extant. The evaluation is to be done over three seasons of field trials. However, certification continues to be voluntary. The emphasis on registration in the new seeds policy ties with the demands of the Plant Variety Protection and Farmers' Right Act passed in 2001. This Act provides for plant breeders rights, which require extant and new plant varieties to be registered on the basis of the characteristics relating to novelty, distinctiveness, uniformity and stability.

Besides regulating quality, the government had also controlled imports and exports of seed. The seeds policy of 1988 allowed limited imports of commercial seeds. Curbs were removed from imports of seeds of vegetables, flowers and ornamental plants. The new seed policy of 2002 allows imports and exports of seeds of all crops. However, all imported seed is also required to go through the process of registration.

While there is considerable demand, the unavailability of quality seeds and planting material has plagued the growth of horticultural sector. There is no legislative framework for quality assurance of vegetatively propagated plants. Many states have brought out necessary legislation but the enforcement has been difficult in the absence of appropriate quality standards.

As a part of Seeds Policy, the Government has been encouraging the private sector to expand the size of seed business, with a view to increasing the seed replacement rate in the country. The Government has been providing capital subsidy for establishment of seed processing plants. Around 400 private companies took advantage of this provision and reportedly 7.5 million quintals of seed processing capacity and 3 million quintals of seed storage capacity were created in the private sector.

Under the New Policy on Seed Development (NPSD), two major initiatives were taken in 2013–14:

1. 100% foreign direct investment was permitted under the automatic route.
2. With thrust on creating seed banks, a Seed Rolling Plan for the period up to 2016–17 was put in place for all the states for identification of good varieties for the seed chain and the agencies responsible for production of seeds at every level. A programme of National Seed Reserve is being implemented through 22 agencies that include NSC, State Seeds Corporations and Agriculture Departments of four states.
3. In 2015, Cotton Seed Price Control Order was issued to control the prices of Bt cotton hybrid seeds. This was done under the Essential Commodities Act, 1955. Cotton is the only transgenic crop approved in India.
4. India is also maintaining a SAARC Seed Bank to help the members of SAARC.
5. Seed business is regulated under the Consumer Protection Act and farmers (as users of seed) are treated as consumers.
6. A new seed bill is being discussed at various levels which aims at ensuring the farmers to get high quality seeds while the seed industry experiences 'ease-of-doing-business'. It may provide for compulsory registration of varieties before the seed is sold to the farmers.

Protection of Plant Varieties and Farmers' Rights (PPV and FR) Act, 2001

For implementing the provisions of the PPV & FR Act, the government has established the PPV & FR Authority in November 2005, Fourteen crops viz., rice, wheat, maize, sorghum, pearl millet, chick pea, pegion pea, green gram, black gram, lentil, field pea, kidney bean, cotton and jute were initially notified for the purpose of registration under this Act. There are plans to extend its operation and coverage to forestry and aromatic and medicinal plants.

Up to the end of March 2019, the PPV and FR Authority has received 16532 applications for registration of varieties. These are either extant varieties (notified earlier under Seeds Act, 1966), new varieties; VCKs (varieties of common knowledge) or EDVs (essentially derived varieties). Of these, 592 applications were filed during 2018–19, which include 78% by farmers, 15% by private seed companies and 7% by the public agencies. After due process of verification, the Authority issues certificates. So far (up to end of March 2019), the Authority has issued 3623 certificates that include 1597 (44%) for the farmers' varieties.

WTO AND SEED INDUSTRY

Soon after the WTO came in existence, the accord on Trade Related Aspects of Intellectual Property Rights (TRIPS) had raised a whole range of apprehensions in the country. To fulfil the TRIPS obligations, the Indian Parliament passed a bill for protection of plant varieties and farmers' breeding rights. TRIPS agreement does not require us to patent seeds. However, it was essential to establish an effective system for the protection of plant varieties, seeds and other forms of propagation material as obliged under the WTO's Plant Breeders' Rights (PBR). With such measures, the participation of private sector in the Indian seed industry got some impetus.

In recent years, especially after Indian agriculture became a part of the world's production and trade regime, the seed industry in India has moved into new directions. The most significant of the post-WTO developments are the emergence of biotechnology as the central focus of agricultural research and the expanding clientele of the genetically modified or transgenic seeds. At present, transgenic seeds are the monopoly of few multinational companies (MNCs) such as Monsanto, Dow-Chemicals, Du-Pont, Novaritis, Zeneca, BASF and Aventis.

This situation is unlikely to remain so for long, as ICAR, Agricultural Universities, Department of Bio-Technology and other research institutes have already made considerable headway in developing transgenic seeds. IARI has developed transgenic seeds of a number of crops, viz., brinjal, tomato, cabbage, rice, mustard, cowpea and potato. Further, transgenic seeds are not free of controversies, especially from the view point of human health and environmental jeopardy. The most recent example of this is Bt brinjal, which is surrounded by controversies. However, Bt Cotton has been a sucess story of the benefits of transgenic seeds in the world as well as in India.

PLANT PROTECTION CHEMICALS

Diseases and pests attack crop plants and decrease yields, and are sometimes responsible for total crop failure. Weeds compete with crop plants for nutrients. It has been shown that farm produce worth ₹ 2.5 trillion (2.5 lakh crore) is lost to insects, pests, diseases and weeds every year. Plant protection chemicals are used to either prevent their attack or control them. These inputs are in the nature of an insurance against possible yield losses. The introduction of new crop varieties in farming has increased the importance of this input because irrigated high-yielding crops provide a conducive environment for weeds, insects and pests. Unlike seed, this input has to be used at different times of crop growth. Plant protection measures begin with soil treatment before sowing, and continue up to disposal of the produce, either in the form of market sale or use as seed during the next crop season. The use of chemicals requires a high level of technical knowledge in the selection of the chemical, the method of its use, the care to be taken in handling it, and an appropriate dosage and frequency of treatment. There are three kinds of pesticides (insecticides, fungicides and weedicides) and three other kinds of chemicals (rodenticides, fumigants and plant growth regulators) commonly used in agriculture.

PRODUCTION AND CONSUMPTION OF PESTICIDES

Pesticides industry started in India with the import of BHC in 1952–53. Indigenous production of pesticides began with the establishment of BHC and DDT plants in 1954. By 1958, India was manufacturing five basic pesticides having production of over 5000 tonnes. The number of indigenously manufactured pesticides increased to 42 in 1977. Later, some more pesticides were added to the range. The pesticides industry has witnessed a steady progress since then. Bulk of the production had been of insecticides. Pesticides in India are manufactured by private sector as well as public sector companies or corporations. At one time, there were 80 large units manufacturing pesticides and their formulations, apart from 500 small units. Production of technical grade pesticides in India was 16,583 tonnes in 1974–75, which increased to 88560 tonnes in 1999–2000 against the installed capacity of around 1.10 lakh tonnes. Today, India is the largest manufacturer of basic pesticide chemicals among the South Asian and African countries, next only to Japan and is among the 10 top producers in the world. India accounts for 2% of pesticide usage in the world. At present, 40 companies are making technical grade (generic) pesticides and there are around 800 formulators (some of these are defunct). Nearly 80% of pesticides sold in India are generic.

The consumption of pesticides in the country was initially very low, so much so that, by the end of the First Five Year Plan (1955–56), only 2350 tonnes were consumed in India (Table 6.11). Later there has been a rapid increase in the area covered by the plant protection measures. The consumption of pesticides increased from 8.62 thousand tonnes in 1960–61 to 45 thousand tonnes in 1980–81 and to 75.03 thousand tonnes in 1990–91. However, since the nineties, a greater emphasis is being given to biological and integrated pest management (IPM). The use of chemical pesticides, therefore, has not increased further. In 2017–18, the consumption was 58.16 thousand tonnes. Out of the total plant protection chemicals used in India, insecticides account for nearly 80%, fungicides 12% and herbicides account for 5%. There is a considerable variation in the use of pesticides from state to state and from crop to crop. The state-wise consumption of pesticides indicates that 33.6% of total pesticides consumption was in Andhra Pradesh, 16.2% in Karnataka, 15.2% in Gujarat and 11.4% in Punjab. The crop-wise pesticide consumption indicates that about 50% is used in cotton, 18% in paddy, 14% in vegetables and 2% in sugarcane. The studies have shown that by every one rupee spent on plant protection, on an average, crop worth about rupees five is protected which otherwise would have been lost due to pests and diseases.

Table 6.11: Consumption of pesticides in India (technical grade material)

Year	Total consumption (thousand tonnes)	Consumption per hectare of gross cropped area (gms)
1955–56	2.35	16
1960–61	8.62	56
1970–71	24.31	147
1980–81	45.00	260
1990–91	75.03	433
1995–96	61.26	328
2000–01	47.04	254
2005–06	39.77	206
2010–11	55.54	281
2015–16	54.12	273
2017–18	58.16	293

Source:
1. Indian Agriculture in Brief, Various Issues, Directorate of Economics and Statistics, Ministry of Agriculture, Government of India, New Delhi.
2. Agricultural Statistics at a Glance, Various Issues, Directorate of Economic and Statistics, Ministry of Agriculture, Government of India, New Delhi.

The domestic production of pesticides in India is less than the demand for them. The gap is met through imports. While assessing the use of pesticides and demand for them, several questions regarding the level of use are being raised. The consumption of pesticides per hectare of gross cropped area in India increased from 16 gms. in

1955–56 to 147 gms in 1970–71 and further to 433 gms in 1990–91. The average per hectare consumption of pesticides decreased to about 328 gms in 1995–96. Since then, it has remained below 300 gms per ha. At this level, the consumption of pesticides in India is lower than in advanced countries like the U.S.A. (1600 gms/hectare), Europe (2000 gms/hectare) and Japan (10790 gms/hectare).

But lower per hectare pesticide use in India vis-a-vis other countries should not be a worry. What is important is not the demand and consumption of chemical pesticides, but the demand for control of pests and diseases to save the potential crop or grain loss. Due to the hazards of use of pesticides, the emphasis is now shifting to Integrated Pest Management (IPM) rather than the use of chemicals. The IPM involves a combination of biological and agronomic practices to control the diseases and pests. Nevertheless, the gross demand for pesticides and associated plant protection equipments as also for the technical skills to use them is likely to expand as farmers are moving towards market-oriented farming system.

REGULATORY FRAMEWORK

The pesticides sector in India is mainly regulated by the Insecticides Act, 1968 and the Insecticides Rules 1971 (with amendments subsequently made there in). These regulate the imports, manufacture, sale, transport, distribution and use of insecticides/pesticides.

There is a Central Insecticide Board and Registration Committee (CIB-RC) with headquarters at Faridabad. Each manufacturer, formulator or importer has to register under the provisions of the Act. There are 260 pesticides approved for use in India. There are some class 1 pesticides allowed in India and a few of these are heavily used. It may be mentioned that the World Health Organization (WHO) classifies pesticides, according to acute toxicity as extremely hazardous (class 1a) and highly hazardous (class 1b). Many of these are banned in different countries.

There are quite a few spurious, fake, duplicate, sub-standard, adulterated or counterfeit pesticides sold to the farmers, which need to be detected and corrective measures taken. There are 70 pesticides testing laboratories with a testing capacity of 75100 samples. There are, in addition, two central pesticides labs (capacity of 1600) and two regional labs (capacity of 2000). The pesticides in the market are continuously reviewed and notified as banned, restricted and under review on the grounds of public health as well as harmful effects on soil/microorganisms.

The other aspect of regulation of pesticide use is the monitoring of chemical residues in food stuffs. There is a Pesticides Residue Monitoring Network (PRMN) with ICAR as the leader. There are nine organizations or Ministries and 25 accredited laboratories, which regularly monitor the residues, following the standard international protocols. Maximum Residual Limits (MRLs) are prescribed by the Ministry of Health and FSSAI. Broadly, the results show that though residues are detected in many samples, these exceed the MRLs only in (say) 2% cases. Pesticide residues occur mainly when the time gap (waiting period) between chemical application and harvesting is less than the prescribed period.

In addition, there are two other legal/regulatory instruments related to pesticides. One is Destructive Insects and Pests (DIP) Act 1914. The other is Plant Quarantine Order, 2003 (with subsequent amendments), which is meant to regulate the imports to safeguard against exotic pests. This aspect is looked after by the Directorate of Plant Protection Quarantine and Storage.

MARKETING OF PESTICIDES

For regulating the manufacture, import, sale and use of pesticides, the Insecticides Act, 1968 is operative in the country. Under this act, there is a provision of registration of pesticides with the Government of India (Ministry of Agriculture). Every formulation has to be registered after making a formal application in this regard to the government.

For marketing of pesticides, the manufacturers or formulators appoint distributors for each region. These distributors are mostly located in urban centres and are either in the cooperative or private sector. They arrange to supply the plant protection chemicals to farmers through a network of dealers. The number of dealers varies from chemical to chemical and area to area. These dealers are either in the private sector, including agro-service centres; agri-clinics and agri-business centres or in the cooperative sector. The lack of technical know-how on the part of the farmers or dealers and the complementary requirement of such equipment as sprayers or dusters make the marketing of plant protection chemicals a difficult and skilled job. Its misuse may endanger human life.

There are around 1.79 lakh pesticides sale points (outlets) in India. Nearly 90% of these are private outlets, 7% are operated by cooperatives and remaining by the government agencies or NGOs. Government departmental outlets are in 11 states that include Himachal Pradesh, Mizoram, Nagaland, Tripura, Sikkim, Uttarakhand, and NCR Delhi.

The identified marketing channels for pesticides in a study conducted in Haryana state[1] are depicted in Fig. 6.2. Amongst all the identified marketing channels for pesticides, the channel involving manufacturer, distributor and retailer is the most common channel.

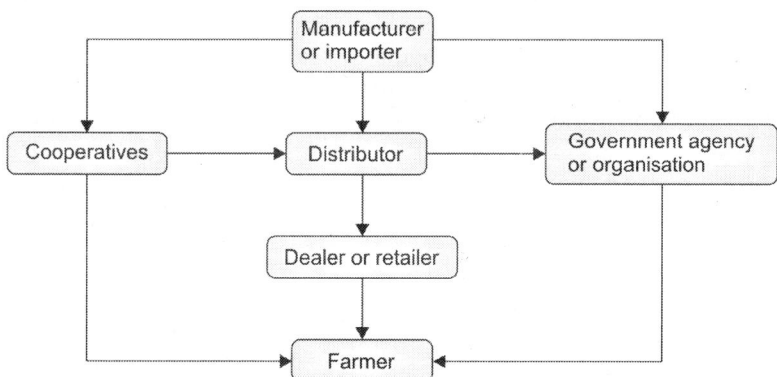

Fig. 6.2: Pesticides marketing channels

Some of the problems faced by the farmers in procuring plant protection chemicals are:

1. The number of pesticides/insecticides depots are inadequate. Each depot covers 10 to 15 villages. Farmers have to travel long distances to get their requirements of plant protection chemicals. This increases the cost of material and results in the wastage of farmer's time. Most of the time, the demand gets blunted.

2. There is a short supply of the pesticides of a particular brand in the market because of insufficient production.

3. Chemicals available from the existing depots are not stocked in adequate quantities, which results in loss of the crop when there is a severe attack by insects/pests.

4. Farmers are not familiar with the insecticide/pesticide which is required for a particular insect/pest/disease, and dosage and techniques of its use. Spraying and dusting machines needed for the use of insecticides and pesticides are costly and beyond the means of an ordinary small farmer.

5. The cost of plant protection measures is very high because of the high prices of insecticides and pesticides.

6. There is, moreover, the non-availability of spraying and dusting equipment on custom service basis.

7. Farmers are often confronted with the problems of supply of poor quality or adulterated lot of insecticides/pesticides. There are dubious labels on packs. Spurious pesticides market is estimated to be around ₹ 3,000 crore. Spurious pesticides lead to crop losses of ₹ 1 lakh crore annually; and

8. Plant protection chemicals of many companies are supplied in large sized packages, which are not convenient for small farmers or vegetable growers.

The following measures are suggested for improving the distribution and marketing of insecticides/pesticides:

1. Private entrepreneurs including agri-clinics and agri-business centres may be encouraged to market insecticides and pesticides, and they should be given adequate technical and financial support, so that they may increase the number of distribution outlets in villages.

2. Cooperative organizations for the marketing of insecticides/pesticides should be revitalized. Farmers Societies should be entrusted with the task of running input depots, including those of pesticides, so that they may become available in time to the members of these societies.

3. At least one sales point in each village and market, including sub-market, should be established to ensure easy availability of insecticides and pesticides.

4. At each sales depot, either in the private or cooperative sector, technical guidance should be provided to the buyers of the chemicals. Sprayers and dusters should also be provided on custom-hire basis to the buyer of the plant protection chemical.

5. More emphasis should be given to biological control and IPM techniques rather than increasing the use of chemicals.

6. The extension education efforts be directed at minimising the crop residues in the product.

7. Efforts should be made for checking the malpractices of pesticide dealers by conducting surprise checks on stocks, quality and prices of insecticides and pesticides charged by the dealers,

8. Only standard or genuine companies should be encouraged to manufacture and supply pesticides to the farmers to check crop losses and prevent accidents in the use of insecticides or pesticides.

PRICES

The prices of pesticides and insecticides are determined by normal demand and supply forces. For the pesticides as a group, the nature of market is very close to a situation of oligopoly or monopolistic competition. Apart from the regulatory measures, public intervention in the domestic market for pesticides existed in the form of explicit subsidies given by the government to certain sections of farmers such as marginal and small farmers and those belonging to scheduled castes and scheduled tribes. As these subsidies were given to meet part of the cost of pesticides in the cultivation of certain crops like oilseeds and pulses, they result in the increase in the quantity of pesticides demanded. The factors such as expansion in irrigation facilities and increase in area under high-yielding varieties also shifted the demand schedule for pesticides and insecticides.

ELECTRICITY

Agriculture, in a way, is an energy-conversion industry. However, electricity is an important energy input in farming, and is made available by the non-farm sector. It is used mainly to energise pumpsets and tubewells, for lifting irrigation water. Threshers and chaffcutters are also operated by electricity. Most agro-based and rural industries use electricity as the source of power.

SUPPLY, DEMAND AND MARKETING

The government or the public sector companies are responsible for the production and distribution of electricity in India. There is no intermediary or middleman between the producer of electricity and consumers (farmers). In other words, the public sector has monopoly control over the supply of electricity. However, now production has been opened to the private sector also but distribution remains under public control.

One of the peculiar phenomenon of this input is that demand for electricity has often outpaced supply. The generation of electricity and laying of transmission lines are highly capital-intensive. Therefore, the total supply has remained constrained by the availability of investable funds for this purpose. This apart, the allocation of the available supply to agriculture and rural sector vis-a-vis other sectors like industries, domestic sector, commerce and railways was determined by the government keeping in view the needs of different sectors within the overall macro-development goals of the country.

Recognizing the importance of electricity for agriculture as well as for other development needs like domestic use, industries and railways, multi-dimensional tremendous progress in electricity sector has been made. First is the rapid expansion in installed capacity for power generation. The installed capacity has multiplied by around 215 times during the last 70 years. Second is harnessing of renewable energy sources (RES). Though thermal power continues to be the main source, the share of RES has gone up. In 2019, out of total installed capacity, the share of thermal power is 64%, of RES is 22%, hydropower is 13% and nuclear is only around 1%. Third, the private sector is contributing significantly now, with a share of 46% in total power capacity. The share of central sector is 24% and of the state sector is 30% in total installed capacity. And the fourth dimension is that the peak deficit (short fall in supply vis-a-vis peak hour demand), which at one time was very high, has declined to negligible level. For example, in 2018–19 it was less than 1%. The progress of electricity generation by different sources of power is shown in Table 6.12. The available supply has grown from mere 6.6 billion kWh in 1950 to 289.4 billion kWh in 1990–91 and further to 1418.9 billion kWh in 2016–17.

The pattern of effective demand for electricity as judged from its actual consumption has undergone a structural change during the last 70 years as shown in Table 6.13. Out of the total electricity consumed in India, the share of agriculture sector has increased

Table 6.12: Pattern of electricity generation in India (utilities)

(billion kWh)

Year	Hydel	Thermal + RES	Nuclear	Total
1950–51+	2.5	2.6	–	5.1 (1.5)
1960–61	7.8	9.1	–	16.9 (3.2)
1970–71	25.2	28.2	2.4	55.8 (5.4)
1980–81	56.5	61.3	3.0	120.8 (8.4)
1990–91	71.7	186.5	6.1	264.3 (25.1)
2000–01	74.5	408.1	16.9	499.5 (55.0)
2010–11	114.4	704.3	26.3	845.0 (120.9)
2016–17	122.4	1075.6	37.9	1235.9 (183.0)

Figures in brackets are for non-utilities. + On calendar year basis.

Source: Economic Survey—Various Issues, Ministry of Finance, Government of India, New Delhi.

Table 6.13: Pattern of electricity consumption in different sectors in India.

(percent utilization)

Year	Agriculture	Domestic	Commercial	Industrial	Rly. and others	Total
1950–51+	3.9	12.6	7.5	62.6	13.4	100
1960–61	6.0	10.7	6.1	69.4	7.8	100
1970–71	10.2	8.8	5.9	67.6	7.5	100
1980–81	17.6	11.2	5.7	58.4	7.1	100
1985–86	19.1	14.0	5.9	54.5	6.5	100
1990–91	26.4	16.8	5.9	44.2	6.7	100
2000–01	26.8	23.9	7.1	34.0	8.2	100
2010–11	20.5	24.3	8.7	36.8	8.3	100
2017–18	18.1	24.2	8.5	41.5	7.7	100

+ On a calendar year basis.

Source: Economic Survey — Various Issues, Ministry of Finance, and Reports of CEA, Ministry of Power, Government of India, New Delhi.

from 3.9 recent in 1950–51 to around 18% in recent years. The electricity consumption in agriculture increased manifold from 162 GWh in 1950 to 204293 GWh in 2017–18.

The supply of electricity in rural areas of the country can be assessed by the number of villages electrified and the number of pumpsets/tubewells energised in the country. The situation at different points of time during the last 70 years is shown in Table 6.14.

Till 1947, only about 1500 villages (0.53%) had been electrified in India. Though with the introduction of the First Five Year Plan in 1951, rural electrification did receive impetus, but by April 1969, the number of villages electrified increased to only 12.8% of the total. To give boost to the supply of electricity in rural areas, the Rural Electrification Corporation (REC) was established at the national level in 1969. It is an autonomous body and is registered as a Company under Companies Act of 1956. Subsequently, State Electricity Boards were established at the state level. The REC promoted and financed Rural Electricity Cooperatives and State Electricity Boards. REC also finances rural electrification projects on area basis. Since the establishment of REC, the supply of electricity in rural areas got accelerated. By the end of March

Table 6.14: Supply of electricity in rural areas of India

As on	Progressive number of villages electrified		Progressive number of pumpsets energized ('000)
	Number	Percentage of total villages	
April 1, 1951	3,061	0.53	21
April 1, 1961	21,754	3.78	198
April 1, 1969	73,939	12.84	1089
April 1, 1980	249,799	43.76	3449
July 31, 1991	481,956	83.20	8992
March 31, 1999	498,775	86.00	11946
March 31, 2009	617,174	99.00	NA
March 31, 2019	597,464	All	21,000

Source: Indian Agriculture in Brief—Various Issues, Directorate of Economics and Statistics, Ministry of Agriculture, Government of India, New Delhi and Rural Electrification Corporation, New Delhi.

1999, 86% of the villages were supplied with electricity. Later, a special campaign was launched for acceleration in rural electrification. As a result, by April 28, 2018, all the census villages (597464) have been electrified.

Apart from reaching the electricity to the villages, special attention was given during the last five years to provide electricity to all the rural households and left-out urban households. It is a historic achievement that by the first week of January 2019, all rural (212 million) and urban households have been provided with electricity connections. It should be noted that about 25.3 million poor households were provided free electricity connections.

As a result of increase in the network of the supply of electricity in the rural areas, there has been tremendous increase in the number of energised irrigation pumpsets and tubewells. The number of pumpsets energised increased from mere 21 thousand in 1951 to 11.9 million in 1999 and further to 21 million by April 2019.

Till 2002, the state electricity boards were solely responsible for generation and distribution of electricity among end users except a small segment under the purview of private distribution companies. After the enactment of Electricity Act 2003, the electricity sector got a boost and several reforms were undertaken. A Central Electricity Authority (CEA) was put in place. Several states reformed the power sector. Separate public sector units (PSUs) or companies were established for generation, transmission and distribution of electricity. Measures were taken to even out supply-demand mismatches (integrated power transmission grid). Central Electricity Regulatory Commission (CERC) and State Electricity Regulatory Commissions (SERCs) were created, with powers to, inter alia, grant inter-state and intra-state licences. The CERC issues regulations regarding the trading margins for inter-state trade in electricity (power purchase agreements).

It needs to be noted that India has not only 'one-nation-one-grid' but also (a) India has turned from a power-deficit to power surplus country, (b) it exports electricity to Nepal and Bangladesh while it imports surplus electricity from Bhutan, (c) it has also expanded and improved transmission system, and (d) provided last mile connectivity to left out villages and households. However, the issues that need attention are (a) control over losses in transmission, (b) achievement of a target of zero-cuts, and (c) supply of electricity to the farmers during day hours for a greater duration.

PRICING OF ELECTRICITY

The government or the public sector so far has the monopoly control over the supply or distribution, therefore, the prices of electricity usually called electricity rates are fixed by the government. The present situation, by and large, is that most states have autonomous organizations or companies dealing with generation (G), transmission (T) or distribution (D) functions. An SERC oversees the pricing of electricity, keeping in view the cost of G, TD and/or purchase price plus cost of T & D. Despite all these arrangements, the final decision on the price to be charged from farmers rests with the state government. Under this scenario, the difference between the price charged from farmers and cost of GTD (or import cost plus T&D) is paid by the state to the state/regional electricity distribution company. A policy of differential pricing has been adopted by the state governments. Different prices are charged in different sectors—business, industry, domestic and agriculture.

Electricity to the agriculture sector is supplied at a lower rate than to the other sectors. Not only this, the rates charged from agriculture as also the domestic sector is less than the economic cost of generation and transmission. Two ways of charging prices (rates) for electricity are in operation—one is the meter basis and the other is the flat rate. In the first method, the farmers are charged on the basis of number of units consumed. However, to ensure a minimum consumption level, minimum charges are prescribed which vary with the horse power of the motor used for lifting the water. In the flat rate system, the rates are linked to the horse power of the motor irrespective of the units of electricity consumed. The rates of electricity under both systems vary from state to state and are revised by the state governments from time to time.

For illustration, the average rates of electricity charged from different categories of consumers during the 1990s are shown in Table 6.15. For the annual behavior of electricity rates for irrigation, see the last table of this chapter. As the average rate charged from farmers and also from domestic users is less than the cost, unless the charges are higher than cost for businesses and industrial users, the distributor companies (DISCOMS) are bound to be in losses. In such cases, the concerned state government has to bear these losses and bail out the DISCOMS. This kind of arrangement, where there is cross subsidization, raises several questions from various quarters. There have been arguments in the literature about the so-called electricity subsidy to the farmers and whether or not it is justified (for details see Acharya, 1997).

Table 6.15: Consumer category-wise sale of electricity and average tariff

Consumer category	1992–93		1999–00	
	Sale (%)	Tariff (Paise per kWh)	Sale (%)	Tariff (Paise per kWh)
Domestic	16.4	77.3	20.0	149.1
Commercial	4.0	165.3	4.9	353.6
Agriculture	29.9	16.1	29.9	24.9
Industrial	35.5	171.5	32.6	350.5
Railway	1.8	206.8	2.3	410.6
Outside state	2.1	91.0	3.2	121.1
Others	10.3	–	7.1	–
All consumers	100.0	105.4	100.0	207.6

Source: Annual Report of Working of SEBs, Government of India, Planning Commission, New Delhi.

DIESEL

Diesel is used as a fuel for tractors and some pumping sets. Diesel market in India is characterized by oligopolistic situation, for there are a few main suppliers and a large number of buyers. The public sector companies which supply diesel oil are Hindustan Petroleum, Bharat Petroleum and Indian Oil Corporation. Now, a private company viz; Reliance is also in the field. These companies stock their products at various locations. From these locations, diesel oil is supplied to filling stations for supply to farmers. In some areas, it is sold to farmers through dealers, who keep the oil in drums. The selling price of diesel is fixed by the government for different regions of the country. The actual price realised from farmers varies from place to place because of the differences in transportation cost and octroi charges; but the difference is marginal.

Quite often, the demand for diesel outpaced its supply. During such acute scarcity periods, the government directly intervened and regulated its supply to farmers through the permit system. The Civil Supplies Departments of State Governments issued through the District Collectors, a permit to each farmer according to his requirements of diesel for pumpsets and tractors. On the basis of these permits, farmers obtained diesel from local agents or filling stations. Often, however, they had to wait for their supplies for quite a long time, or buy their requirements in the black market, which thrived during periods of scarcity. The Ministry of Agriculture monitored the supply of diesel oil to the farm sector. However, now the supply of this is liberalised and prices are determined by the Consortium of these corporations based on the international prices. The situation has considerably improved in recent years. However, the retail prices are still under the control of the government, as far as public sector companies are concerned. During the last 15 years, the prices of diesel have increased at a much higher rate than that of other inputs. In October 2014, diesel prices were decontrolled. Now, these fluctuate frequently, sometimes on daily basis (for annual behaviour of diesel prices, see the last table of this chapter).

FARM MACHINERY

The importance of farm machinery has considerably increased over the years. These have replaced the human and animal power in performing farm operations. In 1970–71, out of the total farm power in Indian agriculture, nearly 61% was sourced from bullocks (45.4%) and human labour (15.4%). Now, their combined share has declined to mere 10% (5% each).

Farm machines are used to perform certain operations with speed and accuracy—operations which may not be possible with human and bullock labour. The machines used for farming are tractors, pumpsets, sprayers/dusters, combine harvesters and threshers. Tractors are used to prepare land to receive seeds and to transport the produce. They are also used as source of power in operating other machines—irrigation pumps, winnowers, threshers, sprayers and dusters.

The estimated mechanization level of farm operations is 40% for tillage, 30% for seeding/planting, 37% for irrigation, 35% for plant protection, 48% for threshing of wheat, and 5% for threshing of other crops (CIAE, quoted in State of Indian Agriculture-NAAS 2009).

The demand for farm machines is a derived demand. Machines are manufactured mostly in the private or corporate sector. The manufacturers market their products through authorised agents/dealers and sub-dealers. Most of the sales depots are located in cities or towns. The State Agro-Industries Corporations have also entered the market for the sale of certain farm machines and their spare parts. Annual investment in farm machines is ₹ 500 billion, including ₹ 200 billion in agroprocessing machines.

PUMPSETS

The demand for pumpsets (including submersible pumps and diesel engines) depends on the level of biochemical technology, the demand pattern for labour and energy, the relative prices of human labour, fuel and oil vis-a-vis pumping sets, the expansion in canal irrigation and the exploitation of ground water. There are 600 manufactures of pumps in India. Manufacturers have to project the demand and adjust their production schedule accordingly. By its credit policy, the government has been encouraging the purchase of pumpsets by farmers. The market for pumpsets is, characterized by monopolistic competition. Different makes of pumpsets are available; and each manufacturer tries to popularise its product by price differentiation and sales promotion activities.

The selling price of pumpsets is fixed by the manufacturer. The dealers or agents have to sell the pumpsets at the fixed prices. The expansion in aggregate demand and the supply of pumpsets can be judged from the number of electric or diesel-operated pumpsets installed by the farmers in India. As shown in Table 6.16, the number of pumpsets was only 87 thousand in March, 1951 and increased to 187 lakh in 2000. The number has gone up further in recent years. Every year, between 2.5 lakh and 5.00 lakh pumpets are being energised, increasing the cumulative number of energised pumpsets. By March 2019, the cumulative number was 3.0 crore.

Table 6.16: Cumulative number of irrigation pumpsets in India

('000)

At the end of March	Diesel pumpsets	Electric pumpsets	Total
1951	66	21	87
1961	230	198	428
1969	721	1089	1810
1980	2553	3449	6002
1991	4850	9100	13950
2000	6480	12220	18700
2019*	9000	21000	30000

* Approximate

Source: Government of India, Agriculture in Brief, 23rd and other Editions, Ministry of Agriculture, New Delhi, CMIE Publications and Industrial Reviews.

The increase in prices of electricity has been at a rate lower than that of diesel. Obviously, the farmers prefer electric operated pumps over diesel engines owing to both lower cost and higher energy use efficiency. As such, the number of electric operated pump sets has increased faster than diesel engines. Even though the use of electric pumps is economical, the frequently interrupted power supply in some remote areas has made diesel engines the only option for these farmers.

TRACTORS

Tractors are mainly used to prepare land for sowing and to transport the produce. They are also used as source of power in operating other machines like irrigation pumps, winnowers, threshers, chaffcutters, the power sprayers/dusters. The nature of the demand for tractors originates from the need to perform the operations of land preparation and transportation and from the requirement of the services performed by other machines.

Tractors were first introduced in Indian farming in the early twenties. The manufacture of tractor in India commenced in 1960. Prior to 1960, they were being imported.

In 1961, Eicher Tractors Ltd and Tractors and Farm Equipments (TAFE) Ltd started manufacturing tractors with foreign collaborations. Later, other industries started manufacturing tractors with foreign know-how, such as Gujarat Tractors Ltd (1963), Escorts Ltd (1966), International Tractors (India) Ltd (1966), and Hindustan Machine Tools Ltd (1977). Punjab Tractors Ltd started production with indigenous technology in 1974. Many more units started manufacturing tractors since then with indigenous and foreign know-how.

Though the indigenous production of tractors increased to a level of 5714 tractors per year in 1965–66, the supply was not sufficient to meet the demand. Therefore, imports of tractors continued even after the commencement of the indigenous production. Up to 1966–67, the country had to import around two to three thousand tractors each year. During the next five years, Indian farming underwent a technological revolution in major wheat growing areas which spurted the demand for tractors.

Tractors were imported from Czechoslovakia; the U.K., the erstwhile USSR, West Germany, Rumania, Bulgaria, Yugoslavia and the German Democratic Republic. The increasing level of imports of tractors which created the problem of non-availability of spare parts and service facilities, had a deleterious effect on the growth of the indigenous tractor industry. In February 1973, the Union Government, therefore, decided to ban the import of tractors altogether, except of those coming under the World Bank assistance scheme.

Since then, the annual production of tractors had been continuously going up as shown in Table 6.17. In 1990–91, 1,39, 233 tractors were manufactured in the country. In 1992, the industrial licensing to manufacture tractors was abolished. The tractor production in the country further increased to over 6.19 lakh by 2012–13. At present, there are 20 units engaged in the manufacturing of tractors. These include 16 domestic and four global companies manufacturing in India. Of these, five companies account for 80% of total tractor production. These are M&M, TAFE, ITL-S, Escorts and J-D. These units have a licensed capacity of over three lakh tractors per year. Presently, India is the largest manufacturer of tractors in the world, accounting for one-third of the global production. India is now exporter of tractors.

The trend in effective demand for tractors in the country can be assessed through the data on sales of tractors. The sales have increased from 30,229 tractors in 1974–75 to 65,101 in 1980–81, 1,39,831 in 1990–91 and 2,54,825 in 2000–01. During 2012–13, 5.9 lakh tractors were sold in the country. From 2013–14 to 2017–18, 34.08 lakh tractors were sold, with 7.97 lakh sales during 2017–18.

Tractor is a durable resource which provides flow-service in the farm sector. Therefore, the number of tractors demanded by the farm-community as a whole during a given year/period, depends inter alia on the inventory or stock of tractors already available in the farm sector. The estimated stock of tractors in the country at different points of time, as shown in Table 6.18 has increased at a rapid rate. In 2018, around 94 lakh tractors were being used in the country. The number of tractors per 1000 ha of net sown area is around 67.

Table 6.17: Production, imports and sale of tractors in India

(number)

Year	Production	Imports	Sale
1961–62	880	2,997	NA
1965–66	5,714	1,989	NA
1970–71	20,104	13,300	NA
1975–76	33,302	1,100	NA
1980–81	70,007	Nil	65,101
1985–86	75,550	Nil	76,886
1990–91	1,39,233	Nil	1,39,831
1995–96	1,91,329	Nil	1,91,497
2000–01	2,61,609	Nil	2,54,825
2007–08	3,45,172	Nil	3,47,000
2012–13	6,19,000	Nil	5,90,672
2017–18	NA	Nil	7,97,000

NA = Not available.

Source: Government of India, Economic Survey, Various Issues, Ministry of Finance, New Delhi; Indian Agriculture in Brief, 28th Edition, Government of India, New Delhi, 2000, and Ag. Stat. at a Glance, 2018.

Table 6.18: Estimated number of tractors in use in India

(thousand number)

Year	Cumulative number	Increase over the 5 years period
1951	8.6	–
1956	21.0	12.4
1961	31.0	10.0
1966	54.0	23.0
1971	143.0	89.0
1976	321.9	178.9
1981	573.0	251.9
1986	923.0	350.0
1991	1468.0	545.0
1996	2182.6	716.6
2001	2640.0	457.4
2008	3500.0	860.0*
2013	5990.0**	2490.0
2018	9398.0**	3408.0

* Increase over 8 year period, ** our estimate

The Government of India had been regulating the supply and prices of tractors. In order to make available indigenous tractors to farmers at reasonable prices, the government issued the Tractors (Price Control) Order in March, 1967. The selling prices of tractors were fixed from time to time on the basis of the recommendations of the Bureau of Industrial Costs and Prices. But the fixation of prices became meaningless because of highly inflationary conditions. The Tractors (Price Control) Order was, therefore, withdrawn in October, 1974 and was replaced by a system of parametric surveillance. Under this system, the government issued guidelines to the manufactures for fixing the prices of their tractors. Any price increase by the manufactures was, subject to scrutiny by the government to check whether the increase fell within the parameters laid down by it. This system ensured a fair return to the manufactures and induced them to go in for the maximum utilization of capacity. This indirect price control system was withdrawn by the government in 1976, except in respect of three preferred models—MF 1035 (35 hp), TAFE 504 (50 hp) manufactured by M/s Tractors and Farm Equipment Ltd. and Ford-3000 (46 hp) manufactured by M/s Escorts Tractors Ltd.

Till the liberalisation of Indian economy, the Tractors (Distribution and Sale) Control Order applicable to only a preferred model prescribed the procedure for registration of orders with dealers for purchase of new tractors. On the basis of the registered demand for tractors in different regions, quota were allotted to respective dealers and farmers had to wait for their turn. The order forbade any individual to buy more than one tractor during the course of a year and banned its subsequent sale within two years, except under specified circumstances.

POWER TILLERS

The supply as well as the demand for power tiller is growing in the country. In early seventies, as the indigenous production fell short of the demand, the supply was augmented by imports. But since then, the indigenous production increased at the levels to meet the demand. During the year 1995–96, while the production of power tillers was 10239, the purchase/sales were little less (Table 6.19). The demand has considerably increased during the last 25 years. In 2017–18, the sales were as high as 52000. As the intensity of cropping in several areas specially in the irrigation command areas is increasing at a rapid rate, the demand for power tillers is likely to expand in the years to come. The government had been providing excise duty exemption to encourage production and sale of power tillers.

Table 6.19: Supply and demand for power tillers in India

(number)

Year	Availability			Sales
	Indigenous production	Imports	Total	
1970–71	1387	70	1457	NA
1975–76	2142	–	2142	2066
1980–81	2125	–	2125	1991
1985–86	3706	–	3706	3754
1990–91	6228	–	6228	6316
1995–96	10239	–	10239	10048
2000–01	NA	–	NA	16018
2010–11	NA	–	NA	55100
2015–16	NA	–	NA	46000
2017–18	NA	–	NA	52000

Source:
1. Government of India, Indian Agriculture in Brief, 28th Edition, Directorate of Economics and Statistics, Ministry of Agriculture, New Delhi, 2000, p. 63.
2. Government of India, Economic Survey, Various Issues, Ministry of Finance, New Delhi, and Agricultural Statistics at a Glance, 2018.

Power tillers have been found to be quite useful by rice and sugarcane growing farmers in the states of Tamil Nadu, Andhra Pradesh, Kerala, Karnataka, West Bengal, Bihar and Maharashtra. Power tillers are becoming popular specially in low land rice fields and hilly terrains.

OTHER FARM EQUIPMENTS

Farm equipments can be conveniently grouped into four categories.

1. Manually Operated Tools (seed-cum-fertilizer drills, seed drills, chaff cutters, wheel-hand hoes, sprayers, threshers)
2. Animal Operated Implements (wooden ploughs, steel ploughs, disc harrows, cultivators, seed drills, seed-cum-fertilizer drills, levelers, wet-land pudlers, animal carts).
3. Equipments Operated by Mechanical and Electric Power (power operated sprayer/dusters, mould board ploughs, cultivators, disc harrows, seed-cum-fertilizer drills, planters, levelers, potato diggers, trailers, paddy threshers, wheat threshers, maize shellers, chaff cutters, combines)
4. New machines, which include self-propelled paddy transplanter, sugarcane harvester, self-propelled reaper, baler, coconut tree climber, aero blast sprayer and laser land leveller.

All these have recorded steady growth. After liberalization, manufacturing of these got a big boost in Haryana, Punjab, Rajasthan, Madhya Pradesh and Uttar Pradesh. Combines are mainly manufactured in Punjab. At present, several models of self propelled combines are under manufacturing. The Bureau of Indian Standards (BIS) has prescribed over 500 standards on agriculture machinery.

TREND IN MECHANIZATION

The mechanization of farm operations takes place in phases. The first phase is the mechanization of those activities that require high power inputs and low control (tillage, transport, water pumping, threshing etc). The second phase in mechanization includes those operations which require medium levels of power and control like seeding, spraying and intercultivation. And the last phase is operations requiring high degree of control and lower power inputs (transplanting, planting of vegetables and harvesting of fruits). The process of farm mechanization in India has followed the same general trend as observed worldwide. In recent years, high power tractors, enhanced capacity machines (rotary tillers, high clearance sprayers etc) and laser levelers are witnessing increasing demand in the country.

REPAIR SERVICES

The demand for repair and maintenance service for farm machines has increased in the recent past because of the increase in the level of mechanization in Indian agriculture. The value of repair service market has increased from ₹ 8309 crore in 2011–12 to ₹ 15527 crore in 2016–17. These account for around 4.8% of total purchased inputs in crop production. The worth of a machine depends on the service and repair facilities available for it. In the absence of these facilities, farm machines remain idle, and there is a wastage of money and time in getting them repaired in city workshops. The supply of these services do not exist in the villages; where they do exist, they are inadequate. Farmers have frequently to travel long distances even for minor repairs of their machines or for the purchase of spare parts. As a result, there is a wastage of time, delays in performance of operation, and an increase in the cost of production. Since the supply falls short of demand, the workshop-owner charges a higher than normal price for his services. They take advantage of the farmer's ignorance of the technical aspects of machine operations. Moreover, farmers are usually in a hurry to get the machine repaired as soon as possible, and this encourages the workshop owners to charge exorbitant prices.

The service sector is completely unorganized. It is a sort of locally, monopolized business of the workshop-owners. Most of the services and repair centres that have come up in the private sector do not possess the necessary facilities for providing an efficient service. A good number of persons who handle the job are self-centred.

The supply of organized service facilities for repair and spare parts, therefore, calls for urgent action. The setting up of agro-industries centres in rural areas is one step in bridging the gap between demand and supply. However, their existing number is inadequate. At many such centres, spare parts are not available. The service is delayed because the demand is greater than the supply. Dealers, too, have their own workshop for the repair and servicing of farm equipments and machines; but they are located only in cities or towns.

It is, therefore, suggested that

1. Workshops for the repair and servicing of farm machinery should be established in rural areas so that farmers do not have to go for petty works to the urban areas;
2. Training of rural youths in servicing and repairing trade and their subsequent financing should be taken up on a large scale; and
3. The supply of tractors, sprayers, dusters, threshers and other equipments on custom-hire basis should be made available by establishing agro-service centres in villages and encouraging the opening of agri-clinics and agri-business centres by agricultural and engineering graduates.

MACHINE HIRING SERVICES

Small and marginal farmers or even some medium landholders are not able to buy the more efficient costly machines. Government agencies, Krishi Vigyan Kendras, private sector companies and cooperative organizations are helping in various ways to enable farmers to acquire the services of farm machines. Several entrepreneurs have emerged in rural areas to cater to their needs. For example, there is a Krishi Kisan App to help farmers benefit by way of placing orders for hiring of machines, apart from the links to seed hubs, weather advisory and demonstration of new technology. Farmers as well as all service providers (custom hiring centers) have been brought on the same portal. Farmers can locate the nearest centre, see the photo of the machine and negotiate the charges/prices and place orders (like hiring a cab). There is another app called 'CHC-Farm Machinery'. Young tech graduates are taking the lead and finding or evolving the innovative ways to help farmers, while helping themselves.

▌ HUMAN LABOUR

Human labour is an important input used in production of all agricultural crops. Some agricultural crops, viz., paddy, sugarcane, vegetables and cotton require more human labour compared to most cereals, pulses and oilseed crops. Farmers depend upon human labour for preparation of land, sowing, transplantation, weeding, harvesting, threshing and transportation. Human labour is also needed for the performance of post-harvest management operations. The availability of human labour in many areas becomes a constraint in peak seasons (harvesting of rabi and kharif crops) and in areas where crops like paddy, cotton, tea and vegetables are grown on large areas.

TREND IN SUPPLY

1. The rate of population growth has decelerated, resulting in the decline in the growth rate of persons in working age.
2. Labour force participation rate of women has shown secular declining trend over the past four decades due to, among other things, increasing enrolment in educational streams at different levels.
3. An increasing number of persons from cultivator households are joining the ranks of wage labour or self-employed in non-farm activities.
4. The number of persons commuting to urban areas for work has steadily increased and rural to rural migration has also emerged as an important form of migration.
5. An increasingly larger proportion of rural labour is getting educated and would be looking for jobs of different kinds, out of agriculture and in urban areas/non-farm activities.

TREND IN DEMAND

1. In agriculture overall, there has been a decline in demand for labour on account of mechanization, and a decline in the average size of holding, coupled with increase in small and marginal holdings.
2. Diversification within agriculture from food to commercial or high value crops has led to change in the nature and quality of labour demand.
3. With emergence of agriculture as an year-round activity in many parts of the country, there has been an increase in the demand for labour.
4. Introduction of contract farming has also led to a change in the demand for labour, at least of the qualitative form.
5. There has also been an increase in the demand for labour, often of the skilled variety, in off-farm activities due to commercialization of agriculture.
6. The faster expansion of the non-farm activity has led both to an increase and a change in the nature of demand for labour.
7. MGNREGA has, in recent years, raised demand for labour, often in competition with agriculture.

WAGE RATES

The price of human labour is the wage rate. The wage rate of supplying human labour for performance of agricultural operations varies from region to region and within a region is based on the cropping pattern followed by the farmers and time of need. The state governments also prescribe minimum wages payable to agricultural labourers by the farmers. State governments have also launched schemes of insurance and accident relief to workers engaged in agricultural operations. Farmers should keep the knowledge of these schemes for deriving advantages in the case of the accidents occurring on their farms. The assured employment of 100 days a year being provided under National Rural Employment Guarantee Scheme has created the problem of availability of labour for farm work in some areas and also impacted the wage rates.

The growth rate of rural wages has accelerated from 2.5% per annum during 1983–1993/94 to 3.4% per annum during 1993/94 to 2011/12. Considering the entire period of 1983 to 2011/12, agricultural wages have grown faster (3.3% per annum) than non-agricultural wages (2.8% per annum) (Papola, 2014). The National Floor Level Minimum Wage (NFLMW), which was ₹ 35 in 1996 and ₹ 137 till May 2015, was raised to ₹ 160 per day in June 2015 and further to ₹ 176 per day from June 2017. However, state governments prescribe minimum wage above this level. For example, Rajasthan state has fixed minimum wage for unskilled agricultural worker at ₹ 225 in May 2019.

FARM CREDIT

Credit is one of the key parameters for increasing output and resource productivity in agriculture. With the technological breakthrough in Indian agriculture, the demand for credit has increased very substantially. In order to sustain and accelerate this technological change in agriculture, the availability of adequate credit and its use in the most efficient way is extremely important.

Credit to farmers is supplied both by institutional and non-institutional financing agencies. The share of institutional financing agencies (commercial banks, cooperative banks, the private and foreign banks) in financing the farm sector is increasing continuously. As on March 31, 2019, 52489 rural branches of scheduled commercial banks, supported by 5.41 lakh Bank Correspondents (BCs) are providing credit and other banking services to the farmers at village level. In addition, over one lakh primary agricultural cooperative/credit societies (PACS), more than 360 district level central cooperative banks (DCBs), 33 state cooperative banks (SCBs), apart from land development banks are serving the agricultural sector in extending short, medium and long term loans to the farmers. All banks (commercial, cooperative and regional rural banks) have issued Kisan Credit Cards (KCCs) to the farmers for increasing their access to institutional credit. Total numbers of KCCs issued till the end of March 2018 is reported to be over 17.67 crore.

The credit supply to the farm sector during the last 40 years from institutional agencies is shown in Table 6.20. During the year 2017–18, institutional agencies supplied ₹ 1168503 crore to the farm sector as credit. For the year 2020–21, it is targeted to provide ₹ 15 lakh crore as institutional credit for the agricultural sector. The share of commercial banks in the total credit supplied to the farm sector was 75%. The cooperatives and regional rural banks accounted for 13% and 12% of total farm credit, respectively.

At the national level, the supply of farm credit in terms of its total magnitude, purpose and price (interest rate) is monitored by the government, but, at the local level, various commercial and cooperative banks engage in some measure of competition. Easy access to banks, simplified procedures and associated technical guidance are the sale promotion tactics used by these agencies to increase their advances.

The price of credit is the interest. The rate of interest on borrowing varies with the agency, the purpose and the period of loan. The farmers are supplied credit by institutional agencies at a concessional rate. Small and marginal farmers are supplied credit at a still lower rate under the differential interest rate scheme. The price as well as the target of specific quantum of supply of credit are fixed by the government. The interest rate structure has now been liberalised. However, the government still intervenes in terms of volume of credit and interest rate, including interest rate subvention in the form of credit policy to be followed by the banks.

Table 6.20: Flow of institutional credit to agriculture

(₹ in crore)

Year	Cooperative banks				Commercial banks			Regional rural banks			Grand total
	Short term	Medium and long term	Total	Short term	Medium and long term	Total	Short term	Medium and long term	Total	
1980–81	1526	600	2126	517*	746*	1263*	–	–	–	3389
1990–91	2822	1151	3973	2173*	2897*	5010*	–	–	–	8983
2000–01	16528	4190	20718	13486	14321	27807	3245	974	4219	52744
2010–11	64527	5578	70105	216773	115933	332706	37808	6160	43968	446779
2017–18	138348	12041	150389	497078	380077	877155	119546	21413	140959	1168503

* Inclusive of credit supplied by Regional Rural Banks. A dash 'indicates break-up not available.
Source: Economic Survey—Various Issues, Ministry of Finance, Government of India, New Delhi, Agricultural statistics 2014, 2018, Ministry of Agriciltue, GOI.

The Government has been prescribing norms of advancing certain minimum percent of its total advances to what is called the priority sector, apart from interest subvention for specified categories of borrowers. The guidelines in this respect are issued usually at the beginning of the financial year. Presently, apart from farmers, micro and small enterprises, and food and agro processing units are included in the priority sector. The proportion earmarked for the priority sector is 40%. For quite some time, there was a distinction between direct and indirect lending for agriculture, but now this has been done away with. The minimum norm for agriculture (including allied activities) is 18% of total lending. Within agriculture, a minimum percentage of total lending (8%) has to go to small and marginal farmers. The lending to micro enterprises has to be at least 7.5%. At least 10% lending has to go to the weaker sections. For different groups of banks, the periods for fulfilling the credit targets are separately defined. The interest subvention is also linked to timely repayment, i.e. those who repay the loan in time, get extra interest subvention.

TREND IN PURCHASED INPUTS

Market orientation of Indian agriculture can also be viewed by looking at the changes in purchased inputs and their prices.

VALUE OF PURCHASED INPUTS

Central Statistical Organization (CSO) works out regularly, the costs of inputs, as a part of National Accounts Statistics. Their estimates for the crops sector for 2011–12 and 2016–17 are shown in Table 6.21. The value of purchased inputs went up from ₹ 2.1 lakh crore in 2011–12 to 3.24 lakh crore in 2016–17.

Cost items	2011–12		2016–17	
	₹ (crore)	%	₹ (crore)	%
Seed	29112	13.9	47217	14.6
Organic manure	21318	10.1	33317	10.3
Fertilizers	41825	19.9	51632	15.9
Repair charges	8309	4.0	15527	4.8
Fodder for animal labour	31244	14.9	47204	14.6
Irrigation charges	4158	2.0	4695	1.5
Marketing charges	38366	18.3	60001	18.5
Electricity	8915	4.2	35759	11.0
Pesticides, etc.	1895	0.9	2214	0.7
Diesel	24684	11.8	26199	8.1
Total	209826	100.0	323765	100.0

Table 6.21: Value of purchased inputs (at current prices)

Source: National Accounts Statistics, CSO, Government of India.

INPUT PRICE INDEX

The index numbers of wholesale prices of different agricultural inputs during the last 25 years are shown in Table 6.22. There has been considerable increase in the prices of electricity and diesel oil during this period, but contained in recent years.

Table 6.22: Index numbers of wholesale prices of agricultural inputs in India

Year	Fertilizers	Electricity for irrigation	Pesticides	Tractors	Diesel (H.S.)	Non-electrical machinery
Base 1993–94 = 100						
1995–96	129.7	134.0	117.1	119.4	108.8	118.4
2000–01	159.1	206.3	123.7	151.9	239.8	144.9
2005–06	175.1	300.8	150.2	179.4	446.8	189.6
Base 2004–05 = 100						
2011–12	137.2	136.8	116.0	137.9	167.8	123.7
2012–13	151.1	170.9	122.2	142.7	192.7	123.0
Base 2011–12 = 100						
2014–15	120	106	121	108	105	NA
2016–17	118	104	117	114	78	NA
2017–18*	118	105	115	114	89	NA

* October to April

Source: Reports of CACP

REFERENCES

1. For more details see:
   (a) Acharya, S. S. (1997), "Input Subsidies in Indian Agriculture: Some Issues", in Policies for Agricultural Development, edited by V.S. Vyas and Pradeep Bhargava, Institute of Development studies, Jaipur and Rawat Publications, pp. 87–120.
   (b) Acharya, S.S. (1997), "Agricultural Price Policy and Development: Some Facts and Emerging Issues", Indian Journal of Agricultural Economics, Vol. 52, No. 1, Jan.–March, 1997, pp. 1–47.
   (c) Acharya, S.S. (2000), "Subsidies in Indian Agriculture and their Beneficiaries", Agricultural Situation in India, Special Number, August, 2000, pp 251–60.
2. Grower, R.K. and M.S. Luhach, "Study on Marketing of Pesticides in Haryana", Indian Journal of Agricultural Marketing, Vol. 20 (1), January–April 2006, p. 49.
3. Venkatesh, P. and Suresh Pal, "Impact of Plant Variety Protection on Indian Seed Industry", Agricultural Economics Research Review, Vol.27, No.1, Jan–June 2014,
4. Papola, T.S., "Economic Diversification and Labour Market Dynamics in Rural India", 12th Prof. G. Parthasarthy Memorial Lecture, AERC, SPU, Vallabh Vidya Nagar (Gujarat), Dec. 5, 2014.

7

Government Intervention and Role in Agricultural Marketing

Efficient marketing is a prerequisite in the development process of any economy. The basic objectives of an efficient marketing are to ensure remunerative prices to the producers and a reduction in marketing costs and margins, to provide commodities to consumers at reasonable prices, and promote the movement of surpluses for economic development. There are many imperfections in the marketing system for agricultural commodities. To protect the interests of the various segments of society, government intervention in the market mechanism is necessary.

▌ WHY SHOULD GOVERNMENT INTERVENE?

At the time of Independence (and even before), India's food supply (even staple food, i.e. cereals) was considerably short of minimum requirement. More than half of the population was living below poverty line meaning thereby that majority of the population had very low food-purchasing power. Further, there was no foreign exchange to import food. Therefore, major immediate goal at that time was to increase the production of food grains for food security. However, on the other side, nearly 80% of the farmers were illiterate and small or tiny landholders and most of them were tenant farmers. Moreover, the then existing agricultural marketing system was extremely inefficient. It was felt that, along with other development initiatives, agricultural marketing system should also receive priority attention. Let us first look at the characteristics of traditional agricultural marketing system.

▌ CHARACTERISTICS OF TRADITIONAL AGRICULTURAL MARKETING SYSTEM

As early as in 1928, the Royal Commission on Agriculture pointed out that the then existing system did not meet the requirement of an ideal marketing system. Important characteristics of the traditional marketing system for agricultural commodities have been discussed below. Some of these still exist, though efforts are under way to improve them.

1. Heavy Village Sales of Agricultural Commodities

A majority of farmers sold a large part of their produce in villages, which resulted in low returns for their produce. There is a difference in the price prevailing at different levels of marketing, i.e. the village, the primary wholesale market, the secondary wholesale, and retail levels. The extent of village sales varied from area to area, commodity to commodity, and also with the status of the farmer. The village sale had been 20 to 60% in foodgrains, 35 to 80% in cash crops and 80 to 90% in perishable commodities. This practice is very common even now in some villages. The factors responsible for village sales are:

(a) Farmers were indebted to village moneylenders, traders or landlords. They were often forced either to enter into advanced sale contracts or sell the produce to them at low prices.

(b) Many villages were not connected by roads. Adequate transport means were not available even in villages connected by roads. It was difficult to carry the produce in bullock or camel carts to market which were often situated at long distances.

(c) There was a small quantity of marketable surplus with a majority of the farmers because of the small size holdings. Farmers with low surplus prefer to sell in the village.

(d) Farmers were hard pressed for money to meet their social and other: obligations, and were often forced to sell their produce right in the villages.

(e) Most of the perishable products need to be marketed in the villages because of their low "keeping" quality and the non-availability of quick transport means.

(f) Many farmers disliked city markets mainly because of their lack of knowledge about prevailing market practices, the possibility of theft or robbery in transit and problems faced by them for selling their produce in city markets.

(g) The information on the prices obtaining in the nearby primary and secondary wholesale markets was not readily available to the farmers.

2. Post-harvest Immediate Sales by Farmers

A majority of the cultivators tend to sell their produce immediately after the harvest at the low prices prevailing at that time. Because of substantial supplies, the markets are glutted in the post-harvest season. Traders often take advantage of this situation. About 60 to 80% of the foodgrains are still marketed in the first quarter of the harvest season.

The reasons for the existence of post-harvest immediate sale of the produce by farmers are:

(a) Poor retention power of the farmers arising out of their pressing need for cash to repay their debts and meet their cash needs for the payment of land revenue, purchase of items of basic necessity, and for meeting their social obligations—all of which are conducted primarily in the off-season.

(b) Inadequate storage facilities available in the villages, either private, public or cooperative.

(c) Fear of loss of the produce by fire, theft and other calamities.

(d) Lack of entrepreneurship and the low risk-bearing ability of the farmers.

(e) In surplus producing states like Punjab and Haryana, most of the marketed surplus of wheat and paddy/rice is sold by the farmers to public agencies at the support prices. As the support prices during a marketing season (year) remain constant, it is advantageous to the farmers to sell in the post-harvest season.

(f) Returns to storage, on point to point basis, are not necessarily positive.

3. Inadequacy of Institutional Marketing Infrastructure and Lack of Producers' Organizations

Farmers were disorganised and marketed their produce individually. Because of their low bargaining power, they had to deal with traders having a strong organization. They could not, therefore, insist on a reservation price for their produce. Producers watched the auction of the produce as silent spectators and were exploited by traders.

4. Multiplicity of Market Charges

The cost of marketing of produce continues to be very high for agricultural goods compared to that of the products of other sectors. A large number of market charges—commission, brokerage, weighment, hamali, karda (impurity charge), dhalta (excessive moisture charge), muddat (charge for making cash payment), darmada (charity for goshala, pigeons, waterhut) etc.—were paid by the farmers. Some of these charges had no relationship with the farmers. The rates of many charges were very high. Further, every state had imposed several cesses or taxes. As a result, the marketing cost of produce was high which adversely affects the realisation of the farmers.

5. Existence of Malpractices

Many malpractices—deduction of unauthorized market charges, spurious deductions, unfair weighment of the produce, unhealthy sale method, taking of karda even for clean produce, taking away a part of the produce as sample by bidders, bungling in accounts, taking 1/2 to 1 kg extra for every quintal of foodgrains, arbitrary deduction for religious and charitable purposes, etc.—were common in the marketing of agricultural products. This resulted in an increase in the real cost of marketing of produce. As a result the producer's share in the consumer's rupee was very low.

6. Lack of Reliable and Up-to-date Market Information

There was no reliable channel for the communication of price information to producer-farmers, who were isolated in remote villages. In the absence of this information, farmers depended on the hearsay reports which they received from village merchants. Village merchants did not reveal the correct price information prevailing in primary and secondary wholesale markets because of their personal interest in buying the produce from the farmers at low prices. In such an uncertain situation farmers sold the produce right in the villages at low prices.

7. Low Marketable Surplus of a Large Variety of Products

Most Indian farmers grow a number of crops, both in the kharif and the rabi seasons. Not only this, they also produce a number of varieties of each crop. That is why the products available for sale with these farmers were many in number, though the quantity of each was very small. When farmers reached the market with a small quantity of the produce, they received very little attention from traders, who are interested in the farmers having a large quantity of the produce for sale. Moreover, marketing charges per unit of the product were higher for a small lot than for a big lot. The tendency to grow a large number of crops among the farmers existed because of the risk in farming, the adoption of diversified farming practices, and the priority given to growing the crops required for home consumption.

8. Absence of Grading and Standardization of Produce

A large number of farmers had little knowledge of the practice of the grading of the produce prior to its sale. They usually mixed up superior and inferior quality products to make a single lot. As a result, they got a lower price for their produce. Sometimes, farmers were penalised by traders for the existence of even a small percentage of poor quality produce in the lot. In the absence of grading, there was the practice of deliberate adulteration of products by traders as well as farmers during the marketing process. These practices include the addition of water in milk, chillies, cotton and wool, the adulterating of better quality foodgrains with inferior quality grains, the addition of sand in wheat and clay pebbles in groundnut. These practices lowered the market prestige and consumers lost confidence in the quality of the product.

9. Absence of Quick Transport Means

Farmers usually carried their surplus to the markets or nearby towns by bullock or camel carts. As a consequence, they had to spend considerable time in marketing their small surplus. This also resulted in considerable deterioration in the quality of the product due to spoilage in case of perishable and spillage in non-perishable commodities.

Further, farmers usually carried their surplus to the markets in loose form as packing materials were either not available or were unaffordable. As a consequence, the farmers suffered heavy losses during loading, transportation and unloading.

10. Strong Associations of Traders and Market Functionaries

While farmers did not have organizations for safeguarding their interests while selling the produce, traders and other market functionaries had their strong associations or unions. As such the practices prevalent in the trade circles were usually biased against the farmers resulting in low prices, high cost of marketing and inconveniences to the farmers at the time of sale.

▌FORMS OF GOVERNMENT INTERVENTION

In the interest of public welfare, the government intervenes in the marketing system. The extent of intervention depends on the objectives of the government and the extent of defects and malpractices prevailing in the system. Government intervention may be direct or indirect; and it may take any one or a combination of the following forms:

1. The framing of rules and regulations for the protection of the interest of some sections of the population. These may relate to the grant or restriction of monopolies, restriction on the activities of traders, licensing and market regulation.
2. Creation of marketing infrastructure such as storage and warehousing, transportation and communication facilities, credit facility, grading and standardization.
3. Administration of prices at different levels of marketing—guaranteeing minimum support prices to producers, providing commodities at fair prices to consumers, and fixing the rates of commission charged by commission agents.
4. Influencing supply and demand by regulating import, internal procurement and distribution.
5. Establishment of organisations, to provide services to the farmers for performing certain marketing functions; and

6. Promotion of farmers' cooperatives or organizations for agricultural marketing and agro-processing.

In India, the Government has intervened in the marketing system of agricultural commodities in various ways at different points of time since 1930's. Efforts put in by various organizations and the intervention by the Government as well as the creation of awareness among the farmers towards marketing of farm products brought out changes in the farmer's marketing practices, as well as in the agricultural marketing system of the country.

The efforts made by the government for creation and/or promotion of infrastructure facilities like roads/railways; storage and warehousing, grading and standardization, cold storage and cool chains, agro-processing, credit delivery, input markets, and promoting alternative marketing channels have already been discussed in earlier chapters. In this chapter, we focus on (a) national and state level organizations; (b) agricultural produce market regulation, (c) food security measures like buffer stocking and public distribution of food grains, (d) agricultural price policy and its implementation; (e) quality control; (f) marketing related legislative measures; (g) promotion of alternative marketing channels; and (h) key changes in marketing policy related to sugar and rice.

▌ DIRECTORATE OF MARKETING AND INSPECTION (DMI)

On the recommendation of the Royal Commission on Agriculture, 1928, and the Central Banking Enquiry Committee, 1931, the Central Marketing Department was established in India in 1935. The office of the Agricultural Marketing Adviser to the Government of India was established in Delhi on January 1, 1935. The aim was to build up a body of knowledge of the marketing of agricultural commodities and suggest measures for the promotion of an orderly and efficient marketing system for farm products. At the same time, similar departments were set up in the erstwhile provinces and princely states. Later, on the recommendations of the Patel Committee, the Inspection Directorate of the Ministry of Food was merged with the Central Marketing Department, and was renamed as the Directorate of Marketing and Inspection (DMI).

The responsibility for the compilation of statistics and for the dissemination of market news by the Central Department was shifted to the Directorate. This Directorate of Marketing and Inspection functions as an attached office of the Ministry of Agriculture (earlier with Rural Development), Government of India, with its head office at Faridabad (Haryana) and a branch head office at Nagpur (Maharashtra). It is headed by the Agricultural Marketing Adviser (AMA), who is assisted by Joint AMAs, Deputy AMAs, Directors and other technical staff. The Directorate of Marketing and Inspection maintains a close liaison between Central and State Governments through its Branch Head Office at Nagpur, regional offices and sub-offices spread all over the country. The regional offices are located at New Delhi, Kolkata, Mumbai, Guntur, Chennai, Bhopal, Guwahati, Lucknow, Chandigarh, Kochi and Jaipur. The regional offices are headed by Deputy Agricultural Marketing Advisers (Dy AMAs). The DMI also has a network of Regional Agmark Laboratories with central Agmark Laboratory at Nagpur as an apex laboratory.

The Directorate of Marketing and Inspection implements the agricultural marketing programmes of the union Government under the supervision and control of Union Ministry of Agriculture. It aims at achieving integrated development of marketing of agricultural and allied products in the country. The DMI currently has five divisions namely administration; market development and reforms; market research and information network; quality control; and vigilance.

The Directorate of Marketing and Inspection was originally assigned the task of implementing the Agricultural Produce (Grading and Marking) Act, 1937; Cold Storage Order, 1980 and Meat Food Products Order, 1973. However, latter two do not exist now. Besides these, the Directorate of Marketing and Inspection is actively associated in rendering assistance, guidance and advice to the State Marketing Departments in regard to the framing of legislation and subsequent implementation of agricultural produce markets Acts for bringing about regulation of agricultural produce markets. The Directorate of Marketing and Inspection undertakes programmes of market research and surveys, training of market personnel, extension and publicity. Between the Head Office, Faridabad and Branch Head Office, Nagpur the sharing of responsibility in respect of these functions is as follows. The Head Office at Faridabad looks after: (i) administration of the Agricultural Produce (Grading and Marking) Act, 1937, rules and other related matters; (ii) administration of Meat Food Products Order, 1973 (till it was repealed); (iii) administration of Cold Storage Order (till it was rescinded in 1977); and (iv) market regulation, development and extension. The Branch Head Office at Nagpur is responsible for: (i) market research, design and planning; (ii) training programmes in agricultural marketing; and (iii) livestock marketing cell.

The details of the specific functions of the Directorate of Marketing and Inspection have been mentioned in the following paragraphs:

MARKET RESEARCH AND COMMODITY SURVEYS

Market research, surveys and investigations were the first basic functions entrusted to the Directorate at its inception in 1935. The programme was designed to conduct in-depth study of the marketing system so as to help planned development thereof. The Directorate has conducted surveys and brought out a number of commodity survey reports covering 120 agricultural, livestock and forest products. The reports include a survey of the prevailing system of marketing of each commodity in the country, the costs of marketing, and the defects in the marketing system. This information facilitated the planning and undertaking development works by focussing the attention of all concerned on the defects. In addition to commodity surveys, the Directorate has conducted surveys and brought out reports on the specific functions and institutions concerned with agricultural marketing, such as cold storage, regulation of markets and cooperative marketing.

The Directorate of Marketing and Inspection has published 330 study reports till recently. The details of studies undertaken and reports published are as under:

Group of commodities	Number of studies published
1. Food grains	21
2. Fruits and vegetables	27
3. Edible nuts and spices	15
4. Oilseeds and oil	12
5. Fibres	10
6. Fish and fish products	4
7. Milk and milk products	4
8. Livestock and livestock products	19

9. Essential oils	6
10. MPDC—Physical planning and post-harvest management	90
11. Others	86
12. Tribal markets	6
13. Reports revised from time to time	30
Total	330

The Directorate also provided research grants to eminent scholars and institutions for undertaking probe-oriented quality research in the field of agricultural marketing.

MARKET EXTENSION

A marketing extension cell was established in the Directorate of Marketing and Inspection in 1962. The main function of this cell is to undertake extension and publicity activities to educate farmers, traders and consumers about various voluntary and regulatory measures brought in to force for the improvement of the marketing system. This work of the extension cell is carried out through various publicity media such as documentary films, cinema slides, exhibitions and distribution of literature.

The marketing extension cell of DMI participates in national and international exhibitions, besides, organizing exhibitions of its own to create awareness about Agmark. Consumers' educational programmes are conducted in collaboration with colleges and voluntary women organizations with practical demonstrations on detection of adulteration in edible products of daily use through simple tests. Also, a quarterly journal 'Agricultural Marketing' is brought out containing popular and research articles on marketing aspects of agricultural products.

STATUTORY REGULATION OF MARKETS AND MARKET PRACTICES

Though the regulation of markets is a state subject, the Directorate of Marketing and Inspection renders guidance and advice to all the State Governments in framing market legislation and its enforcement. To bring about uniformity in the marketing practices over space, the government has set up regulated markets under the Agricultural Produce Market Act. The Act provides for the sale of agricultural produce by open auction in the presence of an official of the Market Committee, the payment of the value of the produce by the buyer immediately after the purchase without any deductions for muddat, the use of standard weights and measures, the sale of clean and graded produce, and the deduction of only the authorised market charges from the value of the produce. More than 7465 markets in the country were brought under the purview of this Act. The Directorate also extends central assistance for the development of selected regulated markets in the country.

The DMI continuously reviewed the status of Agricultural Produce Markets Regulation programme and based on experiences, formulated revised Model Act and circulated to state governments in 2003. Subsequently, the provisions were further revised and another new Model Act was circulated in 2017 (more details in the subsequent section on Regulation).

PROMOTION OF GRADING AND STANDARDIZATION

To secure for farmers the prices which are commensurate with quality, grading is essential. To promote the production and marketing of quality goods and maintain uniformity, the Agricultural Produce (Grading and Marking) Act was passed by the government in 1937. The products graded under this Act are marked with the AGMARK label. The important functions of this cell are:

(a) To draw up grade specifications for various agricultural commodities for internal consumption as well as for exports;
(b) To test the working of these grades;
(c) To supervise the grading work; and
(d) To exercise effective quality control.

The Directorate of Marketing and Inspection has formulated grade specifications for around 220 agricultural commodities (see details in Chapter 4).

TRAINING OF MARKET PERSONNEL

The Directorate of Marketing and Inspection imparts training to different marketing personnel so that they may effectively carry out the various activities associated with the marketing of agricultural products. The training technology adopted by the DMI consists of lecture-cum-group discussions, study tours, case studies, audio-visual-aids, demonstrations and guest faculty lectures. The kind of training courses organised by the Directorate of Marketing and Inspection are shown in Table 7.1 and include the following:

Table 7.1: Trainings in agricultural marketing conducted by Directorate of Marketing and Inspection, Ministry of Agriculture, Government of India, Nagpur

Name of training programme	Venue	Intake capacity	Duration	No. of sessions in a year
1. Diploma Course in Agricultural Marketing	Nagpur	10/15	6 months	One
2. Training Course for Market Secretaries	(a) Lucknow (UP) for Hindi Medium (b) Hyderbad (AP) for English Medium	20 at each centre	3 months	One
3. Training Course for Grading Supervisors	Nagpur	10	15 days	One
4. Training Course for Graders	(a) Hubli (Karnataka) (b) Salem (Tamil Nadu) (c) Chandigarh (Punjab) (d) Pune (Maharashtra)	Decided by concerned authority	1 month	At the discretion of concerned authority
5. Training in Grading of Cotton	Cotton Classing Centre, Nagpur	10	4 weeks	Two
6. Market Intelligence and News Service Course	Nagpur	12	2 weeks	One
7. Market Extension Service Course	Nagpur	12	2 weeks	One
8. State Level Capsule Training Programme	Nagpur/State HQ	Decided by concerned authority	3–5 days	As per requirement of concerned authority

Source: Information Brochure on Training in Agricultural Marketing; Directorate of Marketing and Inspection, Ministry of Agriculture, Government of India, Nagpur.

1. *Diploma Course in Agricultural Marketing:* This course was first organised in 1956 for the middle level personnel of the marketing organisations of different states. The duration of this course which was 11 months is now reduced to six months and is offered at Nagpur. The programme seeks to mould the trainees into general purpose marketing men who could efficiently implement various development programmes in agricultural marketing.

2. *Training Course for Market Secretaries:* This course was started in 1957 for the managers including supervisors of the regulated markets. It is oriented to equip the market secretaries to administer the regulated markets more efficiently. The course is of three months duration, and is offered at Hyderabad and Lucknow.

3. *Training Course for Grading Supervisors:* This course aims at imparting necessary training to the supervisory staff of commercial grading centres run by the market committees and cooperative societies. This course, which was started in 1963, is of a fortnight duration and is offered at Nagpur.

4. *Training Course for Graders:* This course was started in the year 1962. This was originally a combined course for graders, assessors and supervisors at lower level for a period of three months. The programme has been decentralised from 1970 and now is organised in the states of Maharashtra, Tamil Nadu, Punjab and Karnataka in the regional languages. The duration of this course is one month.

5. *Training in Grading of Cotton:* This training is given to persons engaged in grading of special agricultural crop, such as cotton. It is of four weeks duration and organized at Nagpur.

6. *Training Course in Market Intelligence and News Service:* This training course is intended to improve and strengthen the market intelligence and news services in India. The duration of the course is two weeks and is being organised since 1986 at Nagpur.

7. *Training Course in Market Extension Service:* This training course aims at improving the techniques in dissemination of market information and to imbibe consciousness amongst the consumers about the graded and certified products. This training programme is of two weeks duration and was introduced in 1986 at Nagpur.

8. *Capsule Training Programme:* With a view to promoting participative management, quality of work output and organisational performance, a capsule training programme has been initiated from 1991. This programme is of 3 to 5 days duration and is offered at Nagpur and state headquarters.

In addition to these, short-term courses on demand are also being conducted from time to time. These include training programmes on grading of rice, forest produce, tendu leaves and lac for the grassroots level functionaries working in the concerned organizations. Trainings on analytical methods, pre-shipment inspection, and quality control are also organised by DMI. The DMI has so far trained more than 20000 marketing personnel.

MARKETING RESEARCH AND INFORMATION NETWORK (MRIN)

For many decades, DMI has been involved in collecting and disseminating information on prices (daily, weekly, monthly) as also the arrivals in different markets of the country as a part of 'market intelligence cell'. However, since March 2000, the DMI is implementing the ICT based central sector scheme of MRIN with the following objectives:

(a) To establish a nation wide information network for speedy collection and dissemination of market data for their efficient and timely utilization,

(b) To facilitate collection and dissemination of information related to better price realization by the farmers,

(c) To sensitize and orient farmers to new challenges in agricultural marketing by using ICT as a vehicle of extension,

(d) To improve efficiency in agricultural marketing through regular training and extension for reaching region specific farmers in their own languages, and

(e) To provide assistance for marketing research to generate marketing information for its dissemination to farmers and other market functionaries to create an ambience of good marketing practices in the country.

The information network of this new scheme covers market related information, infrastructure related information, price related information and promotion related information. The project is executed with the technical support of NIC. All important wholesale markets have already been linked to the central AGMARKNET portal (http://agmarknet.gov.in). The portal provides information on daily prices and arrivals for more than 300 commodities and 2000 varieties from these markets. State agricultural marketing boards are also on the portal. National atlas of agricultural markets on a GIS platform is also displaying the availability of entire marketing infrastructure in the country. In addition, commodity profiles are also available on the portal. The efforts are going on to improve the content and extend the coverage of the portal.

AGRICULTURAL MARKETING INFRASTRUCTURE

The DMI has also been contributing to creation of agricultural marketing infrastructure. In 1960s to 1980s, when agricultural produce markets were regulated, DMI oversaw the scheme for construction of market yards and sub yards. It was also implementing and overseeing the implementation of rural godowns scheme (GBY) of government of India since 2001. Since 2004, DMI has been implementing the scheme for development and strengthening of agricultural marketing infrastructure, and grading and standardization (AMIGS).

During the XII five year plan, GBY and AMIGS were subsumed into one sub-scheme namely Agricultural Marketing Infrastructure (AMI) which is a sub-scheme of integrated scheme for Agricultural Marketing (ISAM) with effect from April 1, 2014. There are two components under AMI viz; (a) Storage Infrastructure and (b) Marketing Infrastructure other than storage. The objectives of the scheme also include training of farmers, market functionaries and entrepreneurs on various aspects and financial strengthening of state agencies and other stakeholder organizations with provisions of specified subsidies. The subsidy component is higher for NE and hill states/UTs. The private and cooperative markets for agricultural products are also eligible for assistance under the new AMI scheme being implemented by DMI.

PUBLICATION OF JOURNAL

The Directorate has been publishing a quarterly journal of 'Agricultural Marketing' which contains research papers in addition to statistics on different aspects of marketing and news of the Directorate.

STATE AGRICULTURAL MARKETING DEPARTMENTS

Marketing Departments or cells were set up in the States as counterparts of the Central Marketing Department. Initially, the structure of the State Departments varied from

State to State, and their status range from that of a full-fledged department to a cell under the Agriculture Department. However, all the States now have a marketing department/cell to look after the marketing problems of farmers. Many states have separate Ministry of Agricultural Marketing. The primary objectives of the State Department of Agricultural Marketing are to survey the markets in the State and to prepare a programme for their regulation. Administrative control over the market committees is also exercised by the State Directorates of Agricultural Marketing. These State Departments have also been entrusted with the responsibility of gathering market intelligence and grading of agricultural commodities under AGMARK. They get the necessary guidance from the Central Directorate of Marketing and Inspection.

With increasing role of agricultural marketing in the economic development of the state and the increasing activity of market regulation, State Agricultural Marketing Boards were set up in States and Union Territories. These State Agricultural Marketing Boards look after the regulation of markets and bring about an effective level of coordination in the functioning of the regulated markets at the State level. The market regulation scheme received momentum after the establishment of State Agricultural Marketing Boards.

However, with the set up of the State Agricultural Marketing Boards, the State Agricultural Marketing Departments or cells in many states were merged with the Boards and the functions hitherto carried out by the Departments or cells were transferred to the State Agricultural Marketing Boards. But the Governments of India insisted the state governments for the establishment of separate Agricultural Marketing Directorates in the States in addition to the State Agricultural Marketing Boards. Following these recommendations, separate agricultural marketing directorates were established initially in 10 states. These states are Andhra Pradesh, Gujarat, Karnataka, Maharashtra, Madhya Pradesh, Rajasthan, Tamilnadu, Uttar Pradesh, West Bengal and Delhi. In Rajasthan, Agricultural Marketing Directorate was established in 1980 and the functions of market regulation, AGMARK and commercial grading, market intelligence, and budget and administrative work of the market committees were assigned to the Directorate. Later, almost all states established the Departments of Agricultural Marketing, apart from state Agricultural Marketing Boards.

STATE AGRICULTURAL MARKETING BOARDS (SAMBs)

The necessity of a central organization to supervise and provide guidance to market committees was realized with the establishment of a network of market committees. Agricultural Produce Market Regulation Acts were modified to make the provisions for the establishment of State Agricultural Marketing Boards, which were established in 24 States and one Union Territory viz., Rajasthan, Punjab, Haryana, Uttar Pradesh, Madhya Pradesh, Karnataka, Assam, Gujarat, Himachal Pradesh, Andhra Pradesh, West Bengal, Maharashtra, Orissa, Tamil Nadu, Bihar, Jharkhand, Meghalaya, Nagaland, Tripura, Goa, Uttarakhand and Delhi. It has a statutory status in 15 states and advisory status in others. There is a variation in the organization, functions and composition of the Board from State to State. The chairmanship of the Board also differs from State to State.

The Composition of State Agricultural Marketing Boards (SAMBs) has been modified in the revised Agricultural Produce Market Regulation Acts, after the Government of India circulated the model Act incorporating contract farming, direct marketing and setting-up of private and cooperative markets in 2003. As an illustration, the composition of Rajasthan State Agricultural Marketing Board (RSAMB), as per the Agricultural Produce Marketing (Development and Regulation) Act, 2006, is as under:

1. Chairman is Minister of Agriculture or Minister for Agricultural Marketing.
2. Six to 10 members elected amongst chairmen of Market Committees of the State.
3. Five official members including (a) Agricultural Production Commissioner; (b) Secretary, Department of Agriculture/Agricultural Marketing/Cooperation or his representative; (c) Agricultural Marketing Adviser (AMA), Government of India or his nominee; (d) Representative of NABARD; and (e) Registrar, Cooperative Societies of Rajasthan.
4. Chief Executive Officer is Director/Managing Director of RSAMB, appointed by the State Government.

There is a fund called the Market Development Fund, which is administered by the Board. All the receipts of the Board (contributions received from market committees, loans, grants, etc.) are credited and all the expenditure (on improvements and on the regulation of agricultural markets in the State, giving aid to financially weak market committees, paying salary and allowances to its employees, meeting legal expenses and the construction cost of the office and various infrastructures), incurred by the Board is debited to the Market Development Fund.

The main functions of the State Agricultural Marketing Board are:

1. To carry out the training of officers and staff, create facilities for grading and standardization, construct market roads and approach roads to the markets, construct market yard and sub-yards, establish and maintain the Board office and others as specified;
2. To tender advice to the government on the functioning of market committees and on improvements in agricultural marketing as and when referred to; and
3. To frame bye-laws, help in the functioning of market committees, and supervise their operations.

The administrative structure of the State Agricultural Marketing Board consists of the chairman, secretary; and following wings, each entrusted with a particular function:

1. Administrative wing;
2. Accounts wing;
3. Technical (Engineering) wing; (with field offices);
4. Town Planning (Architect) Cell;
5. Computer and Data Analysis Section;
6. Quality Control Section;
7. Publicity and Training Cell; and
8. Post-harvest and Export Promotion Cell

COUNCIL OF STATE AGRICULTURAL MARKETING BOARDS (COSAMB)

The COSAMB, an apex body of the State Agricultural Marketing Boards was established in February, 1988. The need for such a body was felt to coordinate the activities of State Marketing Boards, specially those connected with credit mobilization, central assistance for market development and some common problems. The COSAMB made a great headway in achieving the objectives for which it was incorporated. Under one umbrella of COSAMB, the State Agricultural Marketing Boards have augmented their activities in achieving the objectives.

The COSAMB was established with a view to:

1. setting up a common forum of national stature to build a think-tank and to share the know-how and experience in the cause of strong and efficient agricultural marketing in the country;
2. providing assistance to the member States and to credit institutions to fix norms for arranging credit from the bank and mobilising central assistance for development;
3. setting up a common organization to hold seminars, workshops; exhibitions and to educate and inspire various functionaries in achieving a sound system of agricultural marketing;
4. creating a national non-political autonomous body to pursue policies and issues at the national level;
5. setting common libraries and build a stock of literature;
6. establishing contacts with other institutions and organizations in the country and abroad for the latest developments in the field of agricultural marketing and to build a data bank;
7. undertaking and assisting professional, technical, management and infrastructural consultancy services in the working of agricultural marketing;
8. assisting and advising the Central Government and the State Governments or members in setting up training centres; and
9. seeking representation of COSAMB in various committees of the central and state governments.

REGULATION OF AGRICULTURAL PRODUCE MARKETS

Under the traditional system of marketing of the agricultural products, producer sellers incurred a high marketing cost, and suffered from unauthorized deductions of marketing charges and the prevalence of various malpractices. To improve marketing conditions and with a view to creating fair competitive conditions, the increase in the bargaining power of producer-sellers was considered to be the most important prerequisite of orderly marketing. Most of the defects of, and malpractices under, the then existing marketing system of agricultural products have been more or less removed by the exercise of public control over markets, i.e. by the establishment of regulated markets in the country.

DEFINITION

A regulated market is one which aims at the elimination of the unhealthy and unscrupulous practices, reducing marketing charges and providing facilities to producer-sellers in the market. Any legislative measure designed to regulate the marketing of agricultural produce in order to establish, improve and enforce standard marketing practices and charges may be termed as one which aims at the establishment of regulated markets. Regulated markets have been established by State Governments and rules and regulations have been framed for the conduct of their business.

The establishment of regulated markets is not intended at creating an alternative marketing system. The basic objective has been to create conditions for efficient performance of the private trade, through facilitating free and informal competition. In regulated markets, the farmer is able to sell his marketed surplus in the presence of several buyers through open and competitive bidding. The legislation for the establishment of regulated markets does not make it compulsory for the farmer to sell his produce in the regulated market yard. Instead, voluntary action on the part of the farmers to take advantage of such a market is assumed. The basic philosophy of the establishment of regulated markets is elimination of malpractices in the system and assignment of dominating power to the farmers or their representatives in the functioning of the markets.

OBJECTIVES

The specific objectives of establishment of regulated markets are:

1. To prevent the exploitation of farmers by overcoming the handicaps in the marketing of their products;
2. To make the marketing system most effective and efficient so that farmers may get better prices for their produce, and the goods are made available to consumers at reasonable prices;
3. To provide incentive prices to farmers for inducing them to increase the production both in quantitative and qualitative terms; and
4. To promote an orderly marketing of agricultural produce by improving the infrastructural facilities.

HISTORY OF MARKET REGULATION

The need for the regulation of markets arose from the anxiety of the British rulers to make available supplies of pure cotton at reasonable prices to the textile mills in Manchester (UK). The first regulated Karanjia cotton market was established as early as in 1886 under the Hyderabad Residency Order. The first legislation was the Berar Cotton and Grain Market Act of 1897. Under the provisions of this Act, the British Resident acquired the authority to declare any place in the assigned district a market for the sale and purchase of agricultural produce, and to form a committee to supervise the regulated markets. The 1897 Act became a model Act for legislation in other parts of the country. The Indian Cotton Committee, appointed by the Government of India in 1917, recommended the regulation of cotton markets on the lines of the Berar Act. In 1927, the then Government of Bombay province was the first to enact the Bombay Cotton Market Act. This was the first law in the country which attempted to regulate markets with a view to evolving fair market practices.

The Royal Commission on Agriculture, in its report submitted in 1928, recommended the regulation of market practices and the establishment of regulated markets in India in view of the chaotic conditions that obtained in the agricultural produce markets. Its recommendations were subsequently endorsed by the Central Banking Enquiry Committee, 1931. In 1935, the Government of India established the office of the Agricultural Marketing Adviser (Directorate of Marketing and Inspection) under the Ministry of Food and Agriculture to look into the problems of the marketing of agricultural produce. The Directorate recommended to the State Governments that markets be regulated to safeguard the interests of the farmers and to remove the prevalent malpractices in the markets. In 1938, the Directorate of Marketing and Inspection prepared a model Bill and circulated among the States.

Since then, State Governments have enacted legislation for the regulation of markets in their States. These are: The Hyderabad Agriculture Market Act, 1930; the Central Provinces Cotton Markets Act, 1932; the Madras Commercial Crops Market Act, 1935; the Agricultural Produce Markets Act of Bombay, of Punjab and Mysore, 1939, of

Madhya Bharat, 1952, of Orissa, 1957 and of Rajasthan, 1961. In the mean time, Andhra Pradesh adopted the Madras Act, Gujarat and Maharashtra States inherited the Bombay Act and Delhi and Tripura passed legislation on the lines of the Bombay Model Act. A list of the Agricultural Product Market Acts, initially enacted, in different states is given in Appendix 7.1

Regulation of markets for agricultural products was stressed by several Committees and Commissions from time to time. The important ones are the Banking Enquiry Committee, 1931; The Congress Agrarian Reforms Committee, 1947; The Rural Marketing Committee of the National Congress, 1948; The Planning Commission, 1952; The All India Rural Credit Committee, 1954; and the Agricultural Production Team of the Ford Foundation, 1959.

The Review Committees on the working of the Agricultural Produce markets from time to time have recommended for bringing all the agricultural products including horticulture, forestry and livestock under the purview of Agricultural Produce Market Acts.

PROGRESS

Though the establishment of regulated markets was started during 1930s, the programme got momentum only after the independence. The number of regulated markets which before the commencement of First Five Year Plan (April 1951) was only 236, increased to 715 in March 1961, 5766 in April 1986; 7161 in March, 2001; 7521 in March, 2005 and further to 7566 in March, 2006. Their number decreased to 7465 in March, 2007 and 7157 in March, 2010 due to merger or deregulation of some markets as in Bihar state. In most of the states where state Acts are in place, all the wholesale markets were functioning under the regulation programme (Table 7.2).

The statewise status of regulated markets is presented in Table 7.3. The progress of regulated markets was not uniform in all the States. There was appreciable number of regulated markets in the States of Andhra Pradesh, Gujarat, Haryana, Himachal Pradesh, Karnataka, Maharashtra, Rajasthan, Orissa, Punjab, Tamil Nadu, Uttar Pradesh and West Bengal. The number of regulated markets was negligible in the States of Arunachal Pradesh, Manipur, Meghalaya, Nagaland and Sikkim. Market regulation was not enacted in two States (Jammu & Kashmir and Manipur) and three Union Territories (Andaman & Nicobar Islands, Dadra & Nagar Haveli and Lakshadweep). In the State of Kerala, four regulated markets in the Malabar area were established by the then Madras State under the Madras Commercial Crops Market Act, 1933. It may be mentioned that Bihar had repealed the Act in 2006.

The regulated markets are categorised as super 'A' class, 'A' class, 'B' class, 'C' class and 'D' class markets according to their income from market fee. For example, at one time, in Rajasthan, regulated markets having annual income of ₹ 2.5 crore and above from market fee were categorised as super 'A' class, those having income between ₹ 1.5 to 2.5 crore as 'A' class, income of above ₹ 75 lakh but less than ₹ 1.5 crore as 'B' class, income of ₹ 25 to 75 lakh as 'C' class and those having income of less than ₹ 25 lakh per annum as 'D' class markets. This Classification reflected the size of the market transactions.

On an average, each regulated market in the country serves an area of 459 sq. km. The situation varies from State to State in terms of area covered per regulated market. The average area served by a regulated market in north-east and hill states is large (Table 7.4). In this connection, it may be mentioned that the National Commission on

Table 7.2: Progress of market regulation in India (up to 2010)		
Year ending	Number of regulated markets	Regulated markets as percent of total wholesale assembling markets (7293 in 2001)
March 1951	236	3.23
March 1956	470	6.44
March 1961	715	9.80
March 1966	1012	13.87
March 1976	3528	48.37
March 1980	4446	60.96
March 1986	5766	79.06
March 1990	6217	85.25
March 1991	6640	91.05
March 1992	6738	92.39
March 1993	6752	92.58
March 1994	6809	93.36
March 1995	6836	93.73
March 1996	6968	95.54
March 2000	7077	97.04
March 2001	7161	98.19
March 2005	7521	100
March 2006	7566	100
March 2007/2010	7465/7157	100

Source: Directorate of Marketing and Inspection, Ministry of Agriculture, Government of India, Faridabad.

Agriculture (1976) had recommended that the facility of a regulated market should be available, in general, within a radius of 5 km. Accordingly, the country needs 41838 markets. There is also a wide variation in the average number of villages served by a regulated market in each state. On all-India basis, each regulated market serves 81 villages. There are as many as 27738 markets spread all over the country. Of these, 21221 are primary rural markets. The remaining 6517 are wholesale assembling and terminal markets.

The number of commodities under regulation also varies from State to State; but they include almost all the important agricultural commodities such as foodgrains, oilseeds, fibre crops, commercial crops, fruits and vegetables, forestry produce and livestock products. There is also wide variation across states in infrastructural facilities in the regulated markets. The facilities in many markets are less than what should have been.

Top 10 states of India according to market density, area served and population served by each regulated market are shown in Table 7.5.

IMPORTANT FEATURES OF REGULATED MARKETS

Under the provisions of the Agricultural Produce Market Act, the state government gives notice of its intention to bring a particular area under regulation by notifying the market area, principal assembling market, market yard and sub-market yard, if any, under the principal regulated market. The meaning of these terms is explained in the following paragraphs.

Table 7.3: Number of wholesale, rural primary and regulated markets in India (as on 31 March, 2007/2010)

States/Union Territories	Number of markets			Number of regulated markets		
	Wholesale	Rural primary	Total	Principal	Sub-market yards	Total
1. Andhra Pradesh	312	577	889	312	577	889
2. Arunachal Pradesh**	6	24	30	–	–	–
3. Assam	344	831	1175	20	204	224
4. Bihar*	325	1469	1794	95	415	510
5. Jharkhand	28	592	620	28	170	198
6. Goa	4	24	28	1	7	8
7. Gujarat	209	129	338	196	214	410
8. Haryana	284	185	469	106	178	284
9. Himachal Pradesh	42	30	72	10	32	42
10. Jammu & Kashmir***	26	8	34	–	–	–
11. Karnataka	498	730	1228	146	352	498
12. Kerala***	348	1014	1362	–	–	–
13. Madhya Pradesh	237	1321	1558	235	256	491
14. Chhatisgarh	2	1132	1134	73	106	179
15. Maharashtra	880	3500	4380	294	586	880
16. Manipur***	20	98	118	–	–	–
17. Meghalaya	101	112	213	2	–	2
18. Mizoram**	10	105	115	–	–	–
19. Nagaland**	11	86	97	–	–	–
20. Orissa	398	1150	1548	45	269	314
21. Punjab	437	–	437	145	292	437
22. Rajasthan	426	312	738	125	301	426
23. Sikkim	7	12	19	1	–	1
24. Tamil Nadu	300	677	977	277	15	292
25. Tripura	84	554	638	21	–	21
26. Uttar Pradesh	584	3244	3828	245	342	587
27. Uttarakhand	20	29	49	25	31	56
28. West Bengal	279	2925	3204	43	641	684
29. A & N Islands***	–	–	–	–	–	–
30. Chandigarh	1	–	1	1	–	1
31. Dadra and Nagar H***	–	–	–	–	–	–
32. Daman & Diu	–	–	–	–	–	–
33. Delhi	30	–	30	9	14	23
34. Lakshadweep***	–	–	–	–	–	–
35. Pudducherry	8	–	8	4	4	8
Total (on 31/03/2007)	6261	20870	27131	2459	5006	7465
As on 31/03/2010	6517	21221	27738	2404	4753	7157

* Bihar Agricultural Produce Marketing (Regulation) Act repealed from 1st September, 2006.
** APMR Act not yet implemented. *** APMR Act not enacted.

In Bihar and West Bengal, sub-yards include cold storages and hence figures of total regulated markets and wholesale markets are not comparable. All principal regulated markets are wholesale markets where as sub-markets yards may or may not be a wholesale market as these also include some of the Rural Primary Markets notified for regulation.

Source: Directorate of Marketing and Inspection, Ministry of Agriculture, Government of India, Faridabad.

Table 7.4: State-wise details of area served by regulated markets (as in 2007)

State/Union Territory	Geographical Area ('000 sq. km)	Regulated Markets No. (31-3-07)	Area covered per market (sq. km)	No. of markets required*
1. Andhra Pradesh	275.0	889	309	3501
2. Arunachal Pradesh	83.7	0	0	1066
3. Assam	78.4	224	350	998
4. Bihar	94.1	510	185	1198
5. Jharkhand	79.7	198	402	1015
6. Goa	3.7	8	463	47
7. Gujarat	196.0	410	478	2495
8. Haryana	44.2	284	156	565
9. Himachal Pradesh	55.7	42	1326	709
10. Jammu & Kashmir	222.2	–	–	2829
11. Karnataka	191.8	498	385	2441
12. Kerala	38.9	–	–	495
13. Madhya Pradesh	308.3	491	628	3924
14. Chhatisgarh	135.1	179	755	1719
15. Maharashtra	307.7	880	350	3916
16. Manipur	22.3	0	0	284
17. Meghalaya	22.4	2	11214	285
18. Mizoram	21.1	0	0	268
19. Nagaland	16.6	0	0	211
20. Orissa	155.7	314	496	1982
21. Punjab	50.4	437	115	641
22. Rajasthan	342.2	426	803	4356
23. Sikkim	7.1	1	7096	90
24. Tamil Nadu	130.0	292	445	1655
25. Tripura	10.5	21	499	133
26. Uttar Pradesh	238.5	587	406	3036
27. Uttarakhand	55.8	56	997	711
28. West Bengal	88.7	684	130	1130
29. A & N Islands	8.2	–	–	105
30. Chandigarh	0.1	1	114	1
31. D & N Haveli	0.5	–	–	6
32. Daman & Diu	0.1	–	–	1
33. Delhi	1.5	23	64	19
34. Lakshadweep	0.03	–	–	0
35. Puducherry	0.5	8	62	6
INDIA	**3287.2**	**7465**	**440****	**41838**

* Market requirement (as recommended by National Commission on Agriculture) is one regulated market in the radius of 5 km.

** 459 in 2010

Source: Directorate of Marketing and Inspection, Ministry of Agriculture, Government of India, Faridabad.

Table 7.5: Top ten states of India based on different parameters of market regulation (other than UTs and NE states)

Highest market density per 1000 sq. km. of geographical area	Lowest area served per regulated market (in sq. kms)	Lowest population served per regulated market (lakh persons)
1. Punjab	Punjab	Punjab
2. West Bengal	West Bengal	Haryana
3. Haryana	Haryana	Andhra Pradesh
4. Bihar*	Bihar*	Jharkhand
5. Andhra Pradesh	Andhra Pradesh	Karnataka
6. Assam	Maharashtra	Maharashtra
7. Maharashtra	Assam	Chhattisgarh
8. Karnataka	Karnataka	Orissa
9. Jharkhand	Jharkhand	Assam
10. Uttar Pradesh	Uttar Pradesh	West Bengal

* The situation pertains to the period before the Act was repealed in 2006.
Source: Compiled from data presented in Table 7.4.

1. *Market Area:* The area from which the produce naturally and abundantly flows to a commercial centre, i.e. the market, and which assures adequate business and income to the market committee.
2. *Principal Assembling Market:* It is the main market which is declared as a principal market yard on the basis of transactions and income generated for the market committee.
3. *Market Yard:* This is a specified portion of the market area where the sale, purchase, storage and processing of any of the specified agricultural commodities are carried out.
4. *Sub-Market Yard:* It is the sub-yard of the principal assembling market. This is a small market and does not generate sufficient income to be declared as a principal assembling market.

The physical structures in the market yard can be grouped in four categories depending on their need to buyers/sellers and type of functions/services provided by them to different categories of persons in the market.

Market Yard–Structures

Structures needed for sale and purchase of agricultural commodities	Structures meant for the convenience of buyers and sellers	Structures providing amenities to buyers or sellers	Miscellaneous structures
1. Platforms 2. Storage godowns 3. Shops 4. Drying floors	1. Farmer's rest house 2. Vehicle and cart parking place 3. Animal shed 4. Water huts and water troughs 5. Sanitary facilities 6. Electricity and light 7. Chowkidar hut or security post	1. Canteen 2. Input selling shops 3. Bank 4. Post office	1. Firefighting equipment 2. Watchman and staff quarters 3. Internal roads 4. Boundary of the yard

The important features of regulated markets are as follows:

1. Classification of Regulated Markets

Regulated markets have been classified on the basis of income as well as the produce transacted.

(a) Based on Income

Based on income of the market committee, the regulated markets are of five types—Super, A Grade, B Grade, C Grade and D Grade. The income range for this classification varies from state to state. As mentioned earlier, at one time, in the state of Rajasthan, the regulated markets having income of more than ₹ 2.5 crore per annum were categorized as Super category markets, those having income between ₹ 1.50 and 2.50 crore as 'A' Grade markets, between ₹ 75 lakh and 1.50 crore as 'B' Grade, from ₹ 25 lakh to 75 lakh as 'C' Grade and those having income less than ₹ 25 lakh per annum are categorized as 'D' Grade markets. Based on this income criteria, the number of regulated markets in the state of Rajasthan varied across the years.

(b) Based on the Major Commodities Transacted

Based on the major produce transacted in the regulated markets, the markets are classified as foodgrain markets, fruits and vegetable markets, flower markets, wool markets, cotton markets and so on.

2. Method of Sale

In regulated markets, the sale of agricultural produce is undertaken either by open auction or by the close tender method. In recent years, e-auction or online auction has been introduced in some markets. These methods of sales ensure a fair and competitive price for the produce and prevent the cheating of farmers by market functionaries. By these methods, the sale is carried out under the supervision of an official of the market committee, and the signature of the buyer are taken as soon as the auction is over. The business is done during fixed hours. Generally, different commodities are auctioned at different times of the day. Bidding is kept open to all authorized traders, and the highest bidder is given the produce. The farmer has the option to refuse to accept the sale even at the highest bid, if he or she so desires.

3. Weighment of Produce

Weighment of the produce is done by a licensed weighman with standard weights and a platform scale. This eliminates short weights and malpractices which arise out of weighing with a hand balance. Electronic weighing balances/machines have also been installed in some markets. In some markets, weighbridges have been installed.

4. Grading of Produce

The produce in the regulated markets is expected to be sold only after grading; but because of the absence of facilities, such as space, funds for the employment of technical experts for grading and the purchase of grading equipment, the grading scheme has not been fully implemented in all the markets in the country. There is, therefore, a need for the extension of grading facilities to all the markets in the country. However, traders have their own conventional grading system, which they invariably follow.

5. Market News Service

In regulated markets, there is an arrangement for a proper and correct dissemination of market prices through various media, such as loudspeakers and notice-boards. However, the dissemination of price information is not perfect in most markets because of the non-existence of these facilities. Sometimes, farmers are unable to take advantage of this facility either because of their illiteracy or the non-availability of information when they require it. Nevertheless, market yards also provide the information on national portal AGMARKNET which can be accesed on the website.

6. Market Charges

With the regulation of markets, such market charges as darmada, charity, karda, dhalta and muddat, were abolished, while the rates of other market charges, such as commission, brokerage, hamali and weighing charges, were specified in proportion to the extent of the service rendered by different middlemen. Later, most of the market charges payable by sellers have been transferred to the buyer; the sellers have to pay only for the activities undertaken prior to the sale of the produce, i.e. for transportation, octroi and hamali only. There is, thus, some saving for the producers. The extent of saving varies from market to market and from commodity to commodity.

A large number of studies were conducted during the eighties to estimate/assess the benefits of market regulation programme. The effect of market regulation on the reduction of market charges and the savings to the farmers as per a study by Dantwala in 1991 is presented in Table 7.6. These estimates are based on a survey conducted by the Directorate of Marketing and Inspection, Government of India, New Delhi, covering 496 markets of the country.

Table 7.6: Effect of market regulation on reduction of market charges and savings to farmers

Market charges	Average market charges payable per ₹ 100 worth of produce (₹)		Savings to farmers	
	Before regulation	After regulation	₹	Percent
Commission	1.69	1.13	0.56	33.1
Weighment	0.38	0.15	0.23	60.5
Hamali	0.56	0.20	0.36	64.3
Brokerage	0.29	0.19	0.10	34.5
Charity	0.12	0.06	0.06	50.0
Miscellaneous	0.95	0.32	0.63	66.3
Total	3.99	2.05	1.94	48.6

Source: Dantwala, M.L. (1991), quoted in Indian Journal of Agricultural Marketing, Conference Special, 2000, p. 58.

7. Payment of the Value

It is obligatory on the buyer to make prompt payments for the purchased produce without deductions of any muddat. In unregulated markets, the payment of the price of the produce is made by the buyer after 3 to 15 days of the sale of the produce in accordance with 'ocal market rules. In the past, if the seller insisted on cash payment, the buyer would deduct muddat (cash discount) to the extent of ₹ 0.50 for the produce worth ₹ 100 while making the payment. This practice has been stopped with the regulation of markets. Nevertheless, informal agreements between seller and trader or commission agent continue.

8. Licensing of Market Functionaries

All the market functionaries, from the hamals to traders working in the regulated market, have to obtain a licence from the market committee; after paying the prescribed fees, to carry on their business. The licensed traders have to keep proper records and maintain accounts in accordance with the bye-laws of the market committee. This facilitates the exercise of a proper control on the accounts of the traders and the collection of the correct amount of the market fee by the market committee. Any violation of rules by middlemen may lead to the cancellation/suspension of the licence by the market committee or to the filing of challans in a court of law.

9. Supervision

The day-to-day functioning of regulated markets is supervised by the officials of the market committee, i.e. the secretary, auction clerks, and other staff. The administrative decisions are taken by the nominated/elected market committee.

10. Market Committees

The market committee consists of representatives of all sections, i.e. farmers, traders, cooperative marketing societies, cooperative or commercial banks, autonomous bodies (panchayat samiti and municipal board of the area), and governments officials. Prior to the establishment of regulated markets, the rules for the conduct of the business in the market were framed by traders without any consideration for the interests of other groups, i.e. farmers and consumers. The composition of the market committee varies from state to state.

Even within a state, there is always a minor change when the State Acts are amended. However, the broad structure of the market committees in different states is as shown in Table 7.7. The number of farmers is generally more than half. The other category members may be local legislators or parliamentarians or representatives of local banks, warehousing corporations, local area development boards or non-trader market functionaries. There is also some marginal variation in the composition of committees according to the category of the market (Super, A, B, C and D).

The market committees are the main instruments under the Market Act. The committees are corporate bodies comprising the representatives of various groups. A common feature of the Acts passed by the State Governments is that the number of farmer-members is more than that of other interest groups. The procedure of the election of non-official members of the market committee differs widely across States. While the Act of Karnataka provides for the direct general election, those of Gujarat and Maharashtra provide for indirect election through cooperative marketing and credit societies. In some other states, the Acts provide for the election through the members of the village panchayats.

11. Finances

To meet its administrative expenditure, and in order to create infrastructural facilities in the market area, the market committee gets funds from the following sources:

1. Market fee on the produce brought for sale in the market yard/market area. The rate of this market fee varies from State to State. In the past, the rate of market fee

Table 7.7: Composition of market committees in different states/UTs (as in August 2008)

States/UTs	Total members	Farmers	Traders or Market function-aries	Coope-ratives	Local Adminis-tration	State govt.	Others
Andhra Pradesh	15–20	10–12	2–3	1	1	1	–
Assam	15	7	3	1	1	1	2
Arunachal Pradesh	9	2	2	1	1	3	–
Gujarat	17	8	4	2	1	2	–
Haryana	12–20	7–9	3–6	–	2	–	–
Himachal Pradesh	9–16	5–9	3–6	–	–	1	–
Karnataka	15	9	1	2	2	1	–
Meghalaya	13	5	3	2	1	1	1
Maharashtra	18	10	3	1	2	2	–
Madhya Pradesh	8–20	2–14	2	1	1	1	1
Nagaland	11	6	3	0–1	0–1	2	–
Orissa	17	7	4	–	3	3	–
Punjab	12–20	7–9	3–6	–	2	–	–
Rajasthan	15	7	2	2	–	1	3
Tamil Nadu	15–16	8	3	–	–	3	1
Uttar Pradesh	14–16	5–7	–	–	1	–	3
West Bengal	15	5–7	2	1	1	2	2
Tripura	12	6	1	1	2	2	–
Chandigarh	12–20	7–9	3–6	–	2	–	–
Delhi	14	6	3	1	2	2	–
Goa, Daman & Diu	18	–	3	7+1	4–5	2	–
Puducherry	15–16	8	3	–	–	3	1

Source: AMA, Govt. of India, In some states while total number is specified, the composition is discretionary.

was ₹ 0.20 to 0.30 for produce valued at ₹ 100. At present, the rate varies from 1 to 3% of the value of goods. The market fee is collected at a single point in most of the states.

2. Licence fee/renewal fee of market middlemen functioning in the area, and
3. Subsidy from the government.

12. Functions of Market Committees

The functions of the market committee vary from state to state. Its important functions are:

1. To manage the main and sub-market yards and run them in the interest of the agriculturists and traders;
2. To prescribe the hours of trading with a view to controlling and regulating transactions;
3. To issue, renew or withdraw licences of market functionaries with a view to regulating the entry of persons in the market;
4. To fix market charges for various functions and services;
5. To realise market fee from either buyers or sellers;
6. To collect and disseminate market news among the farmers, and buyers;

7. To settle disputes, which arise out of the sale and purchase of notified agricultural commodities, among farmers, traders and other market functionaries;
8. To control and regulate the behaviour of those who enter the market for transactions;
9. To provide facilities for grading and standardization and to take steps for the prevention of adulteration of agricultural produce;
10. To bring, prosecute and defend any suit on behalf of the market committee; and
11. To create amenities in the market yard and sub-yard as well as in the market area which are necessary for the smooth functioning of the marketing of agricultural commodities.

13. Settlement of Disputes

Disputes arising between producer-seller and traders by reason of the quality of the produce, accounts and deductions of unauthorised charges are solved by the sub-committee of the market committee. This avoids the legal complications and unnecessary expenditure. Prior to regulation, no such facility existed. If a farmer was not satisfied, he had to go to a court of law to get his due justice, which involved a lot of expenditure and wastage of time.

14. Eliminating Malpractices

Regulated markets have brought about a general awakening among producer-sellers. This awakening enables them to protect themselves against a number of malpractices which were formerly prevalent in the unregulated markets. Now the sale slips of the produce sold are given to farmers, showing the details of the quantity sold, the rate of sale and deductions, if any. A copy of the sales slip is supplied to the market committee for purposes of checking. The malpractice of the taking away of samples by the bidders has also been stopped.

15. Provision of Amenities in the Market

The market committee provides the amenities required for a smooth and efficient marketing of the produce of farmer-sellers. These amenities encourage the farmers to bring their produce for sale in the regulated market, and check the tendency on their part to sell locally. There is also a check on the tendency to aversion of city markets for one reason or the other.

The various amenities provided by the market committee are:

1. Link roads and culverts in the area;
2. A spacious market yard;
3. Rest house, cattleshed and water troughs;
4. Light, watchman, drinking water and parking space for vehicles in the market yard;
5. Infrastructure facilities, such as banks, canteens and post office in the market yard.

The amenities vary from State to State and from market to market, depending upon the financial status of the market committee. Most of the regulated markets lack these amenities to the extent of the norms that have been formulated. It is essential, therefore, that the government should see to it that the necessary facilities and amenities are created in accordance with the norms that have been laid down.

The amenities and facilities in most of the regulated markets are not available as per the prescribed norms. The requirement of different amenities and facilities varies with the size of the markets, i.e. A, B, C or D class markets. In general, a regulated market should have suitable (sized) auction platforms, drying yards, trader modules (shop, godown and platform in front of the shop), rest house, cattle shed, storage godowns, office building, waterhut, weighing equipment and toilets. In addition, some basic facilities like internal roads, boundary walls and electric lights should also be available.

The number of regulated markets having different facilities/amenities and the extent of their utilization, as observed in 1999, are shown in Table 7.8. Auction platforms are needed in each market for settlement of price of the produce in a congenial atmosphere between buyers and sellers. However, this facility (both covered and open auction

Amenities	Number of markets with facility (%)	Utilization (%)
Common auction platform (covered)	64	84
Common auction platform (open)	67	82
Common drying yards	26	87
Traders modules	63	89
Retailer's shops	28	100
Storage godowns	74	91
Cold storage	9	100
Weighing equipment	85	100
Processing units	7	83
Grading equipment	30	89
Pledge financing	17	93
Bank	42	100
Post office	28	100
Police post	15	85
Security post	42	97
Farmer's rest house	61	89
Agricultural input shop	29	96
Bathrooms	57	98
Toilets	88	98
Canteen	43	97
Drinking water taps	28	100
Loading, unloading and parking	100	100
Internal roads	89	100
Boundary wall	84	93
Electric lights	89	100
Avenue and shed trees	57	98
Seating benches	28	100
Price display boards	61	92
Public address system	34	91
Public telephone system	24	100
Garbage disposal system	84	100
Drainage system	55	98

Table 7.8: Facilities/amenities in regulated markets

Source: Directorate of Marketing and Inspection, Ministry of Agriculture, Government of India (1999).

platforms) existed only in two-thirds of the regulated markets. At that time, only one-fourth of the markets had common drying yards. Traders modules (shop, godown and platform in front of shop) existed in 63% of the markets. The cold storage units for storage of perishable commodities exists in only 9% of the markets and grading facilities existed in less than one-third of the markets. The basic facilities viz., internal roads, boundary walls, electric light, loading and unloading facilities and weighing equipments were available in more than 80% of the markets, Farmers rest houses existed in more than half of the regulated markets. However, there was considerable gap in the available facilities and their utilisation. Since then, considerable attention has been given for creation of these amenities. The situation is much better now.

16. Farmer's Response Regarding Available Facilities in Regulated Markets

It is interesting to know the farmers response regarding the facilities available in the regulated markets. The responses vary from state to state. The results of a response survey conducted in Punjab for three major crops viz., paddy, wheat and cotton are presented in Table 7.9. More than 70% farmers reported that their knowledge about rates of incidental charges is inadequate. Seventy to eighty percent farmers also reported inadequacy of sheds for animals, seating arrangements and arrangement for food and tea.

Table 7.9: Farmer's response regarding different facilities available in the selected markets of Punjab for marketing of paddy, wheat and cotton

Facilities	Adequate	Partially adequate	Inadequate
1. Knowledge about the rates of incidental charges	18.89	10.00	71.11
2. Available space for unloading	47.78	10.67	35.55
3. Labour availability	87.78	10.00	2.22
4. Availability of parking space at the time of unloading	50.00	15.56	34.44
5. Availability of parking space after unloading	45.56	11.11	43.33
6. Overnight parking facility	10.00	6.67	83.33
7. Availability of proper sheds	6.67	20.00	73.33
8. Availability of electricity	41.11	37.78	21.11
9. Availability of drinking water	24.45	36.66	38.89
10. Arrangement of food and tea	5.55	7.78	86.67
11. Seating arrangement	4.45	11.11	84.44

Source: Sukh Sanjan, S.S. Chahal and M.S. Toor (2000), Impact of Market Regulation on Adequacy of Market Infrastructure in Punjab, Indian Journal of Agricultural Marketing, Vol. XIV, No. 3, September–December, 2000, pp. 15–24

17. Other Incentives for Farmers

In addition to providing amenities a scheme of accidental insurance for agricultural cultivators and labourers handling agricultural machines was also introduced by the market committees in many states. The main objective of this scheme is to pay some financial compensation to the farmers or labourers for injuries occurring by accidents while operating agricultural machines. This apart, some schemes for providing implements and storage bins at subsidised rate have also been taken up by many market committees in the regulated markets out of the funds generated by the market committees. In recent years, they have initiated programmes of exploring distant markets for the produce of their area with a view to increasing the realisation of farmers from their surplus produce. Part of the income generated by market committces is ploughed back by the state governments in constructing village approach roads.

ISSUES AND CONCERNS

Though the market regulation scheme has been implemented in the country, its benefits to the different classes were not considered uniform. During the late 1980s and 1990s, a large number of studies (at micro and macro levels) were carried out to bring out the outcome and impact of market regulation programme (first phase of reforms). These have pointed out that it has a high benefit-cost ratio. Nevertheless, the fact remains that many improvements are needed to make it deliver services in an efficient manner. It is a socio-economic reform. It was felt that there is, a need for bringing home its advantages to the cultivators and market functionaries. To make farmers conscious of their rights in the management and control of the market is a sine qua non of the success of the scheme. Farmers should not submit meekly to the practices of market functionaries when these are contrary to their interest. Market functionaries, too, should be convinced that the market regulation scheme has not been designed to eliminate them but to bring to book those unscrupulous traders who defame the entire trading community by their anti-social and dubious activities. It is with the co-operation of all these agencies—producers, consumers and market functionaries—that the market regulation scheme can make the marketing system of agricultural produce an efficient system.

Regulation of primary wholesale markets was taken up as an institutional innovation and construction of well laid out market yard was considered as an essential requirement of effective implementation of the regulation programme. As the programme was a development-cum-legal measure, it took considerable time to extend it on a wider scale. The market regulation brought its impact in terms of providing higher prices, better returns, reduction of market charges and providing amenities at the time of sale of the produce to the farmers in the vicinity of 7465 regulated markets established in the country. The benefits of the regulated markets to the farmers varied in these market areas according to the extent of infrastructural facilities existing in them. Roughly 55% of the cereals and 90% of the commercial crops are traded in these markets. These markets are also financially sound as they earn revenue at the rate of 0.5% (Gujarat) to 4% (Punjab) of the value of produce transacted.

However, these regulated markets are also full of problems on account of a number of reasons. These are:

1. The benefits of regulation in a market vary directly in proportion to the availability or creation of market infrastructural facilities in them. Out of 7465 regulated markets, a higher level of infrastructural facilities has been created only in 2354 main yards and lower level of infrastructural facilities exists in 5111 sub-yards. Further, there are 20,847 rural primary or periodic markets (hats or shandies) in the rural areas where farmers and rural people congregate periodically to sell their surpluses. In the rural markets, only 15% are reported to have physical infrastructural facilities so far. Thus, there is need for extension of market regulation in all the rural markets. The management of these rural markets of the country is in the hands of local bodies. The management as well as the market conditions for sale in these rural markets are far from satisfactory.

2. Emphasis in most of the regulated markets has remained mainly on the construction activities (roads, yards and building) and on collection of market fees. Adequate emphasis has not been given to the effective implementation of the provisions of market regulation in real sense.

3. Several weaknesses and malpractices are still reported in the working of the regulated markets. The important ones are heavy physical losses of quantity of produce during unloading, cleaning, sieving, loading and transportation; excess weighment; and considerable wastage of time of the farmers in waiting for their turn of auction or weighment, or getting payment on account of congestion in the markets.

4. Bureaucratization in the management of the regulated markets exists in many states which resulted in continuation of several malpractices. At one time, more than 80% market committees were superseded and managed by administrators.

5. Most of the regulated markets do not have qualified staff to manage. Further, due to heterogeneity in provisions of different state marketing Acts, the movement of agricultural produce in the country is not smooth.

6. Most producers refrain from bringing their commodities to these regulated markets as they report that there is no additional benefit to them by selling in regulated markets. They feel that the objective of setting up of these state controlled regulated markets have not been fulfilled; and

7. The regulated markets have indirectly led to the monopolization of trade by way of granting licenses to different types of market middlemen; which in turn are not allowing others to enter in the business in market yards. This way, they are able to charge self determined rates from sellers for providing a particular market service.

Because of these problems, the entire scheme was comprehensively reviewed during the 1990s and several steps were taken to improve the situation during the last 30 years.

REVIEW REPORTS

With a view to reviewing the status of agricultural markets regulation programme and intervention in food grains marketing, a series of in-depth reviews of the entire scenario was commissioned by the government of India since the early 1990s. Some of these are briefly shown here.

HIGH POWERED COMMITTEE ON AGRICULTURAL MARKETING, 1992 (GURU COMMITTEE)

In 1992, a High Powered Committee on Agricultural Marketing was appointed by the Government of India under the chairmanship of Shri Shankarlal Guru, the then Chairman of COSAMB and Gujarat State Agricultural Marketing Board to suggest improvements in the functioning of market regulation scheme. The committee submitted its final report on 26th November, 1992. The Committee in its report suggested several changes in the existing market regulation programme and restructuring of the agricultural produce market committees and State Agricultural Marketing Boards. The highlights of the suggestions are as follows:

1. At least 15,000 more markets should be brought under market regulation in addition to the then existing 6840 regulated markets;

2. The Agricultural Produce Market Regulation Act be enacted in those States and Union Territories where it is not in existence.

3. Separate national bank on agricultural marketing be created to cater to the credit needs of the marketing sector.
4. Tobacco, tea, rubber, cardamom, spices and other plantation crops should be brought under the purview of the market committee instead of commodity development boards.
5. High priority for central assistance be accorded to primary markets situated in DPAP, ITDP, HADP and North-Eastern regions.
6. State Marketing Tribunals be created to deal with the legal disputes in the implementation of State Acts.
7. A Central Agricultural Marketing Board be constituted to develop the markets of national importance.
8. A scheme to insure the farmers against price depressions below the support/procurement level be implemented.
9. A pledge finance scheme of credit for the unsold produce in the regulated markets be implemented.
10. The Chairman of the Market Committee/Board should be elected from amongst the elected agriculturist members of the market committees/Boards.
11. The DMI, COSAMB and NIAM should intensify the activities for development of marketing in the north-eastern region.
12. The DMI be restructured to make it render technical advice effectively.
13. Facilities of NICNET be extended to provide timely information for the users of the market.
14. A tableau on agricultural marketing programme be included in the Republic Day Celebrations.
15. In the preamble of the agricultural produce marketing regulation, along with the producer's interest, the interests of consumers should also be included.

The committee, in all, made 89 recommendations relating to various aspects of agricultural marketing like legal framework, marketing organizations, central assistance scheme, marketing credit, post-harvest technology, research and education.

EXPERT GROUP ON AGRICULTURAL MARKETING, 1998 (ACHARYA GROUP)

The Ministry of Rural Areas and Employment of the Government of India, in July, 1998, constituted an Expert Group for suggesting measures to strengthen the Agricultural Marketing Division of the Ministry and to widen its scope to facilitate the development of marketing network as growth centers in rural areas. The terms of reference of the Expert Group were:

1. To examine the integration of the development of agricultural as well as rural marketing infrastructure with various on-going schemes/programmes of the Ministry of Rural areas and Employment.
2. To assess the feasibility of earmarking of some funds from the various schemes/programmes of the Ministry of Rural Areas and Employment for the development of agricultural and rural markets, particularly in view of the absence of any scheme at present for development of markets in the country.
3. To suggest structural changes in the Agricultural Marketing Division of the Ministry with a view to making it a professional division for assimilating and integrating various findings of the studies as well the development plans of agricultural and rural markets not merely for the policy purposes but also to arrange the funding

from external sources, so as to supplement the resources for the development of rural markets in the country.

4. To examine critically the role of Directorate of Marketing and Inspection and the National Institute of Agricultural Marketing (NIAM) particularly in the changing context of the new economic policy for the development of rural marketing in the country; and

5. To make suggestions for checking the fluctuations in agricultural prices in the context of local and global realities.

The Expert Group recommended for setting up of a strong Division of Agricultural Marketing in the Union Ministry of Agriculture and making NIAM as a professional body. The Expert Group, submitted its report in August, 1999, with 38 recommendations on the above aspects. As a follow up, the Government of India decided to transfer DMI, NIAM and their infrastructure to the Ministry of Agriculture with immediate effect.

EXPERT COMMITTEE ON STRENGTHENING AND DEVELOPING OF AGRICULTURAL MARKETING, 2000 (GURU COMMITTEE)

Under the chairmanship of Shri Shankar Lal Guru, (and S. S. Acharya as a member), this committee was appointed in December, 2000 with the following terms of reference.

1. To review the system of agricultural marketing in the country in the context of increasing agricultural production and liberalization of international trade.

2. To examine the organizational set-up and functioning of the different State Agricultural Marketing Boards and Agricultural Produce Market Committees and to recommend measures to make them more effective instruments for providing better infrastructure and services to the farmers, traders and consumers.

3. To make recommendations for promoting pledge finance, direct marketing and alternative marketing systems.

4. To study the requirements of additional investments in infrastructure, supply chain management from farm to the consumer and other facilities for the marketing system for the next ten years and to make recommendations for encouraging public, private and cooperative sectors to make such investments.

5. To examine the requirements of market intelligence for the farmers, exporters, traders and consumers and to make recommendations in this regard;

6. To examine the requirements of market extension, research, and training for the agricultural marketing system and to make recommendations in this regard; and

7. To recommend measures for effectively utilising information technology tools with special reference to E-Commerce and E-Business, for the development of a modern marketing system.

The Expert Committee submitted its report on 29th June, 2001 giving 45 recommendations on the above seven aspects. One of the recommendations of the Expert Committee was to make an investment of ₹ 270,000 crore in agricultural marketing infrastructure during the next 10 years (10th and 11th Five Year Plan), which was based on the detailed exercise done by a Working Group on agricultural marketing infrastructure constituted under the chairmanship of Dr. S.S. Acharya (a member of the Expert Committee).

INTER-MINISTERIAL TASK FORCE ON AGRICULTURAL MARKETING, 2001 (JAIN)

As a follow up of the Expert Group/Committee, an Inter-Ministerial Task Force was constituted by the Ministry of Agriculture, Government of India to operationalise the recommendations of Expert Committee on Strengthening and Developing of Agricultural Marketing under the chairmanship of Shri R.C.A. Jain, the then Special Secretary, Agriculture on 4th July, 2001.

The Task Force identified nine areas on which the road map for development of agricultural marketing was prepared. These are:

1. Legal amendments necessary for efficient marketing
2. Direct marketing
3. Infrastructural development of markets
4. Pledge financing
5. Warehouse receipt system
6. Forward market system
7. Price support policy
8. Information technology in agricultural marketing; and
9. Market extension, trainings and research in identified areas.

Separate Inter-ministerial Working Groups examined each of these areas and assisted the Task Force in formulating its recommendations. The Task Force submitted its report in 2002. Based on the suggestions of the Task Force, a National Seminar was held on 27th September, 2002 under the chairmanship of the Agriculture Minister, Government of India, at New Delhi, wherein all Ministers and Secretaries of Agricultural Marketing of State Governments, and Union Territories were invited for discussion and for chalking out a program of agricultural marketing reforms. The second phase of marketing reforms during the last 15 years have followed the pathway suggested by the Task Force.

HIGH LEVEL COMMITTEE ON LONG-TERM GRAIN POLICY (ABHIJIT SEN)

The committee was constituted by the Union Ministry of Consumer Affairs, Food, and Public Distribution for formulating a long term grain policy for the country on 17th November, 2000. The Committee submitted its report in July, 2002. The Terms of Reference of the Committee required examination of (a) minimum support prices (MSP) and price support operations; (b) role of Food Corporation of India (FCI); (c) functioning of PDS; (d) policies regarding buffer stocks, open market sales and foreign trade; and (e) allocation of grain for rural development and other welfare programmes. Some of the recommendations of the committee are as follows:

1. As continued food self-sufficiency is an indispensable component of national security, India must continue to plan for cereals self-sufficiency.
2. MSP system with open-ended purchase should continue as this system is WTO-compatible.
3. Minimum support prices should be supplemented by variable import and export tariff policy and MSPs should have some statutory basis.
4. We should go back to a universal PDS.
5. FCI, which has performed reasonably well in several areas of food security, should be continued to play its role as central agency in procurement and distribution of foodgrains.

6. The Commission for Agricultural Costs and Prices should be made an empowered statutory body.
7. Price support operations should be made more effective in eastern and central Indian states; and
8. State Corporations and private sector should be encouraged to play more active role in foodgrains trade.

WORKING GROUP ON AGRICULTURAL MARKETING AND TRADE FOR XI FIVE YEAR PLAN, 2006 (ACHARYA)

As a part of preparation of the strategy for agricultural marketing reforms and international trade in agricultural commodities, the Planning Commission (Government of India) constituted a Working Group in 2006 under the chairmanship of Dr. S.S. Acharya to suggest measures in this regard. The Working Group reviewed the entire scenario of agricultural marketing and trade, and made several recommendations. The report, which is available on the website of the Planning Commission, formed the basis for XI Five Year Plan pertaining to agricultural marketing and trade in India. Some of the major recommendations of the working group, which became a part of XI Five Year Plan, are as follows:

1. Urgent need for modernization of existing market yards and sub-yards.
2. Marketing system improvement and conducive policy environment.
3. Strengthening of marketing infrastructure by stepping up of investment.
4. Improvement in marketing information system with the use of ICT.
5. Human resource development in agricultural marketing.
6. Promotion of exports/external trade.
7. Hold regular elections of APMCs and bring professionalism in functioning of regulated markets.
8. Plough-back the market fee for development of marketing facilities,
9. Extend greater flexibility to stakeholders, sellers as well as buyers, to interact in the markets.
10. Promote grading, standardization, packaging and certification in the market area.
11. Ensure transparency in auction system, penalization for arbitrary deductions from farmers' realization, prompt payment to farmers, and hassle-free transactions in the markets.
12. Improve weighing systems by installing bulk weighment system and handling in a time bound manner.
13. Bring uniformity in the state-level tax structure for agricultural commodities.
14. Remove de facto restrictions on movement of goods across state borders by harmonizing state-level taxes and provide hassle free collection of tax at convenient points.
15. Conceptualize whole country as one unified integrated national market
16. Amend ECA to provide marketing and trade restrictions only in exceptional situations.

THE NEED FOR REORIENTATION

Agricultural marketing scenario in the country has undergone a sea change over the last two decades. Some of these changes which have significant bearing on the working of the regulated markets are:

1. Rapid increase in the marketable surplus on account of increase in agricultural production;
2. Specialization in production in different agro-climatic zones;
3. Emergence of relatively new concepts of markets of national importance; and
4. Changes in the consumer preferences and tastes for food products.

These apart, the marketing policy regime for agriculture has undergone considerable changes since 1991 when a programme of economic reforms was launched in the country. The policy regime has changed both in terms of domestic market and external trade. The liberalization of domestic markets and changes in trade policy ought to subserve the objectives of further improvement in food and nutrition security at the macro as well as household level; reduction in levels of poverty; development of backward and less developed areas; reduction in inter-personal and inter-regional disparities in development and overall accelerated growth of the economy. The achievement of these objectives in the liberalized environment poses considerable challenge to various stakeholders particularly the cooperatives and such marketing organizations as Agricultural Produce Market Committees (APMCs) or Krishi Upaj Mandi Samitis (KUMS), State Agricultural Marketing Boards and State Directorates of Agricultural Marketing.[1]

It is in this context that there is a need to redefine the role of APMCs. Their existing activities were laid down in a situation of scarcity syndrome and when the domestic market for most of the agricultural commodities was insulated from the world market and government intervention was pervasive. No doubt, in the initial phases attention was needed on establishment of market yards, creation of various infrastructures in the yards and evolving systems and procedures for regularization of practices in the yards and sub-yards. The emphasis now, however, should shift to other important activities of topical relevance. Some of the activities which should receive increasing attention of APMCs and departments of agricultural marketing include:

1. Creating grading and quality certification facilities and building up goodwill for the produce of their area in distant markets;
2. Encouraging the local farmers, traders and processors to market the produce under their brands and promoting such local brands based on graded produce;
3. Regular surveillance on the price environment for the most important products of their area, analysing the movement in profitability and promoting an aggressive marketing strategy to keep the price-cost ratio for such products favourable to the farmers;
4. Linking the farmers to bulk processors or traders by promoting contract farming;
5. Carrying out marketing costs and margins studies by mapping out the channels for the produce of their respective areas, estimating the farmers share in consumer's rupee and identifying the scope for reducing costs and margins and increasing the farmers realization for motivating them to increase the production;
6. Identifying the varieties of crops of their area, which fetch better prices and arranging and popularising the seeds of such varieties; and
7. In the event of prices of any commodity, which is covered under the price support programme, tending to dip below the support level, either ensuring that an agency is in place to make purchases at the support price, or entering into the market on behalf of the nodal agency to prevent the prices from falling below the support level.

Some of these functions called for changes in the State Agricultural Produce Market Acts. While amending the relevant provisions of the Act, even the word 'regulation' needs to be dropped as it connotes what is really not true.

These apart, the regulated markets in the changed marketing scenario should assist the producers in increasing their production alongside the performance of various functions related to the marketing of agricultural produce. Some of the ways in which the regulated markets can help by taking new initiatives are:

1. Provision of credit facilities to the producer farmers: This facility will increase the freedom of the producer-sellers to dispose of their produce profitably in the market at fair prices. This can be done by arranging alternative credit facilities and providing the facility of loan on the basis of pledge of their produce. This might inject a sense of loyalty among the producer-sellers towards the regulated markets. Adequate number of financial institutions should be made available in the market yards for this purpose.

2. Supply of agricultural inputs to the producer-farmers: The use of agricultural inputs viz., improved seeds, fertilizers, pesticides and implements has increased in agriculture due to adoption of improved methods of farming by the farmers. There is a need to develop channels through which production inputs could be made available to the farmers economically and conveniently. The regulated markets may be the most suitable place for making available these inputs in adequate quantities. The APMCs may encourage manufacturers to set up their depots in the market yards.

3. Provision of transport facilities for carrying their produce to the markets from places of production. Rural marketing scenario is still beset with the lack of adequate and proper transportation facilities at the producers level. The function can be taken up by the market committees which would help in reduction of transport cost in addition to providing the facility as and when needed.

4. Providing market extension services: There is an increasing need for providing market extension services in terms of planning and preparing the agricultural produce for the market and improved methods of cleaning, grading and packaging of the produce. The regulated markets should facilitate and encourage this aspect in coordination with Krishi Vigyan Kendras.

5. Strengthening of market intelligence activities: The importance of a sound market intelligence system can hardly be over-emphasized in any programme of agricultural marketing development. This information not only helps in selling of their produce at favourable prices but also guides the producer-farmers in planning the production well in advance as per need of the economy. The regulated markets all over the country are collecting and disseminating the market information on arrivals and prices of notified agricultural commodities but there is need to accord priority to its dissemination after meaningful analyses. This is partially being done through AGMARKNET.

SECOND PHASE OF MARKETING REFORMS (2000-14)

Based on the comprehensive reviews by the official committees or expert groups commissioned by the government of India and several consultations with all the stakeholders, considerable changes were made and steps taken to improve the marketing system of agricultural commodities. These include (a) formulation and circulation of new model Act in 2003; (b) special provision for liberalizing the trade in fruits and vegetables; (c) acceleration in the working of agmarknet portal; (d) launch of an Integrated Scheme of Agricultural Marketing, (e) launch of a unified agricultural market initiative by Karnataka state, (f) changes in the policy of levy on sugar mills; and (g) abolition of levy on rice millers. Some features of these are discussed in the following paragraphs.

NEW MODEL ACT, 2003 (PROVISIONS AND PROGRESS)

As a follow up of the concerns expressed by the review committees, a New Model Act, called Agricultural Produce Markets Regulation Act, was circulated to states in 2003. It had incorporated considerable changes in the original APMR Act. Some far reaching changes suggested in the Act were as follows:

1. Provision for establishment of wholesale agricultural produce markets by the private sector (private entrepreneurs, cooperatives or voluntary groups/NGOs).
2. Provision for wholesale purchases directly from the farm gate by processors or bulk buyers.
3. Provision for promotion of contract farming (mechanism for facilitating the tie-ups of bulk purchasers/companies with farmers or farmers groups, including prices, input supply, technical guidance and dispute settlement mechanisms).
4. Provision for establishment of farmers or consumers markets by the private sector, where farmers can directly sell to the consumers.
5. Provision for electronic trading (e-trading).
6. Provision of single point levy of market fee across the state (market fee to be paid only once). If a trader buys a produce in one market by paying a market fee, he/she need not pay market fee again in another market of the state, if the produce is shifted there for sale.
7. Provision of single unified trading licence for trade in all mandies across the state. If a trader does business of buying selling in more than one market of the state, he/she needs only one licence for trade in all the mandies.

As per the Model Act of 2003, several state governments have acted positively and amended their original Acts. It may be noted that Kerala, J&K, Meghalaya, Manipur and Bihar do not have any APMR Act. Of the remaining 24 states, 21 have incorporated the provisions like direct purchases from the farm gate and contract farming while 20 states have made provision for establishment of private wholesale markets. Other provisions have also been incorporated by 15 to 18 states as shown in Table 7.10.

FRUITS AND VEGETABLES (SPECIAL PROVISION)

Fruits and Vegetables were also in the list of commodities under the regulation. However, there were several issues and concerns related to these perishables. In the new Model Act of 2003, there was a suggestion to deregulate or delist these farm products, if the state government so desires. Several states took measures to facilities trade in fruits and vegetables. Some of these are as follows:

1. Some states have excluded or denotified fruits and vegetables from the list of agricultural commodities under regulation.
2. Some states have exempted F & V from the market fee or considerably reduced the market fee.
3. Some states permitted the traders to buy F & V from the farmers outside the market yards without any licence.

Table 7.10: Status of adoption of provisions of Model Act, 2003 (as in November 2016)

(number of states)

Provision	Adopted	Not adopted	No act	Total
1. Private wholesale markets	20	6	3	29
2. Direct purchase form farm gate	21	5	3	29
3. Contract farming	21	5	3	29
4. Farmers' markets	16	10	3	29
5. E-trading	16	10	3	29
6. Single point market fee	18	8	3	29
7. Single trade licence	15	11	3	29

Source: DMI website (viewed January 2020)

4. Some states exempted the market fee but imposed some user charges on buyers, if the trade takes place in the market yard.

As a result of these measures, several licensed private markets have come up and these are facilitating trade in fruits and vegetables in several states.

AGMARKNET

During the ninth Five Year Plan (1997–2002), the Directorate of Marketing and Inspection (DMI), Government of India embarked on an ICT project named AGMARKNET with the help of National Information Centre (NIC). AGMARKNET has been started for linking all important APMCs, SAMBs, state Directorates and DMI Regional offices located throughout the country for effective exchange of information on market prices and arrivals of important commodities. AGMARKNET is expected to play a catalytic role in market-led agriculture extension in India. The AGMARKNET scheme was started in October, 2000 and now most of the wholesale markets have been provided electronic connectivity under the scheme.

The major aim of the scheme is to collect and disseminate price and market related information in respect of agricultural commodities. Price related information on the portal includes maximum, minimum and modal price of the commodity and quantity of arrivals of more than 300 commodities and 2000 varieties. Apart from the prices and arrivals, information related to standards and grades; labeling, sanitary and phyto-sanitary requirements; physical infrastructure of storage and warehousing, national atlas of agricultural markets, commodity profiles, and market laws and regulations is also provided in the portal of AGMARKNET.

The scheme of AGMARKNET is fully funded by the Government of India. The scheme has also been given boost by Prasar Bharati (Doordarshan and All India Radio). Users can access website of AGMARKNET (www. agmarknet.nic.in) at any time.

INTEGRATED SCHEME OF AGRICULTURAL MARKETING (ISAM)

In the 12th Five Year Plan, the Government of India merged all the marketing related central schemes of Ministry of Agriculture in a single Integrated Agricultural Marketing Scheme with five sub-schemes.

1. Agricultural Marketing Infrastructure (AMI)—includes rural godown scheme; development and strengthening of agricultural marketing infrastructure; and grading and standardization.

2. Marketing Research and Information Network (MRIN)
3. Strengthening of Agmark Facilities (SAGF)
4. Capital Contribution and Project Facilitation for small Farm Agri-System Development-Agri-Business Development (ABD)
5. National Institute of Agricultural Marketing (NIAM)

The Agricultural Marketing Division of Department of Agriculture and Cooperation (DAC) of Ministry of Agriculture (MOA) is responsible for agricultural marketing policies and coordination. The DMI is implementing AMI, MRIN and SAGF sub-schemes. The small Farmers Agribusiness Consortium (SFAC), an autonomous body under MOA, is responsible for implementing ABD sub-scheme and NIAM looks after training, education and consultancy services.

The objectives of ISAM are as follows:

1. To encourage (through grants) public, private and cooperative investors for creating agricultural marketing infrastructure.
2. To encourage creation of scientific storage capacity and to promote credit to farmers against stored commodities.
3. To promote primary processing and integrated value addition/value chain system so as to link farmers directly to primary processors.
4. To increase awareness of farmers to new agricultural marketing system by use of information and communication technology (ICT).
5. Arrange for information dissemination to farmers and help in direct tie-up of farmers with traders /processors/bulk buyers.
6. To facilitate standardization of grades and create awareness among farmers.
7. To encourage private investment in agricultural marketing and create assured market for farmers and help in tie-up of farmers with traders, processors or bulk buyers.
8. To promote research, education and training in agricultural marketing.

KARNATAKA MODEL OF UNIFIED AGRICULTURAL MARKET

Karnataka is one of the first states in India to initiate comprehensive agricultural marketing reforms in 2013. It brought out a 'Karnataka Agricultural Marketing Policy' document in 2013 (see salient features in Appendix 2 at the end of this chapter). As a follow up, in July 2013, the state enacted Karnataka Agricultural Marketing (Regulation and Development) Act, 2013, which provided for (a) introduction of warehouse-based sales, (b) single unified licence to traders, and (c) establishment of direct purchase centres to buy notified agricultural produce directly from the farmers (by bulk buyers such as food companies). Currently, all APMC mandies in the state operate under unified online market, where traders across the state and outside use online platform for buying 92 agricultural commodities from the farmers.

Rashtriya e-Market Services Private Limited (ReMS) is the special purpose vehicle, created by Karnataka Government and NCDEX spot exchange, for handling the integration of mandies through single online platform. On the unified market, registered traders are able to participate in the auction process, thus leading to better price discovery. The farmers have the choice to sell in any regulated or private market in the state and receive timely online payment. Besides providing a single license system, increased competition, easy and fast trading, and better price discovery mechanism, the platform developed by ReMS provides value-added services, including assaying and grading. The state already has the experience of operating online trading platform in the state. The tested features include computerized lot entry, automated price discovery, electronic weighing, declaration of winner list, MIS Report, inventory tracking and management and e-permit. For meeting the expenses of online platform, ReMS charges 0.2% of the gross sale amount from the buyers and sellers. The sale of perishables like fruits and vegetables is not included in the online marketing platform.

SUGAR LEVY AND MARKETING POLICY

In May 2013, the Government of India, through a notification, rescinded the control over sugar industry, especially on sugar sale in the open market and PDS (existing through Essential Commodities Act and Sugar Control Orders). The main elements of this policy change are as follows:

1. There will be no levy on sugar mills. The mills will not be required to supply 10% of their sugar output at subsidized prices to the government agency for meeting the requirements of sugar under the PDS/welfare programmes.
2. The government has scrapped the sugar release order mechanism through which government controls sugar sales in the open market. This gives freedom to the sugar mills to sell sugar in the open market as they wish.

As regards doing away with levy on sugar mills, it should be noted that the Government has been distributing (around 2.7 million tonnes) sugar under PDS at a subsidized price of ₹ 13.50 per kg (price continuing unchanged since 2003). The levy price paid to sugar mills was considerably lower than then ex-factory price. Total subsidy out go on sugar is around ₹ 2600 crore per annum. The Union Government has asked the state governments to purchase sugar from the open market and the difference between the ex-mill price of ₹ 32 per kg and PDS retail price of ₹ 13.50 per kg (i.e. a subsidy of ₹ 18.50 per kg) for the existing level of sugar allocation shall be reimbursed by the Union Government.

POLICY OF LEVY ON RICE MILLERS

Ever since the launch of a positive agricultural price policy in mid sixties, the levy on rice millers has a remained a part of the policy up to 1990, as a targeted regime of rice procurement was in vogue. For fulfilling the target of procurement, apart from other measures, levy on rice millers was felt necessary. Under the levy system, each rice miller was required to hand over a specified proportion of its rice (out put of paddy milling) at a defined procurement price, before it was authorized to sell the remaining rice in the open market. The levy proportion in different rice growing states varied from 15 to 75%, which was notified by the state governments (under ECA).

Since 1991, the system of announcing procurement price for paddy was done away and only minimum support price (as for other crops) was announced for paddy. The system of fixing procurement target was also given away. It simply meant that government or its agencies will purchase all the quantities offered by the farmers at the MSP, whether it exceeds or falls short of the grains required for PDS and buffer stocks. However, the system of rice levy on millers continued. Its rationale was that it reduced the burden of buying paddy by the FCI at MSP and then getting it custom milled.

Under, the new policy, in 2014, the Government of India asked all the rice producing states to limit the levy on rice millers to 25% with effect from October 2014. Further, the state governments have also been asked to amend their levy orders to abolish the levy system with effect from October 1, 2015.

▌FURTHER REFORMS: FROM REGULATION TO FACILITATION (SINCE 2014)

From 2015, the government's approach to agricultural marketing system had shifted from regulation to facilitation. Also, several steps were taken during the last five years to shift the focus of agriculture development from only increasing the production of food, feed and fibre to simultaneously increasing the incomes of the farmers.

The measures related to acceleration in enhancing the connectivity of farmers to markets (roads, railways, communication); 100% electrification of villages and rural households; improvements in input delivery systems; and facilitation of contract farming have already been discussed in the preceding chapters. The measures related to increase in marketing facilities for perishables (cold storage and cool chains), agro-processing infrastructure and warehousing have also been discussed in the preceding chapters. The decision related to enhancing the floor for support prices, provision of alternatives for ensuring effective implementation of the support prices, and promotion of agri-exports will be elaborated in subsequent sections/chapters. Here, we look at (a) a move towards one-nation-one-market in the form of launch of National Agricultural Market; (b) new agricultural market facilitation Act; (c) recommendations of the Committee on DFI (Dalwai Committee); and (d) suggestions to further speed up the reforms.

COMMON NATIONAL AGRICULTURAL MARKET

The concrete steps have been announced in 2014-15 to move to a common National Agricultural Market for the country. These include the following:

1. The Ministry of Agriculture has issued a comprehensive advisory to states to go beyond the provisions of Model Act and declare the entire state a single market with one licence valid across the entire state and remove all restrictions on movement of agricultural produce within the state.
2. In order to promote development of a common national market for agricultural commodities through e-platforms, the department launched a central-sector scheme for promotion of National Agricultural Market through Agri-Tech Infrastructure Fund (ATIF) and implemented during 2014-15 to 2016-17. ATIF was used to migrate towards a national market through implementation of a common e-platform for agricultural marketing across all states.
3. In July 2015, the union cabinet approved the setting up of a National Agricultural Market (NAM) by integrating all existing AMPCs across the country through an on-line platform. It is a pan-India electronic trading platform, which seeks to network the existing APMCs and other market yards to create a unified national market for agricultural commodities. As a market, it is virtual but has a physical market (mandi) at the back-end, benefitting all stakeholders concerned. The important conceptual features of NAM are as follows:
   (a) The existing APMCs and all private markets can 'plug in' through electronic platform for trading.
   (b) Through a unified platform, NAM directly connects buyers and sellers in the market space, shunning the intermediaries and associated costs.
   (c) The concept will reform existing market by ushering more transparency, and leverage state-of-the-art technology for a well-regulated national market.
   (d) The framework provides for a wide scope for addressing infrastructure gaps in terms of logistics and warehousing.

(e) The physical transaction would continue to be conducted through concerned APMC.

(f) The SFAC, a lead promoter of the online national market, offers a special software to each APMC (mandi), which agrees to join the national network, free of cost. Besides, SFAC will ensure timely delivery and settlement of payment.

(g) For registration with national online market, the concerned state must provide single licence to traders, which should be valid across the physical boundary of the state.

(h) The APMC would have to make a provision for electronic trading and issue licences to traders from any region or state who who would operate through online trading platform.

LAUNCH OF E-NAM

In April 2016, the Prime Minister formally launched the electronic trading platform of E-NAM. The main objective was to electronically link all the APMCs for trading in agricultural commodities, so that farmers get larger choices (buyers) to sell and the traders also get larger choices (sellers) to buy their trading volumes. It is in a sense spatial integration of all physical markets, which can/will reduce transaction costs, give better price signals, and improve price discovery—a win-win situation for farmers, traders and in turn the consumers.

The pre-requisites for E-NAM to progress include several actions by APMCs. The foremost is the provision of e-trading by legal means by either amendment in APMR Act or by notification. Similarly, the provisions for single point levy of market fee and issue of single trading licences are also necessary. In addition, several items of physical infrastructure in terms of computer terminals, software, assaying facilities, on-line payment system, training and sensitization of farmers, traders, and APMC functionaries, etc. were the minimum requirements. Considerable efforts were made/ are being made in all these directions.

The progress so far (January 2020) is quite satisfying.

1. As planned, 585 mandis got integrated with E-NAM by March, 2018. Additional 415 mandis are being linked and number is likely to be 1000 by March, 2020.

2. Cumulative trade on E-NAM up to January, 2020 is reported to be valued at around ₹ 1,00,000 crore. More than 150 commodities, including fruits and vegetables, are being traded.

3. Several states have commenced inter-state trade and it is progressing well. So far, 21 E-NAM mandis of eight states are doing inter-state trade (the states are UP, AP, MP, Uttarakhand, Telangana, Gujarat, Rajasthan and Maharashtra).

4. So far, around 1.65 crore farmers and 1.30 lakh traders or commission agents are registered on the portal.

5. E-NAM website is available in English, Hindi, Gujarati, Marathi, Tamil, Telugu, Bengali, and Odia languages.

6. E-learning module is also available on the portal.

7. With the help of apps, entry to APMC yard gates has been made easy.

8. Farmers can see real time bidding process and assaying certificates.

9. On-line payment facility for farmers is available and they receive SMS for payment.

10. More features are being added to the portal.

11. Many accredited warehouses have been designated as mandis for linking to E-NAM.
12. New APLM Act has been adopted by 24 states, which will hasten the expansion of E-NAM (see more on APLM Act in next section).
13. There is a plan to link Rural Primary/ Periodic Markets (RPMs) to E-NAM. As and when, these are linked, it will be a boon to entire network.

The major challenges for E-NAM are as follows:

1. Adequate attention to creation of physical and institutional infrastructure including manpower (at mandi and RPM level) will be critical.
2. Internet connectivity (un-interrupted) in all the areas will be essential.
3. The E-NAM is a parallel outfit to compete with physical trade in APMC yards. However, it is only for trade negotiations. All physical functions will continue as hitherto. This aspect will have to be made clear and brought home to all the functionaries for smooth expansion of E-NAM.
4. It should be understood that E-NAM will take time to mature as NSE/BSE, except that in E-NAM, it is the trade in physical products, whereas in NSE/BSE, the trade is in intangibles (dematerialized shares of listed companies).

MODEL AGRICULTURAL PRODUCE AND LIVESTOCK MARKETING (PROMOTION AND FACILITATION) ACT, 2017

It was felt that there is a need to shed 'regulation' and move to promotion and facilitation of agricultural marketing system (including E-NAM) to safeguard the interests of all the stakeholders including farmers, consumers, and market functionaries. It is in this context that after consultations with all the stakeholders, particularly state governments, the Union Ministry of Agriculture, circulated the Model APLM Act in 2017. The salient features of the Act are as follows:

1. Declaration of whole state/UT as one unified market area. Within a state or UT, there may be (a) principal market yard(s) or sub-yard(s) managed by the market committees; (b) private market yard(s) or sub-yard(s) managed by a licensed private person; (c) farmer-consumer market (FCM) yard(s) managed by a market committee; (d) FCM yard(s) managed by a licensed person; and (e) electronic trading platform.
2. Full democratization of market committee and state/UT agricultural marketing board (SAMB).
3. Director of Agricultural Marketing (DAM) will carry out regulatory functions, while Managing Director of SAMB will be mandated with development responsibilities.
4. Creation of a conducive environment for setting up and operation of private wholesale market yards and farmer-consumer market yards so as to enhance competition for the farmers' produce.
5. Promotion of direct interface between farmers and processors, exporters, bulk buyers or end users, with a view to reducing the price spread (allowing purchases at the farm gate).
6. Enabling declaration of warehouses, silos, cold storages and other structures as market sub-yards to provide better access or linkages to the farmers.
7. Giving freedom to farmers to sell to the buyers at a place and time of their choice, to whom so ever they wish, so as to get better prices.

8. Promotion of e-trading to enhance transparency in trade operations and integration of spatially separated markets and to improve price discovery.

9. Provisions for single point levy of market fee across the state and unified single trading licence to realize cost effective transactions.

10. Promotion of national market for agricultural produce through provisioning of inter-state trading license, grading, standardization and quality certification.

11. Rationalization of market fee and commission charges. There is a cap on mandi fee of 1% for foodgrains and other non-perishables and 2% for fruits and vegetables. There is also a cap on commission or arhat (2 % to 4 %)

12. Provision of special commodity market yard(s) and market yard(s) of National Importance (MNI).

13. Licence of private market yard, sub-yard, electronic trading and direct marketing (farm gate buying) has been placed at the same level playing field as within the market yards of the APMCs.

14. Livestock markets will also receive equal attention as per the new Act.

As the basic intent of new model Act is promotion and facilitation (rather than regulation), 24 states have already adopted it. Even the states, where there are no APMCs, are joining the E-NAM. The new Act provides for many options for the farmer to sell their produce. Some states have deregulated almost all the commodities, meaning thereby that there is no compulsion on the farmers to sell their produce only in APMCs yards (earlier Acts also provided this option but APMCs were informally imposing this restriction with the limited objective of collecting the market fee). Many private market yards are emerging and many processors/bulk buyers are now allowed to directly purchase at the farm gate. As and when, rural primary/periodic market yards are developed and linked to main market yards, physically or through E-NAM, the scene of agricultural marketing system will completely transform.

PROMOTION OF FARMER PRODUCER ORGANISATIONS (FPOS)

One of the suggestions is to speed up the promotion of FPOs. It is well known that owing to the dominance of small landholders in India, group marketing is the only option for better price realization of their marketed surplus. Farmers' cooperatives have been successful in dairy sector, in input marketing and in credit mobilization and delivery. As regards crop product marketing, while cooperatives have been successful at national and state levels, the experience at district and lower level has not been encouraging, except in the case of sugarcane and that too in some states. It is in this context that emphasis for product marketing has now shifted to farmer producer organizations or companies. For reaching the full benefits (of new model Act, contract Farming, E-NAM and new initiatives in organized food retail sector) to the small landholders, organization of farmers in groups is extremely critical. Already, five thousand FPOs/FPCs have emerged but all are not reported to be very active. The government is targeting to promote 10,000 FPOs in the country. It will be advisable to organize farmers in at least three tiers. The first tier may be local SHG/FPOs consisting of 500 to 1000 farmers (or spread to 500 to 1000 hectares of cultivated land) belonging to five to ten smaller villages or two to three larger villages. The second tier can be a federation of around 20 to 25 primary (tiers-1) FPOs. Further, tiers-2 FPOs may be federated in to a larger FPO or FPC. Some of the FPOs may be commodity based and others may be area or group of commodities based. As and when this kind of farmers' organizational structure emerges, the benefits of marketing reforms are likely to trickle down to marginal and small farmers also in a larger measure even in remote areas.

COMMITTEE ON DOUBLING FARMERS INCOME (DFI), 2017 (DALWAI)

In April, 2016, the Government of India, constituted a committee on Doubling Farmers' Income (DFI) under the chairmanship of Ashok Dalwai. The time horizon for DFI was considered as six years. The committee submitted its report in 14 volumes in August, 2017. Two of these volumes relate to agricultural marketing. Some of the recommendations on agricultural marketing are as follows:

1. Modern aggregation points or centers at panchayat or block samiti level should be set up that can serve as farm gate loading points for onward wholesale market connections. These aggregation points can be facilitated by gram panchayats, local FPOs and PACS. KVKs and SAU's should adopt these local aggregation centers. Small size cold stores may also be needed in some of these centers.
2. For achieving the desired market density, there is a need for developing 10000 wholesale and 20000 rural primary and retail markets (RPMs etc.), and linking these to E-NAM.
3. For the success of E-NAM, there is a need to harmonize various product standards and grading parameters adopted by different agencies (BIS, APEDA and Agmark).
4. There is a need to take measures for upgrading storage godowns, including cold storages, to the standards as defined by WDRA for issuing NWRs Also comprehensive guidelines should be developed to promote warehouse based post harvest loans and e-NWR based trading.
5. Private sector shareholding up to 26% should be allowed in FPOs /FPCs.
6. The Marketing Division in the Union Ministry of Agriculture (DACFW) should be redesignated as Division of Agri-logistics and Marketing, which may help in increasing focus on marketing and agri-logistic functions.
7. The Directorate of Marketing and Inspection (DMI) at the centre (MOAFW) should be renamed as Directorate of Marketing and Intelligence.

▌SUGGESTIONS FOR SPEEDING UP OF MARKETING REFORMS

The first Model Act for agricultural marketing reforms was circulated to state governments in 1950's and early 1960s. It took almost 16 years for even the major states to adopt it. This delay is mainly because the exercise of formulation of state Act (with specifications), framing of rules under the Act, notification for inviting objections/ suggestions, vetting by various state ministries etc. has to be repeated in every state and each step takes considerable time. And this happens even when Model Act itself is formulated after several rounds of consultations/consensus on each finer detail, with the state governments and their agencies. The experience with the second Model Act of 2003 had been almost similar. Even when the central grant for marketing development was linked to the extent of market reforms (in accordance with the second Model Act), not all states could amend their respective state Acts in line with new provisions suggested in the Model Act.

Considering the long time taken in introducing the intended reforms even after consensus had emerged among the states and the centre, it will be a significant step if the following suggestions, being made by us (Acharya, 2006, 2017) and other stakeholders since long, are implemented urgently.

1. The subject of 'Agricultural Marketing' should be shifted from the 'State list' to the 'Concurrent list' by amending the constitution;

2. On the lines of GST Council, an Agricultural Marketing Council should be put in place to arrive at consensus, from time to time, among the states and the centre in the matters related to agricultural marketing.

3. 'Agriculture' should be redefined as 'the science and practice of production, processing, marketing, distribution and trade of food, feed, fodder and fibre, where marketing includes storage and transportation'.

STATE TRADING

So far, we have discussed market regulation and reform scenario, which does not require direct entry of the government or its agencies in direct trading.

One of the responsibilities of the government is to ensure the supply of essential commodities to the people. This may require direct intervention on its part in trading of agricultural commodities.

OBJECTIVES OF STATE TRADING

The objectives of state trading are:

1. to make available supplies of essential commodities to consumers at reasonable prices on a regular basis;
2. To ensure a fair price of the produce to the farmers so that there may be an adequate incentive to increase production;
3. To minimize violent price fluctuations occurring as a result of seasonal variations in supply and demand;
4. To arrange for the supply of such inputs as fertilizers and insecticides so that the tempo of increased production is maintained;
5. To undertake the procurement and maintenance of buffer stock, and their distribution, whenever and wherever necessary;
6. To arrange for storage, transportation, packaging and processing;
7. To conduct surveys and provide the required statistics to the government so that it may improve the conditions of the farmers; and
8. To check hoarding, black-marketing and profiteering.

TYPES OF STATE TRADING

State trading may be partial or complete, depending upon the extent of intervention desired by the government.

1. Partial State Trading

In partial state trading, private traders and government coexist. Traders are free to buy and sell in the market. The government may place some restrictions on them, such as declaration of stocks, limits on the stocks which can be held at a point of time and submission of regular accounts. The government enters the market for the purchase of commodities directly from producers at notified procurement price. It undertakes the distribution of commodities to consumers through a network of fair price shops. In this way, it safeguards the interest of producers and consumers alike, and keeps a check on the undesirable activities of traders. This kind of partial trading is prevelant in India, mainly for rice and wheat.

2. Complete State Trading

This is the extreme form of trading adopted by the government when partial state trading fails to ensure fair prices to producers and make goods available to consumers at reasonable prices. The purchase and sale of commodities is undertaken entirely by the government or its agencies. Private traders are not allowed to enter the market for purchase or sale. Under this form of state trading, the government remains the sole purchaser and distributor of the commodity.

Complete state trading necessitates the outlay of huge finance, and the provision of storage facilities at important production and consumption centres, and calls for appointment of efficient persons so that the purchase and distribution functions of professional traders may be effectively taken over by a governmental agency. In India, complete wholesale trade in wheat was taken over by the government in 1973; but it had to be given up very soon.

EXPERIENCE OF WHOLESALE TRADE TAKEOVER IN WHEAT

On the recommendation of Chief Ministers' conference held in February, 1973, the wholesale trade in wheat and paddy was taken over by the government from the rabi season of 1972–73. It was intended to eliminate the wholesalers, who were considered to be responsible for creating an artificial scarcity by hoarding with a view to raising prices. It was expected that the direct purchase of foodgrains by the government and their subsequent distribution through fair price shops would bring down the prices. The specific objectives of the complete takeover of wholesale trade were to:

1. Eliminate unwarranted profits of middlemen;
2. Ensure remunerative prices to producers;
3. Guarantee an assured supply of foodgrains (wheat and rice) to consumers at reasonable prices;
4. Arrest the price rise; and
5. Exercise effective public control over the marketable surplus of agricultural commodities—an item of essential necessity for the masses.

Under the wholesale trade takeover scheme, public sector agencies like the Food Corporation of India, the Civil Supply Departments of State Governments and Cooperative Marketing Societies were entrusted with the responsibility of purchasing the marketed surplus and its subsequent disposal to consumers through a network of fair price shops.

The scheme of wholesale trade takeover in wheat did not succeed, and was withdrawn immediately. It was planed to purchase 30 to 35% of the total production of wheat in the country during that year; but government agencies could procure only half of the targetted quantity of 8.5 million tonnes of wheat. The reasons of the failure of the scheme were:

1. Very low procurement prices, i.e. ₹ 76 per quintal;
2. Coaxing farmers by disgruntled traders. Traders were the main sufferers when this scheme was introduced; and they undermined the arrivals of wheat in the market;
3. Over-estimation of the marketable surplus in various States;
4. Inconvenient public purchase system resulting in a long wait by farmers for many hours, and sometimes for more than one day for their turn to hand over the produce and get payment for it. Farmers had to travel long distances to sell their produce at

5. Skewed distribution of marketed surplus in favour of big farmers, who have retention power;
6. Slackness on the part of State Governments in implementing the policy because of lack of sufficient and experienced staff capable of handling the work; and
7. Lack of storage facilities with the government for storing the procured foodgrains.

The government realised that takeover of rice trade would be much more difficult than wheat trade due to its operation on a wide area in the country and also due to the existence of surplus regions within deficit states. Hence government gave away the complete wholesale trade takeover. However, partial state trading has continued mainly through price support or open market operations of Food Corporation of India and National Agricultural Cooperative Marketing Federation for rice, wheat and pulses; and through Cotton Corporation of India for raw cotton. Occasionally, partial state trading is also undertaken in the case of onion, potato, pulses etc. for a very short period on a limited scale when their prices shoot up in the market.

PUBLIC DISTRIBUTION SYSTEM (PDS)

India's public distribution system is perhaps the largest national programme in the world, which aims at achieving food security, at all the levels, i.e. macro food security, household food security and individual food security. Over the years, it focused on all the three essential features of food security, i.e. availability, accessibility and absorption. The PDS in India had graduated from macro to micro food security and form staple food security to nutrition security. There has been a distinct shift in approach to food security in mid-1960s when a positive agricultural price policy was launched (see details of price policy in the next section). The PDS can be understood by looking at its three essential interwoven components viz., Buffer Stocks, Procurement and Distribution.

BUFFER STOCKS

The term buffer stock of foodgrains refers to the stock of foodgrains maintained by the government to be used as a buffer to cushion the shocks of fluctuating supply and price, to meet the emergency needs and to meet the situations arising out of serious unexpected shortages resulting from transport bottlenecks, natural calamities like war, flood, famine, earthquakes, and from the influx of refugees. These are strategic stocks.

The main advantages of maintaining a buffer stock are:

1. It helps in the stabilization of prices by counteracting the effects of the activities of speculators and hoarders;
2. It safeguards the producers against low prices, specially during the surplus-production years; and
3. It imparts stability to the country's food economy.

The government enters the market and purchases foodgrains for the maintenance of the buffer stock. This buffer stock can be built either by internal purchases or by imports from foreign countries. It is maintained by the Food Corporation of India.

In mid-eighties, a buffer stock of 10 million tonnes comprising 5 million tonnes of wheat and equal quantity of rice was considered adequate. It should be noted that this buffer stock is over and above the operational stocks. Considering both together, a stock of around 20 million tonnes was considered necessary for a country of India's size. However, the stock, which can be considered optimum, depends on the level of public distribution of foodgrains intended by the government.

The norms of stocks (called buffer norms) required by the government for both buffer and operational (PDS) purposes are revised every five years by a Technical Group. The recent revision has been in January 2015. The norms are prescribed at four points of time during the year, keeping in view the procurement seasons and likely off-takes. The buffer norms revised from January 2015 are shown in Table 7.11. Out of total prescribed stocks, five million tonnes is strategic component and remaining is operational part.

Table 7.11: Norms of buffer stocks of rice and wheat in India (before and w.e.f. January 2015)

(million tonnes)

Date	Rice		Wheat		Total	
	Before	New	Before	New	Before	New
April 1	14.2	13.58	7.00	7.46	21.2	21.04
July 1	11.8	13.54	20.1	27.58	31.9	41.12
October 1	7.2	10.25	14.0	20.52	21.2	30.77
January 1	13.8	7.61	11.2	13.80	25.0	21.41

Source: Agricultural Statistics at a Glance, and DFPD, GoI.

The actual stocks have generally remained above these norms (except few years), depending on several factors like level of support prices, behaviour of market prices, offtake under public distribution system, import/export policies, and international market price situation. Practically, the size of stocks at any point of time is equal to the carryover stocks (from the previous point of time) plus the procurement minus the off-take during the intervening period. Actual stocks of rice and wheat during the last 20 years are shown in Table 7.12. The stocks as on January, 2020 are considerably

Table 7.12: Stocks of cereals in central pool (as on January 1)

(million tonnes)

Year	Rice	Wheat	Coarse cereals	Total
2000	14.18	17.17	–	31.35
2001	20.70	25.04	0.03	45.77
2002	25.62	32.41	0.08	58.11
2003	19.37	28.83	–	48.20
2004	11.13	12.69	0.60	24.42
2005	12.76	8.93	0.60	22.29
2006	12.64	6.19	0.43	19.26
2007	11.98	5.43	0.09	17.50
2008	11.47	7.71	0.00	19.18
2009	17.58	18.21	0.40	36.19
2010	24.35	23.09	0.25	47.69
2016	25.95	23.79	NA	49.74
2020	42.31	32.80	NA	75.11

* Rice stocks include 67% of unmilled paddy.
Source: Agricultural Statistics at a Glance, 2014, FCI website.

higher than the norms for this day. It may be mentioned that the government has been disposing the excess stocks mainly through open market sales scheme (OMSS). Between 2013–14 and 2018–19, the government sold, under OMSS, around five million tonnes of grains, on an average, every year. The current carrying cost of stocks is estimated as ₹ 5000 per tonne per year. Obviously, carrying excess stocks adds to the subsidy outgo on food security. There is a need to manage the timely disposal of excess stocks through open market sales or other means.

PUBLIC PROCUREMENT OF FOODGRAINS

The term public procurement refers to securing foodgrains by the government or its agencies to meet the requirements for the supply to consumers through fair price shops, and to build a buffer stock to meet emergency needs in agriculturally lean years. The prices at which the procurement operation is carried out are referred to as procurement prices.

During the seventies and eighties, the required quantity of foodgrains was procured at the pre-announced procurement prices in the internal market. Till 1971 procurement prices were announced by the government in the kharif and rabi seasons before or at the time of the harvest season on the recommendations of the Commission for Agricultural Costs and Prices after considering the prevailing conditions of demand and supply in the economy. Since 1971, procurement prices were treated as support prices.

In India till 1990–91, procurement price system was in operation for cereals. Procurement prices were announced and the public agencies tried to procure pre-decided or targeted quantities at these prices. If it was felt that targets of procurement would not be met, inter-state or inter-district movement restrictions were imposed which resulted in depressing the prices in surplus regions enabling the public agencies to procure desired quantities. Apart from the imposition of movement restrictions, several other instruments were used at different points of time to achieve the procurement targets. Some of these adopted in India during different periods are as follows:

1. Levy on Farmers

Under this system, it was made legally obligatory on producers to sell a part of their produce to the government at the procurement price. The quantity to be procured by the government was fixed either in proportion to the acreage under the crop in that year, or at a flat rate. The procurement of foodgrains was done directly by the government or through the agents appointed by it. Cooperative marketing societies, the Food Corporation of India and private traders were generally appointed agents for this purpose on a fixed commission basis. In this way, the government procured foodgrains from a large number of farmers throughout the country, everyone contributing a small quantity. Farmers were free to sell the surplus at open market prices. Levy on producers remained in operation only for a short period during the seventies.

2. Levy on Millers

Under this system, the government procures the required quantity of foodgrains from traders and millers instead of producer-farmers. Traders and millers are legally bound to deliver a fixed percentage of the foodgrains purchased/processed by them to the government at the announced procurement prices. They are free to sell the remaining quantity in the open market at the prevailing prices. For many years, sugar was procured from sugar factories by imposing a levy on the quantity of sugar processed by them. Similarly, rice millers in several states are required to hand over a fixed percentage of rice milled by them to the government at a fixed price. (derived from the support price of paddy). The levy on sugar mills was withdrawn in May 2013 and that on rice millers in October 2015.

3. Pre-emptive Purchases

In this method, the government purchases foodgrains in the open market. The government assumes the first right to purchase the grains at a price settled between the trader and the food producer. Practically, this system does not exist.

4. Open Market Purchases

Under this method, no price is announced beforehand. The government or its agency enters the market as a trader and buys the produce after competing with other traders.

5. Monopoly Procurement

By this method, the government acquires monopoly rights for the purchase of food-grains from farmers. Traders are not allowed to enter the market for this purpose. This method postulates an adequate infrastructure on the part of the government in terms of manpower, storage godowns, finance and transport facilities. Monopoly procurement for wheat was introduced in early seventies but was withdrawn after one season.

6. Purchase at Support Prices

The most important method of procurement is purchasing the required foodgrains at the minimum support prices announced by the government. This method of procurement, which is being followed by the government, is a part of the agricultural price policy in operation in the country since mid-1960s (details in the next section).

To ensure the availability of foodgrains for the masses at reasonable prices, the government has been procuring substantial quantities of foodgrains and maintaining large (buffer) stocks of goodgrains. The foodgrains procured are wheat, rice and coarse grains. Wheat and rice are the two main foodgrains, accounting for over 99% of the total procurement. The coarse grains account for the remaining 1%.

The total quantity of foodgrains procurement has been fluctuating from year to year. The Government of India has been procuring foodgrains ever since the beginning of the First Five Year Plan. The total quantity of internal procurement of all foodgrains was 3.477 million tonnes in 1952, 4.031 million tonnes in 1965, 6.714 million tonnes in 1970 and 8.857 million tonnes in 1971.

During the seventies, procurement of foodgrains averaged at 10 million tonnes (8.7% of the total foodgrains production) and ranged between a low of 5.7 million tonnes in 1974 to a high of 13.8 million tonnes in 1979.

The annual procurement has consistently increased since then. It averaged at 17 million tonnes per year during 1985–90, 23 million tonnes during 1990–95, and 25 million tonnes per year during 1995–2000. During the last 20 years, there has been a step-up in the level of procurement of rice and wheat as shown in Table 7.13, Annual procurement averaged at 39 million tonnes during 2000–01 to 2004–05, which increased further to an average of 66 million tonnes during 2014–15 to 2018–19.

Table 7.13: Public procurement of major foodgrains in India		
		(million tonnes)
Year	Wheat (April–March)	Rice (October–September)	Total
1990–91	11.07	11.74	22.81
1995–96	12.33	10.05	22.38
2000–01	16.36	20.82	37.18
2005–06	14.79	27.66	42.45
2010–11	22.51	34.20	56.71
2011–12	28.34	35.04	63.38
2012–13	38.15	34.04	72.19
2013–14	25.09	31.84	56.93
2014–15	28.02	32.04	60.06
2015–16	28.09	34.22	62.31
2016–17	22.96	38.11	61.07
2017–18	30.82	38.18	69.00
2018–19	35.80	44.40	80.20

Source: Economic Survey. Various Issues, and Ag. Statistics at a Glance.

While, the procurement levels of rice and wheat are discussed above, it must be remembered that nutri cereals are also procured under the price support programme, but their levels of purchases are considerably lower than that of rice and wheat. For example, during the last 5 years (2013–14 to 2017–18), their annual procurement varied from 86000 tonnes to 1.23 million tonnes, with an average of 0.45 million tonnes.

Rice accounts for around 60% of the total foodgrains procurement. The procurement of rice is done in the kharif marketing season, covering the period from October 1 to September 30. Wheat accounts for the remaining 40% of the total quantity of foodgrains procurement. The procurement of wheat is carried out in the rabi marketing season, covering the period from April 1 to end of June, when more than 95% of the year's procurement is over. The state-wise contribution to total procurement has considerably changed in the recent years, though Punjab continues to be the leading contributor for both rice and wheat. The first four states in terms of their contribution to national rice procurement are Punjab, un-divided Andhra Pradesh, Chhatisgarh and Orissa, which together contributed around 67% of total procurement during 2018–19. For wheat procurement, Punjab and Haryana continue to be the leading states, contributing 60% during 2018–19 (Table 7.14). Madhya Pradesh has increased its share in wheat procurement during the current decade.

An examination of the factors that influence the size of the procurement of foodgrains reveals that variations in production, as also the difference between procurement prices and open market prices, generally affect the size of the procurement of rice and wheat.

The responsibility for the procurement of foodgrains has been entrusted by the government to the Food Corporation of India, cooperative marketing societies and Civil Supplies Departments. These agencies procure the foodgrains through their own networks. It is very difficult to procure foodgrains up to the targeted quantity during the short marketing season. In many years, the targeted quantity could not be procured because farmers were reluctant to sell to government agencies. The targets could not be achieved because of the following reasons:

States	2000–01		2018–19	
	Lakh tonnes	%	Lakh tonnes	%
RICE				
Punjab	70	32.9	113	25.5
Andhra Pradesh	72	33.8	100*	22.5
Uttar Pradesh	12	5.6	32	7.2
Chhatisgarh	9	4.2	40	9.0
Odisha	9	4.2	44	9.9
Haryana	15	7.0	39	8.8
West Bengal	4	1.9	20	4.5
Tamil Nadu	17	8.0	13	2.9
Others	5	2.4	43	9.7
Total (Rice)	213	100.0	444	100.0
WHEAT				
Punjab	94	57.3	127	35.5
Haryana	45	27.4	88	24.6
Uttar Pradesh	15	9.1	53	14.8
Rajasthan	5	3.0	15	4.2
Madhya Pradesh	NA	NA	73	20.4
Others	5	3.2	2	0.5
Total (Wheat)	164	100.0	358	100.0

Table 7.14: State-wise procurement of rice and wheat in India (marketing year)

(lakh tonnes)

* AP + Telengana (48.1 + 51.9)

Source: Agricultural Statistics at a Glance, 2009, 2014 and 2018, Ministry of Agriculture, Government of India.

1. Farmers preferred to sell foodgrains to traders to fulfil their debt obligations;
2. Traders and consumers purchased directly from farmers at a price higher than the procurement price, which reduced the quantity of market arrivals;
3. Farmers faced a number of difficulties when they sold foodgrains to government agency, such as delayed payment, payment by cheque instead of cash; delay in the market for their turn to deliver and weigh the produce; rejection of the produce under the pretext of poor quality; and
4. Farmers had to travel long distances to reach the purchasing centres.

DECENTRALIZED PROCUREMENT SCHEME (DCP)

The decentralized procurement scheme (DCP) is in operation since 1997. A number of states namely, West Bengal, Bihar, Madhya Pradesh, Chhattisgarh, Uttarakhand, Gujarat, Orissa, Tamil Nadu, Karnataka, Kerala, Andhra Pradesh, Telengana and the Union Territory of Andaman & Nicobar Islands have opted for implementation of DCP. Under this scheme, the designated States procure, store and also issue foodgrains under the Targeted Public Distribution System (TPDS) and welfare schemes. The excess stocks are accepted by FCI for transfer to deficit states. The difference between the economic cost fixed for the state and the central issue price (CIP) is passed on to the State Government as subsidy. The decentralized system of procurement helps to cover more farmers under the MSP operations, improves efficiency of the PDS, provides foodgrain varieties more suited to local tastes and reduces the transportation costs.

DISTRIBUTION OF FOODGRAINS

The crux of India's PDS is the distribution, because buffer stocks and procurement are means to ultimately make available foodgrains for distribution to the targeted families for assured food security. The distribution of foodgrains to the vulnerable sections of society at fair prices during periods of scarcity and rising prices is the ultimate aim of the policy of the procurement and storage of foodgrains by the government. Public distribution of foodgrains by Central and State Government, from 1965 to 2019 has been shown in Table 7.15. Right up to the mid-1970s, the public distribution system used to be fed by a considerable extent by imported foodgrains. For example, in 1966, the imports of foodgrains exceeded 10 million tonnes. The considerable increase in the production of foodgrains in the 1970s and the governments efforts to increase internal procurement were helpful in cutting down the imports of foodgrains. Consequently, during the three years from 1978 to 1980, foodgrains were not imported. During the 1980s and 1990s, the quantity of foodgrains distributed under the public distribution system has been around 15 million tonnes in most of the years. However, during the later period, the scale of PDS has considerably gone up, particularly since 2004-05. During the last 10 years, the quantities of foodgrains released under PDS have been more than 50 million tonnes.

Foodgrains are distributed mainly through fair price shops (ration shops), cooperative societies and by local retailers as agents on specified terms and conditions. The number of fair price shops was 140,402 at the end of 1968. This number rose over

Table 7.15: Public distribution of foodgrains in India				
				(million tonnes)
Calendar year	Wheat	Rice	Other grains	Total
1965	5.939	3.586	0.554	10.079
1970	5.347	3.050	0.444	8.841
1975	7.545	3.211	0.497	11.253
1980	8.819	6.057	0.117	14.993
1985	8.477	7.231	0.091	15.799
1990	6.568	8.659	0.098	15.325
1995	5.567	9.365	–	14.932
2000–01	4.070	7.970	–	12.040
2005–06	17.172	25.082	–	42.254 (9.7)*
2010–11	23.1	29.9	–	53.0
2015–16	22.4	31.3	–	53.7
2017–18	17.6	25.6	–	43.2
2018–19	23.3	33.5	–	56.8

*Additional for welfare schemes

Source:

1. Bulletin on Food Statistics, Directorate of Economics and Statistics, Ministry of Agriculture and Ministry of Food and Civil Supplies, Government of India, New Delhi, Website of Food Department (E book, 2018–19).
2. Economic Survey, Various Issues, Government of India, New Delhi.

time and reached a level of 3,60,000 in July 1991. The population covered by these fair price shops increased from 299.47 millions in 1971 to 800 millions in 1998, i.e. they covered nearly 80% of the country's population. In 2019, the number of fair price shops increased to 5.33 lakh. At these fair price shops, foodgrains are sold to consumers at fixed prices, which are known as issue prices. The issue prices fixed for important foodgrains from time to time have been given in Table 7.16.

Table 7.16: Central issue prices of rice, wheat and nutri cereals

(₹ per quintal)

Period	APL Normal*/A Grade	BPL	AAY	NFSA (AAY + PHH)
RICE				
1-12-97 to 28-1-1999	550/700	350	–	–
29-1-99 to 31-3-2000	700/905	350		
1-4-2000 to 24-7-2000	1135/1180	590		
25-7-2000 to 11-7-2001	1087/1130	565	300**	
12-7-2001 to 31-3-2002	795/830	565	300	
1-4-2002 to 30-6-2002	695/730	565	300	
1-7-2002 to till date	795/830	565	300	
5-7-2013 to till date	795/830	830	300	300
WHEAT				
1-12-97 to 28-1-1999	450	250		
29-1-1999 to 31-3-1999	650	250		
1-4-1999 to 31-3-2000	682	250		
1-4-2000 to 24-7-2000	900	450		
25-7-2000 to 11-7-2001	830	415	200	
12-7-2001 to 31-3-2002	610	415	200	
1-4-2002 to 30-6-2002	510	415	200	
1-7-2002 till date	610	415	200	
5-7-2013 till date	610			200
NUTRI CEREALS				
w.e.f. 16-11-2005	70% of eco. cost	50 % of eco cost	200	–
2008–09 till date	450	300	150	
5-7-2013 till date				100

* Normal for only NE States, J&K, HP, Sikkim and Jharkhand. ** From December 2000.
APL = Above poverty line, BPL= Below poverty line, AAY= Poorest of the poor
NFSA = National Food Security Act, PHH = Priority households.
Source: Dept of Food and PDS, Ag, Stat. at a Glance, 2018.

The prices at which the central government supplies foodgrains to the states for distribution through the fair price or ration shops are called Central Issue Prices (CIPs). Since July 2002 till the NFSA came into effect in 2013, the CIP for wheat and rice have been kept at constant level but differentiated for the poorest of the poor (AAY), poor (BPL), and not-so-poor (APL) families. For wheat, these were ₹ 200, ₹ 415 and ₹ 610 per quintal, respectively. In the case of rice, the CIPs were ₹ 300 per quintal for Antyodaya (AAY) families, ₹ 565 for BPL families and ₹ 830 per quintal for APL families.

The functioning of the public distribution system was criticised on many grounds. Some of these are:

1. The number of fair price shops was inadequate in the sense that consumers had to travel long distances to get their supply and have to wait for long hours;
2. The supply of foodgrains through fair price shops was inadequate and grains were not available every time the card holder visited the shop.
3. The quality of foodgrains supplied through fair prices shops was inferior; and
4. Retailers were not interested in becoming the agents because of the low commission they receive and the number of formalities they had to observe in the maintenance of accounts.

Some essential commodities supplied for distribution through the public distribution system to the consumers were wheat, rice, sugar, edible oils, soft coke, controlled cloth and kerosene. Several measures were taken during the last three decades to improve the functioning of PDS.

REVAMPED PUBLIC DISTRIBUTION SYSTEM (RPDS) (1992–97)

A Revamped Public Distribution System (RPDS) was launched on January 1, 1992 and made effective from June 1, 1992, in selected blocks of drought-prone, desert, tribal and hilly areas of the country. The main features of the revamped public distribution system are:

1. in addition to the essential commodities included in the general public distribution system, it included additional items of daily use like tea, soap, pulses and iodised salt;
2. the items are delivered by the government at the door of the fair price shops so as to minimise the delay and leakage; and
3. a vigilance committee for each village, consisting of local persons is constituted for a watch on the distribution system.

The Revamped Public Distribution System was initially operated in 1775 blocks located in tribals, hilly and arid areas having poor infrastructure. Under this system, foodgrains—rice and wheat were allocated to States/Union Territories at prices lower by ₹ 50 per quintal than the issue prices for general public distribution system.

TARGETED PUBLIC DISTRIBUTION SYSTEM (TPDS) (SINCE 1997)

The Government of India has been continuously reviewing and announcing some changes in the public distribution system and termed the new system as Targeted Public Distribution System (TPDS). The essential features of TPDS when launched in 1997 were:

1. The States would identify the families below the poverty line (BPL) who would be issued 10 kg or more of foodgrains per month per family at prices almost half the normal central issue prices for general PDS;
2. Population above the poverty line (non-poor) now under PDS would continue to receive normal entitlement at the full central issue price;
3. The supply of foodgrains for the below poverty line population at 10 kg per family per month shall be guaranteed to states by the Centre;
4. The states will be free to add to the quantum, coverage and the subsidy from their own resources; and
5. The subsidised foodgrains will also be issued to all beneficiaries under the Jawahar Rozgar Yojana as per guidelines at the rate of one kg. per manday for which food coupons would be issued to beneficiaries for exchange at fair price shops.

Since 2003, the families below poverty line are being supplied 35 kg of foodgrains per family per month at a price which was 45% of the economic cost. In addition, 20 million poorest of the poor families are being supplied wheat and rice at a rate of ₹ 2.00 and ₹ 3.00 per kg, respectively under Antyodaya Anna Yojana (AAY).

Based on continuous reviews of the PDS, the coverage of the food security programmes was extended to adopt a life cycle approach, i.e. covering pregnant women, infants, school children, old-age people and those working under employment guarantee scheme. The subsidized food grains were made available in the following programmes also.

1. Midday meal scheme for school children,
2. Supplementary Nutrition Programme for under-six children, pregnant and lactating mothers, and adolescent girls,
3. Annapoorna Yojana for 65+ people without any source of regular income,
4. Supply of foodgrains to SC/ST hostels and hostels run by Social Welfare Department, and
5. Supply of foodgrains under Sampoorna Gramin Rojgar Yojana (Total Rural Employment Scheme).

These programmes account for around 10% of the total off-take of food grains under PDS. Midday meals scheme covers 120 million children in 11.35 lakh schools. Class I to IV students are provided ration @ 450 cal. + 12 gm protein per day and class V to VIII provided with 700 cal. + 20 gm protein per day. Some states have included class IX and X students also.

RIGHT TO FOOD: NATIONAL FOOD SECURITY ACT, 2013

Indian Parliament passed the National Food Security Act (NFSA) on September 2013. It entails 'right to food' to the people of India. It legally commits the government to provide/distribute/supply food to the defined population. The salient features of the NFSA are as follows:

1. Targets to cover 75% of the rural and 50% of the urban population.
2. Over all, 67% of the country's population or 81.93 crore people are covered by the Act.
3. Rice, wheat, and nutri cereals are made available to the targeted population at a price of ₹ 3 per kg of rice, ₹ 2 per kg of wheat and ₹ 1 per kg of nutri cereals. These prices were fixed for three years in the Act but have been retained so far.
4. Monthly entitlement of individual is 5 kg per month, while that of the antyodaya household (poorest of the poor) is 35 kg per house hold per month.

One of the provisions of the Act calls for PDS network reforms. In pursuance of this provision, a pre-condition for implementation is that all the state governments identify the beneficiaries and link their food distribution network to an online tracking system.

With enactment of NFSA, India has come a long way in the last six years and turned from a leaky and poorly targeted food transfer to a critical social support for needy households and individuals.

1. Digitization of ration cards has been completed in all the states/UTs. Digitization and Aadhaar seeding has helped weed out fake/duplicate ration cards and thus ghost beneficiaries. There are now around 25 crore ration cards and around 805 million people covered under various provisions of PDS.

2. Computerization of supply chain management has been completed in 25 states/UTs and in remaining states, the work is in progress.
3. Out of 5.33 lakh FPSs, around 4 lakh outlets have been automated with electronic point of sale (PoS) devices (However, in some areas, availability of internet is an issue).
4. Transparency portals have been set up in all the states/UTs. Grievance redressal facility is also in place in all states/UTs, either toll free or online.
5. Online allocation of grains is also complete in all the states/UTs, except a few.
6. Interstate portability of ration cards (card holder can claim ration anywhere in the country, irrespective of his/her native state) is presently operational in many states and going to be an all India roll out soon.
7. Several pilot projects of direct cash transfer (DCT) were completed and the results are mixed. The preferences of households for subsidized grain vs DCT vary across households and depend on several socioeconomic characteristics. Nevertheless, the saving in food subsidy from DCT is quite substantial.
8. Protein rich pulses are increasingly being added as a part of PDS in some states. In addition, in some states milk has been included as a part of midday school programme. These measures, as and when scaled-up, will help in improving nutrition security.

ECONOMIC COST OF PDS GRAINS AND FOOD SUBSIDY

The economic cost of wheat and rice distributed under PDS is the sum of purchase price of grain (pooled cost), procurement incidentals and distribution cost. The details for last two years are shown in Table 7.17. For wheat, the economic cost was ₹ 22.98 and ₹ 24.35 per kg during 2017–2018 and 2018–2019. For rice, these were ₹ 32.81 and 34.72 per kg respectively. The PDS issue prices since the launch of NFSA had remained ₹ 2 and ₹ 3 per kg for wheat and rice, respectively. Obviously, the food subsidy has been going up continuously. It increased from 18005 crore in 2001–02 to ₹ 56394 crore in 2010–11 and further to more than ₹ 1.32 lakh crore in 2018–19.

Table 7.17: Economic cost of wheat and rice

(₹ per quintal)

Item	Wheat		Rice	
	2017–18	2018–19	2017–18	2018–19
1. Pooled cost of grain	1588	1684	2291	2501
2. Procurement incidentals	304	318	482	423
3. Acquisition cost (1 + 2)	1892	2002	2773	2924
4. Distribution cost	406	433	508	548
Economic cost (3 + 4)	2298	2435	3281	3472
Annual buffer carting cost	512	501	512	501

Source: FCI, Department of Food and PDS, Economic Survey, Government of India.

HIGH LEVEL COMMITTEE TO REVIEW THE ROLE OF FCI AND RELATED POLICY (2014)

The committee (Shanta Kumar Committee) made several recommendations. Some of these are as follows:

1. FCI should focus its procurement operations in eastern UP, Bihar, West Bengal and Assam where procurement system is relatively weak and farmers resort to

distress sale. Simultaneously, it should exit Punjab and Haryana where state agencies are well equipped to do the procurement operations.

2. Roll out direct cash transfer to curb PDS leakages (grain diversion), start with big cities (1 million plus) and grain surplus states (that may save around ₹ 30,000 crore food subsidy).

3. Encourage private players to create modern storage and transportation infrastructure.

4. Rationalize tax structure on grain procurement to promote private purchases.

5. Enter into agreements with states before procurement season, to avoid announcement of bonuses and limiting the statutory levies to 3 to 4%.

6. Relook at the FSA, defer it in states that are not prepared, and lower the coverage from 67% to 40%.

7. Outsource stocking operations to CWC, SWCs and also to the private sector.

8. Liquidate food grain stocks once they cross the buffer norms.

9. Beneficiaries under TPDS should be given six months' ration at a time, immediately after the procurement season.

10. Government should synchronise its trade policy with MSP policy.

11. Introduce modern silos and hopper wagon system to minimize the use of gunny bags.

QUALITY CONTROL

AGRICULTURAL PRODUCTS

Quality control of agricultural commodities for long has remained the responsibility of the Directorate of Marketing and Inspection. The Directorate has prescribed grade standards for various agricultural products under the Agricultural Produce (Grading and Marking) Act, 1937. Agricultural commodities are graded under this Act on the basis of the specifications laid down under the grade standards. Graded products bear the AGMARK label, indicating the purity and quality of the product. Consumers are benefited when they buy graded products. The details of the mechanics of grade standards for agricultural products and the progress of grading in India have been discussed in Chapter 4.

MANUFACTURED PRODUCTS

Manufactured products are graded in accordance with the standards laid down by the Indian Standards Institution, now Bureau of Indian Standards (BIS) and bear the ISI label. Manufacturers have to use proper ingredients in specified proportions and follow the technique of manufacture given in the standards laid down by the Bureau of Indian Standards. The ISI label is an indicator of the good quality of the product.

INDIAN STANDARDS INSTITUTION (ISI) (1947 To 1986)

Standardization on an organized basis started in India with the establishment of the Indian Standards Institution. This institution, popularly known as the ISI; was set up in 1947 with the active support of the industrial, scientific and technical organizations in the country. The ISI frames standards in consultation with, and as a result of the co-operation of, the community—industrialists, scientists, administrators and the public. Standardization plays a vital role in the industrial development of a country. Apart from helping the commercial movement and industrial exchanges, standards conserve the production effort by reducing the unnecessary variety in products, by ensuring interchangeability, reducing costs and making mass production possible. Thus, standards lead to the best utilization of the human and material resources of a country. The institution operates under an Act of Parliament (ISI Certification Marks Act), under which manufactured items are stamped with the ISI mark of certification. This mark acts as a third party guarantee to the purchaser that the goods bearing the ISI mark have been produced in accordance with the provisions of the relevant Indian standards.

The World Standards Day is celebrated annually on 14th October, for it was on this day in 1946 that the United Nations Co-ordinating Committees decided to set up the International Organization for Standardization. This international organization now has more than one lakh experts from many countries directly involved in its work, and help it to create more and better international standards.

The aims and objects of the ISI are:

1. Preparation of standards for products, commodities, materials and processes on national and international bases;
2. Promotion of the general adoption of the standards prepared by it at national and international levels;
3. Certification of industrial products and assistance in the production of quality goods;
4. Dissemination of information relating to standards and standardization;
5. Conduct of surveys and training programmes for assistance to Indian industries in organizing their in-plant standards activity;
6. Collaboration with international organizations dealing with standardization for promotion of international trade;
7. Imparting training in industrial standardization to scientists and technologists from abroad; and
8. Performing a watching function in regard to the quality of Indian exports.

The Export Inspection Council exempts certain products, such as light engineering products, diesel engines and power-driven pumps, from preshipment inspection if they carry the ISI mark.

The Indian Standards Institution functions through nine Divisional Councils, which are responsible for the work of standardization in their respective fields. These divisional councils are; agricultural and food products, chemical, civil engineering, consumer products, electro-technical, mechanical engineering, structurals and metals, textiles and cargo movements, marine products and packaging.

ISI has set up more than 2,000 technical committees, sub-committees and panels dealing with different subjects with a membership of more than 24,000 experts representing various interests such as manufacturers, purchasers, consumers, scientific, technical and research organizations and government departments. These experts work in an honorary capacity and evolve national standards by consensus. Each standard specification is finalised after an exhaustive process of testing in laboratories, discussion in the committees and circulation to hundreds of interested parties all over the country.

Formulation of standards through consensus of different interests generally ensures their smooth implementation. In addition, Central and State Governments; local bodies and statutory organizations generally adopt standards in their purchases. Some State Governments decided to give preference to ISI certified products while some others have established standards cell for effective implementation of Indian Standards.

ISI also uses different media for public relations and publicity to spread the message of standardization. As a result, more than 90% of Indian Standards have been adopted by various official and non-official organizations. Various promotional and instructional programmes are carried out. The promotional programmes include management conferences and group meetings. The instructional programmes comprise survey, training programmes and seminars.

For effective implementation of national standards and for bringing the advantages of standardization within the reach of the common consumer, the Institution is operating a certification marks scheme under the ISI (Certification Marks) Act. This Act enables ISI to grant licences to manufactures to use the ISI mark on their products. Every licence includes a scheme of testing and inspection which the licencee is required to follow strictly. During the operation of the licence, ISI carries out regular and surprise inspections of the manufactures to make sure that the scheme of testing and inspection is being properly adhered to. Samples of certified products are drawn from the production line and from the open market and tested in independent laboratories. As a safeguard for the consumer the scheme provides for free replacement of ISI marked goods found to be of substandard quality.

The certification scheme was started in 1955–56. The scheme covers a range of products including consumer products and industrial items such as biscuits, infant milk food, ink, cables, conductors, jute products, steel, paints, shoe polish, pressure cookers, aluminium utensils, coffee, electrical appliances, sports goods and water meters. In the rural sector, the ISI has formulated standards for grain storage structures, fertilizers, pesticides, seeds, farm machinery and implements, pumping sets, gobar gas plants and animal husbandry and dairy equipments. Some items in the market, however, do not conform to these standards because they are produced in the small and tiny sectors without any facility or funds for quality testing.

For adoption of national standards to regulate the quality of industrial manufactures, in-plant standardization is an important requirement. Since 1961, ISI has promoted the concept of in-plant standardization through conferences, symposia and training programmes.

The institution initially established a central laboratory at New Delhi and regional laboratories in Mumbai, Kolkata and Chennai for conducting testing of products covered under the certification marks scheme. These laboratories also undertake investigational work covering food, chemical, electrical and mechanical items for the purpose of evaluation of standards. Laboratory personnel from government agencies and industries covering their products under the ISI certification marks scheme avail of the training facilities provided by the Institution in its laboratories.

ISI serves the interests of the country in the field of international standardization by close collaboration with the international organizations such as ISO (International Organization for Standardization) and the IEC (International Electro Technical Commission) for standardization. It is represented on important administrative bodies of these organizations.

The ISI also works in close collaboration with the similar organizations for standardization in other countries of the ECAFE (Economic Commission for Asia and Far East) Region with a view to promoting standardization activities. It actively participates in the work of the Asian Standards Advisory Committee (ASAC).

The ISI has benefited the consumers as well as the manufacturers. It promotes overall economy and brings about the best utilization of human and material resources by bringing the advantages—minimization of wastages, cutting down unnecessary varieties of products, increasing productivity and reducing costs. It protects the consumers through assured quality. It acts as a third party guarantee. The scheme brings to the consumers the benefits of lower price, better quality, more safety and repair services. To manufactures, it helps in adopting, the process of standardization. This reduces wastage, cost of material, cost of production and increases the chances of profits. It has introduced the metric system of weights and measures. The ISI took up the steel economy project involving a comprehensive standardization programme to give a fillip to the steel industry and introduce economies in the use of structural and special alloy steels. The project resulted in a saving of 24% in the use of metal according to an evaluation by NCAER. Another notable achievement is the preparation of the National Building Code streamlining housing construction practices all over the country.

BUREAU OF INDIAN STANDARDS (BIS) (Since 1987)

The Indian Standards Institution has been renamed as the Bureau of Indian Standards (BIS) with effect from April 1 1987. Along with the change in its name, its status and scope of activities have also been enlarged. The Bureau of Indian Standards performs all the functions of ISI as before with greater thrust to consumer protection, improving the level of quality of Indian products, harmonising the standards formulation and the certification/inspection activities in the country by providing a larger network of testing and consultancy services.

The activities of BIS, for the benefit of industry and consumers, include the following:

1. Formulation of standards
2. Mark of ISI under the Product Certification Scheme
3. Certification of foreign manufacturers and imported goods
4. Hallmarking of gold and silver jewellery
5. Eco Mark for environment-friendly products
6. Compulsory Registration Scheme for notified Electronics and IT goods
7. Quality Management System Certification as per IS/ISO 9001 series
8. Environment Management System (EMS) Certification as per IS/ISO 14001 series
9. Service Quality Management System Certification as per ISO 15000/15700
10. Occupational Health and Safety Management System Certification as per IS 18001
11. Food Safety Management System Certification as per IS/ISO 22000
12. Laboratory Testing, Calibration and Management
13. Training of personnel in the field of standardization and management system.

The Bureau has five regional offices that operate through 33 branches in 28 cities. It has network of eight labs of its own, besides 245 BIS recognized labs and 255 government labs of national institutes of repute and eminence. It uses the National Accreditation Board (NABL), which is a constituent of Quality Council of India. The functions of BIS are carried out with the help of 358 technical committees in 15 departments. The departments include one for food and agriculture and the other for water resources.

Over the past six decades, it has built up over 20710 Indian Standards covering products in different sectors like food and agriculture, chemicals, civil, mechanical and electrical engineering, electronics, textiles and many other products. The standards are constantly reviewed and updated to keep pace with technological innovations and the new social needs. The production of small-scale items based on Indian standards provides competitive capability with large-scale sector.

The Bureau has also made special efforts in the sphere of rural development by formulating over 2,079 standards relevant to the rural sector in areas of agricultural inputs like fertilizers, pesticides, agricultural machinery and farm implements, pumping sets, gobar gas plants and also in the sphere of post-harvest technology. The BIS has also formulated 18 standards for water.

The Bureau is the largest certification agency with over 27066 licences in operation for a wide range of products, covering 910 items varying from food products to electronics. It has become an institution of quality assurance for the consumers.

Standards certification is mandatory for items of mass consumption particularly those affecting health and safety of the consumers. Provision of voluntary certification for items such as colour television, control switches, sodium vapour lamps, jute and canvas products, bus and truck tyres and greases also exists.

Some of the other highlights of BIS activities are:

1. BIS has adopted IS/ISO 9000 series of standards. Now BIS quality certification is on the lines of international norms and is acredited by RVA Netherlands.
2. Under the Foreign Manufacturers Certification Scheme, for certifying imported goods, 343 licences at 37 country points are in operation.
3. Under the Management Systems Certification Scheme, 1229 licences are in operation.
4. BIS is also undertaking HACCP certification. HACCP Certification is a process control system designed to prevent microbial and other hazards in food production. It is based on Quality Management System and IS 15000 which is equivalent to CODEX ALI NORM 97/13A.
5. BIS also works as central enquiry point for WTO.
6. BIS had also started integrated milk certification scheme, applicable to organized dairy sector, which includes certification of both product and milk management system. Under this scheme, a certification licence is granted to use ISI mark on pasteurized milk pouch, along with food safety and management system (FSMS) licence of the organization, right from the farmer level to the retail point. In the area of dairying, BIS has formulated 36 standards, out of which nine standards are mandatory. More than 300 licences are operational against these standards.
7. BIS has also formulated Indian standards for CSR (Corporate Social Responsibility), with the objective of offering guidance on CSR compliance by the companies.

ADVANTAGES OF CERTIFICATION MARK SCHEME

The specific advantages accruing to different sectors of economy from the ISI Certification Mark Scheme are the following:

1. To Manufacturers

- Streamlining of production processes and introduction of quality control system.
- Reaping of production economies accruing from standardization.
- Better image of products in the market, both internal and overseas.
- Winning of consumers' confidence and good-will.
- Meeting the preferences of organized purchase agencies of central and state governments, and public and private sector organizations for ISI marked products.

- Independent audit of quality control system by BIS.
- Financial incentives offered by the Industrial Development Bank of India and nationalized banks.

2. To Consumers

- Conformity with Indian Standards by an independent technical national organization.
- Helps in choosing a standard product.
- Protection from exploitation and deception.
- Free replacement of ISI marked products in case of their being of substandard quality.
- Assurance of safety against hazards to life and property.

3. To Exporters

- Exemption from pre-shipment inspection, where admissible.
- Convenient basis for concluding export contracts.
- Elimination of the need for exhaustive inspection of consignments exported from the country, thus saves expenditure, time and labour.

The authorized or licensed ISI mark has three essential components. The first is ISI mark with a standard design in a rectangle having width and height ratio of 4:3. The second component is a product code at the outside top of the rectangle consisting of letters IS, followed by colon and a three to five digit number, which denotes the product category. The third component is printed at the outside bottom of the rectangle, which denotes the manufacturer. It appears as CM/L followed by a seven digit number. For some product categories, grade is also displayed at the bottom of the ISI mark.

CONSUMER PROTECTION

Food products have the distinction of meeting an essential need of all the consumers, irrespective of their economic and social status. Protecting the consumer's interest relating to food products means providing him wholesome, hygienically prepared and pre-tested quality products to enable him to lead a healthy life. The consumers are often cheated through deceptive and defective weights and measures and adulteration.

The doctrine of "caveat emptor", i.e. 'let the buyer beware' has long been the corner-stone of the consumer laws in India and this is virtually not acceptable to the average consumer now. The doctrine put forward in his favour with the growing consumer awareness is "caveat venditor", that is, 'let the seller beware'.

Various Acts were framed by the government from time to time to protect the consumers. Some of the main Acts enacted and statutory orders passed by the Government to subserve the interest of the consumers are:

1. The Indian Sale of Goods Act, 1930.
2. The Agricultural Produce (Grading and Marking) Act, 1937.
3. The Drugs and Cosmetics Act, 1940.
4. The Indian Standards Institution (Certification Marks) Act, 1952 and now Bureau of Indian Standards Act, 1986.
5. The Emblems and Names (Prevention of Improper Use) Act, 1950.
6. The Prevention of Food Adulteration Act, 1954.
7. The Essential Commodities Act, 1955.
8. The Fruit Products Order, 1955.

9. The Sugar Control Order, 1956.
10. The Export (Quality Control and Inspection) Act, 1963.
11. The Vegetable Oil Products (Control) Order, 1967.
12. The Monopolies and Restrictive Trade Practices Act, 1969, and amended in 1984.
13. The Meat Food Products Order, 1973.
14. The Packaged Commodities Order, 1975.
15. The Trade and Merchandise Marks Act, 1958.
16. The Standards of Weights and Measures Act, 1976.
17. The Cold Storage Order, 1964 and 1980;
18. The Consumer's Protection Act, 1986, 1991, 1993, 2002, 2018; and
19. Food Safety and Standards Act (FSSA), 2006.

The last one, i.e. FSSA is the latest in the series, which has subsumed several food related Acts of the past (see details in the next section).

The above list of legislations is quite impressive for the protection of the consumers. In practice the situation appears obscure due to poor enforcement of them. Under the Consumer's Protection Act, there is a provision that a consumer may get his defective goods replaced or price refunded or get compensation for any loss due to the unfair trade practices of the traders.

The Consumer Protection Act gives the consumer the right to be protected against marketing of goods and services which are hazardous to life and property; the right to be informed about the quality, quantity, purity, potency, standard and price of goods/ services; the right to be assured of access to variety of goods and services at competitive prices; the right to be heard and be assured that consumers' interests will receive due consideration at appropriate forums; the right to seek redressal against unfair trade practices or unscrupulous exploitation of consumers; and the right to consumer education about various laws and procedures to prevent malpractices.

The consumer awareness is very important for achieving the objectives of Consumer Protection Act. At present, the level of awareness is quite low among the consumers and more so among the farmers. Therefore, efforts should be made to create more awareness in rural areas through electronic media.

There is a three tier system for redressal of grievances of consumers under this Act.

1. **At district level:** There is a district level consumer forum. The complaints amounting up to ₹ 20 lakh can be presented in these forums.
2. **At state level:** Complaints valuing ₹ 20 lakh to one crore and writ petitions against the decisions of district forums can be filed in state level forums/courts located in state capitals.
3. **At National Level:** The complaints valuing more than ₹ one crore can be filed in the national forum/court which is located at New Delhi.

Many voluntary agencies are working in the country for giving strength to the consumers movement and the protection of consumers. The important ones are Consumers Guidance Society of India, Mumbai 1966; Consumer Council, Vishakhapatnam, 1970; Consumer Education and Research Centre, Ahmedabad; Consumer Action Forum, Kolkata; Karnataka Consumers Service Society, Bangalore, Grahak Panchayats at Mumbai and Pune and Society for Environmental Communications, New Delhi. The Government has established Consumer's Protection Councils at the State and district levels for the protection of the consumers. The work done by these agencies has been commendable in extending the rights of the consumers by keeping the producers conscious of consumer rights and interests. The main contributions of the organizations are in areas of consumer education (providing information about availability of goods, prices and trade practices), product rating (testing of products) and liaison with government and producers of products. The consumers are also being educated through advertisements in various media.

Under the consumers rights movement, which is over-seen/supported by the Union Ministry of Consumer Affairs, 629 district forums, 35 state commissions and one NCDRC are fully functional ensuring fast track disposal of cases relating to consumers' rights protection. There are 25 state consumer help lines and two national help lines. The networking of consumers' forums is through CONFONET. There is a special campaign against misleading advertisements and food wastage in special functions. For consumer education and empowerment, 'Jago Grahak Jago' campaign is very popular.

CONSUMER EDUCATION

World consumer day is celebrated on 15th March every year. On this day, people are sensitized about consumer rights and responsibilities.

Consumer Rights

1. Right to Safety: To protect against the sale of goods and services which are spurious and or hazardous to life.
2. Right to Information: To know the quality, quantity, weight and the price of the good/service being paid so that consumers are not cheated by unfair trade practices.
3. Right to Choose: To be assured, wherever possible, access to a variety of goods and services at competitive price.
4. Right to be Heard: To be heard and be assured that consumers interest will receive due consideration at appropriate fora.
5. Right to Seek Redressal: To seek legal redressal against unfair or restrictive trade practices or exploitation.
6. Right to Consumer Education: to have access to consumer education.

Consumer Responsibilities

1. Obtain full information regarding quality and price before making any purchase.
2. Be careful about false and/or misleading advertisements.
3. Purchase goods having quality marks like ISI/AGMARK, FSSAI, etc. as and where available for safety and quality.
4. Obtain proper receipt/cash memo for purchases made and guarantee/warranty card, duly signed by the seller, wherever applicable.
5. Approach consumer forum for redressal of grievances against sale of defective goods or deficient services or adoption of unfair or restrictive trade practices.

For undertaking education and research relating to consumers, a Consumer Education and Research Centre is working since long.

The Consumer Education and Research Centre (CERC) is a non-political, non-profit organisation situated at Ahmedabad. It is a public charitable trust registered under the Bombay Public Trust Act, 1950. The CERC is recognised as consumer organisation by the Government of Gujarat. This is the consumer organisation recognised as Research Institute by the Central Government on the recommendation of the Department of Science and Technology.

The main objectives and functions of this centre are:

1. To create an enlightened consumer consciousness and public opinion through the mass media;
2. To study analytically and do research on the working of the public utility services;
3. To carry academic programmes for training the workers and leaders for consumer protection;
4. To approach the legislators for lobbying with them for taking up consumer protection issues on the floor of parliament/assemblies
5. To mobilise and motivate people and other voluntary organisations for protection of consumers from various ills in the society
6. To take recourse to court for redressal of grievances of the consumers.
7. To establish a two-way dialogue with the consumer organisations in the country and those of abroad for mutual benefit and support;
8. To set up consumer product testing laboratory for testing and evaluation of the product such as food, pharmaceutical and domestic electric appliances; and
9. To set up consumer library with facilities for increasing the consumers' knowledge.

In recent years, several other voluntary and non-government organizations have come up in the country, which are working in the interest of the consumers.

QUALITY MANAGEMENT IN FOOD

International Organisation for Standardization

International Organisation for Standardization (ISO) and the Codex Alimentarious Commission (CAC) are the two prominent international organisations engaged in standardization of agricultural products.

International Organisation for Standardization (ISO) came into existence formally on 25th February, 1947. The objective of ISO is to promote the development of standards in the world with a view to facilitating international exchange of goods and services and to develop cooperation in the spheres of intellectual, scientific, technological and economic activities. The central secretariat of ISO is located in Geneva (Switzerland). It is an independent non-government international organization. It currently has a membership of 164 national standards bodies. India is a founder member of ISO. Its structure includes a General Council, ISO Council and Technical Management Board (TMB) with several Technical Committees and advisory groups. It also works with bodies like WTO. Through its members, ISO brings together experts to share knowledge and develop voluntary, consensus-based, market relevant international standards that support innovation and provide solutions to global challenges. So far, the ISO has published 22969 international standards covering almost every industry from technology to food safety, health care and agriculture. The ISO is engaged in the formulation of standards for a large number of agricultural commodities, covering spices and condiments, lac, essential oils, cereals and pulses, food products and stimulant foods. The results of ISO technical work are published as International Standards (ISO standards). This work is carried out by a set-up of technical committees covering different commodities traded internationally. One of the ISO Technical Committee (ISO/TC 34) deals with agricultural food products. This technical committee has 15 sub-committees dealing with various food commodities groups such as fruits and vegetables; milk and milk products; meat and meat products; and spices and condiments. The sub-committees have participating members, observer members and liaison members. The ISO standards published for different commodities are subject to periodical review every five years with a view to keeping the standards up-to-date under the working procedure of ISO.

Some most popular standards published by ISO are (a) ISO 9001:2015 related to quality management system, for companies and organizations; (b) ISO 14001:2015 related to environment management system; and (c) ISO/IEC 27001:2013 related to information technology. This family of standards is designed for any size of organization.

HACCP and World Standards

Indian manufacturers need to upgrade the quality of the products by adopting HACCP (Hazard Analysis and Critical Control Point), a food safety system, which is an internationally recognised auditing method. HACCP focuses attention on chemical, physical and microbial hazards.

The rejection of Indian wheat consignments by Iraq, gherkin containers by European Union and grape containers by U.K. and domestic complaints of presence of rat droppings in wheat are some of the examples quoted as non-compliance of food safety norms. It has harmed our business both on the export front and in the local market. By and large, the 'Made in India' label, for long, was considered as sub-standard produce by people of many countries. Therefore, there was a need to change this perception to make a significant dent in the food export market especially in the processing sector. Food processing sector, comprising fruits and vegetables, grains, milk, fish, meat, poultry products, soft drinks, and alcoholic beverages, is one of the largest sector in terms of production, consumption, employment generation and from export prospects. As such adoption of HACCP concept is important. There is a need to generate continuous awareness and run educational programmes for exporters and also have a legislation to ensure safety norms.

This concept was not so important when the food chain was localised and people consumed locally produced fresh harvested or cooked food without prolonged storage. In recent times, food has become a global issue. Good-looking fresh fruits and vegetables might contain hazardous chemicals and bacteria which may cause ill effects on the health of the consumers, instantaneously or at a later date. Quality management of food is, therefore, essential in fruits and vegetables and other processed products from the safety point of view.

A new era in food safety started in 1960s when USA planned to send astronauts in spaceship to moon. For such a mission, it was necessary to ensure that food provided to the astronauts would not cause illness while on board. With this objective, Pillsbury of NASA (National Aeronautics and Space Administration) developed and used HACCP as preventive system for preparation of food for astronauts.

HACCP and Risk Analysis is a modern concept of quality management applied to food items. The concept of HACCP gained recognition and acceptance globally as a system of choice for food safety due to following reasons:

1. To identify food safety hazards for different farm products and their process of production.
2. To accept responsibility for food safety instead of relying upon compliance with official regulation and inspection by food safety inspectors.
3. Necessity of creating awareness among people to realize their role and responsibility for food safety.

4. To improve the design of food products and process for achieving safe food, and
5. To prepare food companies for future HACCP based food safety regulations and trade specifications.

International food safety standards are developed by the Codex Alimentarius Commission (CODEX). This is a joint commission of FAO and WHO and recognizes HACCP based system for food. As per the WTO requirement, only Codex standards are acceptable for international trade. Therefore, Codex-HACCP is minimum international standard for trade among countries in future. Based on this analysis, appropriate action can be taken to ensure that the areas identified as critical control points are kept under control and are not allowed to endanger the items produced.

There are seven principles of Codex-HACCP.

1. Conduct a hazard analysis
2. Determine the critical control points (CCPs)
3. Establish critical limit
4. Establish a system to monitor control of the CCP
5. Establish the corrective action to be taken when the monitoring indicates that a particular CCP is not under control
6. Establish procedure for verification to confirm that the HACCP system is working effectively
7. Establish documentation concerning all procedures and records appropriate to these principles and their application.

Food safety is analysed in terms of hazards and risks. A hazard is the capacity of a thing to cause harm under certain conditions. The probability that a defined harm will occur is the risk associated with that hazard. The hazards may be physical, chemical or micro-biological and can occur at any stage from raw material to the consumption by the consumer.

The benefits of testing food by HACCP are:

1. Avoids human sufferings;
2. Reduces burden from over burdened health care system;
3. Increases the export of food products;
4. Attracts more foreign tourists; and
5. Increases earning potential of citizens.

Apart from Codex and HACCP, other agencies prescribing the world food standards are as follows:

1. FSA: Food Standards Agency of U.K.
2. FDA: Food and Drug Administration of USA
3. ISO: International Organization for Standardization (European Union) - applicable to all countries

For food processing industry, the ISO 22000 is the standard, which was developed after studying the prevalent standards in different countries. All important features of Dutch HACCP standards have been taken into account in ISO 22000.

Codex Alimentarius Commission (CAC)

Codex (CAC) is an inter-governmental body of the United Nations Organization (UNO). It was established by Food and Agriculture Organization (FAO) and World

Health Organization (WHO) in 1963. It develops harmonized food standards, guidelines, and codes of practices to protect the health of the consumers and ensure fair practices in the food trade. The Commission also promotes coordination of all food standards work undertaken by international governmental and non-governmental organizations. India is a member of Codex Alimentarius Commission since 1964. The CAC works through a large number of international committees. For each codex committee, there is parallel shadow committee in India that works for that particular codex committee. The shadow committee means the committee at the national level, constituted by the national Food Authority for reviewing the agenda of the CAC and its subsidiary committee and preparing India's comments or views on various agenda items. The stakeholders or members represented in each committee are various ministries like agriculture and food processing industries, department of animal husbandry, dairying and fisheries, educational institutions, industry associations like CII and FICCI, and experts/scientists in concerned areas.

Food Safety and Standards Authority of India (FSSAI)

FSSAI has been established under the Food Safety and Standards Act, (FSSA) 2006, as a statutory body for laying down science-based standards for articles of food, and regulating manufacturing, processing, distribution, sale and import of food so as to ensure safe and wholesome food for human consumption. Ministry of Health and Family Welfare, Government of India, is the administrative ministry for implementation of FSSA, which came into effect in the whole country in August 2011. The previous Acts/Orders that have been repealed by the FSSA include the following:

1. Prevention of Food Adulteration Act, 1954
2. Food Products Order (FPO), 1955
3. Vegetable Oil Products (Control) Order 1947, 1980
4. Meat Food Products Order, 1973
5. Solvent Extracted Oil, De-oiled Meal and Edible Flour (Control) Order, 1967
6. Milk and Milk Products Order, 1992 and 2009
7. Edible Oils Packaging (Regulation) Order, 1988
8. Some other Orders relating to Food, issued under ECA, 1955

The FSS Act envisages prevention of fraudulent, unfair, unethical and deceptive trade practices which mislead and harm the consumers by enhancing the sale, supply, use, processing, and consumption of unsafe, contaminated and substandard food items. Such practices include wrong expression of standard, quality, quantity, and grade; and giving guarantee of quality without adequate scientific and legal basis. The FSSAI has brought out 26 regulatory notifications and many others are in the pipeline. Some of these are the following for pursing the objectives of FSSA.

1. Food Safety and Standards (Licensing and Registration of Food Business) Regulations, 2011 (and subsequent Amendment).
2. Food Safety and Standards (Packaging and Labeling) Regulations, 2011 (and subsequent amendments).
3. Food Safety and Standards (Food Products Standards and Food Additives) Regulations, 2011 (Part 1 and Part II) plus subsequent Amendments.
4. Food Safety and Standards (Prohibition and Restrictions on Sales) Regulations, 2011 (plus amendments).
5. Food Safety and Standards (Labeling and Sample Analysis) Regulations, 2011 (with amendments).

6. Food Safety and Standards (Contaminants, Toxins, and Residues) Regulations, 2011 (with amendments).
7. FSSAI Logo, which can be used for printing on labels of packaged food products.
8. Food Safety and Standards (Food or Health Supplements, Nutraceuticals, Special Dietary Foods, Novel Foods etc.) Regulation, 2016.
9. Food Safety and Standards (Foods Recall Procedure) Regulation, 2017
10. Food Safety and Standards (Imports) Regulation, 2017.
11. Food Safety and Standards (Non-Specified Food and Food Ingredients) Regulation, 2017
12. Food Safety and Standards (Organic Food) Regulation, 2017
13. Food Safety and Standards (Alcoholic Beverages), Regulation, 2018.

As per the FSS Act, a Food Business Operator (FBO) has been defined as a person engaged in the business of food manufacture, processing, packaging, transportation, distribution, storage and import etc. and includes food services, catering services, and sale of food or food ingredients. These include wholesalers, retailers, hotels, restaurants, clubs, canteens, re-packers and re-labelers. The FSSAI has asked all FBOs to obtain licence/registration certificate under FSS Regulations. It has been notified that operating without a licence will attract a penalty up to ₹5 lakh and imprisonment up to 6 months.

The registration is also necessary for Petty Food Business Operators (PFBOs). A milk producer whose production capacity of the milk and milk products is less than 100kg/liter per day or is handling milk less than 500 litres per day; meat or meat products producer in the capacity for slaughter of maximum 2 large animals or 10 small animals or 50 poultry birds per day; or any other FBO where annual turnover is less than ₹12 lakh is termed as Petty Food Business Operator and is required to be registered. A PFBO is required to register his food business with the registering officer viz:, food safety officer, local panchayat or municipality. The licence fee is ₹100/- for PFBOs. If the Registration Officer does not communicate any decision within 7 days of making an application, the PFBO can start food business without further wait.

While FSSAI is based in Delhi, the states have Food Safety Commissioners. At the district levels, SDMs have been designated as district level officers to issue licences to FBOs under the Act. There are 680 adjucating officers, 651 district level officers for licences and 3144 food safety officers. Within the short period of coming into existence, the FSSAI has made following progress (apart from framing detailed rules and regulations as mentioned earlier):

1. It has come out with FSSAI safety standards for food items.
2. It has finalized 13000 standards for food additives and ingredients in line with global safety standards (codex) in order to do away with lengthy processes of product approval. If FPOs comply with these standards, they would not be required to seek product approval from FSSAI.
3. Most of the states/UTs have started online registration and licensing of FBOs. The states/UTs have granted 7,83,832 licences and registered 31,90,371 FBOs under the Act. The FSSAI has also granted 37,405 central licences.
4. FSSAI has notified that food labels must carry:
   (a) name of food,
   (b) list of ingredients,
   (c) nutritional information,
   (d) declaration about food additives,

(e) complete name and address of manufacturer,

(f) date of manufacture or packaging,

(g) country of origin for imported food, and

(h) instructions for use. For consumers, it has emphasized that "Make Reading Food Label a Habit."

5. It has established food import clearing system (FICS) at 21 points of entry, which is being operated at several locations like Mumbai, Chennai, Delhi, Kolkata, and Cochin.

6. To facilitate imports, guidelines on imported food items meant for 100% export or re export; and also the guidelines regarding imports meant for exhibition and testing purpose have been issued by FSSAI in January-February 2015.

7. It is regularly publishing a bi-monthly news letter, besides making online availability of (a) Food Licensing and Registration Form, (b) Indian Food Code (form for Licensing); (c) Food Product Approval System; and (d) Food Import Clearance System.

8. FSSAI is also keeping track of ground realities. It has assessed that:

(a) 13% of food items failed to meet the standards,

(b) milk is watered down or adulterated with products such as fertilizer, bleach and detergent to thicken it and give it a white frothy appearance,

(c) sweets are laced with soda,

(d) fruits being artificially ripened by using calcium carbonate,

(e) cheap cooking oil being mixed with expensive oil,

(f) tea waste mixed with new tea, and

(g) urea and blotting paper added to thicken the food.

9. FSSAI has a network of 244 primary food testing labs (FTLs), 18 referral FTLs, apart from 45 state labs and 62 mobile FTLs. An online portal has also been launched (INFOLNET).

10. FSSAI has set up a new Central Consumer Protection Authority (CCPA), which is looking after consumers' interests, including prescription of food labels to well inform the public about contents of the item.

11. FSSAI has launched an 'Indian Organic Integrity Data Base' portal, which recognizes both the existing system for organic certification, viz:, National Program for Organic Production (NPOP) of APEDA and Participatory Guaranteed Scheme (PGS-India).

12. FSSAI has also launched 'Food Regulatory Portal' which is very friendly for food businesses. It provides easy access to globally bench-marked food standards and time bound approval of non-specified food products. It also provides cloud-based platform for food sampling and testing. It has online licensing and registration system. It facilitates hassle free food imports.

13. FSSAI has launched a massive programme of training for FBOs, food safety supervisors, food analysts and all other stakeholders.

14. Under the consumer protection Act, 2018, there is a provision of heavy penalties on false claims and advertisements of packed food products, which gives extra powers to FSSAI for action in the interest of consumers.

The greatest challenge which the FSSAI or for that matter, all the consumers face is that, worldwide as well as in India, the food business owners/operators are a part of the system that prescribes or finalizes the food regulations and/or rules.

ECOMARK

To protect and conserve the nature, Government of India (Ministry of Environment) has instituted a scheme to make the products with ECOMARK based on bio-degradability of the product or what is known in general parlance as environmental friendly products. The salient features of the ECOMARK scheme are as under:

1. The scheme is administered by the Bureau of Indian standards (BIS).
2. The product categories identified under the scheme include soaps and detergents; paints; paper; plastics; cosmetics; textiles; batteries; wood substitutes; propellants and aerosols; food items (edible oils including vanaspati, tea, and coffee); electrical and electronic goods; packing/packaging materials; lubricating and speciality oils; drugs; food preservatives and additives; and pesticides.
3. The scheme is operated on a national basis and provides accredition and labels on household and other consumer products which meet certain environmental criteria and quality requirements presented in relevant Indian standards for the product.
4. Products certified as eligible for the ECOMARK carry the ISI Mark for quality safety and performance of the product and licensed to carry the ECOMARK for a prescribed time period after which it is reassessed.
5. Inspections are carried out and samples taken for analysis of any material or substance in relation to which the BIS-ECOMARK has been used as may be necessary for proper implementation of ECOMARK. For this purpose, the standard Mark of Bureau would be a single mark having a combination of ISI Mark and Eco logo.
6. To operate the scheme, BIS has included additional requirements for ECOMARK in the Indian standards.

Mark to Identify Vegetarian/Non-Vegetarian Food Products

The Government of India by an amendment in the Prevention of Food Adulteration Act, 1955 on 4th October, 2001 and 20th June, 2002 has made it mandatory for the manufacturers of food products to put a label indicating whether the food has been prepared using meat and allied products or otherwise. Under this amendment, the packed food products bearing a mark of a big dot in a square in green colour is indicator of vegetarian product and a mark in brown colour is indicator of non-vegetarian food. This amendment is applicable throughout the country.

Mark of FPO

Till the Fruit Products Order was operational, the products carrying a mark of FPO in an oval with two hanging strips (making inverted –V shape) was mandatory on packed containers of fruits and vegetables processed products. This indicated the quality of the product and conveyed that the production of processed fruit products had been carried out under clean and sanitary conditions.

ADMINISTERED PRICES

COMMISSION FOR AGRICULTURAL COSTS AND PRICES (CACP)

Another method of intervention in the market mechanism has been the announcement of different administered prices viz., minimum support prices, statutory minimum prices, fair and remunerative prices, procurement prices and issue prices. These prices are announced for different agricultural crops by the Government of India on the recommendations of Commission for Agricultural Costs and Prices (CACP). This Commission was originally set up in January, 1965 in the name of the Agricultural Prices Commission (APC).

The Agricultural Prices Commission was set up on the recommendations of the Foodgrains Prices Committee headed by Shri L.K. Jha with the aim of advising the Government on price policy of agricultural commodities with due regard to the interests of both producers and consumers. The price policy of the country aims at evolving a balanced and integrated price structure taking into account the overall needs of the economy and with due regard to the interests of both the groups of the economy. While recommending the price, the Commission considers the following aspects:

1. The need for incentives to farmers for the adoption of improved technology and maximization of production;
2. The need for ensuring a rational utilization of land and other production resources; and
3. The likely effect of the price policy on the rest of the economy, particularly on the cost of living of masses and industrial cost structure.

The terms of reference of the Commission were made broad based in March, 1980 with the change in its name to Commission for Agricultural Costs and Prices. Since 1966, the Commission has set up a fairly logical scheme for arriving at the administered prices of farm products. The Commission, till 1991, had been recommending two sets of administered prices viz., minimum support prices and procurement prices. From 1991 onwards, only minimum support prices (MSPs) are recommended by CACP.

The Commission for Agricultural Costs and Prices is not a statutory body. The Commission submits separate reports recommending the prices and price policy for the kharif and rabi season crops, sugarcane, jute and copra. The Central Government, after considering the reports of the Commission and views of the State Government and keeping in view the demand and supply situation in the country, takes decision on the level of administered prices.

The Commission for Agricultural Costs and Prices (CACP), which was instrumental in evolving a balanced and integrated price structure in the country, has been manned by eminent and experienced agricultural economists. A list of the Chairman of CACP at different points of time is given below:

Year	Name	From	To
1965–66	Prof. M.L. Dantwala	February, 1965	March, 1966
1966–70	Dr. Ashok Mitra	November, 1966	February, 1970
1970–75	Prof. Dharma Narain	March, 1970	May, 1975
1976	Dr. Kamta Prasad	January, 1976	July, 1976
1976–78	Prof. Dharma Narain	July, 1976	May, 1978
1978–81	Dr. A.S. Kahlon	November, 1978	October, 1981
1982–83	Dr. Y.K. Alagh	August, 1982	August, 1983
1983–86	Dr. G.S. Bhalla	October, 1983	October, 1986
1987–91	Dr. S.S. Johl	March, 1987	March, 1991
1991–92	Dr. D.S. Tyagi	April, 1991	May, 1992
1992–96	Dr. S.S. Acharya	May, 1992	October, 1996
1996–97	Dr. D.K. Marothia	December, 1996	June, 1997

Year	Name	From	To
1997–00	Dr. Abhijit Sen	June, 1997	Decemer, 2000
2001	Dr. T. Haq	January, 2001	June, 2001
2001–02	Dr. G.K. Chadha	June, 2001	April, 2002
2002–08	Dr. T. Haq	May, 2002	March, 2008
2008–10	Dr. S. Mahendra Dev	May 2008	August, 2010
2011–14	Dr. Ashok Gulati	March, 2011	Feb. 2014
2014–15	Dr. Ashok Vishandass	March 2014	Dec. 2015
2016–	Dr. V.P. Sharma	June 2016	Continuing

MINIMUM SUPPORT PRICES

This is the price fixed by the Government to protect the producer-farmers against excessive fall in price during harvest season in bumper production years. These prices give a sort of price guarantee to the farmers which means that a price not lower than the announced minimum price will be paid to the farmers when they bring their produce for sale in the market. In case the market price for the commodity falls below the announced minimum price due to bumper production and glut in the market, government agencies purchase the entire quantity offered by the farmers at the announced minimum price. The minimum price has been assigned a statutory status in case of sugarcane and as such the announced price is termed as statutory minimum price. There is statutory binding on sugar factories to pay the minimum announced price and all those transactions or purchase at a price lower than this are taken as illegal. Since 2009–10 crop season, the sugarcane price announced by the central governement is termed as fair and remunerative price (FRP).

Minimum support prices for different agricultural crops viz., foodgrains; oilseeds, fibrc crops, sugarcane and tobacco are announced by the Government of India before the start of the sowing season of the crop. This makes it possible for the farmers to have an idea about the extent of price insurance cover provided by the Government for the crop.

The minimum support prices and statutory minimum prices announced by the government for foodgrains (cereals and pulses), fibre crops (cotton and jute), oilseed crops and sugarcane in last 30 years are shown in Tables 7.18 to 7.20.

As can be seen, support prices are given for the marketing year. The marketing year for all kharif crops, including paddy, is defined as October to September, where as for rabi crops it is defined as April to March. In the case of raw jute, the marketing year is July to June, for cotton September to August and for sugarcane it is November to October.

Sesamum and niger seeds were brought under the support price scheme in the mid-nineties. The support prices announced for them during the last 20 years are as follows:

Marketing year	Sesamum	Niger seed
2001–02	1400	1100
2004–05	1500	1180
2005–06	1550	1200
2008–09	2850	2405
2009–10	2850	2405

Marketing year	Sesamum	Niger seed
2010–11	2900	2450
2011–12	3400	2900
2012–13	4200	3500
2013–14	4500	3500
2014–15	4600	3600
2015–16	4700	3650
2016–17	5000	3825
2017–18	5300	4050
2018–19	6249	5877
2019–20	6485	5940

The prices are in ₹ per quintal.
Source: Agricultural Statistics At a Glance, Various Issues, Ministry of Agriculture, Government of India.

Since the marketing year 2018–19, there had been two major developments in the sphere of minimum support price (MSP) policy of the government.

1. The government announced a floor for the level of MSPs. The announcement had been that MSPs for all the crops (covered under the scheme) will not be less than 1.5 times the cost of production. The cost of production was defined as A2+FL. The cost A2 includes all variable costs (all paid-out costs plus imputed value of farm produced inputs, irrigation costs, depreciation on farm buildings and equipments and interest on working capital) and FL is the imputed value of family labour used in the crop production. By definition, A2 +FL include all the costs except imputed rental value of land and imputed interest on own fixed capital. This is for the first time that a floor for the level of MSP has been defined.

2. The other development has been increase in efforts at all the levels to make the price support effective. There had been several occasions in the past when adequate arrangements for purchase of produce at MSP were not in place and farmer had to sell their produce in the market below the announced MSPs. For remedying this situation of price distress sale, the union government, in 2018, announced 'Pradhan Mantri Annadata Aay Sanrakshan Abhiyan' (PM-AASHA), with the objective of ensuring the announced MSPs to the farmers. Three alternatives have been suggested to the state governments under PM-AASHA:

   (a) Price Support Scheme (PSS): Under this, the price support agencies (FCI, NAFED and state agencies) announce the list of purchase centers and make the price support purchases. This system is presently in vogue.

   (b) Price Deficiency Payment Scheme (PDPS): In this scheme, the farmer is free to sell his produce to any trader in mandi yard and obtain a sale slip showing the quantity sold and the price. If the price is less than the MSP, he becomes entitled for receiving the deficit. The deficit is usually calculated by taking the modal monthly price in the market, rather than the individual's sale price. In this method, no price support purchase operations by the government agencies are necessary. This method was tried by the Madhya Pradesh government but several issues emerged. Niti Aayog feels that this scheme will be economical, if rolled out nationwide.

Table 7.18: Minimum support/procurement prices of cereals

(₹/Qtl.)

Marketing year	Wheat	Paddy	Coarse cereals (jowar, bajra, maize and ragi)
1988–89	173	160–180	145
1989–90	183	185–205	165
1990–91	215	205–225	180
1991–92	225	230–250	205–210
1992–93	275 (25)	270–290	240–245
1993–94	330 (25)	310–350	260–265
1994–95	350	340–380	280–290
1995–96	360	360–395	300–310
1996–97	380	380–415	310–320
1997–98	415 (25)	415–455*	360
1998–99	455 (55)	440–470	390
1999–00	550	490–520	415
2000–01	580	510–540	445
2001–02	610	530–560	485
2002–03	620	530–560 (20)	485 (5)
2003–04	620 (10)	550–580	505
2004–05	630	560–590	515–525
2005–06	640	570–600	525–540
2006–07	650 (50)	580–610 (40)	540–555
2007–08	750 (100)	745–775	600–620
2008–09	1000	900–930	840–915
2009–10	1080	1000–1030	840–915
2010–11	1100	1000–1030	840–915
2011–12	1120 (50)	1080–1110	980–1050
2012–13	1285	1250–1280	1175–1520
2013–14	1350	1310–1345	1250–1520
2014–15	1400	1360–1400	1250–1550
2015–16	1450	1410–1450	1275–1650
2016–17	1525	1470–1510	1330–1725
2017–18	1625	1550–1590	1425–1900
2018–19	1735	1750–1790	1700–2897
2019–20	1840	1815–1835	1760–3150

* With effect from 1997–98, MSP is fixed for only two varieties of paddy viz., common and Grade A.
Note: Figures in the brackets indicate additional bonus announced by the Union Government.
Source: Publications of Directorate of Economics and Statistics, Ministry of Agriculture and Reports of CACP, Government of India, New Delhi.

(c) Private Procurement and Stockist Scheme (PPSS): In this scheme, it is envisaged to persuade private sector to make price support purchases from the farmers at MSP. The participating traders or entrepreneurs are later reimbursed all the costs by the government. State governments are persuading the traders to start buying the produce at MSP under this scheme.

Table 7.19: Minimum support prices of barley, pulses, jute and cotton							
						(₹ per quintal)	
Marketing year	Barley	Gram (Common)	Arhar (Common)	Moong and Urad** (Common)	Raw jute W-5* (TD-5)	Cotton	
						Medium	Long
1988–89	135	290	360	360	250	500	600
1989–90	145	325	425	425	295	570	690
1990–91	180	420	480	480	320	620	750
1991–92	200	450	545	545	375*	695	840
1992–93	210	500	640	640	400*	800	950
1993–94	260	600	700	700	450	900	1050
1994–95	275	640	760	760	470	1000	1200
1995–96	285	670	800	800	490	1150	1350
1996–97	295	700	840	840	510	1180	1380
1997–98	305	740	900	900	570	1330	1530
1998–99	350	895	960	960	650	1440	1650
1999–00	385	1000	1105	1105	750	1575	1775
2000–01	430	1015	1200	1200	785	1625	1825
2001–02	500	1100	1320	1320	810	1675	1875
2002–03	500	1200	1320	1330	850	1675	1875
2003–04	500	1220	1360	1370	860	1725	1925
2004–05	525	1400	1390	1410	890	1760	1960
2005–06	540	1425	1400	1520	910	1760	1980
2006–07	550	1435	1410	1520	1000	1770	1990
2007–08	565	1445	1590	1740	1055	1800	2030
2008–09	650	1600	2000	2520	1250	2500	3000
2009–10	680	1730	2300	2520	1375	2500	3000
2010–11	750	1760	3500	3400	1575	2500	3000
2011–12	780	2100	3700	3800	1675	2800	3300
2012–13	980	2800	3850	4300	2200	3600	3900
2013–14	980	3000	4300	4300	2300	3700	4000
2014–15	1100	3100	4350	4350	2400	3800	4100
2015–16	1150	3175	4625	4625	2700	3800	4100
2016–17	1225	3500	5050	5000	3200	3860	4160
2017–18	1325	4000	5450	5400	3500	4020	4320
2018–19	1410	4400	5675	5600	3700	5150	5450
2019–20	1440	4620	5800	5700	3950	5255	5550

* Up to 1988–89, W-5 variety in Assam and later for TD-5 variety in Assam.

** These are MSP for urad. In last 20 years, MSP for moong were higher.

Source: Publications of Directorate of Economics and Statistics, Ministry of Agriculture and Reports of CACP, Government of India, New Delhi

PROCUREMENT PRICES

Procurement price of a commodity refers to the price at which government procures the commodity from producers/manufacturers for maintaining the buffer stock or the public distribution system. These prices are announced by the Government of India on the recommendations of the Commission for Agricultural Costs and Prices before

Table 7.20: Minimum support prices of oilseeds and sugarcane								
								(₹ per quintal)
Markering year	Mustard (FAQ)	Ground-nut (shell)	Soya-bean black	Soya-bean yellow	Sun-flower seed	Copra*	Saf-flower seed	Sugar-cane**
1988–89	430	430	275	320	450	–	415	19.50
1989–90	460	500	325	370	530	1500	440	22.00
1990–91	575	580	350	400	600	1600	550	23.00
1991–92	600	645	395	445	670	1700	575	26.00
1992–93	670	750	475	525	800	NA	640	31.00
1993–94	700	800	525	580	850	2150	720	34.50
1994–95	810	860	570	650	900	2350	760	39.10
1995–96	830	900	600	680	950	2500	780	42.50
1996–97	860	920	620	700	960	2500	800	45.90
1997–98	890	980	670	750	1000	2700–2925	830	48.45
1998–99	940	1040	705	795	1060	2900–3125	910	52.70
1999–00	1000	1155	755	845	1155	3100–3325	990	56.10
2000–01	1100	1220	795	865	1170	3250–3500	1100	59.50
2001–02	1200	1340	795	885	1185	3300–3550	1200	62.05
2002–03	1300	1355 (20)	795 (10)	885 (10)	1195 (15)	3320–3570	1300	69.50
2003–04	1330	1400	840	930	1250	3320–3570	1300	73.00
2004–05	1600	1500	900	1000	1340	3500–3750	1500	74.50
2005–06	1700	1520	900	1010	1500	3570–3820	1550	79.50
2006–07	1715	1520	900	1020	1500	3590–3840	1565	80.25
2007–08	1715	1550	910	1050	1510	3620–3870	1565	81.18
2008–09	1800	2100	1350	1390	2215	3660–3910	1650	81.18
2009–10	1830	2100	1350	1390	2215	4450–4700	1650	129.84
2010–11	1830	2300	1400	1440	2350	4450–4700	1680	139.12
2011–12	1850	2700	1650	1690	2800	4525–4775	1800	145.00
2012–13	2500	3700	2200	2240	3700	–	2500	170.00
2013–14	3000	4000	2500	2560	3700	5100–5350	2800	210.00
2014–15	3050	4000	2500	2560	3750	5250–5500	3000	200.00
2015–16	3100	4030	2540	2600	3800	5550–5830	3050	230.00
2016–17	3350	4220	–	2775	3950	5950–6240	3300	230.00
2017–18	3700	4450	–	3050	4100	6500–6785	3700	255.00
2018–19	4000	4890	–	3390	5385	7510–7750	4100	275.00
2019–20	4200	5090	–	3710	5650	9521–9920	4945	275.00

Figures in the brackets show additional bonus or special assistance.

* For copra, the prices are of calendar year and the lower and upper prices are for milling and ball copra, respectively.

** Statutory minimum price up to 2008–09 and fair and remunerative price (FRP) since 2009–10. These are linked to recovery of 8.5% up to 2004–05 and 9% after that.

Source: Publications of Directorate of Economics and Statistics, Ministry of Agriculture and Reports of CACP, Government of India, New Delhi

the harvest season of the crop. At these announced prices, government procures the foodgrains (wheat, paddy and coarse grains) in the needed quantity either for maintaining the buffer stock or for the distribution through fair price shops.

The necessity of fixing procurement prices arises out of the government acute need to get a part of the available market supply with a view to maintaining the distribution of foodgrains at reasonable prices to the weaker sections of the society.

Procurement prices are fixed generally at a level which is somewhat higher than the level of minimum support prices but lower than the prevailing market prices. The level of these prices is recommended by the Commission for Agricultural Costs and Prices based on the estimated size of the harvest of that crop and intended quantities to be procured. The procurement prices are lower in relation to the actual market prices and as such farmers and traders are not willing to sell their stocks voluntarily to the government. In such circumstances, the government procures foodgrains at the announced procurement prices either by imposing a levy on the farmers, or on the traders or through other methods as discussed earlier.

The Commission for Agricultural Costs and Prices may or may not recommend both of these prices for a commodity at a point of time. Since 1971, minimum support prices for cereals have been the same as their procurement prices, with the difference, that these are announced before the sowing season. As a result, the procurement price itself became the support price at which the government purchased all the foodgrains offered for sale. If there is no levy on farmers or traders or processors in the strict sense of the term, such a price should be referred to as the purchase price instead of the procurement price.

Procurement prices also became the minimum support prices because the government was bound to purchase the foodgrains offered by the producers for sale. The minimum support prices in general are fixed at a level lower than the market price level to avoid the responsibility of purchasing the entire marketed surplus. Beginning with the kharif crops of 1991–92, the system of announcement of procurement prices has been abolished and only minimum support prices for cereals are announced. In the case of other crops, it were only the minimum support prices that were being announced. Now for all foodgrain crops, only the minimum support price system is operative.[2]

MARKET INTERVENTION SCHEME (MIS)

MIS is an adhoc scheme of price support. The scheme covers horticultural and other commodities which are perishable in nature and are not covered under minimum support price scheme. In order to protect the growers of these commodities from making distress sale, in the event of bumper crop production, during the peak arrival period when prices fall to a very low level, the central government implements market intervention scheme for a particular commodity on the request of the state government concerned. Losses suffered are shared on 50:50 basis by the central and state government.

The market intervention scheme so far has been implemented for commodities like apple, kinnow/malta, garlic, orange, galgal, grapes, mushrooms, clove, black pepper, pineapple, ginger, red chillies, corianderseed, isabgol, chicory, onion, potato, cabbage, and castorseed, in the state of Himachal Pradesh, Haryana, Punjab, Andhra Pradesh, Maharashtra, Karnataka, Rajasthan, Gujarat, Kerala, Jammu & Kashmir, Mizoram, Sikkim, Meghalaya, Tripura, Uttar Pradesh, West Bengal, Madhya Pradesh, Andaman & Nikobar Islands, and Lakshadweep.

The difference between minimum support policy and market intervention scheme is as follows:

Particulars	Minimum Support Policy	Market Intervention Scheme
Commodities included	Fixed, presently 24	Not fixed
Regularity	Regularly announced every year	Not regular but ad hoc
Support prices	Decided by the Central Government on the recommendations of CACP	Decided by the Central and individual state government
Quantum of support	All quantity is purchased which is offered by the farmer	Pre-decided, limited quantities are purchased
Applicability	Throughout the country	Specified limited markets of the state
Time of operation	Throughout the year	Specified period
Incidence of losses, if any	Borne by the Union Government	Equally shared between union and state goverment
Infrastructure required for implementation	Large scale	Limited scale

Source: Economic Survey 2000–2001, Ministry of Finance, Govt. of India, New Delhi, p. 97.

For strengthening the MIS scheme, the government of India in 2015 created a Price Stabilization Fund (PSF) with a corpus of ₹ 500 crore. This new MIS+PSF was implemented during 2014-15 to 2016-17. The corpus was kept in a separate bank account maintained by Small Farmers Agribusiness Consortium (SFAC). It had provided interest-free working capital to eligible agencies to buy horticultural products and sell at reasonable prices. To begin with, onion and potato were taken up. The losses, if any, were met from the fund, and profits, if any, were added to it. PSF was managed by a seven-member management committee.

Subsequently, in 2018, the government announced 'Operation Green' (on the lines of erstwhile 'Operation Flood' for milk) with a seed capital of ₹ 500 crore. The Operation Green (OG) includes tomato, onion and potato (TOP). The objective of OG is to replicate the success of Indian milk story by reducing price volatility in TOP and thereby help both farmers and consumers. The litmus test of the success of OG would be if it can contain roller coaster rides of boom and busts in the prices of these three essential farm products. The success of OG depends on organizing farmers into FPOs (for aggregation), investment in logistics (cold storage and reefer vans) and linkages with processors/bulk buyers.

STATUTORY PRICE CONTROL AND RATIONING

Rationing is a measure for controlling the demand of consumers and keeping the rise in demand under check by allotting a limited quantity per capita per time period.

Rationing can be of two types—statutory rationing and modified rationing. When whole of the population in the command area is covered, it is called the statutory rationing. If only a selected section of the population is covered under the rationing system, it is called modified rationing.

Statutory rationing of foodgrains was undertaken in the country before independence (and continued for some time even after that) because of severe shortages. However, with the improvement in the supply situation, informal rationing has been introduced. Statutory rationing was undertaken in places like Kolkata, Mumbai,

Asansol and Durgapur towns only in rice. The main purpose was to keep the increasing demand of foodgrains under control. The quantity of foodgrains is fixed under statutory rationing, depending upon the supply, and is increased or decreased over time.

This is a social measure which was adopted by the government to protect mainly the vulnerable sections of society who could not afford to purchase foodgrains at open market prices because of their poverty. The rich and well-to-do can buy foodgrains in the required quantity and are able to meet their basic needs. The main purpose of rationing is to provide relief to the poor; the other sections are automatically protected under the constitution. However, on moral grounds, the rich are persuaded not to purchase the rationed commodity. At present, we have a targeted PDS and not rationing per se, because foodgrains are available in the market and any one can buy any quantity.

DIRECT INCOME SUPPORT TO FARMERS

Despite all effort by the government, by way of measures such as assurance of MSPs, market intervention scheme, subsidies on farm inputs, and other general programs like PDS at subsidizes prices, and electrification at household level and market linkages through rural roads, the farmers' distress continues to be a serious issue. During 2016 to 2018, 11 state governments declared waiver of farmers' loans amounting to a total of ₹ 212395 crore to ease the farmers' distress. There are several pros and cons of such loan waivers (see NAAS, 2019). In this connection, the option of direct income support to the farmers was discussed and suggested in various forums during the last five years. Direct income support for farmer is also compatible with the agreement on agriculture under world trade regime. Two state governments (Telangana and Odisha) launched such schemes and recently central government also rolled out such a scheme. Some salient's features of these three schemes are as follows:

RYTHU BANDHU SCHEME (RBS) OF TELANGANA

State government of Telangana launched this income support scheme in May, 2018. The objective is to put in a place an alternative to loan waiver, by way of giving the farmers cash before sowing of the kharif and rabi crops, so that they need not depend on loan for procuring critical farm inputs. The RBS mandates to pay ₹ 8000 per acre to farmers in two instalments, every year before kharif and rabi season. Before launching this scheme, all land records were updated and land ownership titles were given to 58 lakh farmers of the state (1.42 crore acres of land). The farmers are free to grow the crops of their choice. The RBS is available to all the land owners (irrespective of size). It does not include landless and share-cropper farmers.

KALIA SCHEME OF ODISHA

In January 2019, Odisha government launched a Krushak Assistance for Livelihood and Income Augmentation (KALIA) scheme. It covers small and marginal farmers (SMF), landless agricultural households (ALH) and vulnerable agricultural households (VAH). It implies that the scheme is not restricted to land owners. This scheme provides for payment of ₹ 10,000 per year in two instalments of ₹ 5000 each to SMF, ₹ 12500 per year to ALH and ₹ 10000 per year to VAH. Those excluded from the scheme are large farmers, those farmers who are tax payers and whose at least one member is in the government job. Total beneficiaries under the scheme are estimated to be 4.5 million small and marginal farmers, 3.0 million ALH and 1.0 million VAH, with a total of 8.5 million households.

PM-KISAN (GOVERNMENT OF INDIA)

In February 2019, the Union Government announced a Prime Minister Kisan Samman Nidhi (PM-KISAN), which is a direct income support scheme for the farmers of India. Originally, the scheme was announced for small and marginal farmers but later extended to all farmers. Under the scheme, there is a payment of ₹ 6000 per year to each farmer (in three instalments of ₹ 2000 each). Up to January 2020, 9.5 crore farmers have been registered under the scheme and are being paid the instalments. The state governments are being persuaded to speed-up the identification of beneficiaries.

▌ GOVERNMENT SPONSORED NATIONAL ORGANIZATIONS AND THEIR ROLE

FOOD CORPORATION OF INDIA

An efficient management of the food economy with a view to ensuring an equitable distribution of foodgrains at reasonable prices to the vulnerable sections of society is essential in the present socio-economic environment of the country. The government felt the necessity of an organization which can act as its main agency for handling foodgrains, acquire a commanding position in the foodgrain trade as a countervailing force to the speculative activities of private traders and, at the same time, work on commercial lines. Towards the end of 1964, Parliament decided to transfer the government's function of trading in foodgrains to the public sector. Legislation was enacted; and the Food Corporation of India (FCI) was born on January 1, 1965.

The Food Corporation of India initially serviced only four States in the southern part of the country. Later, it extended its services throughout the country. Today, the FCI is the unrivalled food marketing agency, serving the interests of both the farmers and consumers. Its market operations prevent the speculative trader from acting against the interest of the farmer by assuring him a remunerative price for his produce. It ensures a prompt and uninterrupted supply of foodgrains to the vulnerable sections of society all over the country. Operationally, the FCI reaches the remotest corners of the country through its vast network of offices and storage centres. Financially, it is one of the largest public sector undertakings in the country. Its authorized capital was increased to ₹ 10,000 crore in November 2019.

FCI is a public sector organization under the Department of Food and Public Distribution of Government of India. Its head office is at New Delhi. For its functioning, the country has been divided into five zones and within each zone, there are regions (24 in all). Each region generally coincides with a state (except some smaller states) and is headed by a Regional Manager. The next lower levels are divisions and district depots. Every district (more than 600) of the country has a Depot Manager and within each district, there are many operating points.

The main functions of the Food Corporation of India are:

1. To procure a sizeable portion of the marketable surplus of foodgrains at incentive or minimum support prices from the farmers on behalf of the Central and State Governments;
2. To make timely releases of the stocks through the public distribution system (fair price shops and controlled items shops);
3. To minimize seasonal price fluctuations and inter-regional price variations in food commodities by establishing a purchasing and distribution network; and
4. To build up a sizeable buffer stock of cereals to meet the situations that may arise as a result of shortfalls in internal procurement and imports.

The Food Corporation of India started its operations in 1965, and limited them to the States in the southern region and to some of the ports in that region. The anticipated growth of the Corporation was temporarily retarded by two successive years of unprecedented crop failures in 1965–66 and 1966–67, serious shortages of foodgrains and grave scarcity conditions in the country. But its activities received momentum with the onset of the green revolution in the country. By March, 1969, it had taken over all the functions of the Directorate General of Food. To-day, the FCI operates in almost all the States of the Indian Union in one sphere or the other on behalf of the Centre, or of the States, or on its own.

The tremendous growth of the organization is the direct result of the staggering increase in the volume of its business. The progress of the FCI in various areas may be assessed from the following:

1. Procurement

The Food Corporation of India undertakes the procurement of foodgrains, mainly cereals, on behalf of the Government of India and State Governments in the States where it has been entrusted with this responsibility either as a sole agency or jointly with other public procurement agencies. It also undertakes massive price support operations for cereals on behalf of the Central and State Governments to protect the interests of the growers. It prevents distress sales by ensuring to the farmers, predetermined support prices. It also handles huge stocks of foodgrains procured by other agencies for the central pool, and utilizes the services of cooperative societies to the maximum extent possible.

The Food Corporation of India purchases cereals from producers during both the seasons, directly or through the agency of cooperatives or purchasing agents, and from rice millers under various arrangements of procurement determined by different State Governments. The quantities procured in different years by the Food Corporation of India are given in Table 7.21.

Table 7.21: Procurement of foodgrains by Food Corporation of India

(thousand tonnes)

Calendar year	Rice	Wheat	Coarse grains	Total
1965	2951	375	705	4031
1970	3043	3183	488	6714
1975	5042	4098	423	9563
1980	5210	5866	102	11178
1985	9568	10355	184	20107
1990	12792	11094	105	23991
1995	9997	12327	–	22324
2000–01*	20824	16356	–	37180
2005–06*	27656	14785	1150	43591
2010–11*	34198	28335	128	62661
2015–16*	34218	28088	260	62566
2018–19*	44399	35795	86**	80280

* Figures are for marketing year (October–September for rice and April–March for wheat)
** For 2017–2018
Source:
1. Govt. of India, Department of Food, New Delhi, FCI website.
2. Govt. of India, Agricultural Statistics, At a Glance, Various Issues, New Delhi.

A glance at the procurement of foodgrains by the Corporation would show the vital and effective role it has played in the national economy. Commercial purchases of some commodities, viz., cereals and pulses, are also made by the Corporation at market prices with a view to supplying them to the defence services.

2. Storage

The provision of adequate and proper scientific storage facilities for foodgrains from the time of procurement till their distribution is another important function performed by the Corporation. Its responsibility for storage has increased with the transfer to it of the responsibility for building up a buffer stock by the government. Foodgrains are stored in godowns which are scientifically constructed for protection against dampness, rats and fungus.

Till the beginning of 1968, there had been a more or less complete ban on the construction of new godowns. With the onset of the green revolution, there was an urgent necessity of augmenting substantially the storage facilities for foodgrains at the production and consumption centres. The Corporation, therefore, launched a crash programme for the construction of godowns. It also encouraged private parties to construct modern foodgrain godowns on a guaranteed occupation basis. Constant and effective inspection and treatment of foodgrains in storage ensures that the stocks are kept in good condition. New and cheap methods for the preservation of stocks have been developed by the technical experts of the Corporation. Storage losses in FCI godowns have been brought down to less than 1% as against its former very high percentage loss (up to 10%). The activities undertaken by the FCI for this purpose are:

1. It has adequate storage capacity—well-built godowns, silos and CAP (Cover and Plinth) located at strategic points near the production and consumption centres and major ports. CAP storages are in large open areas and are scientifically planned to hold thousands of bags of grains under polythene covers. (It includes both owned and hired).
2. The FCI has taken over the construction of silos in order to switch over to the bulk handling and storage of foodgrains in a phased manner. Silos are tall and massive structures with huge storage facilities and facilities for mechanical handling. As a result, the losses arising out of handling are reduced. Handling charges, too, are brought down. The construction of silos has been taken up in Punjab, Haryana, M.P., Bihar, West Bengal and other states. It hires storage space from CWC, SWCs, state governments and private warehouses and silos (details in Chapter 4).
3. The FCI uses air strips, army barracks and former palaces for the storage of foodgrains during the massive procurement season.
4. The FCI has a chain of 138 quality control laboratories which develop quality control measures to ensure the safe storage of foodgrains. In addition, scientists, technicians and workers air, rotate and fumigate stocks at regular intervals so that quality does not deteriorate.

3. Movement

The Food Corporation of India organises swift and massive movement of foodgrains, mainly by rail and road, to ensure timely arrivals in the areas of consumption and of storage. This activity of the Corporation enables it to maintain a steady public distribution system—from the procurement centres and the ports to the areas of consumption and storage without any serious difficulty. It is one of the largest users of the railways. The quantity transported by rail and road during 1966-67 was only 1.238 million tonnes, which increased to an average of more than 40 million tonnes during the last five years.

FCI undertakes movement of cereals in order to (a) evacuate stocks from surplus regions; (b) meet the requirements of deficit regions for NFSA/TPDS and other schemes; and (c) create buffer stock in the deficit areas. The movement plan is prepared every month involving 1906 FCI owned and hired depots/silos, 557 rail heads and 98 FCI's own sidings. Nearly 85% is moved by railways, rest by roads, and small quantity by ocean vessels (Islands) or riverine movement.

4. Imports/Exports

During the last 30 years, the imports of cereals has not been a regular phenomenon. Only in some years, the need for importing wheat arises. Whenever it happens, the Food Corporation of India handles the imported wheat on government account. This responsibility was entrusted to it by the government in 1969-70. The imported foodgrains are speedily despatched to various destinations to avoid congestion at the ports and to augment supplies to the public distribution system. FCI undertakes exports as a part of food aid or donations to other countries and sometimes small commercial exports.

5. Distribution

Another important function of the Corporation is the distribution of procured/ imported foodgrains through nearly 5.33 lakh fair price shops all over India. Foodgrains are issued on the basis of the allocations made by the Central Government. The Food Corporation of India makes foodgrains available to the vast majority of population at reasonable prices. The quantity of foodgrains distributed through public distribution and open sales has varied between 53 and 63 million tonnes during the last 5 years. FCI handles the largest cereal supply chain system of Asia.

6. Processing

The Food Corporation of India also has made notable strides in the field of food processing. It has acted as a pace-setter in the modernization of food processing operations. It had set up 24 modern rice mills in different States to increase the availability of rice and extract oil from rice bran. It had also set up a Paddy Processing Research Centre at Tiruverur in Tamil Nadu in collaboration with the Government of Tamil Nadu and the Union Ministry of Agriculture with a view to evolving a new technology for increasing the outturn of rice at rice mills, better utilization of bran for the extraction of edible oil and proper use of by products. A solvent extraction plant at Sembanarkoil (Tamil Nadu) was also set up for the manufacture of edible and industrial grade oil from rice bran. These have served as models for private interests in this line to set up such mills elsewhere.

The Corporation had set up paddy dryers in Thanjavur district in Tamil Nadu and a maize dryer at Khanna in Punjab to dry the grain and transport it to other districts without any damage by quick sprouting diseases which break out because of high moisture content. The FCI had also set up a solvent extraction plant at Ujjain (M.P.) to process groundnut. The FCI had set up a maize mill at Faridabad (Haryana) to manufacture a variety of maize products. It has set up a dal mill at Lucknow (U.P.) to meet the purchase requirements of the army. The FCI also produces a protein-rich food (Balahar), a midday meal for school children. Balahar is a mix of wheat flour, groundnut meal, vitamins and minerals. These initiatives of FCI encouraged private investment in food processing sector.

7. Consultancy

The Food Corporation of India has also taken up the function of consultancy service, and provides technical and scientific assistance to other public and private undertakings as well as cooperatives in the country and abroad. The consultancy service offers assistance in the modernization of rice and dal mills and other agro-processing units. The service includes the conduct of feasibility and technoeconomic studies, management systems and optimization studies, and market surveys.

8. Management of Levy Sugar

The corporation was also doing the collection and management of levy sugar on behalf of the Government of India, till recently.

COTTON CORPORATION OF INDIA (CCI)

The Cotton Corporation of India as a public sector agency was set up by the Ministry of Commerce, Government of India, in 1970. Now, it is under the Ministry of Textiles. It was done for handling of imports, purchasing domestic cotton to safeguard the interests of growers and consumers, imparting the needed stability to cotton prices in the long run and maintain the supplies to government and private textile mills. The corporation entered in the domestic market in year 1971–72. Since then it is responsible for price support operations of raw cotton in all the states except Maharashtra, where, for long, a state federation had been performing this job.

    The purchases of cotton by Cotton Corporation of India and the Maharashtra State Cooperative Cotton Growers Marketing Federation (MSCCGMF) during 1991–92 to 2013–14 are shown in Table 7.22. The Cotton Corporation of India (CCI) makes price support as well as commercial purchases to provide price support to the cotton growers except in Maharashtra State. During the last 35 years, the Corporation purchased 3.48 lakh bales to 89.39 lakh bales of raw cotton from the farmers. The cotton is purchased mainly from the states of Gujarat, Andhra Pradesh, Madhya Pradesh, Rajasthan, Punjab, and Haryana. In Maharashtra till recently, the state government did not allow the traders to buy cotton from farmers. The Maharashtra State Cooperatives Cotton Growers Marketing Federation (MSCCGMF) was designated as the state agency to buy all the cotton offered by the farmers at pre-decided prices. The MSCCGMF has been purchasing considerable quantities of raw cotton from the farmers during these years. The purchase of cotton by these two agencies together accounted for around 20 to 30% of the marketed surplus of cotton. Purchases of cotton by the CCI have helped in creating an assured market for the cotton growers and increased competition in the market. However in Maharashtra, government monopoly purchase came in the way of efficient functioning of the raw cotton market.

    Cotton Corporation of India undertakes price support purchases (to provide support to farmers) as well as commercial purchases (for supplying to the mills). As shown in Table 7.23, these two kinds of purchases are, usually inverse, i.e. when MSP purchases are more, commercial purchases are less and vice versa. The state-wise purchases by CCI during 2015–16 to 2017–18 are shown in Table 7.24.

Table 7.22: Purchases of raw cotton by CCI and MSCCGMF

(thousand bales of 170 kg. each)

Cotton/crop year (Sept–August)	CCI	MSCCGMF	Total
1991–92	1000	1063	2063
1992–93	1187	1961	3148
1993–94	776	1330	2106
1994–95	843	1100	1943
1995–96	1016	2775	3791
1996–97	1120	3127	4247
1997–98	812	1118	1930
1998–99	430	NA	430
1999–00	482	25	507
2000–01	603	NA	603
2001–02	967	NA	967
2002–03	396	203	599
2003–04	573	328	901
2004–05	2624	170	2794
2005–06	1027	325	1352
2006–07	1455	NA	1455
2007–08	993	NA	993
2008–09	8939	NA	8939
2009–10	759	NA	759
2010–11	1366	NA	1366
2011–12	349	NA	349
2012–13	2335	NA	2335
2013–14	443	NA	443

CCI = Cotton Corporation of India
MSCCGMF = Maharashtra State Cooperatives Cotton Growers Marketing Federation
Source: Agricultural Statistics at a Glance, various issues, Ministry of Agriculture, Govt. of India, New Delhi.

Table 7.23: Purchase of raw cotton by CCI (2010–11 to 2017–18)

('000 bales of 170 kg each)

Year	MSP	Commercial	Total
2010–11	0	1366	1366
2011–12	8	341	349
2012–13	2287	48	2335
2013–14	41	402	443
2014–15	8696	0	8696
2015–16	845	0	845
2016–17	0	125	125
2017–18	390	681	1071

Source: Ag. Statistics at a Glance, 2015, 2018, GoI.

Table 7.24: State-wise purchase of raw cotton by CCI

		('000 bales of 170 kg each)	
States	2015–16	2016–17	2017–18
Punjab	0	0	12
Haryana	0	0	21
Rajasthan	0	0	45
Gujarat	52	8	171
Maharashtra	116	57	115
MP	29	1	19
Andhra Pradesh	40	5	73
Karnataka	0	16	81
West Bengal	*	*	*
Odisha	12	4	26
Telangana	595	34	508
Total	845	125	1071

* Less than 0.5

Source: Agri. Stat. at a Glance, 2018.

JUTE CORPORATION OF INDIA (JCI)

The Jute Corporation of India (JCI) was set up by the Government of India in the year 1971 to implement the policy of price support to jute growing farmers. The main objective of the Jute Corporation is to ensure the jute growers a reasonable price for their produce and save them from exploitation by the middlemen. The Jute Corporation undertakes purchases of raw jute from the growers at the minimum support prices so that the market prices of raw jute do not fall below the MSP level at any time. The main activities undertaken by this Corporation are price support purchases, commercial procurement, sale, export and import of raw jute. However, the Corporation is also involved in the maintenance of buffer stock of raw jute and import of sisal and manila fibres. The purchases of raw jute by Jute Corporation of India, which include price support purchases by cooperatives and commercial purchase by JCI, are shown in Table 7.25.

Table 7.25: Purchases of raw jute by Jute Corporation of India

		(thousand bales of 180 kg each)	
Year (July–June)	Purchases	Year (July–June)	Purchases
1976–77	813	2010–11	34
1980–81	1062	2012–13	365
1985–86	2825	2013–14	190
1990–91	852	2014–15	57
1995–96	Nil	2015–16	6
2000–01	464	2016–17	226
2005–06	141	2017–18	678

Source: Jute Corporation of India, Kolkata and Agricultural Statistics at a Glance, Government of India, New Delhi.

Raw jute is purchased by the Corporation from the states of West Bengal, Assam, Bihar, Orissa, Tripura, Andhra Pradesh and Meghalaya. However, more than two thirds of the total jute purchases is only from the state of West Bengal. The purchases of Jute by Jute Corporation of India from different States during the last 20 years are shown in Table 7.26. The presence of JCI has helped the jute growers in realising better prices for their produce.

Table 7.26: Purchases of jute by Jute Corporation of India in states

(thousand bales of 180 kg each)

Year	West Bengal	Assam and Meghalaya	Bihar	Andhra Pradesh	Orissa and Tripura	Total
1997–98	600	171	134	53	27	985
2000–01	277	113	72	1	1	464
2001–02	159	48	39	–	*	246
2002–03	880	167	177	68	22	1314
2003–04	742	121	198	41	16	1118
2004–05	312	10	26	–	3	351
2005–06	105	25	11	–	–	141
2006–07	308	118	55	2	1	484
2007–08	503	90	124	28	11	756
2010–11	21	11	2	–	–	34
2011–12	108	18	31	–	2	159
2012–13	271	44	41	6	3	365
2013–14	89	66	33	1	1	190
2014–15	41	3	13	0	*	57
2015–16	2	3	0	0	1	6
2016–17	97	82	23	23	1	226
2017–18	399	193	76	0	10	678

* Negligible

Source: Agricultural Statistics at a Glance, Various Issues, Ministry of Agriculture, Government of India, New Delhi.

AGRICULTURAL AND PROCESSED FOOD PRODUCTS EXPORT DEVELOPMENT AUTHORITY (APEDA)

The Agricultural and Processed Food Products Export Development Authority (APEDA) was established by the Government of India under the Ministry of Commerce on February 13, 1986 under the Agricultural and processed Food Products Export Authority Act, 1985. The main responsibility of APEDA is the export promotion of fruits and vegetable products, meat and meat products, poultry products, dairy products, confectionery, biscuits and bakery products, honey, jaggery, sugar and coca products, alcoholic and non-alcoholic beverages, pickles, chutneys, papads, cereals (non-basmati rice) and other processed foods.

The main objectives of establishing APEDA are:

1. To maximise foreign exchange earnings through increased agro-exports for providing higher incomes to the farmers through higher unit value realization;

2. To create employment opportunities in rural areas by encouraging value added exports of farm products; and

3. To implement schemes for providing financial assistance to improve post-harvest facilities to boost their exports.

The APEDA has brought qualitative changes in the agricultural marketing system and environment and, therefore, has increased the credence of the agri-business in the products. The examples of these may be export of grapes and mangoes from Maharashtra and Karnataka; strawberry and mushrooms from Punjab and Haryana; litchi from Andhra Pradesh, Uttar Pradesh and Bihar; and cutflowers from Delhi, Haryana and Karnataka. The efforts made by APEDA had brought significant impact on the growth of exports of its scheduled products.

The APEDA has build links between Indian producers and the global markets. APEDA has undertaken following development programms to achieve the objectives for which it has been set-up in the country:

1. Development of databases on products, markets and services;
2. Publicity and information dissemination;
3. Inviting official and business delegations from abroad and organization of product promotions and visit of official and trade delegations abroad
4. Organization of seminars, workshops and awareness programmes on exports as well as on latest farming practices;
5. Participation in International Trade Fairs in India and abroad and organization of buyer-seller meets and other business interactions;
6. Information dissemination through APEDA's newsletter, feedback series and library;
7. Providing recommendatory, advisory and other support services to trade and industry;
8. Problem solving in government agencies and organisations, RBI, and Customs related to import-export procedures;
9. Offer of financial assistance under various schemes, which seek to promote and develop agro-exports.

The activities for financial assistance from APEDA include:

1. Strengthening of market intelligence and data base through studies and surveys
2. Quality upgradation as per international standards.
3. Development of infrastructural facilities
4. Research and development
5. Development of packing quality
6. Human resource development, and
7. Upgradation of meat processing facilities

India's agricultural exports during 2017–18 were valued at ₹ 2.52 lakh crore or approximately $36 billion.

MARINE PRODUCTS EXPORT DEVELOPMENT AUTHORITY (MPEDA)

Marine Products Export Development Authority (MPEDA) was set up in the year 1972 to undertake the promotional work relating to the export of marine products. The role of MPEDA includes development of off-shore and deep sea fishing, promoting shrimp farming using latest technology, adopting measures required for diversfying export products and export market, modernizing sea food industry, including new-technology for value addition, extending marketing services and assuming quality control in fishery products. It is also vested with the responsibility of marine product industry development. There are 120 EU approved seafood processing units under MPEDA. Sea food industry accounted for exports worth ₹ 47,646 crore during 2017–18.

NATIONAL HORTICULTURE BOARD (NHB)

National Horticulture Board (NHB) was set up by the Government of India in the year 1984 as an autonomous society under the Societies Registration Act, with a mandate to promote integrated development of horticulture in the country. The specific objectives of setting up of NHB are:

1. To promote the development of horticulture industry in the country;
2. To help in coordinating, stimulating and sustaining the production and processing of fruits and vegetables;
3. To establish sound infrastructure in the field of production, processing and marketing with a focus on post-harvest management to reduce losses;
4. To assist in the establishment of growers societies to advance their economic and social status;
5. To provide technological, financial and other assistance to various market organizations; and
6. To provide market information and build data base in horticulture.

The type of infrastructure being promoted by the NHB include grading and packaging centres, pre-cooling units, cold storages, auction platforms and refrigerated transport facilities. The National Horticulture Board has formulated several innovative schemes in tune with the requirements of the industry. These schemes fall in following broad groups:

1. Development of commercial horticulture through production and post-harvest management.
2. Capital investment-subsidy scheme for construction/expansion/modernisation of cold storages for horticulture produce.
3. Technology development and transfer.
4. Establishment of nutritional gardens in rural areas.
5. Market information service for horticulture crops; and
6. Horticulture promotion service.

NATIONAL DAIRY DEVELOPMENT BOARD (NDDB)

The National Dairy Development Board (NDDB) was established in 1965 through an Act of Parliament at Anand (Gujarat) for providing facilities for increasing the production and marketing of milk and promotion of dairy industry in the country. NDDB provides market support to the producers of milk in rural areas and has been instrumental in increasing the supply of liquid milk and dairy products (ghee, powdered milk both whole milk powder and infant milk powder, butter and chocolate) to the urban consumers at reasonable prices.

Over the years, the NDDB has been successful in organising milk procuers' cooperatives and linked them to district level dairy cooperative infrastructure. The NDDB fostered strategies for increasing milk production, research on cattle feed and fodder, cooperative training and education and for the development of an efficient system of handling, processing and marketing of milk and milk products. NDDB has also been instrumental in setting up of infrastructure facilities for milk processing, transportation and marketing of milk and milk products throughout the country. In recent years, NDDB has been organizing farmers producer organizations (FPOs) in those milk producing areas where cooperatives are turning out to be rigid organizations.

NORTH-EASTERN REGIONAL AGRICULTURAL MARKETING CORPORATION (NERAMC)

The Government of India set-up the North Eastern Regional Agricultural Marketing Corporation (NERAMC) in 1982 to help faster development of agricultural marketing in North-Eastern States of the country viz., Mizoram, Manipur, Tripura, Meghlaya, Nagaland, Arunachal Pradesh and Assam. The main activities assigned to NERAMC are to promote domestic marketing, processing, post-harvest handling and exports. The Corporation acts as a central agency to implement the Central Government policies and programmes and coordinate them in these states to ensure overall development of agricultural and allied commodities marketing. The impact of the establishment of NERAMC is visible in the region by the conspicuous development of marketing and processing infrastructure and their integration with other parts of the country.

HIMACHAL PRADESH HORTICULTURAL PROCESSING AND MARKETING CORPORATION (HPMC)

The HPMC was established in 1974 in the state of Himachal Pradesh with the financial assistance of International Development Association. The primary objectives of establishing HPMC were:

1. To modernise the entire post-harvest handling of horticulture produce in the state, and
2. To curb the profiteering tendencies of private traders by introducing an element of competition.

HPMC is a service-oriented organization with a commitment to ensure remunerative returns to fruit growers and supply nutritive quality products at reasonable prices to consumers. The HPMC has contributed substantially to (a) merchandised grading and scientific packaging of fruits; (b) introduction of juice-dispensing machines; (c) manufacture of apple juice concentrates; and (d) development of sound base for apple and other fruit products. HPMC also provides services of cool-chains and marketing outlets for fresh fruits and vegetables. HPMC has its processing plants to produce juices, pulps, and concentrates. HPMC also has a well knit distributor's network and more than 400 selling booths spread all over India for sale of apple juice and other products.

COMMODITY BOARDS

The Commodity Boards are essentially the producer controlled organizations with government support and authority over a broad range of functions starting with production, processing and marketing of the crops. The commodity boards function under the purview of Ministry of Commerce, Government of India. The state governments have little control over these boards. These commodity boards are mainly confined to plantation and commercial crops in India. The commodity boards also promote both internal and external trade of the commodity. Each board deals with a specific commodity or group of commodities.

The important commodity boards established for the specific commodity or group of commodities in India along with their role are discussed below:

1. Tea Board

Tea Board is the premier organization for looking after the growth and development of Indian Tea Industry. It is a statutory organization of the Union Government established under the Tea Act of 1953. The board functions under the direction of the Commerce Ministry of the Union Government. A committee consisting of the representatives of tea producers, workers, exporters, brokers, the state Government concerned and others provides necessary feed back to the board and the decisions are taken accordingly in consultation with the government. Tea Board was not very active in the earlier decades as the main emphasis was on re-plantation programmes. Thereafter, the board started giving emphasis to increase the domestic production and exports. Tea Board advances loans and subsidies for extension, planting, replanting and refilling besides exploring the possibilities in non-traditional areas and assisting small growers. The Board also advances loans to manufacturers. The Board deals with export promotion, market intelligence and participation in fairs and exhibitions.

2. Coffee Board

As per the Coffee Markets expansion Act, 1940 the Indian Coffee Market Expansion Board was set up on December 21, 1940. The powers of this board were broadened and it was empowered to collect and market the estimated exportable surplus. In the subsequent amendments to the Coffee Market Expansion Act, the powers of the Board were extended to all the estates.

The functions assigned to the Coffee Board are:

1. Promotion of sale and consumption in India and elsewhere of Indian coffee;
2. Promotion of agricultural and technological research in the interest of the coffee industry;
3. Assistance to the coffee estates for their development;
4. Securing better working conditions and the provision and improvement of amenities and incentives for workers; and
5. Implementation of the measures enumerated in the Coffee Act relating to the operation of the surplus pool.

Market intervention in coffee has helped in maintaining price stability in the internal market and also increased the total earnings of the producers.

3. Rubber Board

The Rubber Board is a statutory body constituted under the Rubber Board Act, 1947 by the Ministry of Commerce, Government of India. The main functions of this board are to conduct research and training programmes in production and marketing of rubber, extension services and to plan for the welfare of the plantation workers.

4. Tobacco Board

Tobacco Board was established on 1st January, 1976 under the Tobacco Board Act, 1975 by the Government of India. The main functions of the Tobacco Board are regulation of production and marketing of Virginia tobacco, ensuring remunerative prices to growers by purchasing tobacco from the growers, promoting the grading by growers and recommending minimum prices to the government for export of Virginia tobacco, set up of auction platforms, and conduct of scientific research related to tobacco.

5. Spices Board

The Spices Board was established by the Ministry of Commerce, Government of India in 1987 under the Spices Board Act, 1986 at Cochin. The main objectives of establishment of Spices Board are to improve the production and quality of spices and promote export of different spices to earn foreign exchange. For achieving these objectives, the Spices Board is providing package of services oriented to different commodities and areas according to potential/scope of increase in earnings and exports.

The Spices Board covers 52 different spices crops and has made a good impact in improving export marketing of these spices to different countries.

6. Cardamom Board

The Cardamom, the queen of spices, is grown predominantly in the ever green forests of western ghats of Kerala, Karnataka and Tamilnadu. It is an important plantation crop in the domestic as well as in external trade of the- country. The Cardamom Board was established in the year 1966 under the Ministry of Commerce, Government of India to develop various promotional activities such as extension of plantations, improvement in the quality of cardamom and increasing the productivity of cardamom estates.

7. Coir Board

Coir Board was set up in 1953 by the Commerce Ministry of Government of India under Coir Board Industries Act, 1953, for development of coir industries. The Board has set up the Coir Board Research Institute at Allepey (Kerala) for conducting research on different aspects of coir industry.

8. Silk Board

Silk Board was set up in 1949 by the Government of India under the Central Silk Board Act. The main activities of the Board include development of silk industry, export-import of raw silk, increasing the production of silk, and to help in training and research for silk industry.

9. National Meat and Poultry Processing Board

This board was established in 2009 at Delhi.

10. Indian Grape Processing Board

This board was established at Pune.

▌ LEGISLATIVE MEASURES FOR IMPROVING AGRICULTURAL MARKETING

Another important form of government intervention in the marketing system of agricultural commodities is legislative measures. The legislative measures include framing of rules and regulations for protection of the interests of some or all the sections of the population. The legislative measures intended for improvement in agricultural marketing relate to regulation of markets, licensing of market functionaries, putting up of restrictions on the activities of traders, granting or banning of monopolies to one group of persons, regulating trade in domestic and foreign markets and also the protection of the consumers. A number of legislations were enacted from time to time by the central as well as the state governments. However, these were reviewed from time to time and suitable modifications/amendments were made to make them consistent with the changed marketing scenario of the country.

As an example, special mention need to be made of an order of the government in 2002 which made a sharp departure from the past regime of controls. For a long time, there were several restrictions on marketing of foodgrains. In order to facilitate the free trade and movement of foodgrains, the Government of India has issued a control order titled 'Removal of Licensing Requirements, Stock Limits and Movement Restrictions on Specified Food Stuffs Order, 2002 on February 15, 2002. This Order allows any dealer to freely buy, stock, sell, transport, distribute, dispose, acquire, use or consume any quantity of wheat, paddy/rice, coarse grains (jowar, bajra, maize, ragi), sugar, edible oilseeds and edible oils, without a licence or permit. Under this order, the state governments would require prior permission before issuing any order for regulating, by licence or permit, the storage, transport, and distribution of the specified commodities. Another recent important step taken by the government is to integrate several laws relating to food into one integrated modern food law and enacted Food Safety and Standards Act in 2006.

A comprehensive list of legislations enacted in India at different points of time for regulating the agricultural marketing system is given below (though, quite a few of these have been repealed but their salient features must be known):

LAWS AND ORDERS

1. Laws Regulating the Functioning of Agricultural Produce Markets (market charges and facilities) in States and Union Territories of India

The state Agricultural Produce Market Regulations Acts were enacted in different States and Union Territories during 1960 to 1995. Since agricultural marketing is a state subject, each state/UT enacted its own Act under the overall guidance of Directorate of Marketing and Inspection, Government of India. The details of these Acts are given in Appendix 7.1. As discussed in earlier sections a comprehensive review of the impact of these Acts on the agricultural marketing scenario was undertaken during the 1990s and based on these, reforms in these Acts were suggested to the states in 2003. Some states amended their original APMR Acts on the lines suggested in 2003 Model Act. The state-wise status of reforms as in 2006 is also shown in Appendix 7.1. However, several issues are being continuously raised by the stakeholders on the functioning of APMCs. Considering these, it was felt necessary to move away from 'regulation' to 'facilitation'. Accordingly, two new model Acts were prepared and circulated to the states in recent years. These are (a) Agricultural Produce and Livestock Marketing (Promotion and Facilitation) Act (APLM-PF), 2017 and (b) Agricultural Produce and Livestock Contract Farming (Promotion and Facilitation) Act (APLCF-PF), 2018.

2. Laws Regulating Quality, Grading and Standardization

1. The Agricultural Produce (Grading and Marking) Act, 1937: This Act provides for grading and marking of agricultural commodities. The Act authorises the central

government to frame rules relating to fixing of grade standards and the procedure to be adopted for grading the agricultural commodities included in the schedule. The Act of 1937 was amended in 1986. This amendment revised the provisions of the Act, strengthened the same with a view to promoting and protecting the interests of the consumers and made the penal provisions of the Act more deterrent and thus provided more teeth to the Act.

2. The Bureau of Indian Standards Act, 1986 (previously, the Indian Standards Institutions Certification Marks Act, 1952): Manufactured products are graded in accordance with the standards laid down by the Indian Standards Institution established under this Act and graded products, bear the ISI label. The name of Indian Standards Institution has been changed to Bureau of Indian Standards (BIS) under the Bureau of Indian Standards Act, 1986. The status and scope of activities has been enlarged with greater thrust to consumer protection, improving the level of quality of Indian products and providing larger network of testing and consultancy services.

3. The Export (Quality Control and Inspection) Act, 1963: A comprehensive legislation entitled Export (Quality Control and Inspection) Act, 1963: was enacted with a view to providing sound development of export trade through quality control and inspection and matters connected therewith. The Act came into force on January 1, 1964. Under this Act, powers were vested in the Central Government to notify commodities which shall be subject to statutory pre-shipment inspection, decide the system of inspection, quality control and certification for the notified commodities, recognize, adopt or formulate the standards for the notified commodities, prohibit the export of notified commodities, and satisfy the notified requirements. With a view to encouraging the export of agricultural commodities, there has been considerable liberalization of provisions relating to the exports and imports in recent years (for details see chapter 11).

4. The Fruit Products Order, 1955 and 1997: The Fruit Products Order, 1955 was issued under the Essential Commodities Act, 1955 with a view to regulating the manufacture, storage and sale of fruits and vegetable products. It was repealed in 2006.

5. The Meat Food Products Order, 1973: The Meat Food Products Order, 1973 was issued under the Essential Commodities Act, 1955 and was in force since July 15, 1975. This order covers the manufacture of meat food products in small factories as well as large factories under a licensing system that examines all aspects of hygiene. Under the provisions of this Order, the meat factories have their own captive slaughter houses. It was repealed in 2006.

6. The Rice Milling Industry (Regulation) Act, 1938 and amendment; Regulation and Licensing Rules, 1976: This Act regulated modernization of rice mills and safe preservation of stocks. This has now been repealed.

7. Wheat Roller Flour Mill Licensing and Control Order, 1957: This order controlled the purchase of wheat and manufacture and disposal of different kinds of wheat products of the roller flour mills and safe preservation. This order was withdrawn in 1991.

8. The Milk and Milk Products Order; 1992. (Repealed in 2006)

3. Laws Regulating Weights, Measures and Packaging

1. The Standards of Weights and Measures Act, 1958: This Act prescribes compulsory use of metric system of weights and measures in the country.

2. Jute Packaging Material Act, 1987.
3. The Standards of Weights and Measures (packed goods) Act, 1997: This Act is aimed at ensuring packing of goods of the correct weight in the packages.

4. Law Regulating the Storage and Warehousing of Agricultural Commodities

1. The Agricultural Produce Development and Warehousing Corporation Act, 1956: This Act was passed for creation of scientific storage structures in the country in 1956, which was later replaced by the Warehousing Corporation Act, 1962.
2. The Warehousing Corporation Act, 1962—In 1962, the Government of India decided to break up the Act of 1956 into two separate Acts—the National Cooperative Development Corporation Act, 1962 and the Warehousing Corporations Act, 1962. This Act came in to operation on 18th March, 1962. The Act defines the specific functions and the order of operations of central and state warehousing corporations.
3. The Warehousing (Development and Regulation) Act, 2007. This new Act of warehousing addresses the issue of encouraging private investment in warehousing, regulation of warehouses and legalizing the use of warehouse receipts (WRs) as negotiable and exchangeable instrument. The WRs contain information on quantity, quality, grade and type of commodity. The concept is simple. It unlocks stored value by giving finance secured against goods deposited in an independently controlled warehouse.
4. The Cold Storage Order, 1964 and 1980: The Cold Storage Order, 1964 promulgated by the Government under the Essential Commodities Act, 1955 had the objective of ensuring hygienic and proper refrigeration conditions in a cold store, regulating the growth of cold storage industry in a planned manner, rendering technical guidance for a scientific preservation of food stuffs in a cold store, and preventing exploitation of farmers by cold store owners. This order was replaced by a more comprehensive Cold Storage Order, 1980, which was rescinded by the Cold Storage Order, 1997 for allowing free market mechanism for demand based growth of cold storage industry.

5. Consumer Protection Acts

The important Acts in this category are:

1. The Drugs and Cosmetics Act, 1940
2. The Emblems and Names (Prevention of Improper Use) Act, 1950
3. The Prevention of Food Adulteration Act, 1954, amended in 1964, 1976, 1986 (Repealed in 2006)
4. The Drugs and Magic Remedies (Objectionable Advertisement) Act, 1954
5. The Trade and Merchandise Marks Act, 1958
6. The Monopolies and Restrictive Trade Practices Act, 1969
7. The Consumer Protection Act, 1986, 1991, 1993, 2002.
8. The Essential commodities Act, 1955
9. The Food Safety and Standards Act, 2006

6. Laws Regulating Supply, Demand and Prices

1. The Essential Commodities Act, 1955: A large number of commodities were initially covered under the Essential Commodities Act. The list under this Act was pruned drastically in recent years. The commodities which are under the purview of the Essential Commodities Act are cattle feed including cakes and concentrates, coal

including coke, automatic vehicles—parts and accessories, cotton and woolen cloths, medicines, food products including edible oilseeds and oils, iron and steel and products manufactured by them, paper including newspapers, card boards etc; petroleum and petroleum products, cotton lint, fibre and cotton seed; raw jute and jute goods, fertilizers—organic, inorganic and mixed ones; thread manufactured from cotton, exercise books, insecticides, pesticides and such other chemicals; seeds of food crops, horticulture crops, fodder crops, jute and onion.

2. The Forward Market Control (Regulation) Act, 1952: This Act was enacted with a view to regulating forward contracts and dealing with certain other related matters. The FMC was merged with SEBI in 2015.
3. The Prevention of Black Marketing and Maintenance of Supplies of Essential Commodities Act, 1980.
4. The Indian Sale of Goods Act, 1930.
5. Constitutional Amendment Act, 2016 (related to GST).
6. The Prevention of Cruelty to Animals Act, 1960 (and cattle trade and slaughter house regulations in different states).

7. Laws for Setting up of Commodity Corporations

1. The National Cooperative Development Corporation Act, 1962
2. The Food Corporation of India Act, 1964
3. The Cotton Corporation of India Act, 1970
4. The Jute Corporation of India Act, 1971
5. The Horticultural Processing and Marketing Corporation Act (H.P. and J & K State), 1974
6. The North Eastern Regional Agricultural Marketing Corporation (NERAMAC) Act, 1982

8. Laws for Constituting Commodity Boards

1. The Coffee Market Expansion Act, 1940
2. The Rubber Board Act, 1947
3. The Silk Board Act, 1949
4. The Tea Board Act, 1953
5. The Coir Board (Industries) Act, 1953
6. The National Dairy Development Board Act, 1965
7. The Tobacco Board Act, 1975
8. The National Horticulture Board Act, 1984
9. The Spices Board Act, 1986

9. Other Control Orders

1. Solvent Extracted oil, De-oiled Meal and Edible Flour (Control) Order, 1967 (repealed in 2006)
2. Sugar (Movement Control) Order, 1959
3. Sugarcane (Press-mud Control) Order, 1959
4. The Gur (Movement Control) Order, 1963
5. Sugarcane Control Order, 1966
6. Sugar Control Order, 1966
7. The Sugar (Packaging and Marking) Order, 1970
8. The Sugar (Restriction on Movement) Order, 1972

9. The Levy Sugar Supply (Control) Order, 1979
10. Vegetable Oil Products (Standards of Quality) Order, 1975 (repealed)
11. Vegetable Oil Products (Control) Order, 1977 (repealed in 2006)
12. Pulses, Edible Oilseeds, Edible Oil (Storage control) Order, 1977
13. Cotton (Control) Order, 1986
14. Jute and Jute Textiles (Control) Order, 2000.
15. The Edible Oils Packaging (Regulation) Order, 1998 (repealed in 2006)

10. Acts Related to Agricultural Inputs

1. **Indian Seeds Act, 1966:** This Act aims at regulating the quality of seeds sold for the purpose of agriculture and envisages compulsory labelling and voluntary certification of seeds sold.
2. **Insecticides Act, 1968:** This Act provides for compulsory registration of all insecticides. The Act regulates import, manufacture, transport, storage, sales and use of all pesticides and insecticides in the country.
3. **Fertilizer Control Order, 1957, 1985:** This was promulgated under the Essential Commodities Act. There has been many changes in the order in later years. The basic objective of issue of this order is to regulate the manufacture, sale, distribution, price and quality of fertilizers.
4. **The Fertilizer (Movement Control) Order, 1973:** The basic objective of this Order is to ensure equitable distribution of fertilizers in the states of India.
5. **The Seeds Control Order, 1983**.
6. **The Protection of Plant Varieties and Farmers' Rights (PPV & FR) Act, 2001**.

SALIENT FEATURES OF SOME MARKETING ACTS

Though many of these have been amended many times, and some others repealed or withdrawn, the knowledge of their basic features is relevant and important for students, researchers and scholars of agricultural marketing.

Rajasthan Agricultural Produce Markets Act, 1961 (and subsequent Amendments)

The Rajasthan Agricultural Produce Markets Act, 1961 (Act No. 38 of 1961), received the assent of the President of India on 3rd November, 1961. The Act provides for the better regulation of buying and selling of agricultural produce and the establishment of markets for agricultural produce in the State of Rajasthan. It extends to the whole State.

The salient features of the Act are:

1. The State Government may, by notification in the Official Gazette, declare its intention to regulate the purchase and sale of such agricultural produce and in such areas as may be specified in the notification;
2. The government may notify market areas for the purposes of this Act. For each market area, there shall be one principal market yard and one or more sub-yards;
3. The State Government shall establish a market committee (Mandi Samiti) for every market area for the agricultural produce for which it is declared to be in that area. Every market committee shall be constituted as prescribed under the rules and shall consist of 15 members representing various groups. The term of office of market committee shall be three years. The first market committee would be

nominated by the government for a period of two years. Thereafter elections would be held.

4. No person shall make or recover any trade allowance other than the allowance prescribed by the rules or bye-laws in any market area for any transaction in respect of the agricultural produce concerned.

5. The secretary of the market committee or any officer of the government authorised in this behalf may, for carrying out of any of the duties imposed on the market committee under this Act, inspect at all reasonable times all accounts, registers and other documents pertaining to the purchase and sale of notified agricultural produce and enter any shop, godown, factory or other places where such account books or registers or documents or such goods are kept.

6. The market committee (Mandi Samiti) may, in respect of the market area under its management, make bye-laws for the regulation of the business and conditions of trading therein.

7. The notified agricultural commodities for sale and purchase in the market area are given in the schedule under the Act. The State Government, may, by notification in the Official Gazette, add, amend or cancel any of the items of agricultural produce specified in the schedule.

Agricultural Produce (Grading and Marking) Act, 1937

This act provides for the grading and marking of agricultural (and other) produce and extends to the whole of Indian Union.

The salient features of the Act are:

1. The Central Government may, after previous publication by notification in the Official Gazette, make rules for fixing grade designations to indicate the quality of any scheduled article, defining the quality indicated by every grade designation, specifying grade designation marks to represent particular grade designations, and authorising a person or a body of persons subject to any prescribed conditions, to mark, with a grade designation mark, any article in respect of which such mark has been prescribed, or any covering containing, or label attached to any such article.

2. Anyone who marks any scheduled article with a grade designation mark when he is not authorised to do so by the rules made under Section (1) shall be punishable with fine.

3. Whoever counterfeits any grade designation mark or has in his possession any die, plate or other instrument for the purpose of counterfeiting a grade designation mark shall be punishable with imprisonment or with fine, or with both.

4. Extension of application of Act: The Central Government, after such consultation as it thinks fit with the interests likely to be affected, may, by notification in the Official Gazette, declare that the provisions of this Act shall apply to an article of agricultural produce not included in the schedule (or to an article other than an article of agricultural produce) and on the publication of such notification, such article shall be deemed to be included in the schedule.

General Grading and Marking Rules, 1937

1. These rules may be called the General Grading and Marking Rules, 1937.

2. Any person or body of persons desirous of being authorised to mark any article with a grade designation mark shall apply to the Agricultural Marketing Adviser to the Government of India, Faridabad.

3. If, after due enquiry, the Agricultural Marketing Adviser, or any person duly authorised by him in this behalf, is satisfied that it is expedient in the interest of better marketing that the authorisation be granted, and that the applicant is a fit and proper person to receive a Certificate of Authorisation, he shall issue such a certificate to the applicant. Each certificate shall state the name and address of the authorised person or body of persons, the articles to which grade designation marks will be applied, the period for which the certificate will be valid; and the premises at which the grade designation marks may be applied.

4. It shall be a condition of every certificate of authorisation that during the operation of the certificate, the holder thereof shall, at all reasonable times, give access to the premises named therein to any person duly authorised by the Agricultural Marketing Adviser or by the Central Government, and shall afford him facilities for ascertaining that marking process is being correctly performed. The holder of the certificate shall keep a record of the number of packages marked with each grade designation mark. Any certificate of authorization may be cancelled, revoked, modified, or suspended by the Agricultural Marketing Adviser, or by any other person authorized by the Central Government in that behalf, provided that 14 days, notice in writing has been given to the certificate-holder at the address stated on the certificate, and an opportunity has been given to him to show cause why his certificate should not be cancelled, revoked, modified or suspended.

The Cold Storage Order, 1964, 1980, 1997

The Cold Storage Order, 1964, was passed by the Central Government in exercise of the powers conferred on it by Section 3 of the Essential Commodities Act, 1955 (Act 10 of 1955), and came in force on 1st January, 1965. The salient features of the Order were:

1. No person shall carry on the business of storing of foodstuffs in a cold storage except under, and in accordance with the terms and conditions of a valid licence.

2. Every person, desiring to obtain a licence, shall make an application in duplicate to the Licensing Officer, together with the fees prescribed there of and a copy of the blueprint of cold storage building plan.

3. In granting or refusing a licence, the Licensing Officer shall have regard to the following matters:

    (a) The number of cold storage units operating in the locality where the cold storage, in which the foodstuffs are proposed to be stored, is located;

    (b) Refrigeration conditions of the cold storage in which the foodstuffs are proposed to be stored; and

    (c) Any other matter which the Licensing Officer may consider necessary for the purpose.

4. Cancellation of Licence: The Licensing Officer may, after giving the holder of the licence an opportunity to show cause and after giving him three months notice, cancel any licence granted to him for any breach of the terms and conditions of the licence, or for any contravention of the provisions of this order.

5. Every licence shall, unless previously cancelled, expire on the 31st day of December.

The Cold Storage Order, 1980 was promulgated to replace the earlier order of 1964. This order was applicable all over the country except in the States of Uttar Pradesh, West Bengal, Punjab and Haryana which have promulgated their own state orders for this purpose. This order in addition to ensuring hygienic and proper refrigeration conditions in a cold store also has a provision of rendering technical guidance for a scientific preservation of foodstuffs in a cold store and preventing exploitation of farmers by cold store owners. This Order was replaced by another Order of 1997.

Fruit Products Order, 1955

The Fruit Products Order, 1955 was issued under Section 3 of the Essential Commodities Act, 1955. Under the provisions of this order, the manufacture of all the fruit and vegetable products was to be done under the sanitary and hygienic conditions specified for the manufacturing units in accordance with the standards of quality specified for each of the products. Each package of the product is sealed, labelled and bears Food Products Order number as an indication mark.

Under the Act, a licence from the Agricultural Marketing Adviser was to be obtained to manufacture fruit and vegetable products. This licence was issued only if the factory meets the requirement of hygiene and sanitation. The specifications for preservation and the colour that can be used in these products were laid out. The specification for the final products were also laid down and no product which does not conform to the prescribed specifications could be sold in the market. An expert body, the Central Fruit Products Advisory Committee advised the Central Government on these matters. This Order was repealed in 2006.

Meat Food Products Order, 1973

The Meat Food Products Order, 1973 promulgated under Section 3 of the Essential Commodities Act, 1955 became operative in July, 1975. This aims at hygienic production and maintenance of control in slaughter houses, ensuring proper ante-mortem examination of animals and post-mortem inspection of carcasses. The Order prescribed the sanitary and other requirements to be observed in all establishments as well as the conditions to be observed by the manufacturers in handling, storage, transport, packing, marking and labelling of containers, and in the use of flavouring/colouring material and preservatives. Meat food products were subjected to laboratory testing to ensure that they are free from poisonous metals. This Order was repealed in 2006.

Prevention of Food Adulteration Act, 1954

The Prevention of Food Adulteration Act, 1954 aimed at achieving the objective of protection of the consumers against the sale of adulterated or sub-standard articles in the market. The minimum quality standards were laid down for the enlisted food items under this Act. Enforcement staff have powers to draw the samples of food items from the manufacturer to retailing outlet and get them tested by the Public Analyst. In case the sample was found to be sub-standard or adulterated, the lot was seized and prosecution is initiated against the party which is liable to be punished upon conviction. This Act had been a powerful deterrent against the manufacture and sale of adulterated or sub-standard food products.

The Prevention of Food Adulteration Act, 1954 was promulgated with effect from June 1, 1955. The Act was further amended in 1976. The major objective of enacting this Act was to protect the public from harmful and poisonous food. With the passing of this Act, a uniform food law was prescribed for whole of the Indian Union. The Act set out certain basic requirements of food quality such as absence of extraneous matter, preparation and handling under sanitary conditions, use of certified food additives, preservatives and synthetic colours, packaging in sound containers, proper branding and declaration of the quantity of contents in the package. The Act provided for the local authority of food (health) to prohibit the sale of any article of food in the interest of the public health. The Central Committee for Food Standards (CCFS) was functioning since 1941 under the aegis of the Ministry of Health and Family Welfare. The function of this committee was to advise the Central and State Governments on matters arising out of the administration of the Prevention of Food Adulteration Act. The CCFS consisted of the representatives of the central and state governments and certain research institutes, besides representatives of agricultural, commercial, industrial and consumer interests. There were 85 public analysis laboratories in the country which were administered by the central/local bodies/state governments. In the event of the report of the food sample given by a public analyst appearing to be erroneous, there was a provision in the PFA Act to get the second part of the sample reanalysed by another public analyst in the country. This Act was repealed in 2006.

GOODS AND SERVICE TAX (GST)

Imposition of taxes by the government is another form of intervention in the market. The taxes become part of the marketing cost and thus increase the price spread between the farmers and consumers. The GST regime, which is one nation-one tax system, was introduced in India in July 2017. Prior to this, there were several types of taxes levied by the centre as well as by states and local self government (LSG) institutions, and these varied from state to state and place to place. There used to be long queues at check posts on the state borders for carrying the goods from one state to the other and also for entry into a particular municipality or city boundaries. The new GST system had subsumed 17 such indirect taxes and multiple cesses, reducing compliance burden for traders and eliminating cascading effect of taxes. The system provides for one administration, uniform law and procedure, common classification of goods and services, and a level playing field for all producers/traders across the country.

Entire system, including decision on rates for goods and services, is overseen by a GST council, chaired by Union Finance Minister, with all the state Finance Ministers as members. Currently there are mainly three rates viz:, 5, 12, and 18% (of course the other is zero rate). Almost all agricultural commodities like cereals, pulses, oilseeds, fresh fruits and vegetables, milk, curd, Jaggery, and cheese are exempted. The processed and branded food products are either in 5% or 12% category. The rates continue to be revised from time to time by the GST council.

REFERENCES

1. Acharya has made several suggestions in this regard. See for example, Acharya, S.S., "Liberalization of Agricultural Produce Markets and Role of Marketing Organizations", Keynote address delivered at the 11th National Conference on Agricultural Marketing Organized by Tamil Nadu State Agricultural Marketing Board at Chennai, September 17–19, 1997. The address was subsequently published in the Indian Journal of Agricultual Marketing.
2. For a more detailed discussion on evolution of agricultural price policy in India and on formulation of price policy, see Acharya S.S. and N.L. Agarwal, Agricultural Prices; Analysis and Policy, Oxford and IBH Publishing Co. Pvt. Ltd., New Delhi, 1994.

3. Acharya, S.S. Indian Agriculture and Food Security: Current concerns and Lessons, Asian Journal of Agriculture and Development, Vol.7(1), Manila, Philippines, 2010.
4. Acharya, S.S., Food Security and Indian Agricultural Policies, Production and Marketing Environment, Agricultural Economics Research Review, Vol. 22(1), June 2009.
5. Government of Karnataka, Karnataka Agricultural Marketing Policy, 2013.
6. National Academy of Agricultural Sciences (NAAS), "Loan Waiving versus Income Support Scheme: Challenges and Way Forward," Policy Paper 91, November 2019.

APPENDIX 7.1

AGRICULTURAL PRODUCE MARKET (REGULATION) ACTS OF DIFFERENT STATES AND UNION TERRITORIES OF INDIA (AS IN 2006)

State/UT	Title	Remarks (whether reformed)
1. Andhra Pradesh	The Andhra Pradesh Agricultural Produce and Livestock Markets Act, 1966 (AP Act 16 of 1966)	Reformed
2. Assam	The Assam Agricultural Produce Markets Act, 1972 (Assam Act 23 of 1974)	Reformed
3. Arunachal Pradesh	The Arunachal Pradesh Agricultural Produce Marketing (Regulation) Act, 1989 (Act 6 of 1990)	Reformed
4. Bihar	The Bihar Agricultural Produce Market Act, 1960 (Bihar Act 16 of 1960)	Repealed in 2006
5. Gujarat	The Gujarat Agricultural Produce Markets Act, 1963 (Act 20 of 1964)	Reformed
6. Haryana	The Punjab Agricultural Produce Markets Act, 1961 (Act 23 of 1961)	Reformed
7. Himachal Pradesh	The Himachal Pradesh Agricultural Produce Markets Act, 1969. (Act 9 of 1970)	Reformed
8. Karnataka	The Karnataka Agricultural Produce Marketing (Regulation) Act, 1966 (Act 27 of 1966)	Reformed
9. Kerala	The State has not yet enacted the Agricultural Produce Marketing (Regulation) Act	However, Madras Commercial Crops Act, 1933 is in operation in four markets of Malabar Sea area of the state
10. Madhya Pradesh	The Madhya Pradesh Krishi Upaj Mandi Adhiniyam, 1972 (Act 24 of 1973)	Amended in 1979, Reformed
11. Maharashtra	The Maharashtra Agricultural Produce Marketing (Regulation) Act, 1963 (Act 20 of 1964)	Reformed
12. Manipur	No Act	
13. Meghalaya	The Meghalaya Agricultural Product Markets Act, 1980 (Act 1 of 1980)	Action for Reforms
14. Mizoram	The Mizoram State Agricultural Produce Marketing (Regulation) Act, 1996 (Act 11 of 1996)	Reformed
15. Odisha	The Orissa Agricultural Produce Markets Act, 1956 (Act 3 of 1957)	Reformed
16. Nagaland	The Nagaland Agricultural Produce Marketing (Regulation) Act, 1985 (Act 1 of 1989)	Reformed

State/UT	Title	Remarks (whether reformed)
17. Punjab	The Punjab Agricultural Produce Markets Act, 1961 (Act 23 of 1961)	Reformed
18. Rajasthan	The Rajasthan Agricultural Produce Markets Act, 1961 (Act 38 of 1961)	Reformed
19. Sikkim	The Sikkim Agricultural Produce Market Act, 1993 (Act 1 of 1993)	Reformed
20. Tamil Nadu	The Tamil Nadu Agricultural Produce Markets Act, 1959 (Act 23 of 1959)	Reformed
21. Tripura	The Tripura Agricultural Produce Markets Act, 1979 (Act 15 of 1983)	Reformed
22. Uttar Pradesh	The Uttar Pradesh Krishi Utpadan Mandi Adhiniyam, 1964 (Act 25 of 1964)	Action for Reforms
23. West Bengal	The West Bengal Agricultural Produce Marketing (Regulation) Act, 1972 (Act 35 of 1972)	Action for Reforms
24. Chandigarh	The Punjab Agricultural Produce Markets Act, 1961 (Punjab Act 23 of 1961)	Reformed
25. Delhi	The Delhi Agricultural Produce Marketing (Regulation) Act, 1976. (Act 87 of 1976)	Reformed
26. Goa	The Maharashtra Agricultural Produce Marketing (Regulation) Act, 1963 (Maharashtra Act 20 of 1964)	Reformed
27. Pondicherry	The Pondicherry Agricultural Produce Markets Act, 1973 (Act 3 of 1974)	Action for Reforms
28. Uttarakhand	NA	Reformed

Market Regulation Act has not been enacted in Jammu & Kashmir; Andaman & Nicobar Islands; Dadra & Nagar Haveli; Lakshadweep; and Kerala.

APPENDIX 7.2

SALIENT FEATURES OF KARNATAKA AGRICULTURAL MARKETING POLICY, 2013

OBJECTIVES

1. Creation of a market structure that is transparent and equitable, distinguishes quality and variety, disseminates relevant market information to all market participants for level playing field, and ensures fair returns to all stakeholders, with the seller having the choice to decide the time, place and avenue of sale.
2. Reduction in and/or elimination of barriers to participation in markets, to foster competition and efficient determination of price, and linking the primary markets in the state to the national market for the benefit of all stakeholders in the marketing chain.
3. Address the risks associated with clearing and settlement that arise in the course of marketing of produce by the farmer or by the subsequent buyer, through technology solutions or other appropriate means with linkages to financial institutions.
4. Encourage and promote primary value addition through aggregation, grading and packaging at the village level through farmers self help groups, societies,

associations or producer companies to respond to increasing and changing market demand in domestic, regional and international markets and create awareness on adherence to quality standards for better price realization.

5. Enhancing the skill levels of all stakeholders in the marketing system through well designed capacity building intervention efforts, for deriving benefits arising from primary value addition, modern practices adopted in storage, processing, market systems and the like, with the state playing an active role in stakeholder education effort.

6. Improving access to finance to all market participants, scientific storage and preservation of commodities, encouraging investment in infrastructure for market access, and enabling primary value addition to the commodities.

7. Encouraging setting up of new institutions and strengthening of existing ones to provide state-of-the-art facilities to all stakeholders, fostering self-help groups and cooperative ventures for improving the bargaining and holding capacity of farmers to handle the challenges in marketing.

8. Establishment of a progressive regulatory environment, distinct from the operational level, that promotes public and private initiatives to function in tandem for the benefit of all stakeholders, with the government playing an enabling policy making role.

9. Adopting and leveraging technology at all levels for efficient operations of markets that would be critical in realizing the objectives of this policy.

10. All farmer-friendly provisions of the Karnataka Agricultural Produce Marketing (Regulation and Development) Act, 1966 shall be implemented to safeguard the interests of the farmers.

POLICY INITIATIVES/INSTRUMENTS

In pursuit of the above objectives, the key initiatives of the government would be as under (we prefer to call these Policy Instruments):

1. Regulated Markets
   (a) These would adopt technology required for setting up a comprehensive electronic auction system for transparent price determination.
   (b) A state-wide networked virtual market would be established by linking various regulated markets and warehouses, which will be provided with assaying and grading facilities and other necessary infrastructure.

2. Increasing Competition
   (a) Licensing procedure would be simplified and single unified licence would be made applicable.
   (b) Conditions that restrict participation would be removed to increase competition in the auction of the agricultural produce.
   (c) Administrative processes with regard to licensing would be simplified and automated.

3. Private Markets
   (a) Private markets would be encouraged on a level playing field for providing an alternative, while being part of the networked market.
   (b) Farmers and sellers would have choice to sell in any regulated or private market.
   (c) Private markets and regulated markets would compete to provide efficient services.

4. Quality Standards for Demand Creation
   (a) Steps for laying down quality standards and for creating infrastructure for sampling and assaying the produce in markets and warehouses to facilitate quality-based trading.
   (b) Undertake capacity building programmes to create awareness on quality standards and its importance for creating demand in domestic, regional and international markets.
5. Empowering Farmers
   (a) Organize farmers groups for primary value addition, encourage village aggregation, and to enhance bargaining power of farmers.
   (b) Accreditation of warehouses and encouraging warehouse-based sales.
   (c) Create an enabling environment for farmers to avail pledge loans.
   (d) Effective dissemination of market price information
   (e) Simplified processes and online timely payment to the farmers' accounts.
   (f) Enabling farmers to decide when to sell, and at what price, with a right to reject the price offered.
6. Market Development
   (a) Provision of state-of-the-art technology for marketing of produce, through PPP arrangement.
   (b) Facilitate and develop an efficient and effective agriculture marketing information system.
   (c) Establish linkages with secondary markets, encourage producer companies, aggregation of farmers produce and other measures for realization of better prices.
   (d) Linkages with financial institutions to facilitate seamless clearing and settlement mechanism as well as facilitate pledge loans to farmers and others.
   (e) Put all safeguards to ensure timely online payment to farmers.
7. Contract Farming: Promote contract farming and have a single point for registering contract -farming sponsors with procedures for timely settlement of disputes.
8. Infrastructure
   (a) Development of key and strategic agricultural marketing infrastructure.
   (b) Put in place an enabling and conducive environment for private sector and other stakeholders' investment in agricultural marketing infrastructure with appropriate incentives.
9. Regulatory and Legislative Environment

The Karnataka Agricultural Produce Marketing (Regulation and Development) Act 1966 would be reviewed to facilitate the achievement of aforesaid policy objectives, and implementation of proposed action plan.

Source: Government of Karnataka, 2013.

8

Cooperation and Cooperatives in Agricultural Marketing

▌COOPERATION

MEANING

The word cooperation is composed of two Latin words 'Co' and 'opus'. The word 'co' means to work and 'opus' means together. Cooperation thus means working together or jointly. In simple words, the cooperation can be defined as a form of association of people to work together in order to achieve a particular end but the scope of cooperation is very wide. Cooperation is a philosophy, a ruling principle of our social, educational and industrial life.

Cooperation stands for voluntary association of members for some economic and/ or social objective. In brief, it is self help made through effective organization. Cooperation is based on the principle of help of people especially marginalized individuals who are powerless and to provide benefits (economic/social/moral) to them equal to rich, powerful and resourceful persons, through their joint efforts.

Cooperation has the following three elements:

1. Coming together of individuals on voluntary basis;
2. Equality among members irrespective of capital contributed (one vote for each); and
3. An honest economic and social objective.

The cooperative associations resemble the private business corporations in its method of operation. Principally, as a business institution, the cooperatives must succeed on their own merits. It should not seek any patronage but should receive the same fair treatment as accorded to other business concerns by the government. Basically, the cooperatives are the economic organizations where members are owners, operators and contributors of commodities handled and are direct beneficiaries of the savings that may accrue. No intermediary has to gain or suffer at the expense of members of that organization. It must conduct business according to sound business principles.

Cooperation as a form of economic organization differs from capitalism and socialism. Cooperation involves all the basic aspects of capitalism. These are:

1. Private property right in producer and consumer goods;

2. Capital is invested for gain by the investor;
3. Inheritance to the property and rights;
4. Right of free enterprise;
5. Right of contract;
6. Personal liberty; and
7. Guaranteed privileges to the individuals.

As in capitalism, in cooperation also payment is made for the services utilized and profit is distributed among the individuals as in private corporation. However, the cooperation avoids the drawbacks of capitalism, such as waste, exploitation of workers, absence of any voice of workers in management, denial of share to workers in profit, predominance of self interest and existence of excessive individualism.

Cooperation also differs from socialism. There are, however, some similarities in economic organization between socialism and cooperation. These are:

1. Common ownership and operation of production units to some extent;
2. Service and not the profit as main objective; and
3. Capital subordinated to the individuals

On the other hand, cooperation avoids the shortcomings of socialism as:

1. There exists great supervision of individuals;
2. Excessive interference of state in all affairs of production is absent; and
3. There is a lack of self interest.

Thus, cooperation combines the advantages and avoids the disadvantages of both capitalism and socialism.

MOTIVES OF COOPERATION

Cooperation can be studied in respect of its motives from three angles, i.e. social, political or economic motives. There are three basic corresponding motives in these spheres as under:

Social	Political	Economic
1. Freedom	Liberty	Opportunity
2. Justice	Equality	Security
3. Brotherhood	Fraternity	Partnership

PRINCIPLES OF COOPERATION

The important principles of cooperation are:

1. Voluntary organizations: It is voluntary to the individual to join or not to join the organization.
2. Management of the organization: It is based on democratic principle, i.e. one vote of each member irrespective of financial contribution by the individual to the organization.
3. The working is based on the principle of mutual self-help and reliance. The motto of cooperation is "each for all and all for each". In other words, it is help of himself through the help of others.
4. The aim of the cooperation is not to promote merely the economic interest of its members but also for moral and educative values.

5. Honesty is capitalized, i.e. loans to the members are given mostly on personal security.
6. Loans to the members are given mainly for productive purposes.
7. Neutral to caste, religion, and political party to which members belong.
8. Emphasis in working is on person and not to his wealth.
9. Service is the motto of a cooperative society and not earning the profits.

ADVANTAGES OF COOPERATION

The advantages accrued to the members joining the different forms of cooperation can be grouped into four broad groups as under:

1. Economic benefits: These benefits can be in terms of availability of loan at low rate of interest, low marketing cost per unit of output in sale of agricultural produce through cooperative marketing society, etc.
2. Moral benefits: It boosts the moral of the members as it increases among them the habit of savings and thrift.
3. Social benefits: The society as a whole is also benefited. The members plan and carry such activities, which are beneficial for the whole society viz., water management decisions, use of Common Property Resources (CPRs) and joint decisions for removal of social evils.
4. Educative benefits: The members also get benefits in terms of improvement of their understanding power through opening of schools or start of literacy campaigns in the society.

COOPERATIVES IN INDIA

HISTORY OF COOPERATIVE MOVEMENT IN INDIA

The cooperative movement saw the light of the day in 1844 in England. Later, on the guidance of Schultz, F.W. and Raiffeisen, F.W., cooperative credit societies were started in Germany. This has been the model to the present day cooperative credit structure in the world.

In India, the idea of cooperative movement was first conceived by the then Madras Government in 1892. Sir Frederick was appointed to find ways and means for the constitution of agricultural banks in the state. He suggested the formation of Raiffeisen type societies. These societies started to function in the present century to protect the farmers from exploitation by the traders. Further, on the basis of the precise report of Fredrick Nicholson and Duperhex that the cooperatives were considered worthy of every encouragement. Taking this view as well as the recommendations of the Famine Commission, 1901, a pilot bill was passed and an Act was approved by the legislative council in 1904 to meet the credit needs of rural people. As such the Act of 1904 was a pioneer Act in the cooperative credit movement in India.

The Act of 1904 was amended to meet the growing needs of rural credit and the limitations of the Act were removed by passing another Act in 1912. The Act of 1912 gave a fresh impetus to the growth of cooperative movement by including marketing and processing also in addition to rural credit. In 1914, Sir Edward Maclagan Committee suggested many more changes. In 1919, Government took measures under the Reforms Act of 1919 to make the cooperative credit more effective for furtherance of agricultural growth. In 1945, the Saroya Committee and in 1954 Gorawala Committee suggested changes for improvement in the cooperative system for the welfare of the farmers.

Since then, several committees examined the status of cooperatives from time to time and made suggestions for them to play an important role in rural and agricultural development including marketing of farm inputs and products.

Since independence, the cooperatives were considered a part of the strategy of planned economic development. After 1991 (launch of economic liberalization, privatization and globalization), the cooperatives were at the crossroads in the fast emerging scenario of economic liberalization. The role of cooperatives in economic development again was discussed at various forums. It was realized that given the predominance of small farmers, and prevalence of rural poverty and food insecurity, cooperatives will continue to be critical in liberalized economic environment also. However, it was found that cooperative institutions in the country generally suffered from (a) resource constraints, (b) poor governance and management, (c) low operational efficiency, and (d) unviability of many of these. Subsequently, a National Policy on Cooperatives was adopted. The objectives of the policy are to facilitate all round development of the cooperatives in the country. The Government's role in cooperatives should be as a guide and facilitator. The policy envisages necessary support, encouragement and assistance to cooperatives. Further, the policy aims to ensure that cooperatives work as autonomous, self-reliant and democratically managed institutions, accountable to their members.

The cooperative organizations can be informal groups or formal groups of people. The formal cooperative societies (organizations), once registered under the cooperative societies Act, become a separate legal entity, with an identity of its own. The members have a limited liability. The society can do business, buy properties and make contracts in its own name. The registering authorities are in place in all the districts of the country.

Keeping in view the role of cooperatives in agricultural developments, the subject of 'cooperation' is attached to 'Agriculture' and at the national level, the department is named as 'Department of Agriculture, Cooperation and Farmers Welfare'.

A major development in cooperative sector has been the enactment of Multi State Cooperative Societies Act, 2002. In this Act, the laws relating to cooperative societies were consolidated and amended to provide for registration of such societies that span over more than one state.

STRUCTURE OF COOPERATIVES IN INDIA

Cooperation has mostly a three or four tier structure which is pyramidal in shape, i.e. it is broad at the base level and narrows down at the apex or top level. The examples of three and four tier structures of cooperative credit and marketing are portrayed below:

Level	Short-Term Credit
Three Tier Structure of Cooperative Credit	
1. Primary (village)	PACS (Primary Agricultural Credit Cooperative Societies)
2. District	CCBs (Central Cooperative Banks)
3. State	SCB (State Cooperative Bank)
Four Tier Structure of Cooperative Marketing	
1. Village Level	Primary Agricultural Cooperative Societies (PACS)
2. Intermediate Level (mandi/block)	Primary Agricultural Marketing Societies (General purpose or product specific)
3. State Level	State Cooperative Marketing Federation (General purpose or product specific)
4. National Level	National Cooperative Marketing Federation

The structure of cooperation thus resembles a pyramid with large number of societies at village and few societies at district or intermediate level and one society at the state level. The state level cooperative organizations have a national level federation.

SIZE OF INDIAN COOPERATIVE SECTOR

There are 854355 cooperative societies in India. Out of these, 20.8% (177605) are credit societies, which include 97961 PACS, 1085 district/state level cooperative banks and 1562 urban cooperative banks. The non-credit cooperative societies include marketing (agriculture) cooperatives, dairy cooperatives, livestock/poultry cooperatives, fisheries cooperatives, agro-processing cooperatives, sugar/sugarcane cooperatives and multi state cooperatives. There are also consumer cooperatives, housing cooperatives, women cooperatives and labour cooperatives. Nearly 98% of Indian villages are covered by one or the other cooperative organization.

There are 17 national level cooperative federations, 390 state (level) cooperative federations and 2705 district level cooperatives.

The cooperatives account for 29% of India's fertilizer production. Indian Farmers Fertilizer Cooperative (IFFCO) and Krishak Bharti Fertilizer Cooperative (KRIBHCO) are two national level cooperative organizations with significant contribution in agricultural development and fertilizer marketing. In the sugar sector, cooperative sugar factories account for 31% of India's sugar production. The cooperative sector's storage capacity is 22.8 million tonnes.

COOPERATIVE MARKETING IN INDIA

In mid-1960s, when a new agricultural development strategy was launched in the country, to increase the production of foodgrains for improving food security situation, an institution was needed at the village level to stock and make available new seeds and fertilizers to the farmers. Further, as the farmers, in general, were not in a position to pay for the cost of seeds and fertilizers, a mechanism of providing short term credit (crop loans) was also necessary. It is in this context that a movement to organize village level farmers cooperatives was launched in the country. These village level cooperatives were named as Primary Agricultural Credit (Cooperative) Societies (PACS). They have played a very critical role in ushering green revolution in the country. These continue to be a very sound and effective grassroots level institution for agricultural and rural development.

The establishment of cooperative marketing societies was another step which has been taken to overcome the problems arising out of the then existing system of marketing agricultural produce. It was believed that the objectives of economic development and social justice can be furthered by inter alia channelising agricultural produce through cooperative institutions.

In the fifties and sixties private agencies dominated the Indian foodgrains trade. Farmers complained of the marketing system because they got lower prices, due mainly to high marketing charges and the prevalence of malpractices. The efforts of the government to improve the marketing system of agricultural commodities were only partially successful in creating healthy conditions for scientific and efficient marketing. Moreover, the progress of regulated markets was inadequate and not uniform in all areas. The need for strengthening cooperative organization was, therefore, recognized for the marketing of the produce of farmers and for making inputs available to them at the right price and time. The cooperative institutions are expected to function as competitors of private traders in the market. These organizations can pool the produce of the small farmers having a small surplus to market and improve their bargaining power. They could also help government agencies in the execution of the policy decisions bearing on the procurement and distribution of foodgrains and other essential commodities.

MEANING

Cooperative marketing is composed of two words - cooperative or cooperation and marketing. As mentioned earlier, cooperation means a form of association of people to work together to achieve a particular end. Marketing connotes all business activities involved in the flow of goods and services from the point of production until they are in the hands of the consumers. Cooperative marketing is an extension of the principle of cooperation in the field of marketing. It is a process of marketing through a cooperative association formed voluntarily by its members to perform one or more marketing functions in respect of their produce. In other words, cooperative marketing is the marketing 'for the farmers' and 'by the farmers' that aims at eliminating the chain of functionaries operating between the farmers and the ultimate consumers and thus securing the maximum price for the farmers' produce.

A cooperative sales association is a voluntary business organization established by its member patrons to market farm products collectively for their direct benefit. It is governed by democratic principles, and savings are apportioned to the members on the basis of their patronage. The members are the owners, operators and contributors of the commodities and are the direct beneficiaries of the savings that accrue to the society. No intermediary stands to profit or lose at the expense of the members.

Cooperative marketing organizations are associations of producers for the collective marketing of their produce and for securing for the members the advantages that result from large-scale business which an individual cultivator cannot secure because of his small marketable surplus.

In a cooperative marketing society, the control of the organization is in the hands of the farmers, and each member has one vote irrespective of the number of shares purchased by him. The profit earned by the society is distributed among the members on the basis of the quantity of the produce marketed by her/him. In other words, cooperative marketing societies are established for the purpose of collectively marketing the products of the member farmers. It emphasises the concept of commercialization. Its economic motives and character distinguish it from other associations. These societies resemble private business organization in the method of their operations; but they differ from the capitalistic system chiefly in their motives and organizations.

FUNCTIONS OF MARKETING COOPERATIVES

The main functions of cooperative marketing societies are:

1. To market the produce of the members of the society at fair prices;
2. To safeguard the members for excessive marketing costs and malpractices;
3. To make credit facilities available to the members against the security of the produce brought for sale;
4. To make arrangements for the scientific storage of the members' produce;
5. To provide the facilities of grading and market information which may help them to get a good price for their produce;

6. To introduce the system of pooling so as to acquire a better bargaining power than the individual members having a small quantity of produce for marketing purposes;

7. To act as an agent of the government for the procurement of foodgrains and for the implementation of the price support policy;

8. To arrange for the export of the produce of the members so that they may get better returns;

9. To make arrangements for the transport of the produce of the members from the villages to the market on collective basis and bring about a reduction in the cost of transportation; and

10. To arrange for the supply of inputs required by the farmers, such as improved seeds, fertilizers, insecticides and pesticides.

HISTORY OF COOPERATIVE MARKETING

The history of cooperative marketing in India dates back to 1912, when the Cooperative Marketing Societies Act, 1912 was passed. The first Cooperative Society was formed in Hubli in 1915 to encourage cultivation of improved cotton and to sell it collectively. In 1918, the South Canara Planters Cooperative Sale Society was formed in the then Composite Madras Province for joint sale of arecanut. The Royal Commission on Agriculture (1928) stressed the need for group marketing instead of individual marketing. The Central Banking Enquiry Committee (1931) also underlined the need for organised marketing. The XI Conference of Registrars of Cooperative Societies (1934) also emphasised the importance of Cooperative Marketing. In 1945, the Cooperative Planning Committee recommended that atleast 25% of the marketable surplus should be channellised through Cooperative societies within the next 10 years by forming one society for a group of 200 villages.

The All India Rural Credit Survey Committee (1954) brought to light the dismal performance of the then existing marketing cooperatives. In a sample of 75 districts surveyed, there was no cooperative marketing society in 63 districts. In remaining districts only around 1% of the total sale of agricultural produce was done through the societies. The committee suggested the establishment of primary cooperative marketing societies and linking of credit with marketing. The First Five Year Plan (1951–56) laid stress on the establishment of agricultural marketing and processing cooperative societies. In 1958, the National Agricultural Cooperative Marketing Federation (NAFED) was established as the apex body of cooperative marketing. In 1963, the National Cooperative Development Corporation (NCDC) was set up for promoting programmes relating to cooperative societies. The Mirdha Committee (1965) recommended that the membership of agricultural cooperative marketing societies should be restricted to the agriculturists and traders should not be allowed to join agricultural marketing societies.

The Dantwala Committee (1966) stressed the need for co-operation and integration among the various cooperative organisations after reviewing the pattern of cooperative marketing, distribution of inputs to farmers and supply of consumer products. Based on the survey of the cooperative marketing societies in 1968, the Reserve Bank of India recognised that effective linking of credit with marketing was necessary. The All India Rural Credit Review Committee, 1969 also recommended the strengthening of cooperative marketing, with a view to helping the government agencies in the execution of price support programmes.

TYPES

On the basis of the commodities dealt in by them, the cooperative marketing societies may be grouped into the following types:

1. **Single Commodity Cooperative Marketing Societies:** They deal in the marketing of only one agricultural commodity. They get sufficient business from the farmers producing that single commodity. The examples are Sugarcane Cooperative Society, Cotton Cooperative Society and Oilseed Growers Cooperative Marketing Society.

2. **Multi-Commodity Cooperative Marketing Societies:** They deal in the marketing of a large number of commodities produced by the members, such as foodgrains, oilseeds and cotton. Most of the cooperative marketing societies in India are of this type.

3. **Multi-purpose, Multi-commodity Cooperative Marketing Societies:** These societies market a large number of commodities and perform such other functions as providing credit to members, arranging for the supply of the inputs required by them, and meeting their requirements of essential domestic consumption goods. The Primary Agricultural Cooperative Societies (PACS) are multi-purpose cooperatives.

STRUCTURE

The cooperative marketing societies have both two-tier and three-tier structures. In the states of Assam, Bihar, Kerala, Karnataka, Madhya Pradesh, Orissa, Rajasthan and West Bengal, there is a two-tier pattern with primary marketing societies at the taluka level and state marketing federation as an apex body at the state level. In other states— Andhra Pradesh, Gujarat, Haryana, Himachal Pradesh, Maharashtra, Manipur, Punjab, Tamil Nadu, Uttar Pradesh, and Puducherry, there is a three-tier system with district marketing society in the middle (i.e. mandi, district and state level). At the national level, NAFED serves as the apex institution. The pattern of the three-tier structure is as follows:

1. Base Level

At the base level, there are primary cooperative marketing societies. These societies market the produce of the farmer members in that area. They may be single commodity or multicommodity societies, depending upon the production of the crops in the area. They are located in the primary wholesale market, and their field of operations extends to the area from which the produce comes for sale, which may cover one or two tehsils, panchayat samitis or development blocks.

2. Regional/District Level

At the regional or district level, there are central cooperative marketing unions or federations. Their main job is to market the produce brought for sale by the primary cooperative marketing societies of the area. These are located in the secondary wholesale markets and generally offer a better price for the produce. The primary cooperative marketing societies are members of these unions in addition to the individual farmer members. In the two-tier structure, the state level societies perform the functions of district level societies by opening branches at the district level throughout the state.

3. State Level

At the state level, there are apex (State) cooperative marketing societies or federations. These state level institutions serve the state as a whole. Their members are both the primary cooperative marketing societies and the central cooperative unions of the state. The basic function of these is to coordinate the activities of the affiliated societies and conduct such activities as inter-state trade, export-import, procurement, distribution of inputs and essential consumer goods, dissemination of market information and rendering expert advice on the marketing of agricultural produce. The state level cooperative federations are, in turn, federated to the national level federations.

MEMBERSHIP

There are two types of members of cooperative marketing societies.

1. Ordinary Members

Individual farmers, cooperative farming societies and service societies of the area may become the ordinary members of the cooperative marketing society. They have the right to participate in the deliberations of the society, share in the profits and participate in the decision making process.

2. Nominal Members

Traders with whom the society establishes business dealings are enrolled as nominal members. Nominal members do not have the right to participate in decision making and share in the profits of the societies.

SOURCES OF FINANCE

In 1966, the Dantwala Committee estimated a capital base of ₹ 2.00 lakh for a cooperative marketing society. At 2018 prices, it should be at least ₹ 1 crore. The following are the major sources of finance of a cooperative marketing society.

1. Share Capital

Farmer-members and the State Government subscribe to the share capital of cooperative marketing societies. Members may purchase as many shares as they like. They are encouraged to invest sufficiently in the share capital. They are also persuaded to invest their dividend and bonus in the shares of cooperative marketing societies.

2. Loans

Cooperative marketing societies may raise their finance by way of loans from the Central and State Cooperative Banks and from commercial banks by pledging and hypothecation and also by clean credit to the extent of 50% of owned capital.

3. Subsidy

The Cooperative marketing societies get a subsidy from the government for the purchase of grading machines and transport vehicles to meet their initial heavy expenditure. They also get a subsidy for a part of the cost of the managerial staff for a period of 3 years to make them viable. They are also entitled for subsidy on processing equipments, storage structures and cold chains, as applicable to any individual entrepreneur under the schemes of NHB, MFPI and APEDA.

FUNCTIONING

The important functions carried out by the cooperative marketing societies are:

1. Sale on Commission Basis

Cooperative marketing societies act as commission agents in the market, i.e. they arrange for the sale of the produce brought by the members to the market. The produce is sold by the open auction system to one who bids the highest price. The main advantage, which the farmer-members get by selling the produce through cooperative marketing societies instead of a commission agent, is that they do not have to accept unauthorised deductions or put up with the many malpractices, which are indulged in by individual commission agents. As there is no individual gain to any member in the marketing of the agricultural produce through cooperative marketing societies, no malpractices are expected to be indulged in.

This type of marketing is not risky for cooperative societies. But sometimes traders in the market form a ring and either boycott the auction or bid a low price when the produce is auctioned on the cooperative marketing societies shops. These tactics of the traders reduce the business of cooperative marketing societies. Therefore, farmers hesitate to take their produce for sale in the market through cooperative marketing societies.

2. Purchase of Members' Produce

Cooperative marketing societies also enter the market as buyers. A society participates in bidding together with other traders, and creates conditions of competition. The commodities thus purchased by a society are sold again when prices are higher.

This system of the outright purchase of the produce by the society involves the risk of price fluctuations. If the managers of societies lack business experience, they hesitate to adopt the outright purchase system. The National Cooperative Development Corporation recommended that the outright purchase system should be adopted only by a society which possesses the following qualities:

(a) The society has a trained manager, i.e. one who is capable of understanding the intricacies of the trade;
(b) The society is financially sound and has adequate borrowing facilities;
(c) The society is affiliated to a good viable central level society; and
(d) The society possesses processing facilities.

3. Advancement of Credit

Cooperative marketing societies advance finance to farmers against their stock of foodgrains in the godowns of the societies. This increases the holding power of the farmers and prevents distress sales. Generally, societies advance credit to the extent of 60 to 75% of the value of the produce stored with them. The recoveries are effected from the sale proceeds of the produce of the farmer. This function involves no risk to the society. Moreover, it increases the business.

4. Procurement and Price Support Purchases

Cooperative marketing societies act as agents of the government in the procurement of foodgrains and other agricultural commodities at the announced procurement or support prices.

5. Other Functions

The following functions are also carried out by Cooperative marketing societies, depending upon the availability of funds and other facilities:

(a) They assemble the marketable surplus of small and marginal farmers and transport this surplus from villages to the society headquarters for disposal;

(b) They make arrangements for the grading of the produce and encourage producers to sell the produce after grading so that they may get better prices;

(c) They undertake the processing of produce;

(d) They make arrangements for the export of agricultural commodities in collaboration with the State Level Cooperative Marketing Federation and the National Agricultural Cooperative Marketing Federation;

(e) They undertake inter-state trade in agricultural commodities; and

(f) They distribute agricultural production inputs, such as fertilizers, improved seeds, pesticides, agricultural implements, and such essential consumer articles as sugar, kerosene oil and cloth either directly or through PACS.

6. Integrated System of Cooperative Marketing

An integrated programme of cooperative development embracing credit, marketing, processing, warehousing and storage has been formulated. The important features of the integrated system are linking up of credit with marketing, development of agro-processing on cooperative lines and promotion of storage and warehousing.

PROGRESS OF MARKETING RELATED COOPERATIVES

The cooperatives that help farmers (including milk producers, livestock and poultry farmers, fishers etc.) market their produce include, apart from the so called marketing cooperatives, commodity specific cooperatives, inputs related cooperatives (fertilizer/credit cooperatives) and multipurpose cooperatives. For understanding the progress of marketing related cooperatives, a look at the following groups of cooperatives is necessary:

1. Marketing cooperatives
2. Agro processing cooperatives
3. Dairy cooperatives
4. Consumer cooperatives
5. Credit and related cooperatives
6. Other cooperatives

Marketing Cooperatives

The number of primary agricultural cooperative marketing societies (PAMS) in 2017 was 7399 with a membership of 7.41 million. These are basically of two types viz:, general purpose PAMS and specialized or commodity specific PAMS. The number of general purpose PAMS was 2524 and rest were commodity specific PAMS. The commodity specific PAMS include 1336 related to fruits and vegetables, 142 cotton, 253 tobacco, 150 sugarcane, 142 oils, 600 coir, 46 arecanut, 51 rubber, 29 tea and coffee and 44 related to coconut marketing. The membership, and trade volumes of PAMS have considerably gone up, during the last 60 years. It went up from 1.4 million in 1961 to 7.41 million in 2017. These marketing cooperatives also supply consumer goods to their farmer-members and others.

In addition to PAMS, the network includes 378 district level cooperative marketing societies and 49 state level cooperative marketing federations (which include 25 general purpose and 24 commodity-specific state federations). The cooperatives have 22.8 million tonnes capacity godowns for storage of various commodities handled by these.

Agricultural Processing Cooperatives

Cooperatives have played a very important role in agroprocessing. The processing units established in the cooperative sector are of two distinct patterns, i.e. those which are set up by exclusive processing societies such as cooperative sugar factories and cotton spinning mills; and those which are established as adjuncts of cooperative marketing societies such as processing/ rice mills, dairy processing plants, and oil mills.

For the benefit of sugarcane growers, there are 324 cooperative sugar factories in the country, of which 284 are in operation, which provide marketing and price support to the sugarcane growers of their hinterlands. The cooperative sugar factories account for around 30% of the total sugar produced by sugar factories in the country. In northern states, where private sugar factories dominate the cane market, there are several cane growers cooperative societies to manage the supply of cane produced by the farmers to sugar factories. There was also two-fold increase in cotton spinning and processing units in the country up to 1980-81, but the number decreased thereafter.

In 2011 in the cooperative sector, there were 172 cotton ginning and processing units (societies), 212 cooperative dairy plants, 702 grain processing units, 142 cooperative oil mills, 54 food and vegetable processing units, 23 cooperative tea factories, and 60 other cooperative processing units, with a total of 1689 cooperative agro processing units in the country (Table 8.1). Cooperatives account for 39% of total milk processing capacity in India.

Table 8.1: Cooperative agro processing units in India		
	2008	2011
1. Cooperative sugar factories	316	324
2. Cooperative cotton growing and processing societies	164	172
3. Cooperative dairy plants	190	212
4. Cooperative food grain processing units	690	702
5. Cooperative oil mills	139	142
6. Cooperative fruits and vegetable processing units	49	54
7. Cooperative tea factories	23	23
8. Other cooperative processing units	52	60
Total	1623	1689

Dairy Cooperatives

Dairy or milk producers' cooperatives have been the foundation of white revolution in India. There are now 1.91 lakh primary dairy cooperatives in the country, with a membership of 1.7 crore small milk producers. They handle around 51 million litres of milk per day. These primary milk producer cooperatives form the base of the milk pyramid of India, which include 202 cooperative milk unions, 346 district dairy cooperatives, 15 state cooperative dairy federations and a national milk grid. The national milk grid has 700 cities or towns in its loop.

Consumer Cooperatives

Consumer cooperative network is a four-tier structure. At the national level, there is a National Cooperative Consumer Federation of India (NCCF). It has 34 branches or sub-branches. The next tier is the state level, which has 30 State Cooperative Consumer Federations (with 533 branches). The third tier has 512 wholesale/central cooperative consumer stores (with 5154 branches). At the base of the network (4th tier), there are 25843 primary cooperative consumer stores (with 8622 branch outlets).

These apart, in rural areas, 44418 village level primary agricultural cooperative societies or marketing societies are undertaking distribution/sale of consumer goods, with their normal business. Likewise, in the urban areas, cooperative consumer societies are operating around 37000 outlets. There are 9.6 million members of consumer cooperatives.

Agricultural Credit Cooperatives

Agricultural Credit cooperative structure consists of Primary Agricultural Credit Societies (PACS) at the village level, Central Cooperative Banks at the district level and State Cooperative Banks at the state level. The main purpose of establishing PACS was to provide short-term credit to the farmers for acquiring new farm inputs like seeds, fertilizers, pesticides, and farm equipments. Many of these are also involved in output marketing and supply of essential consumer goods. Thus their role in agricultural marketing is quite important and significant.

The number of village level PACS (including LAMPS and FSS) is around 98000, with a membership of around 8 crore. The PACS in India cover 98% of the villages of the country. At the next (district) level, there are 371 Central Cooperative Banks (CCBs), with 13327 branches. Some CCBs cover more than one district. At the state (apex) level, there are 33 State Cooperative Banks (SCBs), with 997 branches. The PACS, CCBs and SCBs, cater mainly to the short-term credit needs of farmers.

For meeting the medium/long term credit needs, there are 60I Primary Cooperative Agriculture and Rural Development Banks (PCARDBs) with 1086 branches; and 13 State Cooperative Agriculture and Rural Development Banks (SCARDBs) with 761 branches (these were earlier called Land Development Banks).

Other Cooperatives

In addition, there are farming cooperative societies (including joint farming societies), irrigation cooperatives, poultry cooperatives, fisheries cooperatives, and tribal cooperatives, which help their members in buying or selling of relevant goods/inputs/products.

The success of cooperative marketing is not universal across commodities, sectors and geographical regions.

1. In Gujarat, Maharashtra, Andhra Pradesh and Tamil Nadu, considerable quantities of the foodgrains are marketed by cooperative societies. In Maharashtra and Uttar Pradesh, 75% of sugarcane, in Maharashtra and Gujarat, 75% of cotton, and in Karnataka 84% of plantation crops are marketed by the cooperative societies.
2. The performance of cooperatives in dairy and sugarcane sectors is noteworthy. Dairy cooperatives present the most successful example of cooperative marketing.
3. The success of cooperatives and transforming the social and economic landscape of Gujarat and Maharashtra state and some other parts of the country is a testimony of the role of cooperatives in agricultural marketing in the country.

4. The role of the cooperatives in improving the marketing environment for farmers have also been quiet significant.
5. However, the cooperatives as a whole account for only 10% of the total quantities of agricultural commodities marketed by the farmers. This share needs to be improved in the light of predominance of small-scale farmers, technological changes in marketing practices and as a long-term solution for improving farmer's price realization.

REASONS OF SLOW PROGRESS OF COOPERATIVE MARKETING

The main reasons of the slow progress are:

1. Cooperative marketing societies are generally located in big markets/ towns and quick and cheap transport facilities are not available for the carriage of the produce from the villages to the societies;
2. Farmers are indebted to local traders and enter into advance contracts with them for the sale of the crop;
3. Farmers are in immediate need of cash after the harvest to meet their personal obligations. They, therefore, sell their produce to local traders; they cannot wait for the time required to move the produce to the mandi;
4. There is lack of loyalty among members to cooperative marketing societies because of their poor education and absence of the cooperative feeling;
5. In some cases rivalries among farmer-members result in indecision, which hampers the progress of the societies;
6. Members lack confidence in cooperative organizations, for many of the cooperative sector enterprises run at a loss;
7. The societies do not act as banks for the farmers;
8. Managers of societies do not offer business advice to members;
9. Societies do not provide facilities of food and shelter to farmers when they visit the market for the sale of the produce;
10. The managers of the societies are often linked with local traders and become impersonal to the needs of a majority of small and marginal farmers;
11. There is lack of sufficient funds with the societies to meet the credit needs of the farmers against pledging of the produce brought for sale. Nor do they make an advance payment of the value of the produce purchased or sold through them;
12. Cooperative marketing societies are not capable of carrying on their business in competition with traders and commission agents, because of the absence of adequate business expertise among their employees; and
13. There is a lack of sufficient storage facilities with the societies. They, therefore, try to dispose of the produce soon after their arrival; a fact which results in lower prices for the farmers.

SUGGESTIONS FOR STRENGTHENING OF COOPERATIVE MARKETING SOCIETIES

1. The area of the operations of the societies should be large enough so that they may have sufficient business and become viable. Most of the societies at present are not viable because of the small volume of their business.
2. Cooperative marketing societies should develop sufficient storage facilities in the mandi as well as in the villages.
3. The societies should give adequate representation to the small and marginal farmers in their organizational set-up.

4. The cooperative feeling among members should be inculcated by proper education by organising seminars and by the distribution of literature.
5. In the selection of the officials of cooperative marketing societies, weightage should be given to business experience and qualifications. After their selection, the officials should be given proper training so that they may deal efficiently with the business of the society. The efficiency should be rewarded, wherever possible.
6. There is a need for bringing about a proper co-ordination between credit and marketing cooperative societies to facilitate the recovery of loans advanced by credit societies, and make available sufficient finance for marketing societies.
7. The societies should-acquire the transport facility to bring the produce of the members from the villages to the mandi in time and at a lower cost.
8. Cooperative marketing societies should diversify their activities. They should sell the produce and inputs, and engage in the construction of storage facilities.
9. Marketing societies, like the private traders, should provide accommodation and other facilities for their members when the latter come to the mandi.
10. The public procurement and public distribution programmes should be implemented through cooperative marketing societies to increase their business; and
11. The cooperatives should be made free from government control and be allowed to emerge as strong business organizations.

NATIONAL LEVEL COOPERATIVE FEDERATIONS

There are in all 17 national level cooperative federations, of which those directly related to agricultural marketing are as follows:

1. National Cooperative Union of India (NCUI)
2. All India Federation of Cooperative Spinning Mills
3. National Cooperative Dairy Federation of India (NCDF)
4. Krishak Bharti Fertilizer Cooperative Limited (KRIBHCO)
5. Indian Farmers Fertilizer Cooperative Ltd. (IFFCO)
6. National Cooperative Consumers Federation Ltd.
7. National Federation of Fishermen's Cooperatives Ltd.
8. National Federation of State Cooperative Banks Ltd.
9. National Cooperative Agriculture and Rural Development Banks Federations Ltd.
10. National Federation of Cooperative Sugar Factories Ltd.
11. National Agriculture Cooperative Marketing Federation of India Ltd. (NAFED)
12. Tribal Cooperative Marketing Development Federation of India (TRIFED)

As illustration, some details of NCUI, NAFED, TRIFED and NCDF have been shown below.

NATIONAL COOPERATIVE UNION OF INDIA (NCUI)

It is an apex organization of Indian cooperative movement. It was established in 1929 (before independence). The main functions of NCUI include (a) promotion and strengthening of cooperative sector; (b) cooperative education and training; (c) network development of cooperative information system; (d) research in cooperatives; (e) international cooperative relations; (f) image building through cooperative parliamentarians' forum, publications and publicity; and (g) cooperative planning. There are 262 members of NCUI, which include 17 national level federations, 149 state level federations and 96 multi state cooperative societies.

NAFED

Organization

The National Agricultural Cooperative Marketing Federation of India (NAFED) is an apex organisation of marketing cooperatives in the country. It deals in procurement, processing, distribution, export and import of selected agricultural commodities. The NAFED is also the central nodal agency for undertaking price support operations for pulses and oilseeds and market intervention operations for other agricultural commodities.

The National Agricultural Cooperative Marketing Federation (NAFED) was established in October, 1958. The State Level Marketing Federations and the National Cooperative Development Corporation are its members. The head office of NAFED is at Delhi. NAFED has four zonal offices and 26 regional offices. NAFED's area of operation extends to the whole country. It has established branches in all the major port towns and capital cities in the country. The membership structure of NAFED is shown in Chart 8.1.

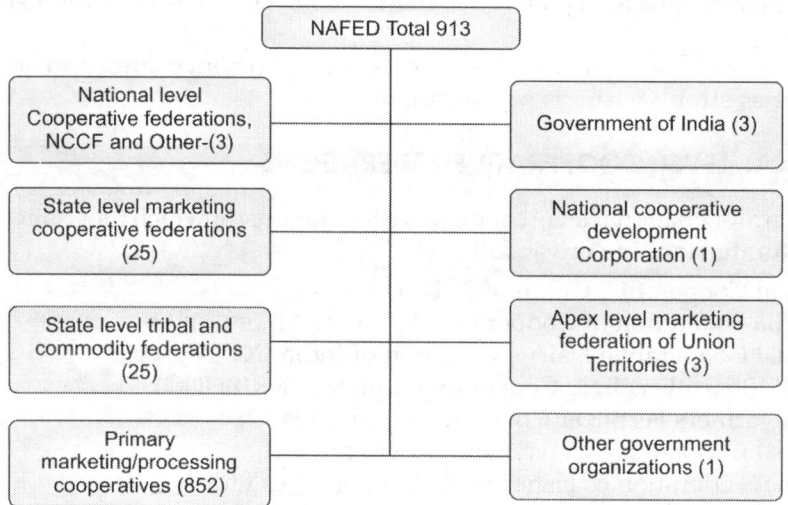

Chart 8.1: Membership of NAFED

Objectives

The main objectives of NAFED are:

1. To coordinate and promote the marketing and trading activities of its affiliated cooperative institutions;
2. To make arrangements for the supply of the agricultural inputs required by member institutions;
3. To promote inter-state and international trade in agricultural and other commodities; and
4. To act as an agent of the government for the purchase, sale, storage and distribution of agricultural products and inputs.

Activities

The NAFED performs the following activities:

1. Internal Trade

NAFED is engaged in interstate trade in agricultural commodities, particularly foodgrains, pulses, oilseeds, cotton, jute, spices, fruits, vegetables and eggs with a view to assuring better prices to the producers. The objectives of internal trade operations are both the market support to farmers and maintaining steady supply of commodities to consumers at reasonable prices. NAFED purchases agricultural commodities through the cooperatives, public sector organisations and state agencies. NAFED supplies grocery items to hospitals, schools, hostels, defence forces and general public.

2. Foreign Trade—Export and Import of Agricultural Commodities

The NAFED exports agricultural commodities, particularly onions (canalised), potatoes, ginger, garlic, nigerseed, sesameseed, gum, deoiled cake of groundnut, soyabean and cottonseed, fresh and processed fruits and vegetables; spices—black pepper, cardamom, turmeric, cuminseed, coriander seed; cereals—rice, barley, bajra, jowar, and ragi and jute bags to various countries including Sri Lanka, England, Mauritius, Australia, Belgium, Canada, Fiji, Hong Kong, Japan, Malaysia, the USA and number of African, West Asian and Gulf countries.

The market intervention undertaken by NAFED has many times helped the growers of such crops as onion, potato, copra, chillies and others in realising reasonable prices even in those years when market prices crashed.

The NAFED also arranges for the imports of pulses, fresh fruits, dry fruits, nutmeg (Jaiphal), mace (Javitri), wetdates and chicory seeds and inputs particularly fertilizers and machinery as and when asked to do so by the Government.

3. Price Support Operations

NAFED is the agency of the government to undertake support price purchases of commodities like groundnut since 1976–77, soyabean and mustard seed since 1977–78, gram, tur, moong and urad since 1978–79 and toria and sunflower seed since 1985–86. Government of India had designated NAFED as the nodal agency for implementing the price support policy for oilseeds and coarse grains during Seventh Five Year Plan period (1985–90). NAFED has standing instructions to intervene in the market when ruling market price falls below the minimum support levels for oilseeds and pulse crops. Since 1991, NAFED has been designated as nodal agency for undertaking price support operations for oilseeds and pulses on a regular basis. Price support operations are taken under PSS, as well as through MIS (market intervention scheme). NAFED had been purchasing potato, onion, eggs, malta, black pepper, chillies, and coriander seeds under MIS. The price stabilization fund, created by the government (under SFAC), is also used by NAFED for price support operations. The procurement of oilseeds under PSS and PSF (oilseeds and pulses) during the last nine years is shown in Tables 8.2 and 8.3. Note that in oilseeds and pulse crops, price support operation are not necessary every year (unlike paddy/rice and wheat where these are necessary every year).

4. Production and Marketing of Agricultural Inputs

NAFED helps the farmers by supplying them seeds, agricultural machinery like harvesting combines, tractors, spare parts and such other inputs as bio-fertilizers. NAFED also imports some of the machines and spare parts from abroad to ensure timely availability of genuine spare parts at reasonable prices. The technical know-

Table 8.2: Procurement of oilseeds by NAFED under Price Support Scheme (2010–11 to 2018–19)*

Oilseeds	Year	Quantity procured (thousand tonnes)
Groundnut pods	2013–14	340
	2014–15	6
	2016–17	211
	2017–18	1047
	2018–19	2
Mustard seed	2014–15	2
	2017–18	49
	2018–19	868
Soyabean	2017–18	72
Sunflower seed	2010–11	1
	2012–13	2
	2013–14	4
	2014–15	4
	2015–16	4
	2016–17	5
	2017–18	7
	2018–19	3
Copra	2010–11	32
	2012–13	75
	2013–14	34
	2016–17	6

* Only for those years when it was more than 500 tonnes.

Source: Agricultural Statistics at a Glance, 2018, Directorate of Economics and Statistics, Ministry of Agriculture, Government of India, New Delhi.

Table 8.3: Procurement of pulses under PSS and PSF

(2010–11 to 2018–19)* ('000 tonnes)

Years	Arhar	Urad	Moong	Gram	Lentil	Total
PSS						
2012–13	16	–	–	–	–	16
2013–14	43	77	–	–	–	120
2014–15	1	7	–	–	–	8
2016–17	196	–	9	–	–	205
2017–18	603	292	407	–	–	1302
PSF						
2015–16	46	5	–	–	–	51
2016–17	908	88	210	61	9	1276
2017–18	258	–	–	–	27	285

* Only for the year when procurement was more than 500 tonnes.

Source: Agricultural Statistics at a Glance, 2018.

how to operate and maintain the machines is also provided to the farmers. The NAFED through its service centres also sells farm tools, agricultural implements and spare parts produced by Krishi Yantra Udyog, Bhiwadi in Rajasthan.

Bio-fertilizers are gaining importance for increasing yields of pulses and oilseeds crops. The NAFED had set up a unit for production of rhizobium culture at Indore (Madhya Pradesh) in 1985 with a capacity to produce bio-fertilizers for 12 lakh hectares a year. NAFED maintains contact with the input supplying institutions such as the National Seeds Corporation and the Fertilizer Corporation of India.

5. Promotional Activities

NAFED maintains expert staff which conducts market studies, collects data and circulates the results among the members. Other promotional activities of NAFED are intensive development of selected marketing societies as pilot centres for cooperative marketing of agricultural produce in each state; improvement in market intelligence services for cooperative marketing societies; conduct of market surveys; training of market personnel; promotion of market regulations and development of infrastructures. It also helps organic certification of organic products. NAFED also offers consultancy services. It has recently launched Bio CNG manufacturing unit under PPP mode, which uses agricultural waste.

6. Developing Cooperative Marketing of Tribal Produce

A separate cell to develop the marketing of produce of the tribal areas (minor forest products) having economic value has been set up with the assistance of NAFED. NAFED arranges market intelligence, establishes better system for auction of tribal produce and develops markets for other commodities like chilgoza, gum, karya etc. Keeping in view the importance of marketing of tribal produce, a separate Tribal Cooperative Marketing Federation (TRIFED) has been set up.

7. Setting of Scientific Storage System

NAFED has set up a cold storage along with an ice factory and a warehouse in Delhi. NAFED has also made pioneering effort in finding ways of developing modern storages. It has set up an onion warehousing complex at Nagapattanam.

8. Processing of Fruits, Vegetables and Others

NAFED has set up a multi-commodity fruit and vegetables processing unit at Delhi and at Vellore (Tamil Nadu). The basic purpose is to develop the processing industry in cooperative sector in a major way so as to make fruits and vegetable marketing and processing to the advantage of the farmers.

NAFED also manufactures jute goods in joint venture with Konark Jute Limited—promoted by Industrial Development Corporation of Orissa.

9. Turnover

The turnover of NAFED, which was ₹ 2293 crore in 2004-05, has gone up to ₹ 20138 crore in 2018-19, showing a remarkable growth.

NATIONAL COOPERATIVE DEVELOPMENT CORPORATION (NCDC)

The National Cooperative Development Corporation (NCDC) was set up in March, 1963 under an Act of Parliament (1962) for promoting, guiding and supporting rural economic activities on cooperative principles. With its head office in Delhi, it has

16 regional offices. The corporation focuses on programmes of promoting, strengthening and developing farmers cooperatives for marketing, processing and storage of agricultural products as also for supply of agricultural inputs and essential consumer goods in rural areas. It tries to equip cooperatives with facilities for promoting income-generating activities including poultry, fishery, handlooms and minor forest products. The corporation supplements the efforts of the state governments in promoting cooperatives.

The NCDC provides financial assistance to cooperative societies through or on the guarantee of the state governments. NCDC provides financial assistance to large number of cooperatives and their activities. This includes:

1. State level cooperative marketing/commodity marketing federations for margin money to raise working capital from banks or increasing marketing and distribution activities.
2. Agro-based processing units including large size oil complexes and small and medium size units for foodgrains, plantation, commercial and horticulture crops;
3. Viable or potentially viable primary agricultural cooperative societies, large size agricultural multipurpose cooperative societies (LAMPS), farmers service societies (FSS), commodity cooperatives and processing cooperatives for construction of godowns;
4. Cooperatives for the establishment of new cold stores and expansion of the existing units as well as for setting up of ice plants;
5. State governments to supplement their resources for share capital participation in the new cooperatives related to processing, repair and custom hiring services;
6. Cooperatives dealing in fruits and vegetables for development of marketing, establishment of processing units, purchase of transport vehicles, construction of storage sheds and retailing;
7. Fishery cooperatives for development of inland and marine fisheries;
8. Cooperatives for dairy development in areas which are outside the programmes of NDDB;
9. Integrated poultry projects including poultry sheds, feed mix units and hatcheries; and
10. Scheduled caste cooperatives for supply of notified commodities, strengthening of share capital base of tribal development cooperative federations, cooperative-marketing societies in hilly areas, rural consumer cooperatives and student cooperative stores.

NCDC has not only provided financial support to the cooperatives but also provided technical guidance to them. The promotional and developmental role of NCDC has led to continuous expansion and diversification of cooperative programmes under its purview.

Cumulative assistance provided by NCDC up to March 2019 was ₹ 124890 crore. An important feature of functioning of NCDC is that it has recorded a loan recovery percentage of 99 during 2018-19. NCDC has also provided assistance to cooperatives to set up cold stores. NCDC has also set up Kisan Call Centres.

Keeping in view the important role that NCDC has played in helping farmers, through strengthening of the cooperatives, the NCDC Act, 1962 was amended in 2002 (vide NCDC Amendment Act, 2002) for further expanding its role. There are two noteworthy changes:

1. It has expended the mandate of NCDC to include foodstuffs, industrial goods, livestock, and services in the programmes and activities, in addition to the existing programmes being implemented by NCDC. The definitions of agricultural produce, industrial goods, livestock and services have also been elaborated.
2. NCDC is now able to provide loan directly to the cooperatives without state/central government guarantee on furnishing of security.

TRIBAL COOPERATIVE MARKETING FEDERATION (TRIFED)

The Tribal Cooperative Marketing Federation (TRIFED) is a national level cooperative federation responsible for marketing development of tribal products and handicrafts. It was established in 1987. The main objective of TRIFED is to serve the interests of the members in more than one state for social and economic betterment of its members by conducting its activities in professional, democratic and autonomous manner through self-help and mutual cooperation for undertaking marketing development of the tribal products. The TRIFED arranges marketing and export of minor forest products produced or collected by the tribals in the tribal-dominated areas and protects the tribals from exploitation by the private traders because of low demand of their products and production in small lots. Its annual sales exceed ₹ 10 crore. TRIFED has 27 members and the membership structure is as follows:

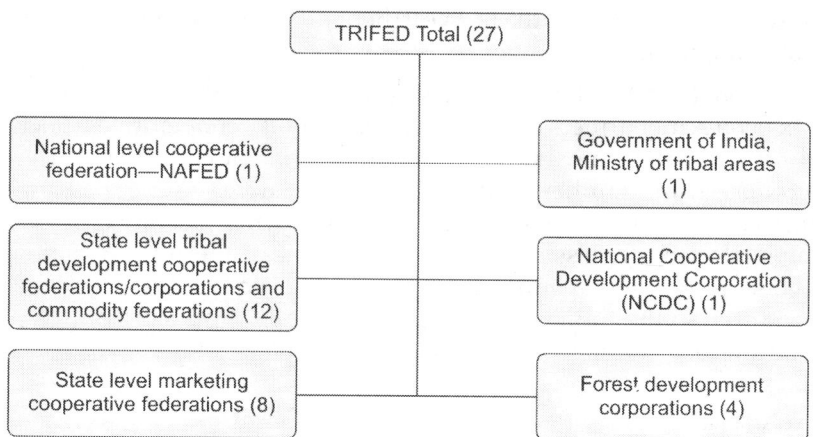

NATIONAL COOPERATIVE DAIRY FEDERATION OF INDIA

With a view to providing marketing support to the milk producers, a sound network of dairy cooperatives has taken shape in the country which has been instrumental in what is called the 'White Revolution' in the country. The network consists of milk producers cooperative societies at the village level, District Milk Cooperative Unions at the district level, State Cooperative Dairy Federations at the state level and National Cooperative Dairy Federation at the national level. Apart from providing market support to the producers in rural areas, this network has been instrumental in supplying liquid milk and dairy products to the urban consumers at reasonable prices.

The main functions of these societies are as follows:

1. Primary Milk Producers Cooperative Societies

The functions of these societies are collection of milk from the farmers and sale to milk unions, sale of cattle feed, help members to increase milk production and provide veterinary and artificial insemination services.

2. District Cooperative Milk Producers Unions

Their functions are organization and supervision of primary milk producers cooperative societies; collection of milk from primary societies for processing and manufacturing of milk products; marketing of liquid milk in their area of operation; distribution of cattle feed; and providing of technical inputs.

3. State Cooperative Dairy Federations

Their functions are production programming, marketing of milk products, assistance in input supply programmes and coordinating bulk purchases. The state level Cooperative Dairy (Marketing) Federations have a vast net work of milk collection, processing and marketing of milk. They manufacture several milk products like butter, ghee, cheese, milk powder and shrikhand. The milk and milk products of these federations are sold as special brand products viz., 'Saras' brand in Rajasthan; 'Sangam' and 'Vijaya' brand in Andhra Pradesh; 'Sudha' brand in Bihar; 'Amul' brand in Gujarat; 'Vita' brand in Haryana; 'Him' brand in Himachal Pradesh; 'Nandini' in Karnataka; 'Milma' in Kerala; 'Sneha' in Madhya Pradesh; 'Gokul' and 'Vikas' in Maharashtra; 'OMFED' in Orissa; 'Verka' in Punjab; 'AAVIN' and 'MAAVIN' in Tamil Nadu; 'Parag' in Uttar Pradesh; and 'Bhagirath' in West Bengal. Some of these like AMUL have become global brand names.

There are more than 1.91 lakh dairy cooperative societies in the country. Nearly 17 million milk producing farmers are the members of dairy cooperatives. These societies have been formed in all the states of the country. However, their concentration is more in the western and southern states. These cooperative societies play an important role in milk marketing in few states of the country. Further, inspite of the existence of large number of milk processing societies, the bulk of the milk is handled by the unorganized sector without commensurate value addition to raw milk. About 35% of the milk produced in the country is converted into milk products through processing and value addition.

▌SOME STATE LEVEL COOPERATIVE MARKETING FEDERATIONS

TILAM SANGH

Recognising the success of the 'Amul Pattern' of dairy cooperatives network, the concept has been extended to the oilseeds sector. Oilseeds growers cooperative societies at the village level and State Cooperative Oilseeds Growers Federations at the state level have been organised in many States. The nomenclature differs from state to state.

In Rajasthan, the State level Cooperative Oilseeds Growers Federation, known as Tilam Sangh was set up in July, 1990. The main objectives of the Federation/Sangh are:

1. To organise the oilseeds growers at grass root level;
2. To augment the production and productivity of oilseeds by providing package of practices;
3. To procure the oilseeds of farmers at their doorstep at premium price;
4. To eliminate the intermediaries and to retrieve the oilseed growers from their clutches;

5. To establish oilseed processing units; and

6. To undertake seed multiplication programme to provide quality seeds to the member producers.

The Tilam Sangh has established eight plants for processing of mustard in different mustard growing areas with a total processing capacity of 1.41 lakh tonnes of mustard. These plants are located at Kota, Fatehnagar (Udaipur); Bikaner, Jalore, Mertacity (Nagaur); Jhunjhunu, Sriganganagar and Gangapurcity (Sawai Madhopur). Around 967 oilseeds growers cooperative societies are working in different mustard growing areas under Tilam Sangh with a total membership of over one lakh members. In addition to mustard plants, Tilam Sangh has also established soyabean and groundnut processing plants. The Tilam Sangh purchases oilseeds from the farmers through the cooperative societies at fair prices and makes available edible oils to the consumers at reasonable prices. With a view to making available improved seeds of oilseed crops, a farm of 40 hectares has also been transferred by the State Government to the Tilam Sangh.

The popularity of edible oils produced by the Tilam Sangh has been increasing continuously among the consumers. Tilam Sangh is supplying 'Tilam brand' edible oils in 1, 2, 5 and 15 litres/kgs packing to the consumers.

RAJFED

RAJFED in Rajasthan, a state level cooperative marketing organization, has been playing a very important role in agriculture marketing. The Rajasthan Cooperative Marketing Federation (RAJFED) was established on November 26, 1957. It is an apex body of marketing cooperatives at the state level. The main aim of establishing RAJFED is to cooperatively handle purchase and sale of agricultural commodities for the benefit of farmers as well as consumers. Following are the main functions of RAJFED:

1. RAJFED purchases the agricultural produce from markets by an open auction method and thus creates condition of competition. Farmer-producers get fair prices of the produce. The commodities so purchased by RAJFED are sold later on when prices are high. In its marketing operations, the RAJFED collaborates with NAFED, Tilham Sangh and other cooperatives. The primary cooperative marketing societies act as commission agents for making the purchases of agricultural commodities on behalf of the RAJFED. RAJFED pays commission to these societies for this work.

2. RAJFED is also engaged in the supply of agricultural inputs such as fertilizers (DAP, Urea, CAN), improved seeds, pesticides, plant protection implements and gypsum to the farmers.

3. RAJFED has established a unit of production of animal feed at Jaipur with a capacity of 12,000 tonnes per annum. The animal feed is distributed through various units of cooperative societies, government departments and private dealers.

4. RAJFED has established a Isabgol Bhushi production plant at Abu Road in 1982 with the help of NAFED with a capacity of 450 tonnes per annum.

5. RAJFED has also established a cold storage and ice-plant at Jaipur.

6. RAJFED works as an important state agency for price support operations whenever need arises.

MARKFED

MARKFED is the state level organization in the state of Punjab. It is a federation of marketing cooperative societies. Its main objective is to help the farmers to secure better prices for their produce by taking care of the market needs and providing agricultural inputs. To achieve this objective, MARKFED's activities include sale of farm inputs (fertilizers, seeds and pesticides); maintenance of godowns; and procurement of agricultural commodities through its member societies. Markfed has also entered the export business and helping establish contract-farming arrangements in the state.

OTHER STATE FEDERATIONS

Almost all states have state level federations of cooperative marketing societies. These may be general purpose federations or commodity specific federations. Some examples of state level federations are state Dairy Cooperative Federations, State Oilseed Federations and State Cotton Federations. State level cooperative marketing federations also exist for marketing and processing of horticultural crops, like HOPCOMS and MAHAGRAPES.

The HOPCOMS has its genesis in the establishment of 'The Bangalore Grape Growers Cooperative Marketing and Processing Society Limited ((BGGCOMS)' on 10th September, 1959 with the main objective of encouraging grapevine cultivation by providing the required inputs, technical know how, and marketing facilities. The society started handling fruits and vegetables apart from grapes since 1965. In 1983, the name of the society was changed as the Bangalore Horticulture Producers Cooperative Marketing and Processing Society Limited (BHOPCOMS) and subsequently in 1987 it became HOPCOMS. The main activities of HOPCOMS are:

- Procurement of fruits and vegetables.
- Marketing of fruits and vegetables.
- Fixing of procurement prices of fruits and vegetables.
- Supply of production inputs to fruits and vegetable growers at reasonable prices, and
- Preparation of juice of fruits and selling them in bottles through retail outlets.

The MAHAGRAPES came into existence in 1991. Mahagrapes is the first organization in Maharashtra state to have the characteristics of both cooperative and a private sector partnership firm. The role of Mahagrapes as a marketing entity itself is a policy innovation. Based on the success of Mahagrapes, Mahabanana and Mahamangoes were set-up in the state subsequently.

9

Market Integration, Efficiency, Costs, Margins and Price Spread

▌MARKET INTEGRATION

MEANING

Integration shows the relationship of the firms in a market. The extent of integration influences the conduct of the firms and consequently their marketing efficiency. The behaviour of a highly integrated market is different from that of a disintegrated market. Markets differ in the extent of integration and, therefore, there is a variation in their degree of efficiency.

Kohls and Uhl[1] have defined market integration as a process which refers to the expansion of firms by consolidating additional marketing functions and activities under a single management. Examples of market integration are the establishment of wholesaling facilities by food retailers and the setting up of another plant by a milk processor. In each case, there is a concentration of decision making in the hands of a single management.

TYPES OF MARKET INTEGRATION

There are three basic kinds of market integration.

1. Horizontal Integration

This occurs when a firm or agency gains control of other firms or agencies performing similar marketing functions at the same level in the marketing sequence. In this type of integration, some marketing agencies (say, sellers) combine to form a union with a view to reducing their effective number and the extent of actual competition in the market. In most markets, there is a large number of agencies which do not effectively compete with each other. This is indicative of some element of horizontal integration. Horizontal integration is advantageous for the members who join the group. Similarly, if farmers join hands and form cooperatives, they are able to sell their produce in bulk and reduce their cost of marketing. Horizontal integration of selling firms is generally not in the interest of the consumers or buyers, because it reduces competition.

The schematic arrangement of a horizontally integrated firm is shown in Fig. 9.1. In this arrangement, there are four firms engaged in buying and selling of foodgrains under the direction of the parent agri-business firm. All the four business firms perform

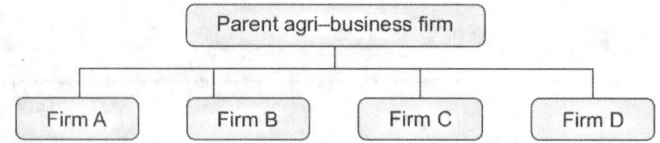

Fig. 9.1: Schematic arrangement of horizontally integrated firms

the same type of marketing function but their locations and areas of operations are different. Cases of such an integration are very commonly found. Frequently a firm will have a central headquarter with a large number of local branches that carry on operations at the local level. Such a network enables the organisation to achieve the economies associated with size of the firm. It also helps the firm to organise some complex types of operations and services which are needed by the local units but individually, they may not be able to perform with ease and/or efficiency.

2. Vertical Integration

Vertical integration occurs when a firm performs more than one activity in the sequence of the marketing process. It is a linking together of two or more functions in the marketing process within a single firm or under a single ownership. For example, if a firm assumes the functions of the wholesaling as well as retailing, it is vertical integration. Another example of vertical integration is a flour mill which engages in wholesale of wheat and retailing activity of flour as well.

The schematic arrangement of a vertically integrated firm is illustrated in Fig. 9.2. In this arrangement a firm is not only engaged in grain purchasing and storage of grains but also owns trucks for transporting the produce from threshing floors/villages to mandi and vice versa. In addition to trading in foodgrains the firm may also be processing the grain for making livestock feed which it sells to the livestock rearers or feed retailers.

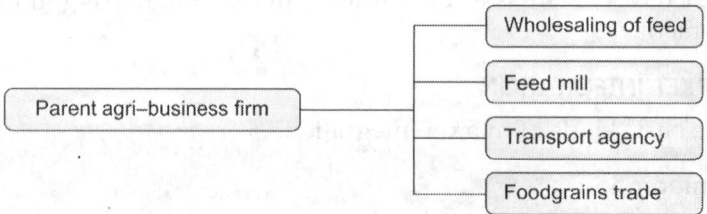

Fig. 9.2: Schematic arrangement of a vertically integrated firm

There have been many reasons for the development of such integrated operations. This type of integration makes it possible to exercise control over both the quantity and quality of the product from the beginning of the production process until the product is ready for the consumer.

Vertical integration leads to some economies in the cost of marketing. A vertically integrated firm has an advantage over other firms in respect of greater market power either in terms of sources of supplies or distribution network. Vertical integration reduces the number of middlemen in the marketing channel. It is of two types, forward or backward, depending upon the stage at which the integration occurs.

1. *Forward Integration:* If a firm assumes another function of marketing which is close to the consumption function, it is a case of forward integration; for example, a wholesaler assuming the function of retailing.
2. *Backward Integration:* This involves ownership or a combination of sources of supply; for example, when a processing firm assumes the function of assembling/ purchasing the produce from villages.

Firms often expand both vertically and horizontally. The modern retail stores are a good example of this. Retailing firms have grown horizontally by expanding either retail stores or a number of commodities they deal in. They have grown vertically by operating their own wholesale, purchasing and processing establishment.

3. Conglomeration

A combination of agencies or activities not directly related to each other may, when it operates under a unified management, be termed a conglomeration. Examples of conglomeration are Hindustan Lever Ltd. (processed vegetables and soaps), Delhi Cloth and General Mills (Cloth and Vanaspati), Birla Group, Tatas, J.K. Group, ITC and NAFED.

The schematic arrangement of a business conglomerate is shown in Fig. 9.3.

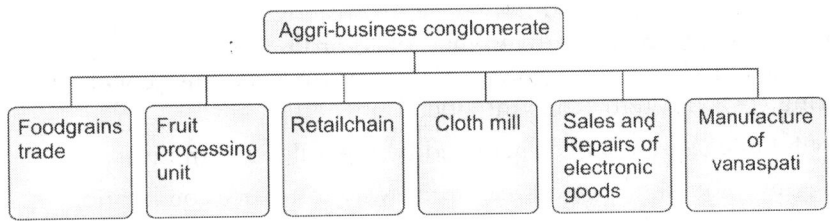

Fig. 9.3: Schematic arrangement of a business conglomerate

The conglomerate is involved in a number of different and frequently unrelated activities. For example, the firm may be dealing in foodgrains trading; processing of horticultural products; cloth milling; selling and repairs of electronic equipments; manufacturer of vanaspati; and a chain of retail outlets Such a conglomeration of activities serves as a means of spreading the risk and helps in expanding the activities to additional markets.

Most of the business firms have some degree of vertical integration, horizontal integration and conglomerate character. The main objective of·such an arrangement is to undertake closely related activities that will permit them to effectively meet the require-ments of their customers. The most common type of integration which exists in our rural markets is that a firm which buys and sells the grains is also engaged in selling of fertilisers, insecticides and pesticides, feed and such other items with the main objective of meeting the multiple needs of their customers, most of whom are farmers.

DEGREE OF INTEGRATION

There are two types of integration; based on degree of integration.

1. Ownership Integration

This occurs when all the decisions and assets of a firm are completely assumed by another firm. The example of this type of integration is a processing firm which buys a wholesaling firm or a retail chain.

2. Contract Integration

This involves an agreement between two firms on certain decisions, while each firm retains its separate identity. When dal mills of an area jointly agree on the pricing of the dals and processed product, it is a case of contract integration. Another example of contract integration is tie up of a dal mill with pulse traders for supply of pulse grains. Contract farming is a typical case of integration between farmers and processors or retailers (chains).

EFFECTS OF INTEGRATION

Integration is an attempt at organizing or co-ordinating the marketing processes to increase operational efficiency and acquire greater power over the selling and/or buying process. Like decentralization, integration in the marketing process may have both advantageous and disadvantageous effects. Whether a particular case of integration is advantageous to society or the individual can be judged by the motive with which it has been undertaken.

The vertical integration of firms may be actuated by the following motives:

1. More profits by taking up additional functions;
2. Risk reduction through improved market co-ordination;
3. Improvement in bargaining power and the prospects of influencing prices; and
4. Lowering costs through achieving operational efficiency.

Horizontal integration may be actuated by the following motives:

1. Buying out a competitor in a time-bound way to reduce competition;
2. Gaining a larger share of the market and higher profits;
3. Attaining economies of scale; and
4. Specializing in the trade.

Horizontal integration in the food industry is limited because of its potential impact on competition.

Conglomeration integration may be actuated by the following motives:

1. Risk reduction through diversification;
2. Acquisition of financial leverage; and
3. Empire-building urge.

By and large market integration reduces marketing costs, particularly the wastage, which is a social loss. But how this saving is shared between farmers, consumers and others, depends on several factors.

MEASUREMENT OF INTEGRATION

The measurement or assessment of the extent of market integration is helpful in the formation of appropriate policies for increasing the efficiency of the marketing process. The measurement or assessment of market integration may be attempted at two levels.

1. Integration among Firms of a Market

The extent of vertical integration in a market may be assessed by counting the number of functions performed by each firm in the market. The extent of horizontal integration may be measured by studying the number of firms performing the same marketing function but operating under one common management.

The result of a study on the existence of vertical and horizontal integration in the marketing of wheat in eight main wheat producing districts of Rajasthan[2] revealed that (Table 9.1) about half of the marketing firms (50.5%) were integrated vertically because they performed two or three functions. Traders in the Sultanpur and Bharatpur markets were relatively more vertically integrated than those in other markets. The ownership of firms of a similar nature by a big firm (Table 9.2) revealed the extent of horizontal integration. The selected markets had very little horizontal integration, for 94% of the firms owned only one establishment.

Table 9.1: Vertical integration of market functionaries

Markets	Total number of firms	Number of functions performed		
		One	Two	Three
1. Sultanpur	23	7 (30.4)	14 (60.9)	2 (8.7)
2. Sriganganagar	280	146 (52.1)	119 (42.5)	15 (5.4)
3. Pali	60	29 (48.3)	20 (33.3)	11 (18.4)
4. Bharatpur	103	38 (36.9)	61 (59.2)	4 (3.9)
5. Tonk	68	41 (60.3)	21 (30.9)	6 (8.8)
6. Sawai Madhopur	46	30 (65.2)	12 (26.1)	4 (8.7)
7. Pratapgarh	72	32 (44.4)	32 (44.4)	8 (11.2)
8. Dausa	104	51 (49.0)	47 (45.2)	6 (5.8)
Total	756	374 (49.5)	326 (43.1)	56 (7.4)

Note: Figures in parentheses indicate the percentage of the total number of firms in each market.

Table 9.2: Horizontal integration of market functionaries

Markets	Total number of firms	Ownership of firms of a similar nature			
		One	Two	Three	More than three
1. Sultanpur	28	27 (96.4)	1 (3.6)	–	–
2. Sriganganagar	293	285 (97.3)	5 (1.7)	2 (0.7)	1 (0.3)
3. Pali	73	69 (94.6)	2 (2.7)	2 (2.7)	–
4. Bharatpur	116	111 (95.7)	4 (3.4)	1 (0.9)	–
5. Tonk	68	62 (91.2)	3 (4.4)	2 (2.9)	1 (1.5)
6. Sawai Madhopur	51	46 (90.2)	3 (5.9)	2 (3.9)	–
7. Pratapgarh	59	46 (77.9)	7 (11.9)	4 (6.8)	2 (3.4)
8. Dausa	118	114 (96.6)	3 (2.5)	1 (0.9)	–
Total	806	760 (94.3)	28 (3.5)	14 (1.7)	4 (0.5)

Note: Figures in parentheses indicate the percentage of the total number of firms in each market.

2. Integration among Spatially Separated Markets

The extent to which prices in spatially separated markets move together or are related to transport costs reflects the degree of integration. A two-way analysis of prices in spatially separated markets may be used to assess the degree of integration.

(a) Price Correlations

One of the indicators of pricing efficiency is the extent of the interrelationship in price movements between selected markets. Uma Lele[3] has defined the interrelationship between price movements in two markets as market integration. The degree of correlation between prices in various markets is taken as an index of the extent to which the two markets are integrated. A higher degree of the correlation coefficient indicates a greater degree of integration; at least in terms of the pricing of the product between market centres and vice versa.

The correlation in the price of a commodity in any two markets is unity under conditions of perfect spatial price integration. Uma Lele[3] is of the opinion that a correlation coefficient of 0.90 or more is a high degree of inter–market price relationship because, in such a case, 81% or more variation in the prices in one market is associated with that in another market, and that the remaining 19% variation may be assumed to stem from transportation, information and data bottlenecks.

Market integration through price correlations may be studied between inter-district village level markets, between primary wholesale markets with corresponding secondary wholesale markets, and among secondary wholesale markets.

A study[4] on foodgrain prices in Rajasthan revealed that about 78% of the price correlation coefficients (168 pairs out of 216 pairs at various levels of markets) were greater than 0.90, showing a high degree of price integration. In 32 market pairs, price integration was perfect. Among crops, integration was found to be very high for gram.

(b) Spatial Price Differential and Transportation Costs

In the absence of statutory restrictions on the movements of goods the differences in prices in two markets should be equal to the transportation cost. If price differences are wider, they simply reveal market imperfections or the existence of disintegrated markets. The observed price differences may also exist because of the differences in the quality of the product and market services. Barring these, the gap between price differentials and transportation cost provides an indicator of market integration.

Spatial price differences and transportation costs may be examined between primary and secondary markets (vertical level) and among secondary wholesale markets (horizontal level).

Prices in primary wholesale markets are generally lower than in secondary wholesale markets. The price gap between the primary and secondary wholesale markets is an indicator of the inadequate flow of commodities between markets and of the efficiency of the system. If the price gap exceeds the transportation cost by a good margin, this is an indicator of poor integration, and vice versa. Similarly, the price gap between secondary markets in relation to transportation cost is an index of the marketing efficiency of price movements and integration between secondary markets.

The results of a study[5] conducted in Rajasthan on foodgrain prices reveal that after accounting for the transportation cost involved in moving the produce between these markets, the spatial price difference between primary and secondary wholesale markets and between various secondary markets was not very high. The study indicated that prices were higher in 67% months in the secondary wholesale markets than in the primary markets, but the price difference outweighed transportation cost in 34% months. In 33% months, the reverse was true. The price difference outweighed transportation cost only in 10% months.

A comprehensive review of these approaches and their merits and demerits are given by Acharya[6] (2001). According to him, the above approaches are very simple but have serious limitations. Therefore, a combination of four alternative empirical procedures have been employed by the researchers for testing of market integration. These are as follows:

1. Correlation Method: Under correlation method, bivariate correlations are estimated between time series of spot prices for a commodity from different market locations. A statistically significant correlation is taken to imply market integration. Examples of studies using this technique are of Lele[7&8], (1967, 1971), Jones[9] (1968) and Acharya[10](1988). However, there are serious problems with this method. A high-level correlation can by no means indicate the integration of markets. Further, markets may well be integrated and yet have low correlations (BIyn[11], 1973, Harriss[12], 1979). Thus, the correlation coefficients can only serve as indicators of likelihood of market integration.

2. Ravallion Procedure: The Ravallion procedure (Ravallion[13&14], 1986, 1987), which is an extension of static correlation method, avoids the main inferential dangers and is able to extract more information about markets than by the statistical correlation method. However, there are also some limitations associated with this method.

3. Co-integration Approach: The co-integration approach does not possess the problems associated with the earlier approaches. The co-integration approach has been used in spatial market analysis by Goodwin and Schroeder[15] (1991), Palaskas and Harriss White[16] (1993), Alexander and Wyeth[17](1994) and Dercon[18](1995).

   It has been argued that all the above approaches rely on price data alone and fail to recognize the pivotal role of transfer costs (Baulch[19], 1997).

4. Parity Bound Models (PBM): Parity Bound Models approach allows transfer costs to vary between periods and thus assumes no specific model of marketing margins. This model can also be estimated by using incomplete time series, which is often the case with price data in many developing countries. One of its limitations, however, lies in the requirement of data on mean transfer costs, which are rarely available. Unless the estimates of transfer costs are accurate, the whole market integration test would be suspect. Further, since only contemporaneous spreads are used in its estimation, it is hard for the PBM to take into account the type of lagged price adjustment postulated by the Granger Causality and Ravallion models. It is, therefore, advisable to estimate the PBM with price data that have been measured at a low frequency (e.g. on a monthly or quarterly basis) in order to allow sufficient time for intermarket arbitrage to occur. However, the estimation of PBM at present appears to be complicated process.

Though none of the techniques available at present is satisfactory, preference is given to co-integration method and results are interpreted with the help of other relevant information about the markets being studied.

CO-INTEGRATION ANALYSIS

Assume that we have month wise prices of a commodity for two (or more) markets X and Y. Four steps are involved in assessing the co-integration (for details see Saxena et al., 2015).

- **First Step:** Checking the stationarity of data—The first step is to examine the stationarity of each individual time series selected for analysis. Augmented Dickey-Fuller (ADF) unit root test (Dickey & Fuller, 1979), Philips Perron (PP) test (Philips and Perron, 1988) and KPSS (Kwiatkowski-Phillips-Schmidt-Shin) test can be used for the purpose.
- **Second Step:** Co-integration analysis—The co-integration depicts long-term relationship between the variables. It means even if two or more series are non-stationary, they are said to be co-integrated if there exists a stationary linear combination of them. Johansen's (1988) multivariate co-integration approach can be used to examine the co-integration between two series. For examining the linkage between two series, two-step Engle-Granger method can be used.
- **Third Step:** Examining long-term causality—For this, Granger causality test can be applied, which examines how one variable explains the latest values of another variable. According to this method, if a variable Y is Granger caused by variable X, it means that values of variable X helped in predicting the values of variable Y and vice versa.
- **Fourth Step:** Estimating Error Correction Model (ECM) for short-term relationship—The co-integration analysis reflects the long term movement of two time series, although in the short-term they may drift apart. Once the series are found to be co-integrated, then the next step is to find out the short run relationship along with the speed of adjustment towards equilibrium, using error correction model (ECM).

CO-INTEGRATION OF INDIAN AGRICULTURAL MARKETS

Co-integration analysis is another mathematical tool to analyze the integration and thereby the efficiency of the marketing system. The results of a comprehensive study by Wilson[20] (2001) have been summarized by Acharya[21] (2003). Wilson analyzed the integration of markets for three important cereals (rice, wheat and jowar) and for rapeseed-mustard and groundnut by using co-integration technique. He used month-end wholesale prices for 18 years for 14 markets for wheat, 36 markets for paddy/rice and 13 markets for jowar. The summary results are presented in Table 9.3.

Table 9.3: Price integration of wholesale markets for foodgrains and oilseeds in India

Crop	No. of markets and prices used	Area	Period	Correlation coefficient exceeded the value in percent market pairs		
				0.6	0.8	0.9
Wheat	10 markets (month-end WSP)	Different	1982–1988	62	31	6
	10 markets (month-end WSP)	states	1992–1998	80	64	42
Paddy	5 markets (monthly WSP)	Different	1981–1988	70	–	–
	15 markets (monthly WSP)	states	1991–1998	65	13	–
Jowar	7 markets (monthly WSP)	Different	1981–1988	–	–	–
	11 markets (monthly WSP)	states	1991–1998	70	32	2
Rapeseed-mustard	8 markets (monthly WSP)	Different	1982–1987	100	93	25
	8 markets (monthly WSP)	states	1992–1997	68	33	32
Groundnut	5 markets (monthly WSP)	Different	1980–1988	100	40	20
	5 markets (monthly WSP)	states	1990–1998	100	70	–

Source: Acharya, S.S. (2001), Domestic Agricultural Marketing Policies, Incentives and Integration, in the book Indian Agricultural Policy at the Crossroads, Acharya, S.S. and D.P. Choudhri (Ed.), Rawat Publications, Jaipur, pp. 184–192.

1. The results for wheat reveal that
   (a) Market adjustment has been fast during both the eighties and nineties;
   (b) The degree of market integration for the nineties has been considerably higher than during the eighties; and
   (c) The markets which demonstrated a higher degree of integration during eighties have certainly become more integrated during the nineties.
2. The results for paddy/rice show that
   (a) The markets for rice also adjust relatively rapidly; and
   (b) The markets during nineties are integrated to a higher degree compared to that during the eighties.
3. The results for jowar reveal that markets exhibit significantly rapid contributing adjustments. However,
   (a) The integration is of much lower degree as compared to wheat and rice; and
   (b) Market integration for jowar during the nineties does not appear to be higher than that during the eighties.
4. The results for rapeseed and mustard reveal
   (a) There is an increasing degree of market integration in the post-liberalization period;
   (b) The integration among markets for rapeseed-mustard also appears to be reasonably high during both the eighties and nineties; and
   (c) The results based on co-integration analysis by Wilson, however, reveal that the rapeseed-mustard markets starting from a very low degree of integration during eighties have become more integrated during the nineties.
5. The prices of groundnut reveal
   (a) A higher degree of co-movement in the nineties compared to that in eighties. The correlation coefficients were greater than 0.7 for 92% market pairs during the nineties as compared to only 38% market pairs during the eighties;
   (b) As regards the groundnut oil, the markets were highly integrated both during eighties and nineties; and
   (c) The results based on co-integration analysis by Wilson reveal that in the case of groundnut markets, integration continues to be low.

MARKETING EFFICIENCY

Marketing efficiency is essentially the degree of market performance. In this sense the concept is broad and dynamic. It encompasses many theoretical manifestations and practical aspects. Broadly, one may look at efficiency of a market structure through the following:

1. Whether it fulfils the objectives assigned to it or expectations from the system at minimum possible cost or maximises the fulfilment of objectives with given level of resources (or costs); and
2. Whether it is responsive to impulses generated through environmental changes and whether impulses are transmitted at all levels in the system.

Expectations from or objectives assigned to the system are of critical importance in assessing the efficiency because various participants have different expectations from the system, which quite often conflict with each other. For example:

1. Farmers expect quick market clearance and higher prices for their produce. They expect the market to buy the products when they are offered for sale at reasonable prices;
2. Consumers expect ready availability of products in the form and quality desired by them at lower prices;
3. Traders and other functionaries expect steady and increasing incomes; and
4. Government expect the system to safeguard the interest of all the three sections and in a proportion which is considered to be fair so that overall long-run welfare of the society is maximised.

DEFINITION OF MARKETING EFFICIENCY

The concept of marketing efficiency is so broad and dynamic that no single definition encompasses all of its theoretical and practical implications. Some of the definitions are given below:

Kohls and Uhl[22]: Marketing efficiency is the ratio of market output (satisfaction) to marketing input (cost of resources). An increase in this ratio represents improved efficiency and a decrease denotes reduced efficiency. A reduction in the cost for the same level of satisfaction or an increase in the satisfaction at a given cost results in the improvement in efficiency.

Jasdanwalla[23]: The term marketing efficiency may be broadly defined as the effectiveness or competence with which a market structure performs its designated function.

Clark[24]: Marketing efficiency has been defined as having the following three components:

1. The effectiveness with which a marketing service is performed;
2. The cost at which the service is performed; and
3. The effect of this cost and the method of performing the service on production and consumption.

Of the three components, the last two are the most important because the satisfaction of the consumer at the lowest possible cost must go hand in hand with the maintenance of a high volume of farm output.

EFFICIENT MARKETING

The movement of goods from producers to consumers at the lowest possible cost, consistent with the provision of the services desired by the consumer, may be termed as efficient marketing. A change that reduces the costs of accomplishing a particular function without reducing consumer satisfaction indicates an improvement in the efficiency. But a change that reduces costs but also reduces consumer satisfaction need not indicate increase in marketing efficiency. A higher level of consumer satisfaction even at a higher marketing cost may mean increased marketing efficiency if the additional satisfaction derived by the consumer outweighs the additional cost incurred on the marketing process.

An efficient marketing system for farm products ensures that:

1. Increase in the farm production is translated into a proportionate increase in the level of real income in the economy, thereby stimulating the emergence of additional surpluses;

2. Good production years do not coincide with low revenues to the producers achieved through effective storage, proper regional distribution and channelising of latent demand; and

3. Consumers derive the greatest possible satisfaction at the least possible cost.

An efficient marketing system is an effective agent of change and an important means for raising the income levels of the farmers and the levels of satisfaction of the consumers. It can be harnessed to improve the quality of life of the masses.

APPROACHES TO THE ASSESSMENT OF MARKETING EFFICIENCY

Traditionally, efficiency of the marketing system has been looked at from the following two angles:

1. Technical or Physical or Operational Efficiency

This aspect of the efficiency pertains to the cost of performing a function. Efficiency is said to have increased when cost of performing a function for each unit of output is reduced. This can be brought about either by reducing physical losses or through change in the technology of the function viz., storage, transportation, handling, and processing. A change in the technique may result either in the reduction of per unit cost (storage cost for a month, transportation cost to a distance of 100 kms or the cost of converting 100 kg of oranges to orange juice) or the increase in the output for a given level of cost.

2. Pricing or Allocative Efficiency

Pricing efficiency means that the system is able to allocate farm products either overtime, across the space or among the traders, processors and consumers (at a point of time) in such a way that no other allocation would make producers and consumers better off. This is achieved via pricing of the product at different stages, at different places, at different times and among different users and hence called pricing efficiency. In simple terms, the pricing efficiency is achieved when following conditions hold:

(a) Price differences between spatially separated markets do not exceed transportation cost;

(b) Intra-year price rise is not more than storage cost; and

(c) Price differences between forms of the product (pulse grain and split dal or wheat grain and wheat flour) do not exceed processing cost.

The pricing efficiency refers to the structural characteristics of the marketing system, where the sellers are able to get the true value of their produce and the consumers receive true worth of their money.

Whenever functions of transportation, storage and processing are performed, cost is incurred, value is added and the product is priced again. The efficiency of marketing is concerned with the extent to which the prices (after these functions are performed) deviate from what the cost of performing these functions warrant. The pricing aspect of marketing efficiency is affected by the extent of competition, dissemination of market information and attitude of the functionaries.

Marketing efficiency in this context may be termed as the pricing efficiency of the marketing system. The relationships between marketing costs and marketing margins and that between gross margins and prices in spatially separated markets between or at different stages of marketing reflect this aspect of marketing efficiency.

The above two types of efficiencies are mutually reinforcing in the long run; one without the other is not enough.

EMPIRICAL ASSESSMENT OF MARKETING EFFICIENCY

Some simple measures to assess the efficiency of the marketing system for agricultural commodities are:

1. Ratio of Output to Input

Conceptually, efficiency of any activity or process is defined as the ratio of output to input. If O and I are, respectively, output and input of the marketing system and E is the index of marketing efficiency; then

$$E = \frac{O}{I} \times 100$$

A higher value of E denotes higher level of efficiency and vice versa. When applied in the area of marketing, output is the 'value added' by the marketing system and 'input is the real cost of marketing (including some fair margins of intermediaries)'. The measurement of 'value added' is not easy. The difference in the price at the farm level (price received by the farmer) and that at the retail level (price paid by the consumers) may be used to measure the 'value added' but it has limitations mainly because of market imperfections. Assuming that degree of imperfection is pervasive, this measure has been used to compare the marketing efficiency of two spatially separated markets, of two commodities or at two points of time. Consider the following examples of marketing efficiency.

Let there be two markets (or channels), A and B, for a commodity. The produce moves in both the markets. Let the marketing costs and value added in these markets be as given in Table 9.4. The efficiency measure is calculated as illustrated in Table 9.4.

Table 9.4: Marketing efficiency of two hypothetical markets		
Particulars	**Market A**	**Market B**
Total marketing cost incurred by all those involved (₹/ton)	1000	1600
Value added measured in terms of difference in the consumer's price and price received by the farmer (₹/ton)	2000	4800
Marketing efficiency (E) (%)	200	300

The obvious conclusion is that market B is more efficient than market A, though marketing cost is higher in market B. But this conclusion may be misleading. It can be argued that consumers were charged three times the actual cost of marketing in market B whereas they were charged only twice in market A. Market B infact could be less efficient than market A. One needs to go into the question of degree of competition in two markets.

2. Shepherd Approach

Shepherd[25] suggested that the ratio of the total value of goods marketed to the marketing cost may be used as a measure of marketing efficiency. The higher the ratio, the higher efficiency and vice versa. This method eliminates the problem of measurement of value added. Consider the following example of working out the marketing efficiency of banana market using the Shepherd's formula:

The banana is sold either to the pre-harvest contractors or directly in the market to the retailer eliminating the pre-harvest contractors. The farmer sold banana to the pre-harvest contractor @ ₹ 1900 per 100 bunches. The retailer's purchase price was ₹ 3270 per 100 bunches. The hypothetical costs and margins on sale of banana are shown in Table 9.5. It can be seen that the marketing efficiency of Channel II is more than that of Channel I. Shepherd's formula does not explicitly take into account the net price received by the farmers in assessing the marketing efficiency. Shepherd's formula assumes that marketing cost itself includes some fair margins of intermediaries. But if the margins retained by the intermediaries are excessive, it is argued that these should not be treated as a part of marketing cost.

Table 9.5: Estimates of marketing costs, margins and marketing efficiency for banana (hypothetical data for illustration)

(₹ per 100 bunches)

Particulars	Channel I (contract sale)	Channel II (direct sale)
1. Marketing costs incurred by producers	–	870
2. Marketing costs incurred by pre-harvest contractor	870	–
3. Margin of pre-harvest contractor	500	–
4. Marketing costs incurred by retailer	320	320
5. Marketing margin of retailer	610	610
6. Total costs and margins	2300	1800
7. Retailer's sale price or consumer's purchase price	4200	4200
8. Net price received by the producer (7 – 6)	1900	2400
9. Shepherd's index of marketing efficiency (7 ÷ 6)	1.83	2.33

3. Acharya[26] Approach

According to Acharya, an ideal measure of marketing efficiency, particularly for comparing the efficiency of alternate markets/channels, should be such which takes into account all of the following:

(a) Total marketing costs (*MC*)
(b) Net marketing margins (*MM*)
(c) Prices received by the farmer (*FP*)
(d) Prices paid by the consumer (*RP*)

Further, the measure should reflect the following relationship between each of these variables and the marketing efficiency (the assumption of "other things remaining the same" is implicit):

1. Higher the (a), lower the efficiency
2. Higher the (b), lower the efficiency
3. Higher the (c), higher the efficiency
4. Higher the (d), lower the efficiency.

As there is an exact relationship among four variables, i.e. $a + b + c = d$, any three of these could be used to arrive at a measure for comparing the marketing efficiency. The following modified measure is, therefore, being suggested by Acharya[26]:

$$MME = FP \div (MC + MM)$$

where *MME* is the modified measure of marketing efficiency and *MC* and *MM* are marketing costs and marketing margins respectively.

A comparison of efficiency measures as worked out by three different methods is given in Table 9.6. The conventional method (*E*) suggests that market C is more efficient than market A which, in turn, is more efficient than market B. Note that the price received by the farmer in market C is the lowest. Hence, this method is not suitable under Indian conditions.

Table 9.6: Measurement of marketing efficiency: Alternative approaches

Particulars	Unit	Market A or Channel A	Market B or Channel B	Market C or Channel C
1. Retailer's sale price or consumer's purchase price (RP)	₹ per quintal	1000	1000	1000
2. Total marketing costs (MC)	₹ per quintal	300	500	300
3. Total net margins of intermediaries (MM)	₹ per quintal	200	100	400
4. Net price received by farmers (FP)	₹ per quintal	500	400	300
5. Value added (1–4)	₹ per quintal	500	600	700
Index of marketing efficiency				
1. Conventional method (E) 5 ÷ 2	Ratio	1.67	1.20	2.33
2. Shepherd's method (ME) 1 ÷ 2	Ratio	3.33	2.00	3.33
3. Acharya's method (MME) 4 ÷ (2 + 3)	Ratio	1.00	0.67	0.43

If marketing margins are not included as a part of marketing cost, the Shepherd's method (*ME*) suggests that market A and market C are equally efficient. Further, these two markets are more efficient than market B. The limitation of this method, as mentioned earlier, is that it does not take into consideration the price received by the farmer.

The limitations of both these methods are taken care by the modified method suggested by Acharya. According to Acharya's method (*MME*), market A is more efficient than market B which, in turn, is more efficient than market C.

Acharya's measure of marketing efficiency can also be stated as:

$$MME = [RP \div (MC + MM)] - 1$$

because $RP = FP + MC + MM$.

In using these methods for comparing the marketing efficiency of alternate channels, it must be recognised that:

1. if marketing cost is defined in a broader sense to include margins of intermediaries also, there is no difference in the conclusion arrived by using Shepherd's formula and Acharya's modified method; and
2. the time, place and form of the commodity at the beginning and end of the channel should be the same in all the channels/markets which are being compared.

In real life situations, the place and form of the commodity are not the same in different channels. Therefore, enough care is needed in deriving conclusions about the efficiency of alternative marketing channels/system. We discuss these aspects in the following sections.

MARKETING COSTS AND EFFICIENCY

Generally, high marketing costs and margins are considered to be indicators of inefficiency in the marketing process. But this is not always true. The fact that a major part of the consumer's rupee is spent on marketing costs does not always mean that something is wrong with the distribution system. A number of factors may operate to cause a high proportion of marketing costs, without any reflection on the efficiency of the marketing system. These are as follows:

1. Place of Production

The geographical localization of production brings about a change in the marketing costs. The example in Table 9.7 illustrates the effect of location on the cost of marketing. Let there be two areas—A and B; one near the market and other away from the market. The marketing cost in area A is 30% of the buyer's price; but it is 40% in area B. Nevertheless, the higher marketing cost in area B is not a reflection of the inefficiency of the marketing system. The distance to which the commodity is transported is higher in the case of B, which results in lower net price accruing to the farmer.

Table 9.7: Effect of geographical location on production and marketing costs

(₹)

Particulars	Production areas	
	A (near the market)	B (away from the market)
Net price for farmer/unit	0.70	0.60
Cost of marketing/unit	0.30	0.40
Buyer's price (economic cost for the buyers)	1.00	1.00

2. Time of Production

Food articles are made available throughout the year, which is possible only by storage. This adds to the cost. The cost of goods sold in the off-season is higher than that in the peak season.

To understand this point, consider the case of wheat. The wheat is generally harvested in the end of March or first fortnight of April. Threshing and winnowing also take time. Assume that the produce is ready for sale by the middle of April. A farmer in Hisar sold his produce in April, 2019 at a price of ₹ 2000 per quintal. If this lot of wheat moves in its normal marketing channel and reaches a consumer in Delhi within say one week, the items of marketing costs are unloading, market fee, weighment, commission etc. in the Hisar Market, and many such items in Delhi market including wholesaler's and retailer's margin. Suppose all these items and transportation cost add up to ₹ 500 per quintal. This means that the consumer in Delhi pays a price of ₹ 2500 per quintal in April, 2019.

If in the month of March, 2020, a retailer in Delhi sells wheat at a price of (say) ₹ 2789 per quintal, it may be justified and may not indicate inefficiency of the marketing system. Some body in the marketing channel has stored this wheat for about 11 months. As the market rate of interest during 2019–20 was around 14% per annum, even if there was no cost of any other item, the interest alone on the value of one quintal of wheat amounts to about ₹ 257 if stored by the farmer and around ₹ 321 if stored by the retailer. Taking a middle position about the stage in the marketing channel where the wheat was stored for about 11 months, the total marketing cost of wheat sold in March, 2020 amounts to about ₹ 789 per quintal (289 + 500). In this case, the increase in the marketing cost does not reveal decrease in the marketing efficiency. This has been illustrated in Table 9.8.

Table 9.8: Effect of sale in different seasons on marketing costs and consumer's price

(₹ per quintal)

Particulars	Sale to the consumers in	
	Post-harvest season	Lean season
Price received by the farmer	2000	2000
Cost of marketing	500	789
Consumer's price	2500	2789

3. Form of Product

The cost of marketing of a processed product is higher than that of a raw product. For example, the cost of marketing tomato juice (which includes cost of processing tomato) is higher than that of fresh tomatoes. But the high marketing cost because of processing is not an indicator of the inefficiency of the system. This changed form adds to the satisfaction of the consumers demand.

The farmer's share, whether it is higher or lower, in consumer's rupee is a measure of marketing efficiency. But there has been a decrease in the farmers share over time. This need not mean a decrease in the efficiency of the system. With the rise in the standard of living and income of the consumers, there is an increased demand for more processed materials and better marketing services, which increase the marketing costs and reduce the farmer's share. This decline in the farmer's share over time is not an adequate indicator of low marketing efficiency.

The levels of marketing costs and margins, or the share of the producer in the price paid by the consumer, should be used as measures of marketing efficiency with some care. A lower marketing cost expressed as a percentage of the consumers' price is not always an indicator of higher marketing efficiency, or vice versa. If that were so, one would conclude that the marketing system in India is more efficient than the system in the USA. In the USA, producers get less than 20% of the price paid for bread by consumers, whereas the Indian farmer gets more than 60% of the price paid by the consumers for his wheat. Even if we include the cost of grinding wheat into flour and then making bread, the share of the Indian farmer will still be much higher than that of the USA farmer.

According to Kohls[27], it is fallacious to take the relative size of marketing margin as an important indicator of marketing efficiency. According to him, the important thing is not the size of the share but rather the total return which is received by agricultural producers from the sale of their products. Higher marketing costs and a more prosperous agriculture are compatible ideas. Dantwala[28] has pointed that it is not necessarily true that the system that provides the distributive service at the lowest cost is the most effective. Good and honest marketing may actually cost more than a slipshod and corrupt one.

OVERVIEW OF MARKETING EFFICIENCY IN INDIAN AGRICULTURE

In a report on Agricultural Marketing in India as a part of the Millennium Study of Indian Farmers prepared by Acharya (2003), based on the review of the past studies it was summarised that:

1. The marketing efficiency in India was low in 1920s as reported by Royal Commission on Agriculture (1928) and Central Banking Enquiry Committee (1931) because of the existence of various malpractices in the trade of agricultural commodities. The situation improved somewhat after the establishment of regulated markets in the country.
2. The degree of marketing/pricing efficiency is relatively low for fruits, vegetables, flowers and other perishable products due to inadequate infrastructure to handle all such perishable products.
3. Pricing efficiency is high for foodgrains and oilseeds even in spatially separated markets as price movements were not higher than transportation costs in more than 80% market pairs. Similarly, inter-seasonal price rise is not greater than storage cost.
4. The improvement in marketing efficiency depends on the competitiveness of the markets. In enhancing the competitiveness of markets, information has a crucial role and this information is not equally accessible to all the market participants. In this connection, encouragement needs to be provided to generate and host useful portals, databases and websites on agricultural marketing.

EFFICIENCY OF INDIAN FOODGRAINS MARKETING SYSTEM

The Indian foodgrains marketing system may be said to be an efficient one because of the following reasons:

1. Existence of Competitive Conditions

The existence of competitive conditions and the desire to maximize profits are the main forces which induce firms to operate efficiently. The following characteristics reflect the competitive conditions in the Indian foodgrain markets:

(a) Market Structure: There is a large number of competitive buyers of foodgrains in the markets.
(b) Market Behaviour: The functionaries in the market reveal no, or very little, tendency to collusion.
(c) Market Performance: Traders in the market function at lower profits and perform various activities at the least possible cost.

2. Technological Advancement

There has been a rapid technological advancement in every field of the market process which has resulted in the reduced cost of market performance. New technological developments in the areas of transportation, processing, storage and market news have resulted in cost reduction and increased marketing efficiency.

3. Low Wages

The marketing system is labour-intensive and the wage rates of labour are lower in India than those in developed countries. The performance cost of market operations is, therefore, low in India.

4. Low Consumer's Income

Consumers' income in India is low. They do not, therefore, prefer the processed form of the products and other market services. As a result, the marketing costs in India are low.

The degree of efficiency of the Indian foodgrain markets varies from market to market within the country. Markets may be said to be efficient if they fulfil the following conditions:

(a) price differentials for the same quality of products in two spatially separated markets should not exceed the cost of transportation;

(b) Price differentials for the same quality of product in the same market over time should not exceed the cost of storage; and

(c) Price differentials for different forms of products in the same market should not be more than the cost of processing them.

For some markets, the above conditions do not hold good because of the following reasons:

(a) **Imperfect knowledge** on the part of buyers and sellers of the demand and supply and other factors associated with the products;

(b) **The existence of monopolistic conditions.** Sometimes, there may be a single firm or a group of firms following some of the tactics for increased profits either by monopolising the situation or by price discrimination;

(c) **Interference of government in free market transactions:** Government interference in free market transactions in the economy in the form of restrictions on foodgrain movements, food zones and rationing which result in increasing the price differentials that are greater than the differentials in transportation, storage and/ or processing costs, which ultimately reduce the efficiency of the marketing system. However, the present mix of interventions in India in the form of price support to the farmers and subsidized distribution of cereals to the targeted groups of consumers is consistent with the national goal of food security for all.

In order to function with maximum efficiency, agricultural markets must satisfy most of the conditions of perfect markets. This is so because the requirements of an efficient marketing system coincide with the requirements of perfect markets. The conditions of perfect competition are:

1. Existence of large number of buyers and sellers in the marketing system without collusion or agreements;

2. Perfect knowledge about market conditions and the logical utilization of that knowledge by all;

3. Homogeneity of the product; and

4. Free mobility of the buyers, sellers and products in the economy.

The second phase of agricultural marketing reforms currently on going, as discussed in earlier chapters would further improve the marketing efficiency of Indian agricultural markets.

▌MARKETING COSTS, MARGINS AND PRICE SPREAD

Market functionaries or institutions move the commodities from the producers to consumers. Every function or service involves cost. The intermediaries or middlemen make some profit to remain in the trade after meeting the cost of the function performed.

In the marketing of agricultural commodities, the difference between the price paid by consumer and the price received by the producer for an equivalent quantity of farm produce is often known as farm-retail spread or price spread. Sometimes, this is termed as gross marketing margin. The total or gross margin includes:

1. The cost involved in moving the product from the point of production to the point of consumption, i.e. the cost of performing the various marketing functions and of operating various agencies; and
2. Profits of the various market functionaries involved in moving the produce from the initial point of production till it reaches the ultimate consumer. The absolute value of the marketing margin varies from channel to channel, market to market and time to time for a product.

CONCEPTS OF MARKETING MARGINS

There are two concepts of marketing margins.

1. Concurrent Margins

These refers to the difference between the prices prevailing at successive stages of marketing at a given point of time. For example, the difference between the farmer's selling price and retail price on a specific date is the total concurrent margin. Concurrent margins do not take into account the time that elapses between the purchase and sale of the produce.

2. Lagged Margins

A lagged margin is the difference between the price received by a seller at a particular stage of marketing and the price paid by him at the preceding stage of marketing during an earlier period. The length of time between the two points denotes the period for which the seller has held the product. The lagged margin concept is a better concept because it takes into account the time that elapse between the purchase and sale by a party and between the sale by the farmer and the purchase by the consumer.

The method of calculating lagged margins is based on the same principle as that involved in the first in–first out method of accounting. However, it is difficult to obtain data on time lags between purchase and sale with a view to maintaining continuous series of marketing margins.

IMPORTANCE OF STUDY OF MARKETING MARGINS AND COSTS

Studies on marketing margins and costs are important, for they reveal many facets of marketing and the price structure, as well as the efficiency of the system.

1. The magnitude of the marketing margins relative to the price of the product indicates the efficiency or otherwise of the marketing system. It refers to the efficiency of the intermediaries between the producer and the consumer in respect of the services rendered and the remuneration received by them. While comparing the efficiency of the marketing system by means of marketing margins over space or time, the difference in the value added to the product through various services/functions is taken into account;
2. Such studies help in estimating the total cost incurred on the marketing process in relation to the price received by the producer and the price paid by the consumer. The cost incurred by each agency in different channels and the share of each agency

in the cost are revealed. This knowledge ultimately helps us to identify the reasons for high marketing costs and the possible ways of reducing them; and

3. The knowledge of marketing margins helps us to formulate and implement appropriate price and marketing policies. Excessive margins point to the need for public intervention in the marketing system.

ESTIMATION OF MARKETING MARGINS AND COSTS

Regular monitoring of marketing margins at regional levels are essential for the formulation and successful implementation of marketing and price policies. A study of marketing margins should include an estimation of the producers' share in the consumer's rupee, the cost of marketing functions and the margins of intermediaries. Marketing margins and costs vary from commodity to commodity, and depend on the amount of processing involved and the market structure for handling of the commodity. Even for the same commodity, the margin may vary from place to place and time to time. A number of factors, such as the method of assembling, the location of the market and the mode of transportation, influence marketing costs and margins. The method of sale, weighment and other facilities, too, affect the marketing costs. Because of a lack of standard grading in agricultural commodities, it is very difficult to make valid comparisons of price data. Adequate precautions have, therefore, to be taken when comparing marketing margins for commodities under different situations.

Inspite of these difficulties, various studies have been conducted in India to study marketing margins and costs with a view to assessing the farmers' share in the consumer's rupee and suggesting measures for improvements in the marketing system. These studies have used different approaches, and vary considerably in their depth.

Three methods are generally used in the computation of marketing margins and costs.

1. Chasing of Lot Method

A specific lot or consignment is selected and chased through the marketing system until it reaches the ultimate consumer. The cost and margin involved at each stage are assessed. The difficulties or limitations of this method are:

(a) It is difficult to chase the movement of a lot from the producer to the ultimate consumer.
(b) Most of the lots lose their identity during the process of marketing, because either the product gets processed or the lot gets mixed up with other lots.
(c) There is no assurance that the lot selected is representative of the whole product.

This method is appropriate for such perishable farm commodities as fruits, vegetables, and milk, because the lag between the time the commodity enters the marketing system and time of its final consumption is very small.

2. Sum of Average Gross Margins Method

The average gross margin at each successive level of marketing is worked out by dividing the difference of the money value of sales and purchase by the number of units of the commodity transacted by a particular agency. The average gross margins of all the intermediaries are added to obtain the total marketing margin as well as the break-up of the consumer's rupee.

The following formula may be used to work out the total marketing margins:

$$M_T = \sum_{i=1}^{n} \left(\frac{S_i - P_i}{Q_i} \right)$$

where M_T = Total marketing margin; S_i = Sale value of a product for ith firm; P_i = Purchase value of a product paid by the ith firm; Q_i = Quantity of the product handled by ith firm; $i = 1, 2, \ldots\ldots\ldots n$ (number of firms involved in the marketing channel).

This method requires considerable effort in the location and examination of the records kept by the intermediaries. The main difficulties in using this method are:

(a) Traders may not allow access to their account books. It would then be difficult to obtain complete and accurate information. Moreover, some traders often make manipulated entries in their account books to evade sales tax, VAT, GST and even income tax; and

(b) This method necessitates adjustment for the difference between the quantities purchased and sold because a part of the product is wasted during handling.

3. Comparison of Prices at Successive Stages of Marketing

Under this method, prices at successive stages of marketing at the producer's, wholesaler's and retailer's levels are compared. The difference is taken as the gross margin. The margin of an intermediary is worked out by deducting the ascertainable costs from the gross margin earned by that intermediary. This method is appropriate when the objective is to study the movements of marketing costs and margins in relation to prices and cost indices. The main difficulties encountered in the use of this method are:

(a) Representative and comparable series of prices for the same quality at successive stages of marketing are not readily available for all the products;

(b) Adjustment for a loss in the quality of the product at various stages of marketing due to wastage and spoilage in processing and handling is difficult;

(c) The price quotation may not cover the price of a product of a comparable quality; and

(d) The time lag between the performance of various marketing operations is not properly accounted for.

The following general rules may be adopted in selection of the method for calculating marketing margins and costs of various agricultural commodities:

Commodities	Method Recommended
(a) For perishable farm products like fruits, vegetables and milk, where the time lag between the commodity entering the marketing system and the time of final consumption is very small.	Chasing of lot or consignment method.
(b) Commodities which require processing before sale to consumers such as paddy, oil-seeds, etc.	Concurrent margins should be calculated by finding the differences in the prices prevailing on the same date at successive levels of marketing.
(c) Commodities not requiring processing before sale to consumers, such as wheat, maize, bajra, jowar, etc.	By comparing the prices prevailing at successive levels of marketing on the same date either for the same market or for a pair of markets.

Irrespective of the method followed, the following information is required for computing marketing costs and margins:

(a) Data on prices of the same variety and quality of the commodity at different stages of marketing, either for one market or for a pair of markets;
(b) Data on marketing charges in cash or kind;
(c) Cost of transportation of the produce at different levels of marketing;
(d) Cost of processing and estimates of the conversion factor from the raw material to finished products;
(e) Cost of all other operations in the marketing process.

Various measures of the price spread and for the computation of marketing costs and margins, and the procedures followed have been given in the paragraphs that follow.

Producer's Price

This is the net price received by the farmer at the time of first sale. This is equal to the wholesale price at the primary assembling centre, minus the charges borne by the farmer in selling his produce. If P_A is the wholesale price in the primary assembling market and C_F is the marketing cost incurred by the farmer, the producer's price (P_F) may by worked out as follows:

$$P_F = P_A - C_F$$

Producer's Share in the Consumer's Rupee

It is the price received by the farmer expressed as a percentage of the retail price (i.e. the price paid by the consumer). If P_r is the retail price, the producer's share in the consumer's rupee (P_s) may be expressed as follows:

$$P_S = (P_F \div P_r)\ 100$$

Marketing Margin of a Middleman

This is the difference between the total payments (cost + purchase price) and receipts (sale price) of the middleman (ith agency). Three alternative measures may be used.

1. Absolute margin of ith middleman (A_{mi})

$$A_{mi} = P_{Ri} - (P_{Pi} + C_{mi})$$

2. Percentage margin of ith middleman (P_{mi})

$$P_{mi} = \frac{P_{Ri} - (P_{pi} + C_{mi})}{P_{Ri}} \times 100$$

3. Percentage mark-up of ith middleman (M_i)

$$M_i = \frac{P_{Ri} - (P_{pi} + C_{mi})}{P_{pi}} \times 100$$

where P_{Ri} = Total value of receipts per unit (sale price); P_{pi} = Purchase value of goods per unit (purchase price); C_{mi} = Cost incurred on marketing per unit.

The margin thus calculated include the profit of the middleman and the returns which accrue to him for storage, the interest on capital and overhead, and establishment expenditure.

Total Cost of Marketing

The total cost, incurred on marketing either in cash or in kind by the producer seller and by the various intermediaries involved in the sale and purchase of the commodity till the commodity reaches the ultimate consumer, may be computed as follows:

$$C = C_F + C_{m1} + C_{m2} + C_{m3} + \ldots + C_{mn}$$

where C = Total cost of marketing of the commodity; C_F = Cost paid by the producer from the time the produce leaves the farm till he sells it; and C_{mi} = Cost incurred by the ith middleman in the process of buying and selling the product.

Some of the costs are linked with the quantity marketed and some are linked with the value of the commodity. The former is a fixed charge, while latter is a variable one. The actual rates of charges are converted in terms of the weight unit or ₹ 100 worth of produce sold. The ad valorem charges are calculated on the basis of the actual market price for the physical unit or ₹ 100 worth of produce sold.

FARMER'S SHARE AND GROSS MARKETING MARGINS

According to Acharya (2003), the gross marketing margins (GMM) can be broken down into three components viz., cost of performing various marketing functions, statutory taxes or levies payable in the marketing channel, and net marketing margins (NMM) retained by market functionaries.

Marketing cost varies from commodity to commodity and changes overtime and space. Marketing costs depend on the perishability of the commodity, need for cold storage facilities, need for processing before consumption, necessity of storage and transportation, distance for transportation and nature of packaging needed. The marketing costs are, therefore, generally high for fruits, vegetables, flowers, sugarcane and cotton compared to foodgrains. Statutory marketing charges include taxes and levies (sales tax or VAT, market fee, octroi, special duty or cess on commercial crops etc.) which are paid in the process of transactions of commodity at different stages of marketing. The rates of these charges vary from state to state, market to market and commodity to commodity. Most of these taxes and levies are on ad valorem basis and as such their incidence is higher on high value crops. The market players have no control on these taxes and levies as these are of statutory in nature. These statutory charges exert considerable effect on gross marketing margins and farmer's share in consumer's rupee. Net marketing margin (NMM) is the amount retained by different market functionaries. The size of net marketing margin depends on the nature of competition, structure of markets and scale of business. Larger the net marketing margin, greater is the inefficiency of the marketing system.

It is now increasingly realized that higher marketing costs do not always reflect inefficiency of the marketing system. The factors, which cause high marketing costs, could be geographical localization of production away from the markets, necessity of storage from production season to the lean season and involvement of processing function in the marketing process. Under such situations, the size of marketing costs reflects only one side of the coin and the other aspects viz., consumer satisfaction is not given any weightage.

Over the period, gross marketing margins (GMM) decreased in foodgrains and oilseed crops due to better competitive conditions in the trade of these commodities. On the other hand, GMM increased in fruits and vegetables due to the expansion in the markets for these crops and their products. As against this, over the period, however, total cost of marketing in absolute terms have shown an increase due to:

1. Increased necessity of packaging all goods;
2. Increased availability of facilities of transportation, communication and storage leading to long distance transportation and storage from production to lean season of the year;
3. Widening of markets due to liberalization of trade and expansion in size of markets leading to movement of products to distant domestic and foreign markets;
4. Increase in the consumer's income leading thereby to higher demand of processed, packed and branded products;
5. Increase in the general price level in the economy thereby leading to increase in the cost of marketing as many marketing charges are linked to the value of the commodity; and
6. Increase in the statutory marketing charges overtime by the government, which in some cases account for 12 to 18% of the gross marketing margins.

A comprehensive review of Indian literature reveals that studies on price-spread and marketing margins for the period 1960 to 1975 are available for only a few crops (wheat, rice, sorghum, pearl millet, chickpea and groundnut). However, in the later period, i.e. 1975–2000 the studies have covered almost all agricultural products—foodgrains, oilseeds, cotton, fruits, vegetables and flowers. (For a summary of results see Acharya, 2003).

There is ample evidence of large variability of the producers share in consumer rupee as well as marketing margins and costs across the crops and study areas. Disregarding the extremities, the farmers share in consumers rupee has been estimated as 56 to 89% for paddy, 77 to 85% for wheat, 72 to 86% for coarse grains, 79 to 86% for pulses and 40 to 85% for oilseeds. The farmer's share in consumer's rupee for perishable farm products (fruits, vegetables and flowers) is generally lower and varied from 32 to 68%.

The studies in general reveal that the producer's share in consumer's rupee has varied with the marketing channel adopted by the farmers. The DMI studies reveal (Table 9.9) that the costs were higher when farmers adopted private channels in marketing of surplus produce compared to the institutional channels and hence farmer's share was lower when they sell through private channels.

A recent comprehensive analysis of statutory charges/taxes and transport and storage costs of wheat by Ramesh Chand[29] has shown that the mark up over farm harvest price prevailing during post-harvest season in a surplus state (like Punjab) needed to attract private sector in wheat trade is 74% to 126% (Goa) for the month of next March. This implies that for wheat supplied to a consumer in Goa in the month of next March, the share of a Punjab wheat grower (based on the price received in the preceding harvest month of May) in the consumer's price is 44.2%. This also means that the statutory charges and marketing costs (storing wheat from May to next March and transportation from Punjab to Goa included) add up to 55.8% of the consumer's price.

Sale of fruits through pre-harvest contractors is also common in fruit producing areas. The studies on estimates of marketing costs and margins reveal that farmers

Table 9.9: Price spread in private and institutional channels in selected agricultural commodities in India (1982–83)

(percentages of consumer's price)

Commodities	Marketing channel	Farmer's share	Marketing costs	Net marketing margins
Rice	Private	65.0	17.7	17.3
	Institutional	66.0	27.0	7.0
Wheat	Private	65.8	20.0	14.2
	Institutional	66.8	27.5	5.7
Apple	Private	41.9	35.0	23.1
	Institutional	52.2	26.2	21.6
Onion	Private	40.6	35.7	23.7
	Institutional	42.2	36.1	21.7
Groundnut	Private	63.6	19.0	17.4
	Institutional	87.6	11.2	1.2

Source: Directorate of Marketing and Inspection, Ministry of Agriculture, Government of India, Faridabad (1985).

Table 9.10: Price spread in marketing of pear fruits through pre-harvest contractors, August 2004

(₹ per quintal)

Particulars	Channel I (through pre-harvest contractors)		Channel II (through retailer)	
	Delhi market	Amritsar market	Delhi market	Amritsar market
1. Producer's net price	604	604	940	914
2. Producer's cost of marketing	–	–	302	123
3. Contractor's cost	318	145	–	–
4. Contractor's net margin	320	288	–	–
5. Retailer's costs	88	140	88	140
6. Retailer's net margin	270	133	270	133
7. Consumer's purchase price	1600	1310	1600	1310
Percent share (%)				
1. Producer's share	38	46	59	70
2. Producer's cost	–	–	19	9
3. Contractor's net margin	20	22	–	–
4. Contractor's cost	20	11	–	–
5. Retailer's net margin	17	10	17	10
6. Retailer's cost	5	11	5	11
7. Consumer's price	100	100	100	100

Source: Sekhon, Navdeep Singh, A.S. Bhullar and M.S. Sidhu, Production and Marketing of Pear in Amritsar District of Punjab, Indian Journal of Agricultural Marketing, Vol. 20 (2), May-August 2006, p. 13.

receive a lower price when they sell through the contractor. (For illustration, see Table 9.10)

The gross marketing margins in marketing of agricultural products have also been worked out from National Accounts Statistics by Acharya, S.S. (1998). In this approach, difference between the total consumers expenditure on a particular farm product and the value of the output at the farm level has been used to estimate gross marketing margin. Based on an aggregate accounting, the gross marketing margin (GMM) as percentage of consumer's price is 19.2 in cereals, 7.2 in oilseeds, 32.9 in fruits and vegetables, 6.7 in milk and milk products, and 37.2 in sugarcane with an overall average of 19.3% for all agricultural commodities. The estimates are shown in Table 9.11.

Table 9.11: Gross marketing margins for major agricultural commodities in India using aggregate accounting approach based on data for 1986–87

(percentages)

Crop groups/crops	Gross marketing margin
Cereals	19.2
Oilseeds	7.2
Fruits and vegetables	32.9
Milk and milk products	6.7
Sugarcane/sugar/gur	37.2
Overall	19.3

Source: Acharya, S.S., Agricultural Marketing in India: Some Facts and Emerging Issues, Indian Journal of Agricultural Economics, 53(3), July–September 1998, pp. 311–32.

FACTORS AFFECTING THE COST OF MARKETING

Studies on the cost of marketing reveal that there is a large variation in the cost per quintal or per ₹ 100 worth of the produce. The factors which affect marketing costs are:

1. Perishability of the Product: The cost of marketing is directly related to the degree of perishability. The higher the perishability, the greater the cost of marketing, and vice versa.
2. Extent of Loss in Storage and Transportation: If the loss in the quality and quantity of product, arising out of wastage or spoilage or shrinkage during the period of storage or in the course of transportation is substantial, the marketing cost will go up.
3. Volume of the Product Handled: The larger the volume of business or turnover of a product, the less will be the per unit cost of marketing.
4. Regularity in the Supply of the Product: If the supply of the product is regular throughout the year, the cost of marketing on per unit basis will be less than in a situation of irregular supply or supply restricted to a few months of the year.
5. Extent of Packaging: The cost of marketing is higher for the commodities requiring packaging.
6. Extent of Adoption of Grading: The cost of marketing of ungraded product is higher than that of the products in which grading can be easily adopted. However, elaborate grading adds to the cost.
7. Necessity of Demand Creation: If substantial advertisement is needed to create the demand of prospective buyers, the total cost of marketing will be higher.
8. Bulkiness of the Product: The marketing cost per unit weight of bulky products is higher than that of products which are not bulky.
9. Need for Retailing: The greater the need for the retailing of a product, the higher the total cost of marketing;

10. Necessity of Storage: The cost of the storage of a product adds to the cost of marketing, whereas the commodities which are produced and sold immediately without any storage attract lower marketing cost.
11. Extent of Risk: The greater the risk involved in the business for a product (due to either the failure of the business, price fluctuations, monopsony of the buyer or the prevalence of unfair practices), the higher is the cost of marketing.
12. Facilities Extended by the Dealers to the Consumers: The greater the facilities extended by the dealer to the consumer (such as return facility for the product, home delivery facility, the facility of supply of goods on credit, the facility of offering entertainment to buyers, etc.), the higher the cost of marketing.

REASONS FOR HIGHER MARKETING COSTS OF AGRICULTURAL COMMODITIES

Generally, the cost of marketing of agricultural commodities is higher than that of manufactured products. The factors responsible for this phenomenon are:

1. Widely Dispersed Farms and Small Output per Farm: There are innumerable producers of agricultural products, each producing a small quantity. Producers are widely dispersed. Hence the cost of assembling is high.
2. Bulkiness of Agricultural Products: Most farm products are bulky in relation to their value. This results in a higher cost of handling, storage and transportation.
3. Difficult Grading: Grading is relatively difficult for agricultural products. Each lot has to be personally inspected during purchase and sale—a fact which increases marketing costs. The sale or purchase by contract or sample is not easy because an inspection of each lot of the product is required by reason of variation in their quality.
4. Irregular Supply: Agricultural products are characterised by seasonal production. Their market supply, therefore, fluctuates during the year. In times of glut, prices go down and the cost of marketing functions, on value basis, goes up.
5. Need for Storage and Processing: There is a greater need for the storage of agricultural products because of the seasonality of their production. The processing of agricultural products is a necessity because all the agricultural products are not consumed in the raw form. Storage and processing add to the cost of marketing. Losses of agricultural products in storage are also high because of their perishability.
6. Large Number of Middlemen: In foodgrain marketing, the number of middlemen is larger because there is no restriction on their entry in the trade. Contrarily, there are many restrictions on the entry into the trade of industrial products. For example, the cumbersome licensing procedure, high risk and high capital requirement make entry into trade in non-farm goods somewhat difficult. The larger the number of middlemen, the higher the marketing costs.
7. Risk Involved: The risk of price fluctuations is higher in agricultural products. The higher risk leads to higher risk premium, which adds to the marketing cost.

MARKETING COST IN INDIA AND OTHER COUNTRIES

In India, the marketing cost of foodgrains is lower than in developed countries. The farmer's share in the price paid by the consumer is higher in India than in developed countries. The factors responsible for this difference are:

1. Foodgrains are sold in a relatively unprocessed form in India, whereas in developed countries, consumers want them mostly, in a processed and ready-to-eat form. In India, the processing of foodgrains is mainly undertaken at the consumers' level.

Therefore, the cost of marketing is lower, and the farmers' share in consumer's rupee is higher in India.

2. Human labour is relatively cheap in India, a fact which keeps the labour component of the marketing cost lower in India than in the developed countries.

MARKETING COSTS OF FOODGRAINS OVER TIME

Over time, there has been an increase in the marketing cost of foodgrains in India. Some of the factors which have been responsible for this increase are:

1. *Shifting Tendency from Subsistence to Commercialised Farming:* Previously, each farmer used to produce foodgrains needed by him; but now, because of specialization in agricultural production and increasing urbanization, the distance between producers and consumers has increased. The cost of moving foodgrains from producers to consumers has, therefore, increased.

2. *Technological Advances in Preservation and Storage:* Formerly, many food products were consumed only during the season of production. Specialization in production and the evolution of short duration high-yielding varieties have resulted in large-scale production, thereby necessitating their storage. Technological advances in storage and preservation, though have facilitated handling of large volumes but have increased the costs and widened the spread between the producers' and the consumer's prices.

3. *Change in the Form of Consumer Demand:* There has been a change in the consumers' behaviour over time. Consumers now like the product in a processed and ready-to-use form following the increasing impact of urbanization. The desire for attractive packaging and home delivery system, too, has had its influence on consumer demand. Their demand for marketing service has, therefore, increased.

HOW TO REDUCE MARKETING COSTS

There are various ways of reducing marketing costs. No single factor can bring about any perceptible reduction in these costs. However, a combination of factors may bring about a significant reduction in the cost of marketing. Some ways of reducing marketing costs for farm products are:

1. Increase the Efficiency of Marketing

An increase in the efficiency of marketing can be brought about by a wide range of activities between producers and consumers. Some major areas in which improved efficiency may result in a reduction in marketing costs are:

(a) *Increasing the Volume of Business:* By increasing the quantity to be handled at a time, one can effectively reduce marketing costs and increase marketing efficiency. Group marketing by farmers can help in this regard.

(b) *Improved Handling Methods:* The new methods of handling, such as pre-packaging of perishable products, the use of fast transportation means, the development of cold storages and an efficient use of labour are some of the methods by which efficiency may be increased and costs reduced.

(c) *Managerial Control:* The adoption of proven management techniques increases efficiency. By a constant monitoring of costs and returns, the efficiency at each stage in marketing may be stepped up.

(d) *Change in Marketing Practices and Technology:* Changes in marketing practices and technology (such as sale of orange juice instead of orange, retailing food services through super markets, and integration of marketing functions) reduce marketing costs and increase marketing efficiency.

2. Reduce Profits in Marketing

Profits in the marketing of agricultural commodities are often the largest because of the inherent risk at various stages of marketing. The risk may be reduced by:

(a) The adoption of hedging operations, improvements in market news service, grading and standardization; and
(b) Increasing the competition in the marketing of farm products.

A decline in marketing margins and costs generally benefits both the producer and the consumer. Only in very few cases the benefits go either to only producers or to only consumers. Apart from such cases, the gains in the efficiency of marketing practices are shared by both. The extent to which these benefits are shared is determined by the nature or characteristics of the supply of, and demand for, the product. For example:

(a) If the supply and demand curves have the same elasticity, producers and consumers share the benefits equally;
(b) If demand is more elastic than supply (e.g. for farm products in the short run), the producers get a larger share of the benefits; and
(c) If the supply is more elastic than the demand (e.g. of many farm products over a longer period), consumers get a larger share of the benefits.

RELATIONSHIP OF FARMER'S PRICE, MARKETING COSTS AND CONSUMER'S PRICE

The farmer receives what the consumer pays after the various costs of marketing have been deducted. This residual, expressed as a percentage of the price paid by the consumer (retail price), is the farmer's share. The farmer's share may be calculated as follows:

$$FS = \frac{(RP - MC)100}{RP}$$

or
$$FS = \frac{PF}{RP} \times 100$$

where FS = Farmer's share in the consumer price expressed as a percentage; RP = Retail price; MC = Marketing costs, including margins; PF = Price received by the farmer.

The farmer's share in the amount of the consumer's outlay at the retail level is not static and undergoes change with the change in market conditions. An increase in the share is taken as an evidence of increase in the efficiency of the marketing system in favour of the farmer, while a decrease in the farmer's share is taken as evidence of the fact that middlemen retain a larger share. The effect of change in marketing charges or costs on the farmer's share is shown in Fig. 9.4.

In period t_3 (compared to period t_2), the farmer's share in the consumer's rupee has increased because of the reduction in marketing costs and margins. It is evident that all the factors which bring about changes in marketing costs affect the farmer's share as well.

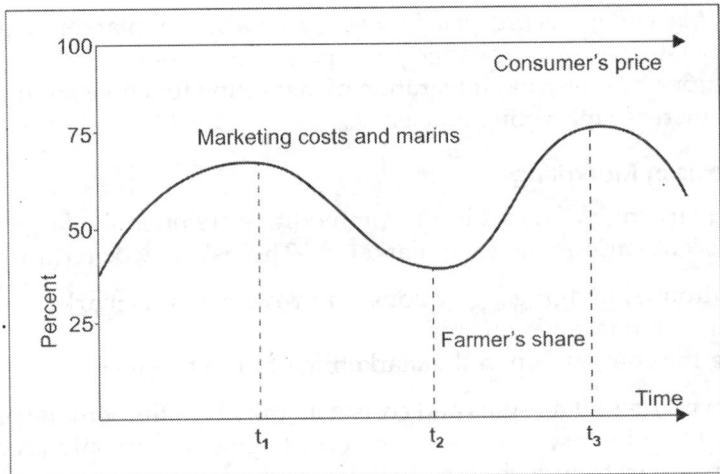

Fig. 9.4: Farmers' share in consumer's Rupee and marketing costs and margins

Several items of the marketing costs are almost sticky, i.e. they do not move up and down with the movement in prices. The basic reason for sticky marketing costs is that many of the items in them are related to the physical volume handled rather than to the value of the product. For example, transport cost, labour cost, weighing cost, storage cost and octroi are charged on the basis of weight.

With any given level of sticky marketing margin or cost, the farmer's share (price received) moves directly with the retail price; that is, if the retail price increases, the farmer's share also increases. But the proportionate change in the farmer's share is more than the proportionate change in the retail price. To illustrate: let the retail price, the marketing costs/margin and the farmer's price be ₹ 100, ₹ 50 and ₹ 50 per unit respectively in period t_1. Suppose, in period t_2, the retail price decreases to ₹ 90 per unit, i.e. a fall of 10%. If the absolute gross marketing margin remains the same, i.e. ₹ 50 per unit, the farmer's price falls to ₹ 40 per unit, i.e. a fall of 20%. In other words, 10% fall in the retail price results in a 20% fall in the farmer's price. This has been shown in Table 9.12.

Table 9.12: Effect of change in retail price on farmer's share

Particulars	Period		Absolute change (₹)	Percentage change
	t_1 (₹)	t_2 (₹)		
Retail price	100	90	10	10
Marketing margin (gross)	50	50	–	–
Farmer's price	50	40	10	20

Another point that emerges from Table 9.12 is that, in period t_1, the price received by the farmer was 50% of the price paid by the consumer but that in period t_2, the farmer received only 44.4% of the price paid by the consumer. To the extent that marketing margins or costs are sticky, the farmers lose more when the retail price decreases.

COMPUTATION OF PRICE SPREAD—ILLUSTRATIONS

An efficient marketing system is a prerequisite for sustaining the tempo of increased agricultural production. This ensures fair returns to the farmers for their efforts. The economic efficiency of the marketing system is generally measured in terms of the price-spread of an agricultural commodity. The smaller the price-spread, the greater the efficiency of the marketing system. This price-spread, besides being influenced by such marketing inputs as storage and transportation, changes with the shifts in the demand for, and/or in the supply, of the product.

The term price-spread has been variously defined and understood according to its usage. Generally, it refers to the difference between the two prices, i.e. the price paid by the consumer and the price received by the producer. A study of the price-spread involves not only the ascertainment of the actual prices at various stages of the marketing channel, but also the costs incurred in the process of the movement of the produce from the farm to the consumer and the margin of various intermediaries.

The following illustrations show the method of calculating the price-spread in foodgrain marketing:

Case I

A farmer, Mr. Bhura (B) comes to Krishi Upaj Mandi, Dausa (regulated market) with 100 bags of wheat (each weighing 100 kg net). He takes the produce to M/s. Jain Brothers (J), a commission agent. Immediately on arrival, Mr. Bhura requests M/s. Jain Brothers to make payment on his behalf to the truck-owner for transporting the produce and for octroi charges. The produce is unloaded from the truck by licensed labourers, who are paid by the commission agent on behalf of the farmer. The produce is put on the auction plat form, and the farmer takes his empty gunny bags. The rates of payments made so far by the commission agent, on behalf of the farmer, are:

1. Transportation charges   @ ₹ 10.00 per bag
2. Octroi   @ ₹ 1.00, per bag
3. Labour for unloading   @ ₹ 3.00 per bag

Now the produce is auctioned and one wholesaler, M/s. Mool Chand purchases this lot at a price of ₹ 2000 per quintal. The commission agent makes the payment to the farmer at the rate of ₹ 2000 per quintal after the deductions shown in (1), (2) and (3) above, which are to be borne by the farmer. The farmer returns home.

Meanwhile, the wholesaler, M/s. Mool Chand, has decided to send this lot to Jaipur market in a hired truck. This wholesaler pays for following items in Dausa market:

4. Cost of gunny bags   @ ₹ 25 per bag
5. Labour charges for filling and stitching bags   @ ₹ 1.00 per bag
6. Weighing   @ 0.25% of the value of the produce
7. Commission to the commission agent   @ 1 % of the value of the produce
8. Market fee to the market committee   @ 1.6 % of the value of the produce
9. Labour charges for loading the bags into the truck   @ ₹ 3.00 per bag
10. Truck transport charges   @ ₹ 10.00 per bag (Dausa to Jaipur)

After arriving in Jaipur market, the wholesaler, M/s. Mool Chand pays for the following items:

11. Labour charges for unloading   @ ₹ 3.00 per bag

The unloading in Jaipur is done at the shop of a commission agent; through him, this lot is sold to M/s. Daulat Chand @ ₹ 2,200 per quintal. The gunny bags are also sold @ ₹ 20 per bag. The commission agent collects the following amounts from the buyer (i.e. M/s. Daulat Chand):

12. Commission	@ 1 % of the value of the produce
13. Weighing charges	@ 0.40% of the value of the produce

M/s. Daulat Chand takes the produce to his shop and, while doing this, he incurs the following expenses:

14. Labour charges for transporting the produce to his shop	@ ₹ 5.00 per bag

Now M/s. Daulat Chand sells wheat to consumers @ ₹ 2,300 per quintal, together with the gunny bags, for which an extra charge of ₹ 16 per bag is realised.

It has been assumed that there is no physical loss during the handling of the produce and that no significant time elapses between various transactions.

Given this information, the marketing costs, the marketing margins and the price-spread in the marketing of wheat may be worked out as follows:

Marketing Costs

(a) Incurred by the farmer, Mr. Bhura

Particulars	Quantity (bags)	Rate (₹/bag)	Amount (₹)
1. Transport charges	100	10.00	1000
2. Octroi	100	1.00	100
3. Labour charges for unloading	100	3.00	300
		Sub total (a)	1400

(b) Incurred by the wholesaler, M/s. Mool Chand of Dausa Market

Particulars	Quantity	Rate (₹)	Amount (₹)
4. Cost of gunny bags (₹ 25 – 20) (his purchase price minus sale price)	100 bags	5.00 per bag	500
5. Labour charges for filling and stitching of bags	100 bags	1.00 per bag	100
6. Weighing charges	₹ 2,00,000 worth of produce	0.25% of the value	500
7. Commission	"	1 % of the value	2,000
8. Market fee	"	1.6 % of the value	3,200
9. Labour charges for loading on to truck	100 bags	3.00/per bag	300
10. Truck transportation from Dausa to Jaipur	100 bags	10.00/per bag	1,000
11. Labour charges for unloading from the truck at Jaipur	100 bags	3.00/per bag	300
		Sub total (b)	7,900

(c) Incurred by M/s. Daulat Chand of Jaipur

Particulars	Quantity	Rate (₹)	Amount (₹)
12. Cost of gunny bags (₹ 20 – 16)	100 bags	4.00 per bag	400
13. Commission on value of the produce	₹ 2,20,000 worth of the produce	1% of the value	2,200
14. Weighing charges	"	0.4% of the value	880
15. Transport charges from market to his shop	100 bags	5.00/per bag	500
		Sub total (c)	3,980

Total marketing cost (a + b + c) = ₹ 13,280

Profits or Net Margins of Traders

Profit of a trader = Receipts (sale value) minus purchase value minus cost incurred

$$A_{mi} = P_{Ri} - (P_{pi} + C_{mi})$$

Profit (net margin) of M/s. Mool Chand of Dausa (in ₹)

$$= ₹ (2,200 \times 100) - (₹ 2,000 \times 100) - (₹ 7,900.00)$$
$$= 2,20,000 - 2,00,000 - 7,900 = 12100$$

Profit or net margin of M/s. Daulat Chand of Jaipur (in ₹)

$$= ₹ (2,300 \times 100) - (₹ 2,200 \times 100) - (₹ 3,980)$$
$$= 2,30,000 - 2,20,000 - 3,980 = ₹ 6,020$$

Total margins for both traders = ₹ 12,100 + ₹ 6,020 = ₹ 18,120

Price Received by the Farmer

Gross price received ₹ 2000 per quintal

Cost borne by the farmer @ ₹ 14.00 per quintal (₹ 1400 for 100 quintals).

Net price received $(P_F) = P_A - C_F = 2,000 - 14.00 = ₹ 1,986.00$ per quintal or

2,00,000 – 1,400 = ₹ 1,98,600 for 100 quintals.

Price Spread

The price spread is as follows:

Particulars	Gross for whole lot of 100 quintals (₹)	Per quintal (₹)	Percent share in the price paid by the consumer
Farmer's share or net receipt of the farmer	1,98,600	1,986	86.3
Marketing cost	13,280	132.80	5.8
Marketing margins—(total for both traders) net profit retained by them after meeting their costs	18120	181.20	7.9
Price paid by the consumer	2,30,000	2300	100.00

Case II

A farmer Mr. Ramu, comes to Krishi Upaj Mandi Samiti, Madanganj, with 500 bags of wheat, each weighing 100 kg net. He takes the produce to M/s. Mehta Brothers—a commission agent. Mr. Ramu requests his commission agent to make the following payments on his behalf:

1. To truck-owner for transporting the wheat   @ ₹ 4.00 per bag
2. Octroi charges   @ ₹ 1.00 per bag
3. To labourers for unloading the produce from the trucks @ ₹ 2.00 per bag

Now the produce is auctioned and a wholesaler, M/s. Phool Chand, purchases the produce at a price of ₹ 1800 per quintal. The commission agent makes the payment to the farmer after deducting the expenses on items (1), (2) and (3) above. The wholesaler incurs the following expenses in the purchase of wheat at Madanganj market:

4. Cost of the gunny bags   @ ₹ 20 per bag
5. Sales tax   @ Nil
6. Labour charges for filling and stitching of   @ ₹ 1.00 per bag
   bags
7. Commission   @ 1 % of the value of the produce
8. Market fee to the market committee   @ 1.6 % of the value of the produce
9. Weighing   @ 0.25% of the value of the produce

The wholesaler decides to transport this wheat to the secondary wholesale market at Jaipur by rail and incurs the following expenses;

10. Cartage to station at Madanganj   @ ₹ 4.00 per bag
11. Railway freight (for the whole lot) @ ₹ 3000
12. Loading and unloading charges   @ ₹ 4.00 per bag
13. Cartage at Jaipur   @ ₹ 4.00 per bag

The wheat of the wholesaler, M/s. Phool Chand, is sold to a retailer, M/s. Padam of Jaipur, through his commission agent @ ₹ 2000 per quintal. The empty gunny bags are purchased by the retailer @ ₹ 16 per bag. The commission agent collects the following amounts from the buyer (M/s. Padam);

14. Commission   @ 1.25% of the value of the produce
15. Labour charges for unloading   @ ₹ 2.00 per bag
16. Weighing charges   @ 0.25% of the value of the produce

M/s. Padam takes the produce to his shop in his own truck and sells it to consumers @ ₹ 2100 per quintal. The empty bags are disposed of by him @ ₹ 14 per bag in the market. For the sake of simplicity, it has been assumed that there is no loss in transit and no significant time lag.

Given this information—producer's price, producers' share in consumers' rupee, absolute margin of the wholesaler and retailer; and marketing costs incurred by producer, wholesaler and retailer; and price spread can be worked out as follows:

Marketing Costs

(a) Cost incurred by the farmer (since the farmer had no money, commission agent paid these and deducted them from the payment made to the farmer).

Particulars	Quantity	Rate (₹)	Amount (₹)
1. Transportation cost	500 bags	4/bag	2,000
2. Octroi	500 bags	1/bag	500
3. Labour charges	500 bags	2/bag	1,000
		Sub total (a)	3,500

(b) Cost incurred by the wholesaler, M/s. Phool Chand, at Madanganj and Jaipur markets.

Particulars	Quantity/value	Rate (₹)	Amount (₹)
4. Cost of gunny bags (purchase price minus sale price)	500 bags	4/bag	2,000
5. Sales tax on value			Nil
6. Labour charges for filling and stitching of bags	500 bags	1.00/bag	500
7. Commission	₹ 9,00,000	1% of the value	9,000
8. Market fee	"	1.6 % of the value	14,400
9. Weighing charges	"	0.25%	2,250
10. Cartage at Madanganj	500 bags	4/bag	2,000
11. Railway freight (total)	500 bags	Fixed	3,000
12. Loading and unloading charges at Jaipur	500 bags	4/bag	2,000
13. Cartage at Jaipur	500 bags	4/bag	2,000
		Sub total (b)	37,150

(c) Cost incurred by the retailer, M/s. Padam.

Particulars	Quantity/value	Rate (₹)	Amount (₹)
14. Commission	₹ 10,00,000 (500 × 2000)	1.25% of value	12,500
15. Weighing charges	"	0.25%.	2,500
16. Labour charges	500 bags	2.00/bag	1000
17. Cost of gunny bags (purchase price minus sale price)	500 bags	2.00/bag	1,000
		Sub total (c)	17,000

Total marketing costs (a + b + c) = ₹ 57,650

Marketing Margins

(a) Margin of wholesaler (M/s. Phool Chand)

$$= P_{Ri} - (P_{pi} + C_{mi})$$
$$= (500 \times 2000) - (500 \times 1800) - 37{,}150$$
$$= ₹ 62{,}850$$

(b) Margin of retailer (M/s. Padam)

$$= P_{ri} - (P_{pi} + C_{mi})$$
$$= (500 \times 2100) - (500 \times 2000) - 17{,}000$$
$$= ₹ 33{,}000$$

Producer's Price

$$P_A = P_F - C_F = (500 \times 1800) - 3500 = ₹ 8{,}96{,}500 \text{ for 500 quintals}$$
$$= ₹ 1793 \text{ per quintal.}$$

Producer's Share in Consumer's Rupee

$$P_S = (P_F \div P_R) 100 = (1793 \div 2100) 100 = 85.4\%.$$

Price Spread

The price spread in this case is as follows:

Particulars	Gross for whole lot of 500 quintals (₹)	Per quintal (₹)	Percent share in the price paid by the consumer
Farmer's share	8,96,500	1,793.00	85.4
Marketing costs	57,650	115.30	5.5
Marketing margins of both the traders	95,850	191.70	9.1
Price paid by the consumers	10,50,000	2,100.00	100.00

LIMITATIONS OF PRICE-SPREAD STUDIES

Price-spread studies conducted both at micro and macro level present enormous difficulties and hence results are not comparable both overtime and space. Some of the difficulties are:

1. There is considerable regional variation in prices of commodities. Further, the varieties grown and marketed in different regions are not comparable.
2. Prices of commodities are not correlated to the recognized quality standards and, therefore, not comparable.
3. The number of intermediaries between the producer and the ultimate consumer or the length of the marketing channel varies from area to area.
4. The authentic data of prices paid by various intermediaries during a period of time are not available. Generally intermediaries do not maintain the accounts. Even if they maintain, access of researchers to such records is impossible.
5. There are divergent methods of handling and transportation followed in different regions which results in large variation in marketing costs, margins and price-spread across commodities and regions.

TERMS OF TRADE FOR AGRICULTURE/FARMERS

As explained in Chapter 2, the concept of Terms of Trade (ToT) was first used to connote the rate at which a country exchanges its exports with imports. In simple terms, it is the ratio of prices of a country's exports (prices received by the country) to the prices paid for imports. Later, the concept was applied to assess or monitor the inter-sector exchanges with in the country. In India, assessment and monitoring of terms of trade for the agricultural sector has been a regular feature when a positive agricultural price policy was launched in the country in the middle of sixties. The terms of trade for the agricultural sector reflect the outcome of the forces of demand and supply, coupled with the impact of various forms of government intervention, as discussed in chapter 7 and others. The rationale for this came from a large number of research studies, which revealed that farmers (especially in developing countries like India) respond to relative prices, rather than just prices received by them. The concept of ToT between agriculture and non-agriculture sectors represents the relative prices.

The terms of trade for the agricultural sector, specifically mean the ratio of prices of agricultural products (prices received by the farmer) to the prices of items purchased by the farmers (prices paid by them). The CACP, while recommending the price policy measures, and the Government of India, while announcing the policy measures, keep in view (take into consideration) the movements in terms of trade for the agricultural sector. When the terms of reference of CACP were revised by the Government in 1980, the movement in ToT for the agricultural sector was added as one of the factors to be kept in view while formulating the agricultural price policy.

There are three approaches to assess the movement in relative prices of agriculture and other sectors (ToT).

1. One is the ratio of indices of agricultural product prices to that of prices of manufactured goods.
2. The second is the ratio of implicit price indicators of the two sectors obtained as GDP deflators (worked out from the value of GDP from agriculture and value of GDP from other sector at current and constant prices).
3. The third approach is what is being used by the Ministry of Agriculture, Government of India. In this approach, for working out the index of prices received by the agricultural sector, the marketed surplus-output ratios of various farm products are used as weights. Further, the indices of price paid by the agricultural sector are worked out separately for (a) farm inputs, (b) items of capital formation/ investment, and (c) items of final consumption. It is in this sense, that this approach is considered to be superior to the first two common/conventional approaches.

The first comprehensive methodology to work out the index of ToT for agricultural sector was developed by the Task Force on Terms of Trade, headed by A.S. Kahlon in 1995 (S.S. Acharya, the then chairman of CACP, was a member of this Task Force). This methodology is being used by the MoA (GoI) for regularly bringing out the indices of ToT for use by the CACP as well as Government of India (Government of India, 1995).

The methodology for working out the indices of ToT involves several parameters like weighting diagram and price quotations, which undergo changes over time. Keeping this dynamism in view, the Ministry of Agriculture constituted a Working Group in 2012 to revisit the methodology being used (as per the Task Force of 1995). The Working Group, chaired by S. Mahendra Dev, has submitted its report in January 2015 (GoI, 2015). Some major recommendations of the Working Group are as follows:

1. The methodology for construction of index of prices paid by agriculture to non-agriculture will continue to be the same as before viz:, the combined index of prices paid for final consumption, intermediate consumption, and capital formation. However, there is a need for construction of indices of ToT separately for farmers and for agriculture (aggregate), because the latter comprises both farmers and agricultural labourers. In the basket for intermediate consumption, hired labour should be included for farmers and not for agriculture.
2. The base period for construction of index of ToT should be shifted from triennium ending (TE) 1990–91 to TE 2011–12.
3. As regards representative prices, wholesale prices (in place of farm harvest prices) should be used for representing prices received by the farmers. For computing indices of prices paid for intermediate consumption and capital formation, wholesale price indices should continue to be used. As regards indices of prices of items of final consumption, rural retail prices should be used.
4. The weights for items of prices received (cereals, pulses, oilseeds, commercial crops, horticultural crops, livestock products and forest produce) for the TE 2011–12 should be taken from the National Accounts Statistics (NAS). Further, in view of

the unavailability of marketed surplus ratios of many products, total value of output can be used in place of value of marketed surplus.

5. The weights for items of final consumption should be taken from monthly per capita consumption expenditure for 2011–12 from 68th round of NSS, for items purchased from the non-agricultural sector, calculated per annum, multiplied with the population in agriculture and then aggregated.

6. The weights for items covered under intermediate goods, should be calculated by taking their value from NAS for the TE 2011–12. The value of hired labour can be taken as the compensation to hired labour in the unorganized sector as reported in the Factor Income Statement of NAS.

7. The weights for items covered under capital formation goods for construction and machinery in total gross capital formation for TE 2011–12 of NAS can be taken (assigned) based on the economy as a whole for the construction sector items and based on WPI for the machinery sector items. The Working Group considered this as the best possible approach given the non-availability of precise data.

8. The combined index of prices paid can be calculated by taking the share of each group (items for final consumption, intermediate consumption and capital formation) in total aggregated expenditure of the agricultural sector.

The Working Group found significant changes in the weighting diagram by using the new suggested methodology. The hired labour, which accounted for 25% of intermediate consumption previously, now accounts for 43%. As per the new methodology, intermediate consumption goods account for 43% of total purchased items.

The limitations of the first two approaches (ratio of prices of agricultural products to the prices of manufactured goods and the ratio of two deflators) can be easily recognized. The present methodology (as per A.S. Kahlon Task Force) and the modified methodology (now suggested by the Working Group) are superior to the two traditional approaches. For comparison, the Working Group's estimates and the estimates of two other approaches, by using a common base period (2004–05) are shown in Table 9.13.

Table 9.13: Index of terms of trade for farmers/agriculture (comparison of three approaches)

(2004–05 = 100)

Year	Working group		GDP deflators or Gross ToT	Ag. product price index as % of manufactured price index
	Farmers	Agricultural sector		
2004–05	100	100	100	100
2005–06	96.6	97.9	102.4	101.0
2006–07	99.1	101.5	104.1	104.0
2007–08	105.0	106.4	109.4	107.1
2008–09	113.8	115.1	113.9	110.9
2009–10	114.0	120.6	123.1	122.7
2010–11	117.2	126.2	127.0	135.7
2011–12	110.8	121.1	126.4	136.5
2012–13	110.8	123.7	127.4	142.5
2013–14	108.8	121.5	NA	153.9

Source: Dev and Rao (2015).

The indices of ToT between agricultural and non agricultural sectors or the percentage ratio of price received by the farmers to price paid by them for the last one decade are shown in Table 9.14.

Table 9.14: Terms of trade between agricultural and non-agricultural sectors (Base: TE 2011–12)

Years	Index of prices received (IPR)	Index of prices paid (IPP)	Index of terms of trade (ITOT)
2007–08	71.45	82.35	86.76
2008–09	81.72	87.06	93.86
2009–10	90.57	92.11	98.33
20010–11	101.10	98.26	102.90
2011–12	108.32	109.63	98.81
2012–13	120.34	119.18	100.97
2013–14	132.85	126.98	104.62
2014–15	140.66	131.51	106.96
2015–16	145.30	136.10	106.76
2016–17	153.80	140.30	109.62
2017–18 (P)	156.26	144.91	107.83

P = Provisional.
Source: Ag. Statistics at a Glance, 2018.

REFERENCES

1. Kohls, R.L. and J.N. Uhl., Marketing of Agricultural Products, Macmillan Publishing Co. Inc, New York, 1980, p. 260.
2. Chauhan, K.K.S. and R.V. Singh, Marketing of Wheat in Rajasthan, University of Udaipur, College of Agriculture, Jobner, 1973, pp. 56–59.
3. Lele, Uma J. "Traders of Sholapur", in Developing Rural India: Plan and Practice, John W. Mellor, Thomas F. Weaver, Uma J. Lele and Sheldon R. Simon, Lalvani Publishing House, Mumbai, 1972, pp. 257–262.
4. Agarwal, N.L., Foodgrain Prices in Rajasthan, Unpublished Ph.D. Thesis, University of Udaipur, Udaipur, 1982, p. 305.
5. Agarwal, N.L., Ibid. pp. 252–253.
6. Acharya, S.S., Domestic Agricultural Marketing Policies, Incentives and Integration, in the Indian Agricultural Policy at Crossroads, Ed. by S.S. Acharya and D.P. Chaudhri, Rawat Publications, jaipur, 2001, pp. 129–205.
7. Lele, Uma, J. (1967), Market Intervention: A Study of Sorghum Prices in Western India, Indian Journal of Farm Economics, Vol. 49, pp. 147–59.
8. Lele, Uma, J. (1971), Foodgrain Marketing in India: Private Performance of Public Policy, Cornell University Press, Ithaca.
9. Jones, W.O. (1968), The Structure of Staple Food Marketing in Nigeria as Revealed by Price Analysis, Food Research Institute Studies, 8:95–124.
10. Acharya, S.S. (1988), Agricultural Production, Marketing and Price Policy in India, Mittal Publications, Delhi.
11. Blyn, G. (1973), Price Series Correlation as a Measure of Market Integration, Indian Journal of Agricultural Economics, Vol. 28, No. 1, January–March, pp. 56–59.
12. Harris, B. (1979), There is a method in my madness or it is vice-versa? Measuring Agricultural Market Performance; Food Research Institute Studies, 17, 197–218.

13. Ravallion, M. (1986), Testing Market Integration, American Journal of Agricultural Economics, 68: 102–109.

14. Ravallion, M. (1987), Markets and Famines, Clarendon Press, Oxford.

15. Godwin, B. and Schroeder, T. (1991), Cointegration Tests and Spatial Price Linkages in Regional Cattle Markets, Amercian Journal of Agricultural Economics, 73: 452–64.

16. Palaskas, T.B. and Harriss-White, B. (1993), Testing Market Integration: New Approaches with Case Material from West Bengal Food Economy, Journal of Development Studies, 30: 1–57.

17. Alexander, C. and Wyeth, J. (1994), Cointegration and Market Integration: An Application to the Indonesian Rice Market, Journal of Development Studies, 30: 303–28.

18. Dercon, S. (1995), On Market Integration and Liberalisation, Method and Application to Ethiopia, Journal of Development Studies, 32: 112–43.

19. Baulch, B. (1997), Transfer Costs, Spatial Arbitrage and Testing Food Market Integration, Amercian Journal of Agricultural Economics, 79: 477–87.

20. Wilson, E.J., Testing Integration of Agricultural Produce Markets in India, in Indian Agricultural Policy at the Crossroads, Ed. by S.S. Acharya and D.P. Chaudhri, Rawat Publications, 2001, pp. 213–238.

21. Acharya, S.S., Agricultural Marketing in India — Millennium Study of Indian Farmers, Ministry of Agriculture, Government of India, New Delhi, 2003.

22. Kohls, R.L. and J.N. Uhl., Marketing of Agricultural Products, Macmillan Publishing Co. Inc., New York, 1980, p. 589.

23. Jasdanwalla, Z.Y., Marketing Efficiency in Indian Agriculture, Allied Publishers Private Limited, New Delhi, 1966, p. 2.

24. Clark, Fred E., Principles of Marketing, 1954, p. 777.

25. Shepherd, G.S., Marketing Farm Products—Economic Analysis, Iowa State University Press, Ames, Iowa, USA, 1965, p. 254.

26. Acharya, S.S. is former Chairman of the Commission for Agricultural Costs and Prices, Government of India. Though retail price, i.e. the price paid by the consumer (RP) can also be used as the numerator without any change in the conclusion, Acharya suggests that under Indian conditions, it is advisable to use farm harvest price (FP) or the price received by the farmer as the numerator.

27. Kohls, R.L., Marketing of Agricultural Products, The Macmillan Publishing Company Inc., New York, 1958, p. 78.

28. Dantwala, M.L., "Problems Before the New Marketing Agencies," Indian Journal of Agricultural Economics, Vol. XIII, No. 2, April–June 1957, p. 183.

29. Chand, Ramesh, Government Intervention in Foodgrain Markets in the New Context, A Report Prepared for Union Ministry of Consumer Affairs, Food and Public Distribution, National Centre for Agricultural Economics and Policy Research, New Delhi, September, 2002.

30. Government of India (1995), Report of the Task Force on Terms of Trade between Agricultural and Non-agricultural Sectors, Ministry of Agriculture, New Delhi.

31. Government of India (2015), Report of the Working Group on Terms of Trade between Agricultural and Non-agricultural Sectors, Ministry of Agriculture, New Delhi.

32. Dev S.M. and N.C. Rao (2015), "Improved Terms of Trade for Agriculture: Results from Revised Methodology", Economic and Political Weekly, Vol. L, No.15, April 11, p.19–22.

33. Dickey, D.A. and Fuller, W.A. (1979), "Distribution of the Estimators for Auto Regressive Time-series with a Unit Root Test", Journal of American Statistical Association, Vol. 74, p. 427–31.

34. Phillips, P.C.B. and Perron, P. (1988)," Testing for Unit Roots in Time Series Regression", Biometrika, Vol. 75 (2), p. 335–46.

35. Johansen, S. (1988), "Statistical Analysis of Cointegration Vector", Journal of Economic Dynamics and Control, Vol. 12 (2–3), p. 231–34.

36. Saxena, Raka, Ranjit Kumar Paul, Simmi Rana, Shikha Chaurasia, Kavita Pal, Zeeshan and Deepika Joshi (2015), "Agriculture Trade Structure and Linkages in SAARC: An Empirical Investigation", AERR, Vol. 28(2), p. 311–28.

10

Training, Research, Extension and Statistics in Agricultural Marketing

A decision on appropriate marketing development strategy, the evolution of proper marketing policy and a choice of policy instruments require a continual flow of advice, information and assessment of the existing marketing system from all those who are engaged and involved in this activity, directly or indirectly. Every system generates impulses as a result of environmental changes, including policy. These impulses have to be observed, recorded, analyzed and interpreted for the decision and policy making. This task is performed usually by administrators and field staff who man the various marketing institutions, scientists and economists, organizations representing farmers, traders and consumers, and the like. The extent to which these persons are able to help in the formulation of better policies depends largely on the training and research network and statistical information system available in a country. This chapter deals with training, research facilities, extension, and statistical organisation as they relate to agricultural marketing, with special reference to the Indian situation.

TRAINING IN AGRICULTURAL MARKETING

Training is an essential adjunct of human resource development. It helps in bridging the gap in knowledge and skills. In addition, training enables the stakeholders to change their attitudes.

The training in agricultural marketing has received attention both at the national and international levels. The Food and Agricultural Organisation's workshop on the Improvement of Agricultural Marketing Training in Asia, held in Thailand in April, 1979, is indicative of international concern in this regard. This workshop was preceded by regional surveys and followed by regional seminars.

Training in agricultural marketing should aim at equipping those who perform various marketing functions with development-oriented perception, practical knowledge, analytical ability and managerial and risk-taking capabilities. Broadly seven target groups can be identified for training in agricultural marketing, viz., farmers, consumers, market functionaries, staff of market committees, chairmen and members of market committees, development policy-makers, programme-planners, and staff of marketing organizations. These stakeholders interact at different levels and at different stages and influence the performance of agricultural marketing system.

For understanding the existing status of training and identifying the training needs, at least three levels of training in agricultural marketing may be identified:

1. University or College Level Training
2. On-the-Job Training
3. Other Training

UNIVERSITY OR COLLEGE LEVEL TRAINING

There is a wide variation in agricultural marketing training programmes among the universities in Asia. The courses and their contents vary from institution to institution. Basically, four patterns may be observed.

1. Till recently agricultural universities and colleges offered a general course in agricultural development for all the graduates at the first degree level. This course partly covered agricultural marketing well. In addition, there were a number of elective courses. Students who opted for agricultural economics as their elective subject were offered a course of three semester credits in agricultural marketing.

   Now, agricultural universities and colleges, under the changed common pattern of I.C.A.R., offer a general course in agricultural marketing for all the students at the first degree level in addition to other courses in agricultural economics discipline viz., farm management, resource economics and agricultural finance and management. The course on agricultural marketing is of three semester credits. Some agricultural universities also offer first degree programmes in agricultural marketing and agribusiness.

2. At the post-graduate level, one course of three semester credits in agricultural marketing and another course of an equal number of credits in agricultural price analysis are invariably offered to all those who desire to take an M.Sc. Ag. degree in agricultural economics. Some agricultural universities offer a specialized Master's course in agricultural marketing. The tendency among M.Sc. students of other disciplines to take agricultural marketing and related courses in agricultural economics as their minor fields is increasing. At the Ph.D. level, these courses are offered at a very advanced level.

3. Many universities in Asia have separate departments of marketing which are a part of the Faculty of Commerce. These departments, however, do not offer specific courses in agricultural marketing. In the Faculty of Economics, agricultural marketing is taught at the B.A. and M.A. level as a part of other courses offered to those who specialise in agricultural economics. Institutes of Management have marketing as an essential component of their course offerings.

4. Apart from general degree programmes, in recent years, the degree and diploma courses in business management (including agri-business) have been initiated by several institutions and universities. Keeping in view the role of agriculture in economic development and emerging scenario of international trade in agricultural commodities, sufficient grinding of prospective managers in the basic concepts, conditions and practices of agricultural marketing system would be quite critical in developing countries.

The XI Five Year Plan Sub-Working Group on Agricultural Economics and Marketing Education and Research (chaired by S.S. Acharya) has emphasized the need for advanced courses, both at B.Sc. Ag. and Master's levels, in agricultural marketing and agribusiness to meet the increasing demand for trained manpower in this field in the years to come (Refer report of the Working Group on Agricultural Research and Education for XI Five Year Plan).

Formal education facilities in agricultural economics, including agricultural marketing, are available in 63 agricultural universities, four deemed-to-be universities, three central agricultural universities and several general universities (with agriculture faculty).

ON-THE-JOB TRAINING

The persons recruited by the state governments to work in agricultural produce markets are usually from the general stream of education. Obviously, these have little exposure to agricultural marketing. It is in this context that the DMI (Government of India) launched a series of short duration courses for agricultural marketing personnel engaged/employed in organizations operated by the government, cooperatives or even the private sector.

The Directorate of Marketing and Inspection has designed various training programmes as per needs of the marketing personnel at different levels. The basic objectives of these programmes are as under:

1. To create an ambiance of good agricultural marketing practices in the country to protect the interests of all market users as well as consumers,
2. To sensitize the agri-marketing personnel to respond to the changing demand trends, challenges of the post-WTO era and globalized marketing environment,
3. To provide up to date information about the on going marketing developments and empower them to formulate policies to further the cause of agricultural marketing in the country.
4. To improve the managerial skills among the marketing personnel,
5. To promote scientific grading and standardization of agricultural commodities both at producer and market level and integrate the same with the export marketing,
6. To improve the efficiency of market intelligence and market news services in the process of market integration.
7. To train the marketing personnel in preparing survey reports on various commodities and functional aspects.
8. To improve the communication skills of the marketing personnel to develop awareness among market users about various market developmental schemes monitored by different departments of Government of India.

The training courses regularly organized by DMI included the following:

1. Diploma Course in Agricultural Marketing
2. Market Secretaries Training Course
3. Grading Supervisors Course
4. Training for grading service at the state level
5. Orientation training for newly recruited officers.
6. Market intelligence and news service
7. Course on marketing extension service
8. Graders training course.

In addition, there are several programmes organized for market functionaries to sensitize these with new developments and schemes related to agricultural marketing. The duration varies from three days to six months. In these courses, the private candidates (who are not employed) are also admitted but they have to bear the cost of entire training.

The details of training programmes and courses conducted by the Directorate of Marketing and Inspection (Government of India) are given in Chapter 7. Occasionally, agricultural universities and other institutes organise refresher courses, summer institutes and workshops for college and university teachers, researchers and officials of the marketing department of the government. The Indian Council of Agricultural Research encourages and financially assists the universities in organising such courses.

OTHER TRAININGS

Though agriculture extension work is the responsibility of State Departments of Agriculture, the training of farmers and extension workers has been assigned to agricultural universities. There is a separate Directorate of Extension Education in all the agricultural universities. There is now a Krishi Vigyan Kendra (KVK) in each of the rural districts of India. But the element of training in agricultural marketing is almost lacking. This is evident from the mix of training programmes and the absence of agricultural marketing specialist among the band of specialists who man these Directorates and KVKs. Problems at the post-harvest stage and their solutions have started receiving attention following the emphasis by the Indian Council of Agricultural Research on research in this area.

The education of farmers on important aspects of marketing such as grading and standardization, storage, processing, market information and pricing of farm products has not received due attention in the programme of farmers training.

One of the most important aspect of training and extension education in agricultural marketing that needs attention is that at all the levels, certain weaknesses exist in the teaching materials or textbooks and teaching methods. As a result, there is an absence of practical knowledge and orientation in marketing. Therefore, this area deserves the immediate attention of policy-makers and administrators in India. At the international level, the FAO has already initiated action in this direction.[1]

The Report[2] of the Expert Committee on Strengthening and Developing of Agricultural Marketing (headed by Shri Shanker Lal Guru) has pointed out that substantial training efforts are needed to upgrade the technical, managerial and organizational skills and knowledge of field and marketing personnel. Under the changing scenario, the following areas of training of market personnel should be given major thrust:

1. Agri-business management
2. WTO and its implications
3. Post-harvest management, including packaging and branding
4. Grading, standardization and quality assurance
5. Information technology
6. Emerging employment potential in agricultural marketing
7. Organic food marketing
8. Marketing extension
9. Entrepreneurship programmes for rural Agri-clinics
10. Training of Trainers (ToT)
11. Special post-graduate course on agricultural marketing in state agricultural universities.
12. Food Processing
13. Food Safety and quality maintenance
14. Food chains and retailing

During the last one decade, two important institutes have been set up in the country to inter alia promote formal education and training in agro and food processing. These institutes are Indian Institute of Crop Processing Technology (IICPT) at Thanjavur (Tamil Nadu) and National Institute of Food Technology, Entrepreneurship and Management (NIFTEM) at Kundli in Sonepat district of Haryana. In addition, recent up-gradation of National Centre for Agricultural Economics and Policy Research (NCAP) to the level of National Institute (NIAP) is also helping in strengthening agricultural marketing training and research in the country.

CAPACITY BUILDING PROGRAMMES

Capacity building has become the buzzword during the last two decades at national as well as international levels. The capacity building means the task of establishing and enhancing human and institutional capacities. This involves strengthening the skills, capacities and abilities of all the stakeholders from policy makers and senior managers to the grass root level functionaries, clients and ultimate intended beneficiaries (farmers). A large number of capacity building programmes in agricultural marketing are being organized by the Agricultural Universities, ICAR Institutes, and international organizations like FAO, IFPRI, ICRISAT and others. As stakeholders at various levels are involved, such programmes are differently named either as workshops, seminars, dialogues or consultations. But their basic intent remains the same, i.e. capacity building on the selected theme.

▌RESEARCH IN AGRICULTURAL MARKETING

IMPORTANCE AND OBJECTIVES

The importance of research has increased after independence and following rapid changes in the economy from the "very little capital investment and few marketing services" stage to the "monetized and surplus" one. Specialization in production has resulted in the concentration of agricultural products in the most favourable areas. On the consumer's side, the pattern of demand has been changing with the increase in incomes and change in life styles. Industrialization and urbanization have resulted in the concentration of population in places that are far from the areas of production, as a result of which food has to move over long distances. The movement of food over time and space has added a new dimension to agricultural marketing.

The need for research in agricultural marketing has been recognised by the planners and policy-makers. Research in this field can contribute to the establishment of facts and the evolution of the policy measures that may be necessary for developing a successful marketing strategy with regard to production, consumption, distribution and pricing.

A smooth functioning of the marketing system is essential for price stability and for proper incentives to producers. In the present context, when agricultural production is on the increase, the marketing system should be suitably altered to sustain the increases by providing efficient and prompt services. This may be done by keeping a constant watch on developments and by anticipating the problems in marketing. The objectives of a research programme in agricultural marketing are:

1. To understand the existing/traditional marketing system;
2. To diagnose the problems confronting the farmers, marketing agencies and consumers in a dynamic context; and

3. To analyze and predict the impact and effectiveness of alternative policy measures in solving these problems.

MARKET RESEARCH IN BUSINESS AND OBJECTIVES

The term market research in the context of a business firm is used in a narrow sense. It refers to the assessment of demand, distribution method and policy. Hanson[3] has defined market research for a firm as the study of consumer demand so that it may expand its output and market its product. The objectives of market research for a business firm are:

1. To discover the people's need of the product which a firm contemplates to manufacture;
2. To examine the reactions of consumers when the product is put on the market with a view to briging improvements in the quality, design and packaging of the product on the basis of their preferences; and
3. To examine the performance of the present system of marketing with particular reference to costs and efficiency.

PROGRESS OF RESEARCH IN AGRICULTURAL MARKETING

The year 1935 is a milestone in the field of agricultural marketing research, for it was then that the Central Marketing Department (now the Directorate of Marketing and Inspection) was established. The Central Marketing Directorate realised that no improvement in the marketing system can be brought about unless the existing system was studied in detail. It, therefore, conducted commodity surveys for the study of the existing-marketing system and brought out commodity survey reports from time to time. These reports paved the way for further research in the applied field of agricultural marketing.

Research in agricultural marketing is also undertaken by universities and other institutes. But very few macro level studies were taken up by these institutions. The research in agricultural marketing, particularly of applied nature, has been meagre and scanty because of the strenuous and time-consuming job of collecting primary data from different market functionaries. Most market functionaries are not ready to part with correct information and data. A lot of scope exists for research in this field. Marketing research in India could not make headway for long, but the situation during the last three decades has changed.

At present, agricultural marketing research in the country is both descriptive and analytical. A description of facts enables us to understand what the system is, but not the why part of it. The structural interrelationships among the various variables of the system have to be understood. There is, therefore, an urgent need for an analysis of the most relevant variables in marketing research. This need is now being addressed by several research institutes.

Research facilities in agricultural marketing are available in 63 agricultural universities, four deemed-to-be universities, three central agricultural universities, and four agricultural faculties of central universities. Besides these, agricultural marketing research, as a component, is available in most of ICAR's 69 institutes, 48 project directorates, 14 national research centres and 57 All-India coordinated projects. The CGIAR institutes, located in India (ICRISAT) or with offices in India, have also been contributing to research on issues related to agricultural marketing. The ICRISAT (with its headquarters at Hyderabad) had adopted 'Inclusive Market-Oriented Development

(IMOD)' as overarching framework for its Strategic Plan 2011–20. In addition, as mentioned earlier, two new institutes viz., IICPT and NIFTEM are conducting research on the theme of crop processing and food processing technology, apart from education and skill development in these areas.

STEPS IN MARKETING RESEARCH

Important steps involved in marketing research are:

1. Identification of the Problem

The first step in research is the identification of the problem. Marketing research problems relate to the estimation of demand and supply, the study of market functions and functionaries, the estimation of marketing costs and margins for individual commodities through various channels, the assessment of marketing efficiency, and the estimation of marketed surplus. The problems may be identified by:

(a) Studying the deviation between the marketing system in terms of its defined norm and its operation in practice;
(b) Discovering the needs of different segments of the economy—for example, producers, consumers, traders and processors;
(c) Reviewing past studies; and
(d) Questioning the existing concepts.

2. Formulation of Hypothesis

The second step in research is the formulation of hypotheses. A hypothesis is formulated for the identified problem to delineate the area of research and raise the questions to be answered from research.

3. Designing Empirical Procedures

An empirical procedure is to be designed to get appropriate data from farmers, consumers, market middlemen and government departments. The data are obtained after developing suitable schedules and questionnaires and pretesting them in the field. Apart from the structured surveys, participatory rural appraisal (PRA) and rapid rural appraisal (RRA) techniques are quite useful in understanding the market practices and perceptions of the farmers, traders and other stake holders.

4. Collection and Analysis of Data

The data for research in agricultural marketing may be collected either from primary sources or from secondary sources. The primary sources include the farmer-producers, market middlemen and consumers. The secondary sources include the published data brought out by different government and semi-government organizations either regularly or occasionally. The collected data are tabulated and analysed by using suitable tools of analysis.

5. Interpretation of the Findings

The results of the analysis should be interpreted to find out solutions to the identified problems of agricultural marketing. It is suggested that before finalising the results and final drafting of the report, interaction with various groups of stakeholders may be held as their feed back proves quite useful in drawing inferences from the analysis of data.

APPROACHES TO STUDY PROBLEMS OF MARKETING OF AGRICULTURAL COMMODITIES

Marketing is a subject which bristles with wide and varied problems. It includes the services and functions of different specialised institutions and middlemen. Different commodities have special marketing problems. Therefore, the results of the study of one commodity may not be applicable to other commodities. Various approaches have been suggested and used to study marketing problems. These are functional, institutional and commodity approaches. Sometimes, the behavioural system approach and the legal approach, too, are considered. No single approach, however, is satisfactory in all respects, for each has its merits and demerits. Therefore, a combination of these approaches is essential if we are to understand all the aspects of the problem related to marketing of a commodity in a better and lucid manner.

1. Functional Approach

Under this approach, the marketing system is studied by considering it as a composite of specialized functions and business activities. It consists of the study of marketing activities which are involved in moving the goods and services through time, space and succession of ownership. The functional approach takes into account the jobs which must be performed in the process of the movement of goods. This approach enables one to evaluate the importance of each marketing function and to suggest improvements (for details of marketing functions, see chapter 4). The advantages of the functional approach in any study of agricultural marketing problems are:

(a) It enables us to make an inter-functional comparison of the marketing costs;
(b) It enables us to make an inter-agency comparison of the cost of performing a marketing function; and
(c) It enables us to make an inter-commodity comparison of the cost of performing various marketing functions.

2. Institutional Approach

The institutional approach to a study of marketing problems implies a study of agencies and institutions which perform various functions in the marketing process. Under this approach are studied the nature and character of various middlemen and other related agencies involved in the movement of the product. The human element receives the primary emphasis.

There are a large number of agencies and institutions which perform various marketing functions in the process of the movement of the goods from producers to consumers. They may be individuals, partnership firms, corporations, cooperatives or government organizations. These agencies vary widely in size and ownership. They get their reward in the form of marketing margins (for details see chapter 5, 7 and 8).

This approach helps us to find answers to the problems of "who does what" in the marketing process, whether the margin of the agency is commensurate with the services rendered, which government regulations are necessary so that their unlawful activities may be curbed, and how to simplify the procedural system.

3. Commodity Approach

Under this approach, the commodity is the pivot around which all institutional and functional details are studied. The problems of marketing differ from commodity to commodity mainly because of the seasonality of production, the variations in its handling, storage, processing, and the number of middlemen involved in them. For example, potatoes are stored in cold storage, while wheat is stored in godowns. Paddy is processed to make rice at the miller's level, while other cereals are processed mostly at consumer's level. Commodity studies lead to repetition and duplication because many commodities, falling in a particular group, have common marketing features.

4. Behavioural System Approach

This approach refers to the study of the behaviour of firms, institutions and organisations which exist in the marketing system for different commodities. The marketing process is continually changing in its organization and functional combinations. An understanding of the behaviour of the individuals is essential if changes in the behaviour and functioning of the marketing system are to be predicted.

5. Legal Approach

The legal approach is another dimension of the study of the system of agricultural marketing. Increasing government intervention through legislation has increased the importance of this approach. The market regulation programme in many states was delayed because of the emergence of legal problems. Various rules and regulations have been introduced to achieve specified goals. Whether or not the purpose is being served? The questions which require a legal approach for the solution of the problems are: Is the purpose being achieved? What modifications, if any, are necessary?

AGRICULTURAL MARKETING RESEARCH INSTITUTIONS

Some major Indian institutions engaged in conducting agricultural marketing research are:

1. Directorate of Marketing and Inspection (DMI)

The Directorate of Marketing and Inspection is the prime agency which studies the marketing problems arising out of the purchase and sale of agricultural commodities in the country at the macro level from 1935 onwards. Based on the micro and macro-level studies conducted in different areas, the Directorate has published a number of commodity survey reports. These reports provide useful research findings on marketed surplus, costs and margins in the marketing of different crops in different producing areas, defects in the marketing system, solutions to these problems and the methods to be adopted to bring about improvements in the marketing system. The Directorate has also undertaken research studies on cold storage, grading, warehousing and market regulations. Though the reports brought out by the Directorate are mainly descriptive, they do serve as a useful base for further analytical research.

2. State Agricultural Marketing Boards (SAMBs)

Most States in India have, by now, established a State Agricultural Marketing Board for the purpose of supervising and advising the functioning of regulated markets. The Board, therefore, is an executive-cum-advisory body. It effects improvements in the regulation scheme, supervises the functioning of regulated markets and advises market committees and the State Government on related matters. A research cell functions under this Board. This cell collects data on various aspects of marketing; but the data are not subjected to any scientific analysis and theoretical scrutiny. In some States, the State Agricultural Marketing Board gets the research work done by agricultural universities or other institutions. For this purpose, it renders financial assistance to them.

3. Indian Council of Agricultural Research (ICAR)

The Indian Council of Agricultural Research encourages research in agricultural universities and its own institutions on agricultural marketing by financially assisting them in their research projects. The Council had a scientific cell for agricultural economics, marketing and statistics, which helped in the selection of research projects for assistance by the Council, and advised on the direction which the overall research and training policy in this area may take. However, it does not exist now. The ICAR has also established a National Centre for Agricultural Economics and Policy Research (NCAP) at New Delhi which undertakes, inter alia, the research on areas related to agricultural marketing. This centre has been recently upgraded to the level of National Institute (NIAP).

The NIAP of ICAR has brought out a large number of well-analyzed research reports in several areas of agricultural marketing and trade, including cereals, horticultural commodities, livestock and fisheries. Many of the research studies have become critical inputs in formulation or modification of national policies.

The Indian Council of Agricultural Research has also implemented a World-Bank-assisted mega project titled 'National Agricultural Innovation Project (NAIP)'. Out of the four components of NAIP, two components were closely related to agricultural marketing. These were 'agricultural supply chains' and 'livelihood security options based on technology'. There were more than 70 sub-projects, implemented in various parts of the country by consortiums of five to 10 national level institutions and state agricultural universities. In each one of these 70 sub-projects, which included thousands of farmers, it was demonstrated that how market linkages of farmers can help in faster adoption of new agricultural technologies. These projects were operated in operational research mode. This apart, as mentioned earlier, in all the ICAR institutes, agricultural universities and coordinated/network projects, there is a component of agricultural marketing related research.

4. National Council of Applied Economic Research (NCAER)

The National Council of Applied Economic Research conducts studies on agricultural marketing which are of all-India significance. Among the important studies conducted by the NCAER in the past are Demand and Supply Projections of Agricultural Commodities and Structure and Behaviour of Prices of Foodgrains, Changing pattern of rural livelihoods, and agricultural marketing system in the country.

5. Agricultural and Other Universities

Agricultural universities, too, are engaged in conducting marketing studies through their departments of Agricultural Economics. Many of these studies are in the nature of students' projects submitted in partial fulfilment of their M.Sc.Ag., M.A. and Ph.D. degrees. The faculty members of these universities conduct short-term projects on important problems of marketing in agricultural commodities. The universities implement national or international research projects sponsored by the UGC, the ICAR and other organisations. Many agricultural universities are partners in ICAR's national research projects.

6. Agro-Economic Research Centres (AERCs)

Agro-Economic Research Centres in different States in the country undèrtake projects on the marketing problems confronting farmers in their areas of operation. These centres carry out in-house as well as sponsored studies in the field of agricultural marketing. The research issues of topical interest are jointly identified by AERCs and the Ministry of Agriultural. On each of the identified theme, more than one AERC work together to bing out multi-locational results.

7. International Institutions

International institutions, such as the International Crops Research Institute for Semi-Arid Tropics (ICRISAT), International Food Policy Reserch Institute (IFPRI), International Live stock Research Institute (ILRI), International Water Management Institute (IWMI), the Food and Agriculture Organisation (FAO), The World Bank and Asian Development Bank, have undertaken national-level projects in agricultural marketing for the important crops of the region. The research findings of these studies are of valuable assistance in providing guidelines to the member countries/institutions to further probe into the problems in their areas. In recent years, the World Bank has also undertaken or sponsored several studies relating to marketing of agricultural commodities in India and several other countries. Many such study reports brought out by these institutions, are now available. All these organizations have their country office in India, which can be accessed by the researchers. The CGIAR has emphasized research in its priorities for the period 2005–15. Hence, all CG centres have stepped-up their research efforts in this area. As mentioned earlier, ICRISAT has adopted inclusive market-oriented development (IMOD) as overarching framework for its 10-year strategic plan to 2020. The Asia office of IFPRI has also sponsored several studies on agricultural marketing related issues.

8. National Institute of Agricultural Marketing (NIAM)[4]

The National Institute of Agricultural Marketing (NIAM) is a pioneering national level organization set up by the Government of India on 8th August, 1988 at Jaipur, for undertaking specialized training, education, consultancy and research in agricultural marketing. NIAM is an autonomous body under the aegis of Ministry of Agriculture, Government of India. The following functions have been assigned to NIAM:

(a) Research in agricultural marketing for government, cooperatives and other institutions, both on public funding and by contract;
(b) Training to develop leadership potential, management and technical competence for the operation of the agricultural marketing enterprises and services. It offers and sponsors specialized marketing courses at various levels as are necessary to supplement the existing facilities;
(c) Undertaking development and investment project formulation activities for public, cooperatives and private institutions; and
(d) Undertaking advisory and consultancy services on marketing policies, investment programmes and marketing development strategies and rendering specific advice to marketing enterprises (state, private and cooperatives).

The National Institute of Agricultural Marketing organizes workshops, seminars and trainings in the micro and macro areas of agricultural marketing. The objectives, as per the Memorandum and Rules and Regulations of the Institute, are:

1. To undertake and promote study of applied and operational research in problem areas of agricultural marketing.
2. To impart training to the various levels of functionaries involved in agricultural marketing activities.
3. To conduct research through long-term projects; policy formulation, prepare status papers on leading issues; case studies in specific marketing problems, processing industries and export management which have direct bearing on the national economy.
4. To offer consultancy services to the state and central departments, public sector undertakings, and cooperatives in formulation of projects, preparing master plans for states, export institutions, traders and farmers.
5. To act as a national nodal point for coordination of various research studies and dissemination of technologies relevant to agricultural marketing.
6. To develop promising human resources by providing long term structured courses in agricultural marketing by offering diploma/degree courses.
7. To help state governments to generate self employment for educated youths by exploiting local potential resources.
8. To act as an agent of government to formulate policies on emerging issues in agricultural marketing.
9. To create wide information network in the country in agricultural marketing for the benefit of all concerned to evolve efficient, innovative and competitive marketing process; and
10. To develop as "Centre of Excellence" in the field of agricultural marketing by establishing liaison with international organizations.

The Institute caters to the needs of all State Agricultural Marketing Boards, Commodity Boards, all cooperative sector organizations, and input supplying agencies. The Institute also caters to the training/research needs of processing centers, extension wings of state agricultural universities, state department of agricultural marketing, horticulture, animal husbandry, dairying and non-government organizations involved in processing of agricultural products, produce marketing, farmers and entrepreneurs.

9. Indian Society of Agricultural Marketing (ISAM)

The Indian Society of Agricultural Marketing was set up in 1986 with the following objectives:

(a) To promote the study of social and economic problems of agricultural sector and to improve the technical competence of the personnel working in various spheres of agricultural marketing;
(b) To conduct studies independently or jointly with other allied organisations on the problems of agricultural marketing and publish the findings in the Society's journal or as independent bulletins;
(c) To conduct periodical seminars/workshops or conferences on various aspects of agricultural marketing; and
(d) To maintain close coordination with various academic and government institutions and trading and banking organisations connected with agricultural marketing.

The society publishes a research journal Indian Journal of Agricultural Marketing (Chief Editor: S.S.Acharya) and organises annual conferences to exchange research findings in areas related to agricultural marketing.

10. Indian Council of Social Science Research (ICSSR)

The Indian Council of Social Science Research (ICSSR) also supports research work related to agricultural marketing. Almost in all states, Institutes of Development Research have come up which are supported by ICSSR as well as concerned state governments. Many of these institutes have a programme of development studies which includes research related to agricultural marketing also. In Rajasthan, the Institute of Development Studies, Jaipur (IDSJ) has a strong programme of studies, research and dialogues on agricultural marketing, price policies and related issues.

RESEARCH PROBLEMS/AREAS IN AGRICULTURAL MARKETING

The field of agricultural marketing research is comparatively new. Some of the research problems in agricultural marketing are listed in the paragraphs that follow.

It is difficult to lay down priorities in terms of a commodity or a research problem. There are regional variations in the importance of the commodity as well as the marketing problem. In general, an integrated study, embracing price formation, market structure, marketing margins and market policy for the important commodities of the region, should be undertaken, depending on the objectives of the study.

Area	Specific researchable problems or themes
1. Packaging	(a) Suitability of various packaging materials for different kinds of farm products (perishable and non-perishable).
	(b) Consumer preferences for packaging material and size of packages.
	(c) Benefit-cost analysis of packaging and different packaging materials.
2. Transportation	(a) Estimation of the magnitude of losses in transportation by different methods for different farm products.
	(b) Determination of the most economic method of transportation for different farm products from one market area to another.
	(c) Assessment of the extent of spatial variation in prices and their relationship with the cost of transportation.
3. Storage	(a) Finding the most economical and efficient method of storage for different farm products.
	(b) Estimation of the magnitude of quantitative and qualitative losses during storage.
	(c) Economics of different storage structures, including cold storage.
	(d) Extent of temporal price variations and their relationship with the cost of storage.
	(e) Optimum location of storage structures and their size.
	(f) Impact of NWR system on farmers.
4. Processing	(a) Economics of location and scale of processing units.
	(b) Survey of existing processing industries, estimation of their demand and working out the gaps between requirement and availability.
	(c) Estimation of the cost of processing by different available methods.
	(d) Economics of mini or primary processing at farm gate level.
	(e) Estimation of margins in the processing of different agricultural products.
	(f) Relative economics of processing in the public, private and cooperative sectors.

5. Marketable and marketed surplus	(a) The size of the marketable and marketed surplus of different farm products and their projections over time.
	(b) Relationship between the marketed and marketable surplus with the size of the holding and the prices of the products.
	(c) Examining the selling behaviour of farmers in terms of choice of time, place and agency.
	(d) Examining the pattern of market arrivals and of the prices of the crops, and factors influencing the volume of arrivals.
6. Marketing efficiency	(a) Relative efficiency of the marketing of farm products through different marketing agencies—private trader, cooperative marketing society and public agency.
	(b) The extent of integration in the marketing of farm products and their relationship with market efficiency.
	(c) Study of the market structure, market conduct and performance, and estimation of the extent of market concentration.
7. Marketing costs, margins and price-spread	(a) Estimation of marketing costs, margins and price-spread for different agricultural commodities in different marketing channels.
	(b) Estimation of the relationship in prices—farm (harvest) prices, wholesale prices and retail prices—and the extent of competitiveness in marketing;
8. Market regulation programme	(a) Estimation of the benefits of a market regulation programme for different groups in the minimization of the cost of marketing, profit margin and price-spread.
	(b) Determination of the most optimum size of the market yard.
9. Demand and supply	(a) Projections of the demand for, and supply of, different farm products and gaps in different time periods at regional and national level.
	(b) Projection of the demand for farm inputs at different levels of technology.
10. Prices	(a) Examination of the extent and magnitude of price fluctuations for different farm products and the factors affecting them.
	(b) Examination of the relationship in different types of prices.
	(c) Study of the seasonal behaviour of prices of farm products.
	(d) Examination of the nature of price movements in primary, secondary and terminal markets for different agricultural commodities.
11. Recent innovations in agricultural marketing	(a) Impact of farmers' markets on pricing efficiency.
	(b) Impact of contract farming or contract marketing on marketing efficiency.
	(c) Comparative study of alternative value chains/retail chains on marketing efficiency and impact on traditional traders and farmers.
	(d) Impact of electronic spot markets or exchanges.
	(e) Impact of E-NAM on farmers' income and on marketing efficiency.
	(f) Impact of 100% FDI in food retail on various sectors.

In addition, following areas for research in agricultural marketing need attention:

1. Marketing technology adaptable to Indian conditions like labour intensive technologies without compromising the quality;
2. Dynamic studies of comparative advantage of various agricultural commodities of India;

3. Generating outlook information for prices and production for long, medium and short term decisions;
4. Evaluation of structure, conduct and performance of agricultural markets;
5. Studies on costs and margins of agricultural crops;
6. Marketing studies for poultry, livestock, fisheries and processed products;
7. Impact of entry of organized retail chains and FDI in retail for grocery and food;
8. Assessment of losses in the marketing chain and how to minimize these;
9. Export potential of farm products and how to reach the benefits to small farmers;
10. Estimation of income and price elasticities of farm products;
11. Analysis of various models of group or collective marketing by the farmers; and
12. Analysis of market size and agribusiness of dairy sector, poultry sector, floriculture and spices sectors.

JOURNAL ARTICLES ON AGRICULTURAL MARKETING

Output of agricultural marketing research is brought out as in-house research bulletins, research reports submitted to the sponsorer, or published journal articles. Each one of these has its own relevance and importance. Most of the research scientists/ academicians endeavor to publish their work as journal article. A very incisive analysis, of research articles published in leading two Indian journals and three international journals (on agricultural economics), by Mruthyunjaya (2015) revealed that the theme of agricultural marketing covered 16% (250 out of 1567) of the total articles in international journals whereas it was only 13% (100 out of 766) in Indian journals. One reason for lower percentage share in Indian journals could be that in India, there is an exclusive agricultural marketing journal, which was not included in the study. The journals included in the study were Indian Journal of Agricultural Economics, Agricultural Economics Research Review, American Journal of Agricultural Economics, Australian Journal of Agricultural Economics and Canadian Journal of Agricultural Economics. The study pertained to the ten-year period 2004 to 2013.

█ MARKETING EXTENSION

NECESSITY

Over the years, Indian agriculture witnessed the remarkable change in production from the subsistence economy to the market economy. The credit for this goes to evolution of new technologies by the scientists for increased crop production, role of extension agencies in carrying the technology from lab to farmer's fields and increased investment in agriculture sector for creation of infrastructural facilities. The increased production necessitated handling of farm products and their movement from producing areas to the consuming areas. This in turn demanded efficient marketing system.

Agricultural extension so far has been confined to issues of intensification and diversification of production. Increasing production by itself is no solution unless the increased produce finds a suitable market through agricultural extension work. The country is also facing a new challenge of globalization and liberalization.

The inclusion of marketing component as an integral part of agricultural development in the country is regarded as a backbone both in developing and developed countries. The marketing system needs to continuously adjust to the need of producers and consumers and should aim at minimization of marketing costs and increase the producers' share in the consumers' rupee. This inter alia necessitates providing market extension services, i.e. educating the farmers in planning and preparation of agricultural produce for the market in terms of improved methods in cleaning, packaging, grading; strengthening of market intelligence; and assessment of profitability of alternative enterprises in addition to the motivation of the farmers in increasing production and arranging needed value addition.

For strengthening and developing agricultural marketing system in the country, it is of paramount importance that sustained efforts in the area of marketing training and extension are made. Although agricultural production technology transfer services exist at the village level, the marketing extension service, designed to benefit the farmers and other market functionaries, does not exist. Marketing extension has not been accorded due importance it deserves in the light of shifting agricultural policies towards sound marketing system. Most states in India do not even have separate Marketing Extension cell to undertake regular extension activities. The absence of emphasis on marketing extension has led to sluggish orientation of farmers to agricultural marketing. An effective marketing extension service is the need of the hour. This has assumed even greater importance in the light of fast changing business environment as a result of liberalization of the economy and the potential which export orientation of agriculture provides for improving farmers' incomes.

Though agriculture extension work is the responsibility of State Departments of Agriculture, the training of farmers and extension workers has been assigned to state agricultural universities. There is a separate Directorate of Extension Education in all the State Agricultural Universities. There is a KVK in each rural district of the country. But the element of training in agricultural marketing is very weak. This is evident from the mix of training programmes and the absence of agricultural marketing specialist among the band of specialists who man these KVKs. Problems at the post-harvest stage and their solutions have been receiving attention following the emphasis by the Indian Council of Agricultural Research on research studies and extension education in this area.

At the national level, a small marketing extension cell was set-up in the Directorate of Marketing and Inspection, Government of India in the year 1962. Since its inception, this cell has taken a leading role in spreading the knowledge pertaining to agricultural marketing and other related issues. Various media are being utilized to reach the target groups. The Directorate of Marketing and Inspection also brings out a quarterly journal "Agricultural Marketing", which contains studies and data on certain marketing aspects. This Directorate also brings out commodity reports and bulletins on various aspects of marketing, but there is very little material on marketing education to the farmers.

State Agricultural Marketing Boards have also a training and publicity cell to educate the farmers on problems in the field of agricultural marketing. Periodical publications are also brought out by these marketing boards. The education of farmers on important aspects of marketing such as grading and standardization, storage, processing, market information, pricing of farm products and product planning has started receiving attention in the programmes of farmers training conducted by various organizations.

By and large, up to the 1980s, marketing extension service to educate the farmers, improve their marketing skills, bring change in their attitude and equip them on various intricacies of agricultural marketing was grossly inadequate. There were no arrangements to provide advice on product planning, selection of market outlets and source of market information. Extension activity in the field of agricultural marketing both at the central and state levels was restricted to publicity regarding benefits of regulated markets and various schemes in the area of agricultural marketing launched by the State and Central Governments.

One of the most important aspects of training and extension education in agricultural marketing that needs attention is certain weaknesses that exist in the teaching material or text books and teaching methods. As a result there is an absence of practical knowledge and orientation in marketing. Therefore, this area deserves the immediate attention of policy makers and administrators in India. At the International level, the FAO has been taking action in this direction.

After the economic liberalization in the 1990's, there has been a sea change in the extension approach, which has been decentralized and made more farmer centric and market driven. It was during this period that the Agricultural Technology Management Agency (ATMA) was introduced in 1998, initially in seven states and later extended to other areas. Under ATMA, the farmers' oriented activities include training to farmers, organizing demonstrations, exposure-visits (inter-state and inter-district) and dissemination of information through various traditional and electronic media. It may be mentioned here that M.S. Swaminathan Research Foundation, Chennai has set up Rural Knowledge Centres (RKCs) in Tamil Nadu and Puducherry, which are based on hub and spoke model. The information provided by these centres is location specific.

One very important feature of the agricultural extension during 1990's has been the entry of private sector viz., corporates, NGO's, agro-input industries, contract farming companies, and producers' cooperatives in extension services. Mahindra Subha Labh and, NABARD-led Agri-Clinics and Agri-Business centres by the private individuals (mostly agricultural graduates) are a few to mention. Mahindra Subha Labh Services Limited (MSSL) was formed in 2001 with the objective to provide what the company described as 'Integrated Yield and Profit Solutions'. The company has established, through its franchisee 'Mahindra Krishi Vihar (MKV)', a one step shop for farmers that provides access to quality inputs, machinery, credit, advisory and field supervision services and buy-back of the output at better prices.

AREAS OF EXTENSION EDUCATION IN MARKETING

Extension education in the field of agricultural marketing should include the following aspects:

1. Information on consumer's preferences, internal (domestic) demand and also export demand for farm products, which can fetch adequate financial rewards to farmers efforts.
2. The requirements of qualities of different farm products both for internal and export markets.
3. Importance of grading of farm produce before marketing, grade standards laid out for different agricultural commodities, and innovative techniques involved in grading of farm products.
4. Information on relative prices of different farm products and trends therein.
5. Knowledge on methods of storage and advantages of undertaking processing of farm products.
6. Knowledge of regulated markets and E-NAM and advantages of selling farm products in them.
7. Knowledge of innovative marketing channels viz., Apni Mandi, Rythu Bazars, Uzhavar Sandies, Hadaspar Vegetable markets and Kisan Mandi and advantages of selling the perishable and other farm products in them.

8. Information on various government programmes pertaining to price, support viz.; minimum support prices, statutory minimum prices, procurement prices and market intervention scheme (MIS).
9. Marketing improvement programmes and subsidies provided for various purposes such as on purchase of plastic crates for handling of vegetables, and transportation of fruits and vegetables for the benefit of farming community.
10. Technological changes pertaining to performance of marketing functions; and
11. Knowledge on post-harvest management of different farm products particularly of fruits, vegetables, spices and other perishable products, including livestock products.

EXTENSION METHODS

Appropriate combination of following two forms of extension education methods should be followed in carrying various messages to the farmers:

1. Individual Contact Methods: Farm and home visit, mailed letters, office calls, SMS on cell phones, etc.
2. Mass or Group Contact Methods: Leaflets, Radio, Television, Newspapers, Journals, Local magazines, Puppet show, Film show, Scientific meetings (krishak samwad, krishak meet), and Collaborative training programmes organized in villages with the technical assistance of different departments/organizations such as NIAM, CWC and SWC, NABARD, NAFED, APEDA and Agricultural Universities.

PRIVATISATION OF EXTENSION SERVICES

Extension services, which were mostly public funded worldwide until two decades ago, are increasingly coming under private domain. The transformation of agriculture from a mere subsistence activity to commercialized agri-business in the developed and developing countries and the associated gradual change of technology from a public good to private good has provided incentives for commercial agencies to invest in this sector. The increasing inability of the governments to adequately fund its extension machinery was, however, the real force behind the search for alternative approaches for public funded extension to privatization. Those who argue for privatization of extension education point out the defects in public funded extension as highly inefficient, too beauracratic, lack of accountability, supply driven rather demand driven, and dominated by constraints such as operational funds, trained manpower and transportation bottlenecks. Further, the growing dissatisfaction of the users with the quality of extension services available, increased flow of agricultural surplus, increased specialization among farmers, and technological developments in mass media have led to the emergence of various types of privatization of extension services. Some of the forms of privatization of extension efforts are:

1. Introduction of cost recovery programmes by government extension departments on selected services.
2. Cost-sharing by farmer's groups.
3. Contracting services to non-government organizations.
4. Partial withdrawal of government extension agencies from favourable regions or from high value crops.
5. Increasing involvement of input companies and product marketing companies in transfer of technology.

6. Growing number of private consultants/firms.
7. Rise in number of NGO's and their willingness to implement rural development programmes.
8. Arrangements made by producer's cooperatives or companies to meet their extension and other demands.
9. Involvement of agro-processing companies to provide all types of services to their contract growers; and
10. Promotion of Agri-Clinics and Agri-Business Centres to provide technical advice to the farmers.

Under privatization, it is envisaged to provide expert services and advice to farmers on new technology, market extension services and trends and prices of crop products in markets and also the supply of inputs and farm equipments on hire. Privatization of extension activities would facilitate tailor made extension services beneficial to both farmers as well as entrepreneurs.

In India, NGOs, private sector, private limited companies, corporate bodies, self-help groups and farmers associations are augmenting and supplementing public sector extension efforts to a large extent. Several states have already started taking steps in this regard. To facilitate private agencies to undertake extension programmes on regular basis for the country as a whole, a 24 hour TV Kisan Channel has been launched to educate farmers which can be used by public agencies as well as private extension agents.

One of the major recommendations of the Expert Committee on Agricultural Marketing (2001) was on providing market extension services to assist small and marginal farmers in marketing their produce in the liberalized trade environment. It has been emphasized that a massive programme of marketing extension needs to be launched at the field level wherein extension messages should encompass all important dimensions of agricultural marketing. Some of the suggestions of the committee are as follows:

1. Advise on product planning, i.e. the careful selection of the crops, and varieties to be grown with marketability in mind to the farmers which is very essential to enable them to withstand the competition in the market.
2. Providing marketing information on various aspects of marketing viz., current prices, market arrivals, forecasting of market trend, arrangements available in the market related to storage and transportation, quality standards and post-harvest handling requirements. Farmers also have to be educated/trained in taking appropriate signals from the forward and futures prices. The information should be area specific, crop specific and buyer specific. Every agricultural market should have an extension cell equipped with internet and other audio-visual facilities necessary to educate farmers in various aspects of marketing functions and services.
3. Securing market for farmers: The extension agency should advise farmers on how, when and at what price to sell farm products and how to make contract marketing arrangements with processors, wholesale traders or other bulk buyers.
4. Advice on alternate marketing: Advice on planned marketing strategy to avoid gluts in the small local markets is also necessary. They can be advised to take benefit of warehousing with pledge finance schemes, entering into forward contracts or to go for future trading. A planned marketing strategy can help the farmers in getting higher price and help stabilize local market prices.

5. Advice on improved marketing practices: Farmers need education on improved harvesting methods, standardization and grading, improved packing and handling practices, and appropriate storage methods for profitable marketing of their produce; and

6. Advice on establishing and operating markets: Marketing extension should help rural population to establish and operate markets on their own to save from exploitative elements. Operating within the framework of marketing rules and regulations, the farmers will be able to protect their interests better than when they visit distant wholesale or terminal markets.

During the last 20 years, the scenario has shown signs of significant change.

1. Government of India (various ministries) and state governments have launched and implementing several schemes to help the farmers, farmers groups and others who help the farmers in marketing activities.

2. Several private companies, including input supplying agencies, contracting companies, retail chains, and warehouse owners have come up with special efforts to help farmers link-up with the markets.

3. Several initiatives by SFAC, NABARD, NCDC, IFFCO, NDDB and commercial banks are helping farmers to organize into groups for collective marketing.

4. The State APMR Acts have been amended and the APMCs are encouraging farmers' markets, facilitating direct purchases from farmers and allowing setting up of private or cooperative markets, which are facilitating better linkages of ordinary farmers to the markets.

5. Several markets are now linked to e-auction platform and further emergence of a common national market (E-NAM) for agricultural commodities will revolutionize marketing extension system in the country.

STATISTICS IN AGRICULTURAL MARKETING

The data on various aspects of agricultural marketing are important for policy formulation and research. The non-availability of data has limited the scope of research in this field.

COVERAGE AND AVAILABILITY OF DATA

The requirement of data for agricultural marketing research/studies depends on the objectives of the study. It also depends on the scope and approach of the study, i.e. functional, institutional, commodity, behavioural or legal. The data required may be primary (grassroots level) or aggregates and averages/ratios at the market, district, state or national levels. Following kinds of data are available and can be accessed at relevant levels:

1. Daily prices and arrivals of different commodities in all the regulated market yards (primary wholesale markets and secondary wholesale markets).

2. Number of licensed traders/commission agents, and prescribed commission charges, market fee and/or cess in each regulated market.

3. Quantities of different commodities graded at producer's, seller's or farmer's level in regulated markets.

4. Quantities stored (in and out) in storage godowns, warehouses/silos and cold storage of different commodities and storage charges.

5. Sample information on month-wise wholesale and retail prices for various commodities (including processed products) that are used for construction of wholesale and consumer (retail) price indices.
6. Sample Information on time, place and quantity of commodities sold and prices received by farmers (as a part of cost of production studies).
7. Quantities of various commodities processed and outturn of processed products (collected as a part of Annual Survey of Industries).
8. Quantities and prices of agricultural commodities and agro-processed products exported and imported with prices.
9. Quantities of various commodities procured at MSPs and quantities distributed under PDS (in different schemes).
10. Quantities of various commodities handled by and storage capacity of cooperatives.
11. Information related to E-NAM like number of registered traders and farmers/ FPOs, quantity transacted, assaying charges, prices of each lot etc.

AGENCIES OR SOURCES FOR DATA

1. The obvious source of primary information is the organization at the grassroots level viz:, agricultural produce market committees; individual warehouses, godowns, silos or cold stores; and cooperative marketing organizations.
2. The state level organizations/offices compile the primary information and bring out aggregated information. These are State Agricultural Marketing Boards, State Directorates of Agricultural Marketing, State Directorates of Economics and Statistics, State Warehousing Corporations, State Registrars of Cooperative Societies and State Agro-Industries sections/departments.
3. At the national level, following organizations aggregate, compile and bring out comprehensive reports and also provide disaggregated information on their websites/portals/Annual Reports:
   (a) Directorate of Marketing and Intelligence, Ministry of Agriculture and Farmers Welfare (AGMARKNET)
   (b) Directorate of Economics and Statistics (MOAFW)
   (c) Central Warehousing Corporation (CWC)
   (d) Ministry of Statistics and Program Implementation
   (e) Agricultural and Processed Products Export Development Authority (APEDA) and Directorate General of Foreign Trade (DGFT)
   (f) Ministry of Food Processing Industries (MoFPI)
   (g) Food Corporation of India (FCI)
   (h) National Agricultural Cooperative Marketing Federation (NAFED)
   (i) National Cooperative Union of India (NCUI)
   (j) Cotton Corporation of India (CCI)
   (k) Jute Corporation of India (JCI)
   (l) Fertilizer Association of India

PUBLICATIONS OF MARKET STATISTICS

Market statistics are available in the publications brought out regularly as well as periodically by the concerned departments. Some of the publications in which market statistics are regularly published are:

1. Agricultural Situation in India—Monthly
2. Bulletin on Prices—Weekly

3. Reserve Bank of India Bulletin—Monthly
4. Agricultural Marketing—Quarterly
5. Bulletin on Foods Statistics—Biannual
6. Indian Agriculture in Brief—Annual
7. Commodity Survey Reports—Occasional
8. Newspapers—Economic Times, Financial Express; The Hindu, The Observer and several others—Daily
9. Annual reports brought out by the concerned departments, stating their progress.
10. Economic Survey—Annual
11. CMIE Publications—Different intervals
12. Reports of the Commission for Agricultural Costs and Prices—Annual and Seasonal
13. Indian Journal of Agricultural Marketing—Quarterly
14. Publications of State Agricultural Marketing Boards.
15. Agricultural Marketing Statistics—Annual
16. Agricultural Statistics at a Glance—Annual
17. Fertilizer Statistics—Annual
18. Rural Development Statistics—Annual
19. Annual Reports of Various Ministries

DISSEMINATION OF MARKET STATISTICS

Market statistics, in addition to their publication in bulletins, are broadcast daily by the regional stations of All India Radio. The information on prices is broadcast by the AIR two to three times a day so that farmers may take advantages of it and come to a decision about when and where to market their produce. The information is also telecast in bulletins of Kisan Channel, regional channels of Doordarshan and several other TV channels.

Before its merger with SEBI, the Forward Markets Commission launched a project in association with four commodity exchanges (MCX, NCDEX, NMCE and ICEX), through which real time spot and future prices of various commodities in different markets are provided to farmers. In 2009–10, 183 ticker boards were installed in various mandies of 14 states. In 2010–11, 705 more such boards were installed in various mandies. By now, 8000 ticker boards have been installed through out the country covering all the mandies. The cost of each board is around ₹ 80,000.

Much of the market related information is available on Agmarknet, websites of DMI and concerned ministries and many other websites on the internet.

Problems

There is a gap in the quantity and quality of available market statistics and their requirements for the purpose of finding solutions to problems with a view to formulating a satisfactory agricultural marketing policy. The available statistics fall short of requirements because of the dynamic nature of the subject. Whatever statistical information is available is not properly utilized because of lack of trained manpower and facilities for their analysis in the marketing sections of the government and various Boards. The statistics available from different sources do not meet the researcher's requirements, whose main problem arises out of data gaps and/or non-availability of data for comparable periods and variety. The problem of non-availability has been accentuated by the non-co-operation of market middlemen (traders). Traders often do not maintain complete records or hesitate to allow access to their records. Some of the important problems in market statistics are of the following nature:

1. Incomplete series due to data gaps;
2. Time lag in the publication of the data, which often makes them obsolete specially in view of rapid policy changes;
3. Lack of comparable data over time due to change in concepts and definition;
4. Non-reliability of the collected data because the collector has no research aptitude and interest; and
5. Market statistics of prices are not comparable because of variations in the quality of the product.

These gaps in market statistics should be minimized so that they may be more useful to researchers as well as policy-makers.

NATIONAL AGRICULTURAL MARKET ATLAS (NAMA)

Various Expert Committees, set up after the initiation of economic reforms and establishment of World Trade Organization (WTO), based on reviews of the existing agricultural marketing system, are of the view that there is need for developing infrastructural facilities, identification of new markets, modernization of existing markets, assessment of marketable surplus of different crops, developing of sub-markets, and matching the infrastructural availability with the available output in the country. For all this, an integrated marketing system has been visualized. This encompasses all sort of information at one particular point. This can be very well dovetailed with the planning document of the planners, who have in their fold systematic infrastructure planning to be used by the marketing system. Synchronization of both the plans is extremely useful for translating the avowed objectives in a desirable manner. With this objective, development of an Agricultural Market Atlas is necessary.

National agricultural market atlas is of immense use in identifying the gap between magnitude of production and availability in terms of number and facility of market centres in a particular area. The market atlas is an effective tool to outgrow such infrastructure imbalances. The market atlas is the best tool to sort out the problem with the help of satellite image. An over view of the different geographical contours/tracts can help in taking such decision as to where to transport perishable commodities and where to develop a new road, which can cater to the need of horticultural crops growing farmers or where to develop new markets for these crops.

Now the time is ripe enough to predict with the help of satellite images the output of a produce which will arrive in the markets. The application and use of national market atlas will mitigate many of such problems and will help in prioritizing the category of the markets to be developed viz., primary, secondary, rural, terminal and mega markets.

The Ministry of Agriculture and National Institute of Agricultural Marketing (NIAM) have recognized the importance and launched the initiative of the National Market Atlas and its benefits are visible. The NAMA is a component of Agmarknet with an additional feature of spatial data. It provides the GIS web interface to visualize the daily market scenario on the national map. Users can find interactive dynamic market atlas and national atlas. Information on DMI and NIAM etc. is also given. It guides the users to go to the related links to see information they wish to seek.

▌EMERGING ISSUES IN AGRICULTURAL MARKETING

The rapid development in agricultural research and the introduction of technological innovations, viz., high-yielding varieties, improved agricultural implements, fertilizers and pesticides, have brought about a breakthrough in Indian agriculture. This development in the field of agriculture, popularly known as the green/white/yellow/blue revolutions, has given rise to new problems in agricultural marketing. It is essential to maintain the tempo of these revolutions. The farmer-producers should be assured of a fair price for their produce, failing which they may lose the incentive to increase agricultural production. A fair price for the produce may be assured when there is an orderly marketing system in the country. But an orderly marketing system can be created only when the problems, which have emerged, are effectively tackled. There is an urgent need in the present context for tackling the emerging problems of agricultural marketing more resolutely and efficiently than ever before. The improvement in the domestic marketing system has assumed special significance with the launch of new economic policy in 1991 and opening up of the external trade regime. The important problems which have emerged pertain to the following areas:

1. Increase in Production Levels and Market Arrivals

With increased market arrivals, and in order to enforce strict market regulations, it is necessary that a large number of spacious market yards are developed in rural areas with all the necessary marketing facilities. Without spacious market yards, it is not possible to centralise and effectively supervise the transactions taking place in the area. The development of these spacious market yards is also essential for the performance of various marketing functions, such as grading, cleaning, sieving and weighment of the produce.

Presently, most of the markets do not have spacious market yards and the transactions are carried on in congested areas in the centre of the city and on the roadsides. Recently, some market committees have constructed spacious market yards; but a majority of market committees do not have them because of the paucity of funds and the non-availability of land.

2. Price Instability

Agricultural prices are very unstable and fluctuate violently. These prices fall in the post-harvest months and increase later in the year. This situation has worsened with the increased market arrivals as a result of the emergence of surpluses, especially of wheat and rice and also of pulses and perishables in recent years. The increasing instability in prices adversely affects the income of farmers as well as the tempo of increasing production. There is, therefore, a need for reducing price instability. Several steps may be taken for farmers to get a better share in the consumer's rupee. Some of the steps taken to check price instability are: fixation of minimum support prices of the crops by the government; purchase of the commodities if market prices fall below that level; development of warehousing facilities to check post-harvest sales among the farmers; and making warehouse receipt a tradable negotiable instrument.

3. Market Intelligence

Market intelligence is another problem that has emerged and is an important adjunct of orderly marketing. With the increased marketed surplus and opening up of trade, the importance of market intelligence has increased. Farmers market the produce in the village and nearby assembling centres out of their ignorance of the price prevailing in the nearby primary wholesale, secondary wholesale and terminal markets. Traders take the advantage of the ignorance of the farmer because they have full knowledge of the prices prevailing in other markets. This places the traders in a superior bargaining position.

The Directorate of Economics and Statistics, Government of India, as well as the State Departments of Agricultural Marketing have been collecting data on wholesale and retail prices at various markets and disseminating the information through periodical bulletins issued on the All India Radio and in the form of publications. However, this is not a satisfactory position because the information provided is stale in the sense that, by the time it reaches the farmers, the market prices have changed. Farmers are not able to take advantage of the available intelligence because of their illiteracy. There is, therefore, an urgent need for refinement in the available market intelligence, so that announcements of market information may be made on the expected prices, arrivals, demand and supply. Linking of all markets through Agmarknet is helping, to some extent, in this regard.

There is also a need for the publication of "outlook" reports for each district to help the farmers to decide about their cropping and production programmes. NCAP and TNAU, under the fold of NAIP of ICAR, have initiated action in this regard. This needs to be made a regular permanent feature in all the states.

4. Crop Price Forecasts

While timely availability of information about current prices helps farmers to decide where and when to sell their surplus produce, it does not help them to cover the price risk faced by them. The schemes like MSPs and market intervention (MIS) have also been reported to be only partially effective because the government/ implementing agencies could not make adequate arrangements in time to purchase the quantities offered by the farmers. There are several cases of substantial quantities of pulses, onion, potato, and even wheat and paddy in some areas sold by the farmers below the announced support prices due to delays or short fall in purchase arrangements. Even export/import decision gets delayed due to lack of likely price scenarios in advance and thus both farmers and consumers suffer. It is in this context that generation and dissemination of crop price forecasts well in advance (like that of weather forecast by IMD) is critically important for the country. About 10 years before (late 1990's), this work was taken up by the ICAR as a part of National Agriculture Innovation Project (NAIP). Forecast models were perfected and human resource capacity was built up.

A group of ICAR Institutes and SAU's proved that crop price forecasts can be accurate up to more that 90%. Later (till recently), the work continued under the leadership of National Institute of Agricultural Economics and Policy Research (NIAP) and it was shown that crop prices can be predicted three to six months in advance with 90 to 95% accuracy. But both these five years exercises were done in project modes. Obviously the technical and HR capacity developed is not being used as the project has terminated. It was being argued since long that this work should become a permanent feature of the system in an institutional framework (Acharya, 2017a, 2017b). This is high time that a crop price forecasting system becomes a permanent feature in MOAFW, realizing that the technical and HR capacity already exist in ICAR-NIAP.

5. Grading of Agricultural Commodities

The grading of agricultural commodities has assumed importance in recent years because of the introduction of many new varieties of certain crops specially of wheat, rice and other cereals. There is a big price difference between varieties which arises out of consumer preferences. In the absence of proper grading, both the producers and the consumers suffer. The producers suffer in the absence of grading because they get the same price for the best quality of the produce and for a fair average quality. Grading ensures that producers receive a price which is commensurate with the quality of the produce. At the same time, grading protects the consumers against adulteration. The progress in the adoption of grading by farmers and consumer preference for graded rather than ungraded produce has been poor; and this situation needs to be corrected. Scientific quality testing machines and equipment can solve this problem by rendering a quick and systematic grading service. Facilities for grading must be developed at the farm level and at the market level. In the peak season, because of the accumulation of huge stocks, it is not possible to introduce grading at the market level. That is why the provision of grading at the farm level is important.

6. Storage Facilities

The problem of the storage of farm produce has been accentuated by the increase in the volume of production. Storage is necessary, at the village site to check the tendency of immediate post-harvest sale by the farmers, and at the market level so that the various marketing functions may be performed and advantage may be taken of any price rise. In the recent past, warehouses have been constructed by the Food Corporation of India, the Central Warehousing Corporation, the State Warehousing Corporations, Cooperative Marketing Societies and the government. Individuals also have built-up storage space. The available storage space in these warehouses is still less than the requirement. There is, therefore, a need for the construction of more godowns, specially in rural and hilly/desert areas.

To make farming more attractive and the hard labour of the farming community more remunerative, it is necessary that storage facilities should be extended right up to the producers' level. This is possible only by having a network of warehouses and rural godowns all over the country, which can serve not only as places of storage but as places of orderly transactions of selling and buying, and thus ensure that farmers do not travel long distances to market their produce. Once this system is evolved, a farmer will be in a better position to market his produce and get benefits of hard work. The on going efforts need to be upscaled.

7. Cold Storage, Cool Chains and Processing

The facilities of cold storage, cool chains and processing of perishable commodities (fruits, vegetables and livestock products) are very inadequate. While the production as well as the demand for these products is increasing, the expansion in facilities for their handling and processing has been quite slow. There is a need to create a favourable policy environment for investment by the private sector in these facilities.

8. Quality Testing and Control

The most important marketing problem today is the tendency of adulteration and use of harmful substances in finished products sold to the consumers. The examples are synthetic milk, prohibited colours and preservatives in processed food, and adulterated spices. The chemical residues and contents in several packed and even branded food products have been found to be considerably higher than prescribed maximum limits. The FSSAI has initiated several measures but their impact on the ground is yet to be realized.

9. Marketing of Agricultural Inputs

The marketing procedures of agricultural inputs have to be improved. This is most essential in the present context of the use of new technology. The newly-evolved HYV seeds are more responsive to other inputs viz; fertilizers and pesctides, and arrangements for their timely availability are a prerequisite for the success of the programmes of agricultural development. The quality testing system for inputs also needs improvement. The expansion of quality testing lab network and soil health card scheme may help.

10. Ancillary Facilities in the Markets

The exiting markets lack the ancillary facilities, which include banks, a post office, and shopping centres. Also, the facilities of sorting, grading and packaging in the villages or primary markets are not available. These should be increased to provide full benefits to the farmers.

11. Finance for Market Development

Market development with all the ancillary facilities is highly capital-intensive, and returns on investment in it accrue slowly over a period of years. For market development, a special arrangement for financing is necessary.

12. Education of Farmers

As mentioned earlier, it is equally necessary to educate farmers in marketing of agricultural products so that they may derive necessary benefits of their activities. Without proper education, benefits cannot reach the farmers fully. This is more so in the present context with the availability of increased marketed surplus with producer-farmers and high volatility in the prices of agricultural commodities. The farmers need to be organized into marketing groups. The efforts need to be upscaled.

13. General Dissatisfaction of Farmers

The problems listed above generally reflect the inadequacy of the market infrastructure for handling of the increased volume of production. This inadequacy is reflected in the general discontentment amongst farmers which, if not checked, will thwart the tempo of increasing production. The grievances which have to be dealt with at the local level are: congestion in the markets; delays in cleaning, weighing and auction of their produce; loss arising out of pilferage and excess weighment; delays in payment by traders; absence of procurement agents when market price falls below the support level; and nonavailability of inputs at the right place and time. These apart, several macroeconomic policies adversely affect the farmers in certain regions creating discontentment among the farmers. There is, therefore, a need to sensitize the policy-makers also for reducing the dissatisfaction among farmers.

14. Unabated Avoidable Wastage in Marketing Chain

About 15 years before, we estimated the avoidable wastage in the marketing chain worth ₹ 50,000 crore. Several measures were taken to reduce this wastage. The MoFPI claimed that they were able to reduce these by 20%. Even if this claim is accepted, the current level of wastage now is not ₹ 40000 crore, but it is about 1 lakh crore, because of the rise in prices of the wasted farm products. Therefore, all round ongoing efforts to improve the marketing efficiency must continue with more vigour.

15. Neglect of Rural Supermarkets (Haats)

While encouraging the entry of organized retailers, setting up of farmers' markets, and various schemes/incentives for organizing the farmers into marketing groups (like FPOs, SHGs or cooperatives) are the essential steps in right direction, there is a need to pay attention to our traditional haat system.

Haats are rightly called rural supermarkets, by some, because haats are the nerve centres of economic, social and cultural life of villages. Villagers, producers, sellers and farmers depend on haats not only for the disposal of their produce, but also for the purchase of their needs. At present, around 43000 haats are held in India, of which 70% are continuing from the pre-independence period. Each haat is spread over 5 to 6 acres of land and has 300 to 500 stalls selling anything and everything that villagers need in their daily lives. These are usually held, on an average, 24 km away from the nearest big town. The average catchment area of a large haat is 57 villages and of a small haat is 21 villages. The haats are usually located near market places (52%), bus stops (35%), temples (5%) and others (8%). Before Independence, haats in rural areas were organized by either zamindars or rulers of princely states, but later by the panchayats. These days, the haats are held on the panchayat or common lands. In many states, private parties are increasingly organizing the haats. The participation fee is charged from the stall owners and it is very low. For example in 2010–11, it was, on an average, ₹ 13 per stall per day.

Across the states, the number of haats has been reported to be 10380 in U.P., 4993 in Bihar, 4078 in West Bengal, 3996 in Jharkhand and 3758 in Maharashtra. The average number of visitors is 11900 in large haats and 5600 in small haats. Haats sell almost all kinds of products that are needed by a rural life style from agricultural products to hair pins. Though agricultural products (53%) still have a major share in total number of stalls, manufactured goods (19%) and processed foods (6%) have gradually found entry. The proportion of stalls selling handicrafts and forest products is around 5% each. Each seller, usually tries to visit several haats to increase its sales. On an average, a seller visits at least three haats per week. In Maharashtra, a seller visits even six haats per week. Many companies are using haats to sell their branded products, including FMCG products. Some companies have even contributed to improving the infrastructure in the haats (like raised platforms, providing drinking water and toilets for women) and in turn they got branding rights from the panchayat or haat contractor.

The haat system demonstrates the Indian ingenuity of keeping product prices low, because there are no high shop rentals, no salesmen salaries or investment in display selves and shop interiors. The fee for a stall is very low and that too the same for a poor woman seller or a MNC selling consumer goods. The haat system provides a lesson for high-cost organized retail business operators. There is a need for a well-thought out way forward to support these rural supermarkets which fit well into rural India's culture. These will help farmers (as consumers as well as sellers), rural families, urban companies looking for rural markets, and other stakeholders.

■ REFERENCES

1. For details, sea FAO, Agricultural Marketing Training Improvement in Asia, 1979.
2. Ministry of Agriculture, Report of the Expert Committee on Strengthening and Developing Agricultural Marketing, Government of India, June 2001, p. 38.
3. Hanson, J.L., A., Dictionary of Economics and Commerce. The English Language Book Society, London. 1974, p. 318.
4. National Institute of Agricultural Marketing, NIAM: An Over View, Jaipur, 2000.
5. "Research Problems in Agricultural Marketing in the Changed Context", Seminar Series V, Seminar on Emerging Problems of Marketing of Agricultural Commodities, Indian Society of Agricultural Economics, Mumbai, 1972, p. 322.
6. Acharya, S.S., Agricultural Marketing and External Trade: Infrastructure and Policy Needs for XI five year plan (2007–12), Indian Journal of Agricultural Marketing, Vol 21(2), May–August 2007.
7. Acharya, S.S. and Ramesh Chand, Report of the Sub-group on Agricultural Economics, Marketing and Agribusiness for XI Five Year Plan, Planning commission, Government of India, December 2006.
8. Acharya, S.S and A. Bhatnagar, Report of the working group on Agricultural marketing Infrastructure and Policy Required for Internal and External Trade for XI Five Year Plan, Planning Commission, Government of India, January 2007.
9. Mruthyunjaya, Status and Strategies for Strengthening of Agricultural Economics Research and Education in NARS of India, Agricultural Economics Research Review, Vol. 28(1), Jan–June, 2015.
10. Acharya, S.S (2017a), "Assurance of Minimum Support Prices to Farmers: Need for Crop Price Forecasts for Effective Pre-Emptive Market Interventions", NAAS Research, Education and Technology Forum News, Vol. 17, No. 3, July–September 2017, p. 9–14.
11. Acharya, S.S (2017b), "Effective Implementation of Agricultural Price and Marketing Policies for Doubling Farmers' Income: Doable Priority Actions", AERR, Vol. 30, Conference Number, p. 1–12.

11

External Trade in Agricultural Products

One other important aspect of agricultural marketing is the marketing of agricultural commodities across the nation's borders usually termed as external trade. As far as an individual country is concerned, external trade is done in the form of either exports or imports. Exports provide the market support for the country's surpluses and generate foreign exchange earnings which increases the country's capacity to import other goods, but at the same time in the short run, they reduce the domestic availability of the commodities exported and consequently raise the domestic price level. Imports, on the other hand, though reduce the foreign exchange reserves, augment the domestic availability of goods and if these pertain to the capital goods or inputs, expand or improve the production capacity. Imports of final goods in the short run, depress the domestic prices. The effect of exports and imports of final goods on the domestic price level is such that, in the absence of any public intervention in the domestic market, the producers gain by exports and lose by imports. One other point to be noted in this regard is that the levels of exports and imports of a country are inter-related as the capacity of imports depends on its ability to export.

International trade or marketing essentially involves buyers and sellers of two different countries. Usually the currencies are different and convertibility is quite often not automatic. This apart, depending on the development philosophy, domestic economic levels, natural resource endowments and national objectives like self-sufficiency or self-reliance, there are quite a few barriers-physical, tariff, subsidies etc. that are imposed by the national governments. Therefore, the buying and selling of commodities across national boundaries usually have not been taking place in the framework of free market environment.

TRADE POLICY FOR AGRICULTURE

TILL THE END OF EIGHTIES

Over the four decades of Indian planning, the perception about the importance of external sector in economic development had gone through a number of changes. These changes can be categorized as under:

1. During the first half of 1950s, i.e. the period of first five year plan, the foreign trade was considered to be almost irrelevant for economic development in India and hence export policies did not receive emphasis.

2. During mid 1950s to mid 1970s (1955 to 1975), the foreign trade was seen as a constraint on growth and India followed a moderately outward looking economic policy. During second five year plan (1956–61), agricultural items hardly received any export incentives. Rather several items were under export restrictions. The third five year plan (1961–66) marked a radical departure in the export policy. Various export promotion measures were introduced in the form of fiscal incentives, import entitlement scheme, direct financial incentives and marketing incentives from the government. Following the devaluation of Indian rupee in 1966, many of the export promotion measures were abolished. The promotion of India's exports continued during the fourth five year plan period (1969–74). To facilitate agricultural exports, the fourth plan extended the compulsory quality control and grading under Agmark. This period was also marked by the establishment of organizations aimed at providing services to the exporting community. These include Export Promotion Councils, Commodity Boards, and the Trade Development Authority, which were set up in early 1970's.

3. During late 1970's to early 1980's, the external trade got more prominent place. In 1977, a Task Force on Agricultural Exports, headed by G.V.K. Rao, in its report submitted to the Government of India, criticized the adhocism applied to agricultural trade. The report pointed out that India did not have an independent export policy for agricultural commodities and agricultural export policy during the 1970's remained 'adhoc, short term and mere reaction to the situation'. The report suggested that long term policies should be formulated and frequent changes in export policies should be avoided. Another committee was appointed by Government of India under the chairmanship of P.E. Alexander in 1977 to review the export-import policies and procedures. The Committee in its report submitted in 1978, recognised that the export control measures for agricultural commodities have resulted in supply uncertainties and loss of market share. The Committee recommended for the stability of export policies, advocated more transparency in making these policies, replacement of licensing system by the tariff system, rationalization of export incentives, elimination of multiplicity of incentives, and strengthening of institutional infrastructure for export promotion. Another Committee on trade policies under the chairmanship of Abid Hussain was appointed by the Government of India in 1984 to review the trade policies and suggest rationalization and improvements in these policies. The Committee realized that incentive element in these policies was small and was not sufficient to offset the negative export bias implicit in other domestic policies. The report proposed rationalization of duty drawback schemes, tax concessions, and increased fiscal concessions to increase the relative profitability of exports.

4. Mid 1980's to early 1990's marks the period when India started liberalization of the external sector. It began in mid 1980s but gathered pace only since 1991. However, up to mid 1990s, agriculture remained largely a protected sector in the Indian economy. During this period, the main policy objectives were to ensure stability of domestic prices of agricultural items. The government actively regulated agricultural exports through a variety of measures like export taxes, export ceilings, canalization and export prohibitions. Monitoring of agricultural exports was done on adhoc and short term basis to keep the domestic supply of agricultural goods stable.

5. The procedure of exports/imports was such that there always had been a gap between policy announcement and the final outcome. For example, first of all the Government announced its decision to allow exports up to some ceiling. The submission of applications by exporters, allotment of quota and actual shipment took time, as a result actual quantity shipped by the time of next harvest season, quite often fell short of the originally intended quantity. The same situation held true in the case of imports on account of the time that is required in contracting and negotiating with the buyers, arranging the shipment and off-loading at the Indian ports.

ECONOMIC LIBERALIZATION SINCE 1991

In July 1991, India introduced radical policy reforms in various economic sectors including trade. These include devaluation of rupee in 1991 and making rupee partially convertible. Trade restrictions on agricultural products were left mostly untouched in the 1991 reforms but during subsequent trade policy changes, restrictions on agricultural products were gradually lifted. India signed the Uruguay Round Agreement on 15th April 1994 at Marrakesh. The treaty introduced agricultural trade in the multilateral agreement for the first time. The aim of this treaty was to eliminate physical controls on agricultural trade by replacing them with bound tariff rates. The Agreement on Agriculture came into effect from January 1, 1995 which marked the beginning of a new era of agricultural trade policy in India.

The macro-economic reforms introduced in India aim at liberalization of trade and industry and at progressively moving towards linking the Indian economy with the world economy. The reform package is being extended to the farm sector also. The export of agricultural commodities as also agro-based processed products is being encouraged and controls/restrictions on trade in agricultural commodities have been increasingly relaxed. The physical controls on import/export of agricultural commodities have been replaced by tariffication, which are also being progressively reduced.

As a part of comprehensive programme of macro-economic stabilization and structural adjustment, there had been a move towards a more open trade regime. In this move, there were five key medium-term objectives:

1. Broadening and simplification of export incentive measures and the removal of restrictions on exports;
2. Elimination of quantitative restrictions on imports;
3. Substantial reduction in tariff rates;
4. Decanalization of exports and imports with the exception of a few items; and
5. Moving to a foreign exchange system which is free of allocative restrictions for trade.

A package of trade policy announced in July–August, 1991, incorporated quite a few changes which aimed at strengthening export incentives and eliminating a substantial proportion of import licensing requirements. Imports of raw materials and components were linked to exports and replenishment licensing (REP) system was replaced by a new instrument named Eximscrip. The Eximscrip was issued to exporters at higher rates and was made freely tradeable. With effect from April 1, 1992 the Eximscrip scheme was abolished. The export licensing and minimum export price (MEP) provisions were also liberalised. It was also intended to decanalise the imports/exports with a view to reducing the role of state monopoly agencies in the foreign trade. The Indian rupee was made partially convertible. The custom duties on certain pesticides were reduced and on oilseeds, pulses and seeds of fruits, flowers and of vegetables were abolished.

The meaning and consequence of liberal international trade in agricultural products should be understood and recognised. Even under a regime of liberal international trade, import tariffs/duties are likely to remain in force. The presence of tariffs and duties and the absence of physical restrictions imply that though the domestic prices will still remain different than those in the international market—the difference depending on the rates of tariffs and duties and the exchange rate of Indian rupee in relation to other currencies—but the relative price structure in the domestic market will tend towards that in the international market.

An examination of the domestic prices in India and the price in the international market shows that the prices of wheat, rice and raw cotton in the domestic market had been generally lower than those in the international market and those of sugar and edible oils/oilseeds were higher in the domestic market as compared to those in the international market. The implication of liberalization of trade in these commodities is that prices of wheat, rice and raw cotton would rise and those of sugar and edible oils would decline. But in view of the huge subsidies being given by major producing countries to their farmers, specially in the case of wheat and sugar (by paying higher prices to the producers and selling at lower prices in the international market, the difference of the two being met by government), the international prices are not market determined prices and their levels are prone to the decisions of the governments of such countries. Also, the international prices have been showing very high instability. Therefore, the decisions on imports and exports of agricultural products require careful assessment of the international markets for these products.

EXPORT-IMPORT POLICY, 1992–97

The Government of India announced a new five year export-import policy effective from April 1, 1992 which gave further push to liberalization of imports and intended to give significant boost to exports. Under this policy, the international trade was made free subject to a negative list of imports and exports. But as far as farm products and related goods are concerned, most of them remained a part of the negative list, as per the following details:

Negative List of Exports

1. Permitted, Subject to Licensing: Coconut, copra, seeds and planting materials, cotton seed, vegetable oils, groundnut cakes, rice bran, milk, cattle, camels, chemical fertilizers.
2. Permitted through Canalising Agency: Onion (NAFED), Niger Seed (NAFED/TRIFED), Powdered Milk (NDDB), Ghee (NDDB).
3. Permitted without a Licence but subject to terms and conditions: Basmati rice, non-basmati rice, wheat, barley, maize, bajra, jowar, ragi, HPS groundnut, raw cotton (Bengal desi, Assam comilla, staple cotton, yellow picking), sesame seed, sugar, gram and gram flour, wheat flour, deoiled groundnut cake, deoiled rice bran, VFC tobacco, soyabean extractions, cotton yarn, black pepper, etc.

Negative List of Imports

1. Canalised Items: All fertilizers (MMTC), edible oils (STC, HVOC), seeds of oilseed crops (STC, HVOC), Cereals (FCI).

2. Restricted Items: Livestock, plants, seeds and other materials (licence from the Department of Agriculture).

The import of pulses, raw cashewnut, seeds of vegetables and flowers, plants, tubers and bulbs, flowers etc. were placed in the negative list.

The philosophy underlying these massive trade policy reforms include the following:

1. Trade: Both exports and imports can flourish in a free regime.
2. Trade policy should go far beyond balancing of imports and exports and should lead to better technology, greater investment and more efficient production at home.
3. Liberalization and removal of licensing, quantitative restrictions and other discretionary controls on matters relating to exports and imports are essential to trade policy reforms. This meant fewer governmental restrictions, greater freedom to trade and lesser administrative controls.

The process of pruning the negative list and decanalization has continued in recent years. The process of liberalization of import-exports of farm commodities, which started in mid-nineties got momentum after 1997. Year after year, the negative list was pruned and quantitative restrictions were withdrawn. By 2001, the import policy became quite liberal (Table 11.1).

MAIN FEATURES OF EXPORT-IMPORT (EXIM) POLICY, 2002–07

- Removal of all quantitative restrictions and decanalization of exports (except a few sensitive items) of farm products
- Scheme of Special Economic Zones (SEZ) strengthened
- Major thrust to promote agricultural exports by setting up of Agri Export Zones [AEZs] and by removing export restrictions on designated items (agro and agro-based products)
- Transport subsidy provided for export of fruits, vegetables, floriculture, poultry and dairy products
- Simplification of procedures to further reduce transaction costs
- Widening of the scope of Market Access Initiative Scheme to include setting up of 'Business Centres in Indian Missions abroad for focused market promotion of exports.
- Dereservation from small scale industry provisions of over 50 items including agricultural implements.

The post-WTO trade scenario for Indian agriculture and main factors impacting the trade in the middle of 2000s are summarized in Table 11.2.

FOREIGN TRADE POLICY, 2015–20 AND EXPORT FACILITATION REFORMS

On April 1, 2015, Government of India notified Foreign Trade Policy for the period 2015 to 2020. This was done under Foreign Trade (Development and Regulation) Act, 1992. The policy, inter alia, encompasses (a) legal framework and trade facilitation measures aimed at ease of doing business; (b) Indian trade classification with harmonized system (ITC-HS) of exports and imports; (c) Indian schemes for exports; (d) duty exemption or remission schemes; (e) export promotion capital goods scheme; (f) export oriented units (EOUs), including biotechnology parks; (g) deemed exports; and (h) resolution of quality complaints and trade disputes. Some major trade facilitation reforms that are transforming trade scenario are as follows:

1. Round the clock customs clearance facilities for consignments are now available 24 × 7 at major ports.

Table 11.1: Import policy of important agricultural commodities

(as on 29 June, 2001)

Commodity	Duty on import	Import policy
1. Rice	Up to 80%	• Canalized through Food Corporation of India (FCI) • Rice with 50% or more broken is allowed freely
2. Wheat	50%	• Import is canalized through FCI • Import by roller flour mills was also allowed freely till recently • From December 1, 1999 a duty of 50% was imposed • STC/MMTC/PEC are permitted to import wheat on behalf of roller flour mills
3. Maize, jowar, bajra	50%	• Import is canalized through FCI and PEC • Import of maize for manufacture of poultry and animal feed is permitted freely on actual user condition subject to registration of import contract/letter of credit with NAFED • Import of maize for supply to poultry and animal feed manufacturers and for starch industry up to 50,000 MT each by NAFED has been permitted
4. Oilseeds (except copra)	35%	Import is allowed freely
5. Rapeseed oil, sunflower oil, soya oil, cotton seed oil	35%	Import is allowed freely
6. Crude oil for refining	75%	Import is allowed freely
7. Crude palm oil (edible grade)	75%	Import is allowed freely
8. Refined vegetable oils of edible grade	85%	Import is allowed freely
9. Soyabean oil	38.5%	Import is allowed freely
10. Pulses	5%	Import is allowed freely
11. Vegetables except onion	15%	Import is allowed freely
12. Onion	0%	Import is allowed freely
13. Fruits		
Dates	35%	–
Fresh grapes	25%	–
Apples	50%	–
14. Sugar	60%	Import is allowed freely
15. Cotton		
Cotton	5%	Import is allowed freely except from Pakistan
Cotton wastes including yarn waste	25%	All import contracts shall compulsory be registered with Textile Commissioner
Cotton carbed or combed	35%	
Cotton yarn	20%	

Source: Agricultural Statistics at a Glance, Ministry of Agriculture, Government of India, 2001, and 2002, pp. 155–156

2. All the documents are accepted in electronic format, thereby reducing the need for physical documents. It saves time and cost, and also eliminates physical interface between government agencies and the traders.

3. As a part of port communication system, a centralized web-based message exchange platform provides facilities for e-payments, e-invoices and e-delivery orders, which make maritime trade easy.

Table 11.2: Post-WTO trade scenario for Indian agriculture		

(as in 2006)

Products	Trade scene	Main factors
Rice	Exports adversely affected, import threat	Increased competition from developing countries like Vietnam and Thailand
Wheat	Exports adversely affected, import threat	Low prices and subsidies and support in EU and US
Sugar	Exports adversely affected	Subsidies in EU and USA
Cotton	Exports adversely affected, imports increased	Decline in domestic production and subsidies in USA
Tea, coffee and spices	Exports adversely affected	Competition from Vietnam, Indonesia, Sri Lanka and others
Vegetable oils	Serious import threat	Superior technology in other major producing countries
Palm oil	Sharp rise in imports	Cheap price and close substitution
Oil cakes	Exports adversely affected	East Asia crisis and GM varieties in USA, Argentina and Brazil, subsidies in USA
Horticultural products	Exports increased, more scope	Rising demand for high value and processed food
Dairy products	Imports possible, checked through tariffs	Subsidies in EU, USA and Canada
Meat and meat products	Exports increased, more scope	Preference for low cost and safe products

Source: National Academy of Agricultural Sciences, NAAS, Policy Paper 38, Dec. 2006.

4. The Risk Management System (RMS) at major ports and airports enables selective scrutiny of consignments. More than 80% of these get cleared without inspection.
5. Now Majority of shipping lines are issuing e-delivery orders.
6. RFID enabled e-seals have been introduced for sealing of export containers, which provides flexibility of sealing the cargo without supervision from regulatory authorities.
7. Phase 1 of fourth container terminal has commenced and a capacity of 24 lakh containers has been added at JNPT. Direct Port Entry (DPE) for exports is now availed by 80% of the exporters at JNPT, leading to time saving in exports. Further, adoption of Direct Port Delivery for imports (at JNPT) had increased from 27% to 40%. As a result, dwell time at the port has reduced because the containers can be picked up directly from the port.

AGRICULTURE EXPORT POLICY, 2018

The new agriculture export policy, which was being debated and discussed among the stakeholders, was finally approved by the Union Cabinet. The salient features of the policy are as follows:

1. To help in achieving the target of doubling farmers income by 2022, the export policy aims to increase agri exports to $ 60 billion by 2022 and further to a level of $ 100 billion in next few years.
2. The objective is to make India one of the 10 top agri export countries in the world and strive to double the India's share in world agri exports.

3. The policy recognizes the need for a stable and predictable trade policy regime, with limited state interference.
4. It focuses on agri export oriented production and better price realization for farmers (new APLM Act, new Contact Farming Act and Land leasing legislation will help).
5. The policy emphasizes the diversification of exports by products and destinations.
6. The efforts will be directed to promote perishables and high value commodities and value added products for exports. In addition, novel indigenous, organic, ethnic, and traditional agri products will also be focused items for exports.
7. Special funds have been assigned to set up specialized clusters in different states for specified agri products/groups of products for exports. There is a plan to involve FPOs and link these clusters/FPOs to global value chains.
8. The policy seeks to provide an institutional mechanism for tackling market access barriers and to deal with SPS issues.
9. The policy assures that processed agri products and all organic products will not be under the ambit of restricted category (MEP, export duty and export ban will not be applicable).
10. The policy has advocated for greater involvement of states, improvement in infra-structure and logistics and promotion of R&D for exportable product development.
11. The agri export policy has also suggested for empowered institutions in India on the lines of USFDA and European FSA (Food Safety Authority).

As a follow up, the government of India in September 2019 appointed Agriculture counsellors in 10 countries, which are among the top five export destinations in one or more agri products under 13 categories identified for export push. This is followed by placing experts in these countries to monitor market conditions in the destination countries, interact with traders and exporters in India and assist in resolving the market access issues. The ten countries are UAE, Vietnam, USA, China, Bangladesh, Nepal, Saudi Arabia, Iran, Malaysia and Japan. It may be noted that in 2018, UAE was among the top five buyers in 19 categories and Vietnam was top buffalo meat destination and one of the top buyers of maize, groundnut and poultry products. USA was among the top five countries for floriculture, fruits and vegetable seeds, guar gum and many others.

▌ SHARE OF AGRICULTURAL PRODUCTS IN TOTAL IMPORTS/EXPORTS OF INDIA

In 1960–61, exports of agricultural and allied products (AAP) were valued at ₹ 284 crore which accounted for 44.2% of total value of exports. Over the years, though the exports of AAP (in value term) increased but as the exports of other commodities increased at a rate faster than that of AAP, the share of AAP in the total exports of the country has gone down. In 1990–91, exports were valued at ₹ 6013 crore but they represented only 18.5% of total exports (Table 11.3). After 1990–91, the exports of AAP from India in value terms increased at a faster rate. In 2000–01, exports of AAP were valued at ₹ 28657 crore. During this period the exports of other commodities also increased at a rapid rate. Therefore, the share of AAP in total exports from India was about 14% during 2000–01.

However, during the next decade, the exports of AAP increased manifold to ₹ 113047 crore in 2010–11. The share of AAP exports in total exports was 9.9% during 2010–11. The exports of AAP further went up to ₹ 251564 crore in 2017–18, accounting for 12.9% of total Indian exports.

As regards the imports of agricultural and allied products, there has been an increasing trend in value of imports. The increase has been sharper after the

Table 11.3: Trend and shares of agricultural and allied products in total imports and exports of India

(₹ in crore)

Year	Imports			Exports		
	Total imports	Agricultural and allied products	Share of AAP in total imports (%)	Total exports	Agricultural and allied products	Share of AAP in total exports (%)
1960–61	1122	310	27.6	642	284	44.2
1970–71	1634	424	25.9	1535	487	31.7
1980–81	12549	1142	9.1	6711	2057	30.6
1990–91	43171	1206	2.8	32527	6013	18.5
2000–01	228307	12086	5.3	201356	28657	14.2
2010–11	1683466	51074	3.0	1136964	113047	9.9
2017–18	3001034	152095	5.1	1956515	251564	12.9

Source: Economic Survey 2009–10, 2014–15 and Ag. Statistics at a Glance, 2018, Government of India, New Delhi.

liberalization of imports of edible oils in the mid 1990s. The value of imports which had increased from ₹ 310 crore in 1960–61 to ₹ 1206 crore in 1990–91, jumped to ₹ 152094 crore in 2017–18. However, the share of agriculture imports to total imports has been consistently decreasing. Against a share of 26 to 28% in 1960–61 and 1970–71, the share of agricultural imports to total imports in 2017–18 was only 5.1%.

▌CHANGES IN INDIA'S EXPORT AND IMPORT BASKET

EXPORT BASKET

A number of agricultural commodities are exported from India. The commodities exported from India fall broadly in three categories:

1. Traditional export items: These products are cashew nuts/shelled; castor oil; coffee; raw cotton, cotton waste; fruits, spices, sugar and molasses; tea and tobacco-unmanufactured.
2. Non-Traditional items but uncertain: These items are raw jute; raw wool; gums, resins and lac, essential vegetable oils; and non-essential vegetable oils (excluding castor oil).
3. Non-Traditional items with good prospects: These items are floriculture products; HPS groundnut; oil meals; meat and meat preparations; fresh and processed fruits; fresh and processed vegetables; sesame and niger seeds; shellac; wheat and rice.

The growth in the export of non-traditional new items has been at a rate higher than that of traditional items (Table 11.4). The exports of traditional items (tea, coffee, cashew, cotton, sugar, and tobacco) increased many fold from ₹ 208 crore in 1960–61 to ₹ 2926 crore in 1990–91 and further to ₹ 41165 crore in 2017–18, but their share in total agri exports declined from 73% in 1960–61 to 46% in 1990–91 and further to only 16% in 2017–18. This happened owing to rapid increase in exports of non-traditional items (fish, meat, F&V, processed products, rice and spices) from mere ₹ 30 crore in 1960–61 to ₹ 2230 crore in 1990–91 and further to ₹ 1.76 lakh crore in 2017–18. The share of non-traditional items in total agri exports went up from mere 11% in 1960–61 to 35% in 1990–91 and further to 70% in 2017–18. The relative importance of various

Table 11.4: Exports of agricultural and allied products from India

(value in ₹ crore)

Commodity	1960–61	1970–71	1980–81	1990–91	2000–01	2010–11	2017–18
Tea and coffee	131	173	640	1332	3161	6364	11642
Oil cakes/meals	14	55	125	609	2045	11070	7043
Tobacco and products	16	33	141	263	871	3985	6022
Cashew	19	57	140	447	1883	2853	5978
Spices	17	39	11	239	1619	7887	20085
Sugar and molasses	30	29	40	38	511	5687	5323
Raw cotton	12	14	165	846	224	13162	12200
Rice	–	5	224	462	2943	11586	50308
Fish and fish products Preparations	5	31	217	960	6367	11917	47646
Meat and meat products	1	3	56	140	1470	8929	27233
Furits and vegetables	6	12	80	216	1609	6768	16874
Other processed foods	1	4	36	213	1094	13552	13745
Others	32	31	182	952	4785	9287	27465
Total	284	487	2057	6317	28582	113047	251564

Source: Economic Survey, Various Issues. Ministry of Finance, Government of India, New Delhi, Ag. Statistics at a Glance, 2018.

commodities in total agricultural and allied products exports has substantially changed during the last 25 years.

During the last two decades, agricultural exports have received special attention since it is in this area that there exists great potential for raising farm incomes, tackling unemployment problem and earning foreign exchange. The impetus for accelerated growth in agricultural exports is envisaged through increased infrastructure support and by building up conducive policy environment. Some of the measures undertaken in this connection include, market determined exchange rate policy, lowering import duties on capital goods, particularly machinery necessary for food processing, easier availability of credit for exports, removal of restrictions on export of agricultural products and several concessions to export-oriented units.

Some of India's major export markets for various agricultural commodities and major competitors are shown in Table 11.5.

IMPORT BASKET

India has been importing a large number of agricultural commodities, including cereals, pulses, edible oils and sugar. The major change in the import basket occurred since 1990's, when India became almost self-sufficient in cereals and imports of edible oils were liberalized. Since then, vegetable (edible) oils and pulse imports have dominated India's agri imports basket. As shown in Table 11.6, imports of edible oils which accounted for 26.7% and 15.5% of total agri imports during 1990–91 and 1995–96, jumped to almost 50% since 2000–01. Pulse imports were 39.2% of total agri imports during 1990–91 and ranged between 11 and 19% during the last 15 years. Edible oils and pulses together accounted for around two-thirds of total agri import values during the last 10 years.

Table 11.5: Major export markets and competitors for agricultural commodities from India

Products	Major export markets	Major competitors
1. Basmati rice	S. Arabia, Kuwait, UAE, UK, USA, B'desh, Indonesia, Iran, Philippines	Pakistan
2. Non-basmati rice	Countries in Sub-Saharan Africa	Thailand, USA and Vietnam
3. Wheat	B'desh, Yemen, Kenya, Turkey and Netherlands	Canada, USA, Australia and Argentina
4. Tobacco	Russia, Yemen, B'desh, Vietnam, Sri Lanka and Nepal	Brazil, USA, Zimbabwe, China and Argentina
5. Spices	East Asia, West Europe, West Asia and North Africa	Thailand, Indonesia, Brazil, China, Sri Lanka, Vietnam, Nepal, Bhutan, Spain, Mexico and Morocco
6. Cashew	Australia, Germany, Hong Kong, Netherlands, Singapore, UK, USA and UAE	Brazil, Vietnam, Mozambique, Ivory Coast and Guinea Bissau
7. Niger seeds	USA, EU	Myanmar
8. Oil meals	Republic of Korea, Singapore, Indonesia, Philippines and Japan	USA, Argentina and Brazil
9. Sugar	Indonesia, Pakistan, Sri Lanka, EU, USA and Russia	Cuba, Brazil, Thailand, Australia and France
10. Fruits	Middle East, UK, France, USA and Netherlands	Chile, Pakistan, Philippines, Columbia, S. Africa, Australia and Israel
11. Vegetables	Sri Lanka, Russia, USA, UK, UAE, Germany and S. Arabia	China, Turkey, Thailand, Philippines, Israel and S. Africa
12. Meat and meat products	Malaysia, UAE, Mauritius, Jordan, Turkey and S. Arabia	China, Thailand, Rep. of Korea and Israel
13. Floriculture	USA, Netherlands, UK, Germany, Japan and Italy	Kenya, Israel, S. Africa, Netherlands and Denmark
14. Tea	Europe, USA, Russia, Saudi Arabia, UK	Sri Lanka, Kenya
15. Coffee	Europe, USA, UK, Russia, UAE	

Source: Ministry of Commerce, Government of India (Annual Reports).

Table 11.6: Some aspects of India's agri imports basket

Year	Vegetable (edible) oils		Pulses		Total agri imports (crore)
	Values (₹ crore)	%	Value (₹ crore)	%	
1990–91	322	26.7	473	39.2	1206
1995–96	2262	15.5	686	4.7	14566
2000–01	5977	49.5	498	4.1	12086
2005–06	8961	41.7	2476	11.5	21499
2010–11	25919	50.7	7512	14.7	51074
2015–16	68677	48.9	25619	18.3	140289
2017–18	74996	49.3	18749	12.3	152095

Source: Agriculture Statistics at a Glance, 2018, Government of India.

▌ STATUS OF IMPORTS/EXPORTS OF SOME AGRICULTURAL COMMODITIES

In this section, we look at the status of exports from India and imports of some non-traditional items like cereals, pulses and raw cotton. Oilseeds and sugarcane are important cash crops in the country. There cannot be international trade in sugarcane but sugar is an important internationally traded commodity and the market for sugar affects the interests of cane growers and sugar factories as also of the consumers. Therefore, we also look at international trade in sugar. In the case of oilseeds also, the major trade is in their derivative-edible oils, and oil meals, hence we look at the trade in edible oils also.

CEREALS

Ever since independence, though self-sufficiency in cereals was on the top of agenda of the development planning, India has remained consistent net importer of rice up to 1972 and that of wheat till recently. The year 1966 marks a turning point in the trend of imports of cereals in India. During this year, the net import of cereals at 10.35 million tonnes was the highest since independence. As the quantities of imports and exports vary from one year to the other, it is appropriate to look at the aggregate of imports and exports for different five years periods. The aggregates of imports and exports of wheat, rice and other cereals for the period from 1955 to 1999 are given in Table 11.7 and for 2000–01 to 2017–18 in Table 11.8.

Table 11.7: India: Imports and exports of cereals during 1955 to 1999

('000 tonnes)

Particulars	1955 to 1959	1960 to 1964	1965 to 1969	1970 to 1974	1975 to 1979	1980 to 1984	1985 to 1989	1990 to 1994	1995 to 1999
Rice									
Import	2030	2601	2956	577	374	925	1158	187	42
Export	146	6	27	117	550	2478	1654	7658	16665
Net import	1884	2595	2929	460	(–)176	(–)1553	(–)496	(–)7471	(–)16623
Wheat									
Import	10697	20422	28671	12170	13390	6494	2011	3065	5277
Export	2	–	1	24	697	108	452	275	1782
Net import	10695	20422	28670	12146	12693	6386	1559	2790	3495
Other cereals									
Import	165	71	4431	1871	713	2000	–	–	211
Export	9	–	24	35	9	108	15	242	130
Net import	156	71	4407	1836	704	1892	(–)15	(–)242	81
Total cereals									
Import	12892	23094	36058	14618	14477	9419	3619	3252	5530
Export	157	6	52	176	1256	2694	2121	8175	18577
Net import	12735	23088	36006	14442	13221	6725	1048	(–)4923	(–)13047

Source: Compiled from "Bulletin on Food Statistics" 1987–89, Ministry of Agriculture, Government of India, New Delhi and updated from other sources.

Table 11.8: India: Imports and exports of cereals during 2000-01 to 2017-18			
			('000 tonnes)
Particulars	**2000–01 to 2004–05**	**2005–06 to 2009–10**	**2013–14 to 2017–18**
Rice			
Import	15	1	7
Export	16898	19949	57045
Net export	16883	19958	57038
Wheat			
Import	5	8037	7957
Export	13235	793	9752
Net export	13230	(–)7244	1795
Other cereals**			
Import	231	296	1154
Export	2078	11387	12363
Net export	1847	11091	11209
Total cereals			
Import	251	8234	9118
Export	32211	32129	79160
Net export	31960	23795	70042

* Negligible
** Includes cereal preparations
Source: Agriculture Statistics at a Glance, 2009, 2014 and 2018, Ministry of Agriculture, Government of India.

As regards rice, India became net exporter in the middle of 1970s. But the real spurt in rice exports occurred in early 1990s when liberal export policies were launched. Rice exports (five year aggregate) increased from around half a million (496 thousand) tonnes during 1985–89 to 7.5 million tonnes during 1990–94, 16.6 million tonnes during 1995–99, 16.9 million tonnes during 2000–05, 20 million tonnes during 2005–10 and 57 million tonnes during the recent five year period (2013–18).

The situation for wheat has not been very encouraging. Taking five-year block, India has been net importer of wheat except during the 2000–01 to 2004–05 and recent five year period (2013–18), when India was net exporter of wheat. However, a redeeming feature of India's cereal (staple food) economy has been that, taking all cereals together (including cereal preparations) India has emerged as net exporter of cereals since the first half of 1990s. The net export of cereals aggregated to 4.9 million tonnes during 1990–94, 13 million tonnes during 1995–99, 32 million tonnes during 2000–05, 24 million tonnes during 2005–10 and 70 million tonnes in the recent five year period of 2013–18.

PULSES

In order to augment the domestic availability, India has been importing pulses throughout the last four decades. The imports substantially increased from 1983–84 onwards. The country's imports stood at 1.73 lakh tonnes in 1980–81 but increased to 4.31 lakh tonnes in 1985–86. In, 1990–91, the country imported 6.63 lakh tonnes of pulses valued at around ₹ 473 crore (Table 11.9). During the 1990s, the country imported around five lakh tonnes of pulses annually. The imports spurted during the last decade, when India imported 2.7 million tonnes of pulses during 2010–11 and more than 5 million tonnes during 2015–16 and 2017–18. The import bill during 2015–16 was ₹ 25619 crore. The import basket of pulses, though varies from year to year but it generally includes moong, pigeon-pea (tur/arhar) and Kabuli gram. Moong is imported from China, Thailand and Burma. Pigeon-pea is imported from Kenya, Tanzania and Burma and Kabuli gram is imported from Australia, Turkey, Tanzania and Burma. Although the pulses have been under OGL (Open General Licence) but the contracts for import of pulses had to be registered with the NAFED. Measures such as reduction of import duty on pulses (for example, from 35% to 10% w.e.f. Nov. 1989) and removal of cash margin on opening letters of credit for imports of pulses have been used in the past with a view to encouraging imports and consequently increasing the availability of pulses in the domestic market. The duty on imports of pulses further brought down to 5% and later the imports were freely allowed without duty.

Table 11.9: Import of pulses in India		
Year	Quantity ('000 tonnes)	Value (₹ in crore)
1976–77	12	NA
1980–81	173	NA
1985–86	431	189
1990–91	663	473
1995–96	491	686
2000–01	350	498
2005–06	1696	2476
2010–11	2778	7512
2015–16	5798	25619
2017–18	5607	18749

Source: Agricultural Statistics at a Glance, 2009, 2018.

OILSEEDS/EDIBLE OILS

Due to rapid increase in the demand for edible oils in the last five decades and despite the increase in production of oilseeds, the rate of growth in supply of edible oils in the country did not keep pace with the growth in demand. It is in this scenario that the country had to augment the domestic supplies through the import of edible oils. Table 11.10 shows the imports of edible oils in the country in various years.

In 1980–81, the country imported 16.33 lakh tonnes of edible oils valued at ₹ 677 crore. The import of edible oils decreased in the later years. However, it was still at a considerably higher level of 10.36 lakh tonnes in 1985–86. Later, with a view to maintaining a favourable price environment for the oilseed growers for encouraging them to adopt new technology and increase the production, the imports were kept restricted. The imports came down to 5.26 lakh tonnes in 1990–91. However, during the last 30 years, the consideration for the growth of production of oilseeds and the objective of helping oilseed growers, which are mainly resource poor and dry land farmers, were relegated in the background. The imports of edible oils were placed under OGL and duties reduced to as low as 10 or 20%. The imports, therefore, went

Table 11.10: Import of edible oils in India		
Year	Quantity ('000 tonnes)	Value (₹ in crore)
1960–61	31.1	4
1970–71	84.7	23
1980–81	1633.3	677
1985–86	1036.4	735
1990–91	525.8	326
1995–96	1062.0	2260
2000–01	4267.9	6093
2005–06	4288.1	8961
2010–11	6905.4	29860
2015–16	15643.7	68677
2017–18	15361.0	74996

Source: Government of India, Indian Economic Survey, Various Issues, and Agricultural Statistics at a Glance 2014, and 2018.

up to 10.62 lakh tonnes valued at ₹ 2260 crore in 1995–96. In subsequent years, these increased manifold. During 2017–18, India imported more than 153 lakh tonnes of edible oils valued at ₹ 74996 crore.

While India imports edible oils, it exports considerable quantities of oil cakes/meals. The details of quantities and values of oil cakes exports are given in Table 11.11. With the increase in international prices of oil cakes/meals, the country has earned substantial part of the edible oil import bill from exports of oil cakes/meals. For example, during 2010–11, against the import value of edible oils at ₹ 29860 crore, the earning from export of oil cakes was ₹ 11070 crore. However, in later years, the import value of edible oils far outpaced earnings from exports of oilmeals.

Table 11.11: India: Exports of oil cakes/meals		
Year	Quantity ('000 tonnes)	Value (₹ in crore)
1960–61	433.8	14
1970–71	878.5	55
1980–81	886.0	125
1990–91	2447.8	609
2000–01	2417.8	2045
2005–06	5976.0	4875
2010–11	6936.9	11070
2015–16	2056.4	3600
2017–18	3570.8	7043

Source: Agricultural Statistics at a Glance, 2009, 2014, 2018; and Economic Survey, 2009–10, 2014–15.

Apart from oil meals, India also regularly exports sesamum seed, niger seed and groundnut (HPS). For example, during 2010–11, the export value of these oilseeds was ₹ 850 crore, which increased to ₹ 8176 crore during 2015–16. In 2017–18, India exported 11.5 lakh tonnes of these oilseeds valued at ₹ 7573 crore.

SUGAR

India has been exporting sugar in most of the years. Nevertheless, in some years massive imports had to be done to meet the demand in the domestic market. The domestic market for sugar has often remained under control though the extent of control has varied over time. In some years, though it was completely decontrolled but the decontrol was short lived. The data in Table 11.12 show that from 1974–75 to 1978–79, the country's export of sugar fluctuated between 2.02 lakh tonnes and 10.21 lakh tonnes.

Table 11.12: India: Import and export of sugar (1974 to 2009)			
			('000 tonnes)
Yeat (Oct–Sept)	Import	Export	Net export
1974–75	–	924	924
1975–76	–	1021	1021
1976–77	–	312	312
1977–78	–	202	202
1978–79	–	863	863
1979–80	180	290	110
1980–81	215	61	(–)154
1981–82	–	383	383
1982–83	–	422	422
1983–84	64	706	642
1984–85	1217	20	(–)1197
1985–86	1626	54	(–)1572
1986–87	951	26	(–)925
1987–88	71	28	(–)43
1988–89	–	28	28
1989–90	242	35	(–)207
1990–91	–	207	207
1991–92	–	583	583
1992–93	–	397	397
1993–94	460	55	(–)405
1994–95	674	41	(–)633
1995–96 (Apr–March)	150	864	714
1996–97	2	667	665
1997–98	347	173	(–)174
1998–99	900	13	(–)887
1999–00	1181	13	(–)1168
2000–01	30	339	309
2001–02	26	1456	1430
2002–03	41	1662	1421
2003–04	74	1201	1127
2004–05	933	109	(–)824
2005–06	559	321	(–)238
2006–07	1	1643	1642
2007–08	1	4685	4684
2008–09	386	3332	2946

In the following sugar season India imported about 1.80 lakh tonnes but exports of sugar were more than the imports. However, in 1980–81 sugar season, India imported 2.15 lakh tonnes of sugar which exceeded exports by 1.54 lakh tonnes. Between 1984–85 and 1986–87, the country had to import 37.94 lakh tonnes of sugar. In 1989–90, sugar season also, the imports of sugar were more than the exports. During 1990–91 to 1992–93, the availability of domestic production was comfortable and import was not needed. In fact, country could export sugar in these three years. However, India imported (net) 10.38 lakh tonnes of sugar during 1993–94 and 1994–95. India was net exporter in nine years and importer in five years between 1995–96 and 2008–09. Taking 14-years period as a whole, India has been net exporter of sugar, with exports exceeding the imports by 11.8 million tonnes during the entire period or 8.3 lakh tonnes per year.

The import-export position of sugar during the last five years is shown in Table 11.13. During 2013–14, India's net export of sugar was around 1.6 million tonnes. Though imports continued in the following three years, exports in each year far exceeded the imports. During 2017–18, imports were more than imports. However, considering all the five years together, India's net exports were 3.67 million tonnes with an average of 0.73 million tonnes per year. The average net export earnings were ₹ 2672 crore per year.

Table 11.13: India: Imports and exports of sugar (2013 to 2018)

Quantity in '000 tonnes
Value in ₹ crore

Year	Imports		Exports		Net exports	
	Q	V	Q	V	Q	V
2013–14	881	2287	2478	7178	1597	4891
2014–15	1539	3668	1955	5329	416	1611
2015–16	1943	4038	3844	9825	1901	5787
2016–17	2146	6869	2544	8660	398	1791
2017–18	2403	6036	1758	5266	(–)645	(–)770

Source: Agricultural Statistics at a Glance, 2014, 2018.

RAW COTTON

India has been traditionally an importer of cotton. But lately it has emerged as an exporter. The data in Tables 11.14 and 11.15 show that during the last 40 years, it has exported raw cotton in all the years. During the 10 year period up to 1995–96, net exports of raw cotton from India averaged at 6 lakh bales. However, since then, the situation has considerably changed. During 1999–00 to 2004–05, India imported substantial quantities of raw cotton. However, the situation in the later period reversed. As shown in Table 11.15, during the last 12 years (2005–06 to 2017–18), India exported substantial quantities of raw cotton (including cotton wastes) and remained net exporter.

FRUITS AND VEGETABLES

India exports seeds of fruits and vegetables; fresh fruits; fresh vegetables and processed fruits and vegetables. There has been considerable increase in the exports of all these

Table 11.14: India: Import and export of raw cotton

(1978–79 to 2002–03)
(lakh bales)

Year (Sept–Aug)	Import	Export	Net export
1978–79	0.3	2.7	2.4
1979–80	–	5.5	5.5
1980–81	–	7.0	7.0
1981–82	0.5	3.7	3.2
1982–83	–	7.0	7.0
1983–84	–	3.5	3.5
1984–85	0.5	1.8	1.3
1985–86	–	4.5	4.5
1986–87	–	14.3	14.3
1987–88	3.0	0.4	(–)2.6
1988–89	2.3	1.0	(–)1.3
1989–90	–	12.8	12.8
1990–91	0.1	22.0	21.4
1992–93	3.1	3.8	0.7
1993–94	0.2	17.5	17.3
1994–95	4.8	4.2	(–)0.6
1995–96	4.1	2.0	(–)2.1
1996–97 (Apr–Mar)	0.2	15.9	15.7
1997–98	0.6	9.3	8.7
1998–99	3.4	2.5	(–)0.9
1999–00	14.0	0.9	(–)13.1
2000–01	12.5	1.7	(–)10.8
2001–02	22.8	0.5	(–)22.2
2002–03	11.8	0.6	(–)11.2

Source: Office of the Textile Commissioner and Directorate of Cotton Development (Quoted in Reports of CACP) and Agricultural Statistics at a Glance, Various Issues.

during the last three decades. The value of exports went up from ₹ 915 crore during 1995–96 to 15051 crore during 2017–18 (Table 11.16). There is vast scope to further increase their exports.

OTHER AGRICULTURAL PRODUCTS

The trend in exports of broad groups of agricultural products has already been shown before. Further, exports/imports of cereals, pulses, edible oilseeds, sugar, raw cotton, fruits and vegetables have been seen earlier. These apart, some other agricultural products also need explicit understanding in terms of their importance in export basket of India. These are dairy products, poultry products, floriculture products and guargum. The exports of these products during the last 10 years are shown in Table 11.17. Though, there are inter-year fluctuations, by and large, there is an increasing trend in the exports (values) of all these products. Guargum meal has emerged as an important export item. Guar crop, a kharif crop of semi-arid or dry land areas, is the source of guargum. Once used only as green vegetables or cattle feed, guar is turning

Table 11.15: India: Imports and exports of raw cotton (2003–04 to 2017–18)

	Import		Export		Net export	
	Quantity ('000 tonnes)	Value (₹ crore)	Quantity ('000 tonnes)	Value (₹ crore)	Quantity ('000 tonnes)	Value (₹ crore)
2003–04	253	1570	180	942	(–)73	(–)628
2004–05	192	1136	87	423	(–)105	(–)713
2005–06	99	704	615	2904	516	2200
2006–07	81	663	1162	6108	1081	5445
2007–08	136	912	1558	8865	1422	7953
2008–09	212	1690	458	2866	246	1176
2009–10	171	1241	1358	9537	1187	8296
2010–11	58	624	1886	13160	1828	12536
2015–16	233	2566	1347	12821	1114	10255
2016–17	500	6339	996	10907	496	4568
2017–18	469	6307	1101	12200	632	5893

Source: Agricultural Statistics at a Glance, 2009 and 2014, 2018; and Economic Survey, 2009–10 and 2014–15.

Table 11.16: India: Exports of fruits and vegetables

(₹ in crore)

Year	F & V seeds	Fresh fruits	Fresh vegetables	Processed F & V	Total
1995–96	41	230	297	347	915
2000–01	63	386	457	785	1691
2005–06	93	1121	920	1415	3549
2006–07	122	1414	1547	1362	4645
2007–08	142	1447	1478	1376	4445
2008–09	120	1945	2454	1810	6329
2009–10	145	2269	2942	1912	7268
2010–11	185	2174	2546	1815	6720
2011–12	288	2558	2907	2714	8467
2012–13	348	3295	3288	3088	10019
2013–14	417	3646	5384	3332	12779
2014–15	427	3160	4666	3627	11880
2015–16	529	4191	5237	3767	13724
2016–17	523	4974	5791	3921	15209
2017–18	671	4913	5298	4169	15051

Source: Agricultural Statistics at a Glance, Various Issues, Ministry of Agriculture, Government of India, New Delhi.

out to be like gold. Guargum is white free-flowing powder, derived from guar splits. It is widely used by global oil and gas industry owing to its unique binding quality. In 2012–13, India exported guargum meal worth ₹ 21287 crore. In later years, however, it went down.

Table 11.17: India: Exports of some other agricultural products

(₹ in crore)

Year	Dairy products	Poultry products	Floriculture products	Guar gum
2004–05	459	282	223	689
2005–06	795	313	301	1094
2006–07	497	314	653	1126
2007–08	960	430	340	1126
2008–09	1130	414	369	1339
2009–10	797	373	294	1133
2010–11	1217	314	296	2939
2011–12	648	458	365	16524
2012–13	2325	495	423	21287
2013–14	4408	567	456	11735
2014–15	2169	651	461	9478
2015–16	1677	769	483	3234
2016–17	1701	530	547	3107
2017–18	1955	552	507	4170

Source: Agricultural Statistics at a Glance, 2014, 2018.

INDIA'S SHARE IN WORLD AGRICUTURAL TRADE

There are two ways of looking at India's share in the world trade for major agricultural commodities. One way is to see the share of India in the total value of exports in the world market at international prices and the other is look at India's share in the total physical quantity of exports/imports in the world. Let us look at in value terms.

As per the international trade statistics of UNO, the values of total world exports and India's share for 12 agro commodity groups are shown in Table 11.18. The world agri exports increased from $ 151 billion in 1980 to $ 866 billion in 2016. The share of India's agri exports, in the world total, went up from 1.2/1.3% in 1980/1990 to 1.7% in 2000, 2.3% in 2010 and further to 3.1% in 2016. Apart from this positive development, there are three other changes in India's agri export performance that need to be noted. One, India continues to maintain a leading share in exports of rice and spices. India's share in rice exports, which was only 3.7% in 1980, increased continuously and reached 24.5% in 2016. India's share in spices exports in 2016 was 17.1% of total world export of spices. Two, India has been able to continuously increase its share in the exports of meat and fishes groups of commodities. And three, India's traditional export item (tea), which accounted for 27.7% of world exports in 1980, has continuously lost its share in export basket. In 2016, India's tea exports accounted for only 8.4% of world tea exports.

BALANCE OF TRADE AND EXTENT OF GLOBALIZATION

The balance of trade analysis is usually done at the country level for the economy as a whole. But, it is also interesting to look at it at the sectoral level. The difference between the value of exports and value of imports related to a particular sector reveals the balance of trade. The difference in agricultural exports over agricultural imports, shown

Table 11.18: Trend in India's share in world trade (exports of agri. products)

(US $ million)

Code group	Commodity groups	1980 World	1980 India	1980 India share %	1990 World	1990 India	1990 India share %	2000 World	2000 India	2000 India share %	2010 World	2010 India	2010 India share %	2016 World	2016 India	2016 India share %
01	Meat	17832	67	0.4	34118	77	0.2	44690	324	0.7	112000	1821	1.6	131100	3975	3.0
03	Fish	12258	242	2.0	32847	521	1.6	50875	1391	2.7	101800	2403	2.4	134800	5499	4.1
042	Rice	4355	160	3.7	3995	254	6.4	6411	654	10.2	20300	2296	11.3	21700	5316	24.5
04	Other cereals*	37634	41	0.1	41319	31	0.1	47164	129	0.3	74000	840	1.1	125600	780	0.6
05	Veg. + fruits	24018	259	1.1	50225	400	0.8	68355	856	1.3	180700	2338	1.3	239700	3269	1.4
06	Sugar	16183	46	0.3	14236	21	0.1	13866	118	0.9	45500	1096	2.4	51200	1769	3.5
071	Coffee	12979	271	2.1	8659	148	1.7	11559	264	2.5	29600	558	1.9	36700	844	2.3
074	Tea	1631	452	27.7	2650	585	22.1	3087	431	14.0	7200	720	10.0	8400	705	8.4
075	Spices	1072	156	14.5	1415	109	7.7	2541	261	10.3	6000	927	15.4	10300	1766	17.1
08	Animal feed	10322	164	1.6	15603	336	2.2	20295	469	2.3	57600	2067	3.6	70100	775	1.1
12	Tobacco	3423	151	4.4	17860	145	0.8	21628	147	0.7	35100	879	2.5	44100	1012	2.3
22	Oilseeds	9487	30	0.3	10477	83	0.8	14388	244	1.7	56400	911	1.6	72600	1352	1.9
	Total of above	151194	2039	1.3	233404	2710	1.2	304859	5288	1.7	7260200	16856	2.3	866300	27062	3.1

Source: Economic Survey, Vol. 2, 2017–18, A-126-9, Government of India, * = other than rice

in Table 11.19, reveals that, except in 1960–61, the balance of trade of Indian agriculture has been positive and shown remarkable growth during the last five decades. The value of agricultural exports over imports which was only ₹ 63 crore in 1970–71, increased to ₹ 4807 crore in 1990–91, ₹ 16571 crore in 2000–01 and further to ₹ 61973 crore in 2010–11. It sharply increased further to ₹ 99469 crore in 2017–18.

Table 11.19: Balance of trade and extent of globalization of Indian agriculture

(₹ in crore)

Year	Imports	Exports	Balance of trade (E-I)	Total trade imports + exports	Trade as % of ag. GDP
1960–61	310	284	(–)26	594	8.2
1970–71	424	487	63	911	4.9
1980–81	1142	2057	915	3199	6.4
1990–91	1206	6013	4807	7219	4.3
2000–01	12086	28657	16571	40743	8.0
2010–11	51074	113047	61973	164121	10.8
2017–18	152095	251564	99469	403659	14.6

Source: Economic Survey 2014–15 and Agricultural Statistics at a Glance 2014, 2018.

The degree of globalization has been measured by the share of total trade (exports plus imports) in agricultural GDP. The degree or extent of globalization in the Indian context needs to be seen from 1991 onwards because this marks the period when economic liberalization began in the country. The analysis clearly reveals that the extent of globalization of Indian agriculture (total agriculture trade as % of agricultural GDP) increased from 4.3% in 1990–91 to 8% in 2000–01, 10.8% in 2010–11 and further to 14.6% in 2017–18.

PROSPECTS OF AGRICULTURAL TRADE

The prospects for India in the international marketing of agricultural commodities appear as follows:

FOODGRAINS

In the case of foodgrains, the cultivation of basmati rice is already expanding and its export is under OGL. But in the case of non-basmati rice and wheat, although the production level is keeping pace with the demand and a situation has also developed where the country has consistently remained surplus over say a five years period. However, it ought to be recognised that some countries like USA and EC are highly subsidising the trade in wheat and till such time that distortion in world market gets corrected, price signals arising in the international market cannot be treated as free market price signals. Therefore, the country should judiciously use import tariffs to safeguard the interests of cereal producing farmers. However, basmati rice should be exported as much as possible and even in the case of non-basmati rice, maize and wheat, India should enter the export market as a reliable seller.

The spurt in the export of non-basmati rice, after the lifting of the ban on rice exports in 2011, clearly shows that India can further push rice exports. However, the main issue with exports of non-basmati rice is the water use in rice production. Wheat and maize are other key cereals which provide huge potential for increasing exports. The main concern is the huge demand-supply gap in pulses and increasing dependence on pulse imports. The country has enough potential to reduce the import-dependence of pulses. Efforts are going on in this direction.

COTTON

Cotton is one among the agricultural products which has the largest possi–bilities for higher exports. India has a comparative advantage in the production of cotton vis-a-vis other countries. The unit cost of production of cotton in India is lower than that in many other countries. This apart, there are some other factors which provide an advantage to this country.

Mechanical harvesting of cotton results in picking of all the bolls-mature, half mature or immature—from a plant as mechanical pickers operate on the principle of sucking. Also trash content in mechanically harvested cotton is more than that in the manually harvested cotton. The countries where sophisticated ginning machines and high quality cleaners are not available, prefer manually picked cotton. In India, cotton picking is manually done. Therefore, Indian raw cotton has advantage over mechanically picked cotton of such other countries like USA, Russia and Australia in terms of maturity, purity, strength and spinning quality. This factor can be exploited if ginning factories are modernized. This apart, extra-long staple cotton remains under-spinned in most of the Indian textile units. Suvin is perhaps the best quality extralong staple cotton in the world. While most of the Indian mills can hardly spin this variety up to 120 counts, those in Japan and Australia can spin Suvin cotton up to 160 or 180 counts. Therefore, the country should expand the cultivation and increase export of extra-long staple cotton as much as possible even if it requires import of medium staple cotton to meet the demand for domestic industry. With Bt cotton revolution in the country, the production and exports are increasing, which need to be sustained.

SUGAR

The world sugar market is considered as one of the most distorted international commodity markets. Although international price of sugar is generally lower than that in the domestic market but it should be kept in mind that world sugar is highly subsidised by major exporting countries like USA, Cuba and Brazil. International price of sugar remains lower than that in India not because India's cost of production is higher but because of the element of subsidy. As far as real cost of production is concerned, India is a competitive producer of sugar. The major issue in sugar exports is the phenomenon of sugar cycles which has become a recurrent phenomenon. The recent decision of the Government to do away with levy on sugar mills and decontrol the sugar release mechanism is a step in the right direction. However, the system of State Advised Prices prevalent in some states comes in the way of maintaining a healthy balance between the interests of sugarcane growers and processors.

PROCESSED FOODS

Although exports of processed foods based on fruits, vegetables, fish and meat, have picked up in the last decade, but still there is ample scope for their expansion. What is needed is upgradation of processing and packaging technology. Foreign collaboration and equity support can reduce both technological and financial constraints in increasing the production and exports of processed foods. Setting up of Agri Export Zones (AEZs) and mega food parks are the steps in the right direction to promote exports of processed foods.

The processed food exports are overseen by two apex level agencies, namely (i) APEDA—Agricultural and Processed Products Export Development Authority; and (ii) MPEDA—Marine Products Export Development Authority. The Ministry of Food Processing Industries (MFPI) is the nodal government entity proactively involved with the food processing industries.

OTHER COMMODITIES

For other high value commodities like tea, cashew, coffee, spices, HPS groundnut, dairy products, poultry products, organics, ethnic spices and flowers, there is a need to improve the quality and to recapture the markets or explore, some new markets. Notwithstanding the expansion in domestic demand for tea, the country can produce higher quality tea for exports. In the case of cashew, though India has lost the US market, the policy of exporting cashew to the erstwhile USSR countries should be reconsidered. In the case of HPS groundnut, India should try to produce and export bold 40/50 counts grade which is demanded in the international market. In this regared, APEDA has taken several initiatives in recent years.

As mentioned earlier, spurt in imports of edible (vegetable) oils, since the mid-1990s is a major concern. There is a need to speed up the current efforts to reduce India's dependence on imported edible oils, which is currently around two-thirds of the domestic demand.

PROMOTING AGRICULTURAL EXPORTS FROM INDIA

We have seen in the preceding sections that (a) exports of agricultural and allied products (APP) have been increasing at a rapid rate during the last decade; (b) balance of trade for agricultural sector has been positive and agri-exports have exceeded agri-imports by considerable margins; (c) the share of India in world agri-exports has been increasing; and (d) the prospects of further expansion in agri-exports are quite positive.

We have also seen that Government of India approved and launched a new agri export policy in 2018 with a view to doubling agri exports to $ 60 billion by 2022 and further $ 100 billion in next few years. The Key features of the policy are (a) predictable and stable export environment with limited state interferences; (b) speeding up of reforms launched in the form of new APLM (promotion and facilitation) Act, 2017; Contract Farming (Facilitation) Act, 2018; Promotion of FPOs/FPCs; and liberalized land leasing norms; (c) focus on cluster-based agri export oriented production and linking these clusters to global value chains; (d) attention to high value and value added agri exports; (e) removal of processed agri products and organic products from the restricted category; (f) provision of institutional mechanism for tackling market access barriers and dealing with SPS issues; and (g) greater involvement of states and improvement of infra, logistics and R&D for exportable product development. During the last two years (2017–18 and 2018–19), India's agri exports have crossed $ 38 billion mark and several measures are on way to further boost this level.

Currently six commodity groups (marine products, rice, meat, spices, fresh fruits and vegetables and sugar) account for nearly 80% of India's agri exports. In most of these, India is globally competitive in most of the years. Another feature of India's agri export basket is that the processed products account for 22%, mini-processed products 62.1% and fresh products account for 15.9% of total exports.

The current promotional programmes and conducive policies provide prospects of further accelerating the growth of agricultural exports.

1. Agriculture and Processed Products Export Development Agency (APEDA) monitors and implements promotional programmes for most of the agricultural exports (except for commodities like tea, coffee, spices and tobacco, for which separate commodity Boards exist). APEDA monitors a large number of product categories and custom lines.

2. The thrust is and should be on eyeing two groups of markets viz:, neighbouring countries which provide advantages of transshipment costs and volumes; and developed country markets which ensure higher realization.

3. Currently, around 22% of total agri-exports is processed foods including frozen vegetables, ethnic snacks, chutney, pickles, and R-T-E foods like Rajma and chhola, which cater to Indian diaspora. There is huge potential to expand this component. Government's decision to dereserve some of these R-T-E foods from MSMEs will help in promoting their exports.

4. India is already known as low-cost producer. This advantage can be exploited if quality and residue level maintenance can be ensured.

5. Basmati and non-basmati rice, poultry products, chillies and turmeric powders are emerging opportunities for exports.

6. Maize, for food, feed and starch, is also an emerging potential item for export enhancement.

7. Fiscal incentives and investment in mega food parks with state-of-the-art infrastructure and processing facilities; encouragement and promotion of cold value chains (network of cold storages and reefer vans); and permission of 100% FDI in agri and food marketing will help in expanding exports of value-added agri products.

8. India is regular exporter of basmati rice. Even when a ban on export of non-basmati (NB) rice was imposed, basmati exports were allowed with minimum export price (MEP). Sometimes, the level of prescribed MEP becomes an irritant. This apart, pesticide residues also become an issue. For example, in September 2012, US bound basmati rice consignments were rejected due to the presence of banned pesticide residues. All the stakeholders need to remain cautious to avoid such situations as these harm our market potential. A new basmati variety (Pusa 1718) is set to sustain India's aromatic long-grained rice exports through higher yield and ability to fight bacterial blight disease.

9. The state governments' policies are also quite important in promoting or thwarting the exports. For example, the government of Punjab in 2013–14, waived the taxes and levies on basmati paddy, with the condition that it will have to be processed within the state. This policy helped exports of basmati rice by (a) shifting area from common rice to basmati rice (which uses less water); (b) promoting local processing industry; and (c) reducing the domestic cost of rice production.

10. A related policy decision of the Government of India to ask all the rice producing states to do away with levy from October 2015 is helping rice exports.

11. In recent years, buffalo meat has also emerged as one of the key drivers of export growth. The tempo has to be maintained.

12. The efforts of APEDA to promote exports of some key products by following a cluster approach is a step in the right direction. Some of the commodities under the cluster approach of APEDA are basmati rice (Punjab and Haryana), buffalo meat (Western UP), grape and grape wine (Nasik region of Maharashtra), pomegranate (Satara and Pune region), dehydrated onions and garlic (Gujarat), poultry meat or eggs (Namakkal) and mango pulp (U.P. and Maharashtra).

13. As a part of export monitoring efforts of APEDA, another noteworthy step is the setting up of internet-based residue traceability software systems that are used for monitoring the fresh produce exported from India to EU and other countries. The systems help in achieving product standardization and facilitating farmers through various stages of sampling, testing, certification and packing. The systems include GrapeNet, AnarNet, TraceNet, MeatNet, BasmatiNet, and PeanutNet. In addition, the APEDA has put in place an end-to-end system called HortiNet for monitoring the exports by barcoding them. It is a web-based architecture that integrates all the stakeholder viz:, farmers, state governments, horticulture departments, testing labs, and exporters. The exported cartons will have barcodes and numbers that will allow traceability back to the farmers.

14. APEDA has also launched a new initiative MangoNet to promote export of mangoes. The mango growers can register on MangoNet for export. The services of BARC and Board of Radiation and Isotope Technology (BRIT) are being used to irradiate mangoes for export. Indian exporters have to give 'hot water treatment' before exporting. The National Plant Protection Organization (NPPO) has made it mandatory to give hot water treatment before giving phyto-sanitary certificates. Mangoes are currently exported to 72 countries and mango pulp to 141 countries.

15. An important aspect of boosting agri-exports is registration of geographical indications (GI). Under the Geographical Indications of Goods (Registration and Protection) Act, 1999, the APEDA has been designated as the custodian of GI rights for farm produce. GI is used to identify agricultural, natural, and manufactured goods that originate from a definite geographical territory. It is granted based on a special quality, reputation and other characteristics. GI registration confers legal protection in India and prevents unauthorized use of registered GI by others. The authorized user has exclusive right to use GI. The registration is valid initially for 10 years and can be renewed. Legal protection helps to boost exports of such products.

16. The ongoing efforts show that the possibilities of increasing agricultural exports are quite encouraging. A report, by Confederation of Indian Industry (CII) and Mckinsey, titled "India as an Agriculture and High Value Food Power House: A New Vision for 2050", brought out in 2013, has shown that agricultural exports are likely to rise many fold by 2030 and reach ₹ 7.72 lakh crore.

The finance commission is currently planning to incentivise states to undertake reforms in the agriculture marketing system so as to boost agri exports and thereby reduce import-dependence. A committee has been tasked to suggest the modalities.

INDIA'S AGRI-EXPORT DESTINATIONS

For fully understanding the prospects and promotion of agricultural exports from India, one has to keep in view the present and possibly new export destinations. India's current agri-export destinations for some commodities are shown in Table 11.20. It

Table 11.20: India's agricultural export destinations

Commodity	Top destinations
1. Non-basmati rice	Nigeria, Senegal, Benin, Cote D'Voire, South Africa, UAE, Guinea, Nepal, Saudi Arabia, Cameroon
2. Basmati rice	Iran, Saudi Arabia, Kuwait, UAE, Yemen
3. Wheat	Bangladesh, Korea, Yemen, UAE, Djibouti, Ethiopia, Indonesia, Thailand, Oman, Tanzania, Somalia, Sri Lanka
4. Coarse cereals	Indonesia, Vietnam, Malaysia, Taiwan, UAE, Bangladesh, Iran, Nepal, Yemen, Oman
5. Buffalo meat	Vietnam, Malaysia, Egypt, Thailand, Saudi Arabia, Indonesia, Iraq, Myanmar
6. Guar gum	USA, China, Russia, Norway, Germany
7. Poultry products	Afghanistan, Saudi Arabia, Indonesia, Germany, Netherlands, Denmark, Oman, Maldives, Liberia, Japan, Vietnam
8. Dairy products	Bangladesh, Egypt, UAE, Saudi Arabia, Algeria, Yemen, Singapore, Iran, Pakistan, Philippines, USA
9. Flowers	US, Netherlands, Germany, UK, Canada, Japan, UAE, Italy, Ethiopia, Australia
10. Groundnut	Indonesia, Vietnam, Malaysia, Philippines, Thailand, Ukraine, Pakistan, UAE, Singapore, Yemen
11. Vegetables	Pakistan, UAE, UK, Nepal, Saudi Arabia, Qatar, Kuwait, US, Sri Lanka, Maldives
12. Fresh mango	UAE, Bangladesh, UK, Nepal, Saudi Arabia, Qatar, Kuwait, Singapore, Bahrain, Canada, Oman, USA.
13. Processed meat	UAE, Myanmar, Qatar, Maldives, Bhutan
14. Sheep and goat meat	UAE, SA, Qatar, Kuwait, Oman
15. Processed fruit and juices	USA, Netherlands, Saudi Arabia, UAE, UK
16. Processed vegetables	USA, UK, Germany, Russia, Australia
17. Fish etc.	USA, SE Asia, EU, Japan, China, Middle East

Source: APEDA and MPEDA.

may be noted that top five or top ten destinations for each export item do not remain the same every year. The information given in the table is general list of destinations during the last 5 to 10 years.

AGRI-EXPORT ZONES (AEZs)

In a fast changing international trade environment and with a view to providing remunerative returns to the farming community in a sustained manner, efforts have been made to provide improved access to the overseas markets of agriculture and allied sectors. In this direction, the Government announced the proposal to set-up Agri-Export Zones in the EXIM Policy 2001–02 for the purpose of developing and sourcing raw materials and their processing/packing leading to final exports. The concept essentially embodies a cluster approach of identifying the potential products and the geographical regions in which such products are grown and adoption of end-to-end approach of integration of the entire process. Under the scheme, the state governments identify the products with export potential which have comparative advantage in the local area. APEDA is the nodal agency of the Central Government to promote setting up of Agri-Export Zones.

The Central Government has sanctioned and notified 60 Agri-Export Zones (AEZs) in 20 states—West Bengal, Uttaranchal, Karnataka, Punjab, Uttar Pradesh, Tamil Nadu, Maharashtra, Andhra Pradesh, Tripura, Jammu & Kashmir, Madhya Pradesh, Bihar, Gujarat, Sikkim, Himachal Pradesh, Orissa, Kerala, Assam, Rajasthan and Jharkhand. Agricultural products covered under these AEZs are Litchi, pineapple, potatoes, onion, garlic, mangoes (Kesar, Chausa, Dusshari, Alphonso), grapes, flowers, apples, vegetables, walnuts, gherkins, wheat, ginger, turmeric, basmati rice and seed spices. Apart from increasing the exports, a substantial amount of direct and indirect employment was generated as a consequence of setting up of these zones. The initial 41 AEZs entailed an estimated investment of ₹ 1142.5 crore, which include ₹ 333.68 crore from various Central Government agencies like APEDA, NHB, Ministry of Food Processing Industry and Ministry of Agriculture; ₹ 168.61 crore from state governments and ₹ 640.24 crore from the private sector.

The progress of all the established AEZs has not been uniform and sluggish in some ones. In view of the sluggish progress of some of the AEZs, the Ministry of Commerce, Government of India constituted a Peer Review Group for the evaluation of the performance of AEZs. The main findings of the Peer Review Group are as follows:

1. There has been low public investment from the central/state government in the AEZs; and
2. There has been an indiscrete proliferation of AEZs without any mid-term review.

Later, YES Banks initiated an exercise of short listing of AEZs. The APEDA is pursuing cluster approach for promoting exports.

CURRENT CUSTOM DUTIES AND BOUND RATES

The structure of custom duties and bound rates for key agricultural commodities are shown in Table 11.21. India's effective custom duties had always remained considerably lower than the bound rates.

EXPORT-IMPORT (EXIM) BANK OF INDIA

Availability of adequate export credit at competitive rates is an important determinant of export performance. To achieve this objective, EXIM bank has been set up in the country. The EXIM bank provides finance at various stages of the export cycle—import finance, product development, production, marketing, pre- and post-shipment as well as for overseas investment. It has a range of export credit programmes such as supplier's and buyer's credit, for Indian companies executing contracts overseas for projects, products as well as services. The EXIM bank also extends guarantee facilities to facilitate Indian companies in executing export contracts and extends lines of credit on its own or on behalf of Government of India to overseas governments or to agencies nominated by them or to financial institutions overseas for encouraging imports from India on deferred payment terms. EXIM bank's import lines of credit provide financing for import of capital goods and related services, particularly for enhancing export production capabilities of small and medium sized export oriented units. EXIM bank, with its medium/long term liability structure, concentrates primarily on medium to long term export finance besides working capital. EXIM bank also provides refinance and rediscounting facilities to commercial banks so that short term exports are not

Table 11.21: Tariff and bound rates on major agricultural commodities

(as on January 25, 2019)

Commodities	Basic custom duty %	Bound duty %
Cereals and pulses		
• Wheat	30	100
• Maize seed	50	70
• Jowar	50	80
• Rice (milled)	70	70
• Pulses	0–60	50/100
• Cereal product	30	150
Livestock products		
• Milk powder	60	60
• Butter and ghee	30	40
• Cheese	30	40
• Fresh milk and cream	30	100
Other products		
• Soyabean oil (crude)	35	45
• Rapeseed mustard oil (crude)	35	70
• Groundnut, sunflower, and safflower oil (crude)	35	300
• Coconut oil (crude)	35	300
• Palm oil (crude)	44	300
• Palm oil (refined), RBD palmolein	54	300
• Sugar	100	150
• Apples	50	50
• Onion	Nil	100

Source: Agricultural Statistics at a Glance, Ministry of Agriculture, Government of India, 2018.

affected by lack of finance. EXIM bank lending has steadily increased over the years. During 2018–19, the bank sanctioned ₹ 38000 crore and at the end of March 2019, its net loan assets were ₹ 93617 crore.

INTERNATIONAL TRADE AGREEMENTS

GATT (THE GENERAL AGREEMENT ON TRADE AND TARIFFS)

The General Agreement on Trade and Tariff (GATT) was a multinational treaty to liberalize world trade. It took effect on 1st January, 1948 and ended when 117 member states signed the Uruguay Round of negotiations in Marrakesh, Morocco on 15th April, 1994. GATT's administrative structure in Geneva was succeeded by the World Trade Organization (WTO) under the Uruguay Round agreement.

GATT established a code of conduct for international trade, based on the principle that the trade should be conducted without discrimination, tariffs should be reduced through multilateral negotiations, and member countries should consult each other to overcome trade problems. GATT centre operated jointly with the United Nations Conference on Trade and Development to assist developing nations in promoting their exports.

Under GATT, a total of eight round of talks of trade negotiations brought about phased reductions in tariffs and other trade barriers. The prolonged eighth round of talks began in September, 1986 at part del Este, Uruguay. In this round the participants agreed to expand the negotiations to include banking, investment, intellectual properties and telecommunications. The talks were concluded in December, 1993 and resulted in the far reaching trade liberalization in the history. Trade in the agricultural commodities was included in the agreement for the first time.

The Uruguay Round launched over 1986–94 was the most ambitious so far. This round also established the World Trade Organisation (WTO), the successor to the General Agreement on Trade and Tariff (GATT). It brought international trade rules to areas previously excluded or subjected to weak rules (agriculture, textiles and clothing), services, Trade Related Investment Measures, (TRIMS), Trade Related Intellectual Property Rights (TRIPS) and strengthened the dispute settlement mechanism. Despite these achievements, the global trading system faces major challenges. Against these challenges, the ministerial conference in Doha in November, 2001 adopted the Development Agenda, which calls for a more coherent approach to trade and development and puts the needs and interests of the developing countries at the heart of the WTO's work program.

WORLD TRADE ORGANISATION (WTO)

WTO is an international body to supervise and encourage international trade. The Uruguay Round of trade talks concluded in 1994 resulted in setting up of the World Trade Organisation (WTO) to take over the functioning of GATT for encouraging multilateral trade in goods and services. The WTO began functioning on 1st January, 1995. The Agreement on Agriculture (AoA) under WTO requires clear understanding for visualizing its implications.

AGREEMENT ON AGRICULTURE (AOA) UNDER WTO

The provisions under AoA can be understood to consist of five broad groups:

1. Market Access Commitment
2. Reduction Commitment for Aggregate Measure of Support (AMS)
3. Reduction Commitment for Export Subsidy
4. Sanitary and Phyto-Sanitary Measures (SPS)
5. Trade Related Intellectual Property Rights (TRIPS)

Market Access

The provisions under market access commitment include the following:

(a) Tariffication of all non-tariff barriers (like converting quantitative restrictions to import duty)
(b) Reduction of all tariffs in a time bound framework as follows:

Countries	Period	Reduction %
Developed	6 years	36
Developing	10 years	24
Less developed	–	0
Those with BoP problem	–	0

(c) If imports of foreign goods to the domestic market is less than 3% in the base period (1986–88), it must be brought to 3% and to further raise it to 5% in the implementation period.

(d) If dumping is proved, the countries will have the freedom to increase the import duty.

Aggregate Measure of Support (AMS)

The aggregate measure of support for a country's agriculture is the sum of product specific and non-product specific subsidies. If AMS in the base period (1986–88) is more than the permissible limit, it should be reduced by the following amount during the implementation period:

Country	Permissible AMS (% of GDP)	Reduction commitment if exceeds the limit (%)
Developed	5%	20%
Developing	10%	13%
Less developed	NA	–
Those with BOP problem	–	0

Under the agreement, the supports (AMS) are classified into three groups for the purpose of reduction commitment, depending on their effect on distortion of trade.

1. Green Box Support/Subsidies: These are not subjected to reduction commitment as these are considered as non-trade distorting supports. The examples are research, training and extension; pests and disease control; government stock holding for food security; and support in times of natural calamities or disasters.
2. Blue Box Support/Subsidies: These groups include direct payment to farmers for income support, income insurance or other payments that are not trade distorting. These are also exempt from reduction commitment.
3. Amber Box Support/Subsidies: All other forms (not included in green and blue box) of supports are considered as trade distorting and are subjected to reduction commitment. Subsidies on agri exports and government purchases at minimum supports prices are counted as amber box subsidies.

Export Subsidies

The reduction commitment for export subsidies requires that (a) the developed countries would reduce it by 36% in six year; and (b) the developing countries would reduce it by 24% in 10 years.

Sanitary and Phyto-Sanitary Measures (SPS)

The SPS provisions of AoA require all exporters to employ international standards relating to sanitary and phyto-sanitary conditions. In the case of default, the importing countries are allowed to prohibit imports from defaulting countries.

TRIPS

Trade related intellectual property rights include copyrights, trade marks, geographic indications, industrial designs, and patents. According to AoA, all the countries are required to provide for arrangements for protection of plant varieties. The developing countries were given a period of five years to evolve such arrangements.

In addition to these, there are quite a few other inter-governmental agreements under WTO, which include the following:

(a) Technical Barriers to Trade (TBT)
(b) Anti Dumping Measures.
(c) Subsidies and Countervailing Measures
(d) Safeguard Mechanisms.

IMPLICATIONS FOR INDIA

India's concern and response to AoA (under WTO) have been as follows:

1. India has been maintaining quantitative restrictions (QRs) on import of 825 agricultural products as on 1st April, 1997. Under the provisions of the Agreement, such QRs were to be eliminated. India had sought to remove them in three phases within an overall time frame of six years, i.e. up to 31st March, 2001. These QRs have since been replaced with appropriate tariffs.

2. The Agreement also imposed constraints on the level of domestic support provided to the agricultural sector. In India's case, it may have, in future, some implications on minimum support prices given to farmers and on the subsidies given on agricultural inputs. However, the Agreement allows us to provide domestic support to the extent of 10% of the total value of agricultural produce. Our support to the Indian farmers continued to be less than permissible limit for about two decades. Further, about 94% farmers are small and marginal farmers, which are exempted from reduction commitment.

3. Disciplines on export subsidy did not affect us as India was not providing any export subsidy on agricultural products.

4. When UR agreement was signed in 1994, developed countries had very high support levels for agriculture. Despite restrictions imposed under the agreement, most developed counties have retained high levels of support for agriculture by shifting most of their subsidies to the 'Green Box' forms of support, not restricted by the agreement. In contrast, owing to fiscal constraints, developing countries had much lower levels of support for agriculture when the agreement was signed. Due to the constraints imposed by the Agreement, the level of support for agriculture in developing countries has barely been allowed to increase.

   Green box subsidies in developed countries, not directly linked to levels of production, have helped their farmers to innovate, invest and increase productivity. The WTO does not restrict green box types of support. However, in small holder dominated agriculture in developing countries, administrative constraints limit the possibility and feasibility of using green box subsidies. Green box measures such as decoupled payments—lump cash transfer to farmers—are not feasible in most developing countries.

   For developing countries, price support to the farmers and consumer subsidies, combined with public procurement and stockholding are the main instruments. Under UR agreement, price support was considered trade distorting and was classified under 'Amber Box'. Amber box subsidies (for countries without negative AMS), were limited at the minimum level, defined as 5% of current value of output for developed countries and 10% for developing countries.

   Under the current rules, the difference between prices offered by a public procurement program and price prevailing in 1986–88 is multiplied by the total output (potentially eligible for support, and not the actual procurement) is used to calculate the level of support. Since current price levels are much higher than the 1986–88 reference price, any price support results in the violation of the minimum restriction.

5. At the Bali Ministerial conference (of WTO) held in December 2013, the issue was raised by India and other developing countries with a view to addressing the food security concerns by suitably changing the relevant rules. Specifically, following suggestions were made:

   (a) Reduce disparities in permissible agricultural support levels between developed and developing countries.

   (b) Update reference prices (presently pegged at 1986–88 nominal prices) to account for global inflation in food prices.

   (c) Compute levels of price support using actual quantity procured rather than total quantity eligible for procurement.

   (d) Provide exemption from restrictions for procurement on grounds of food security or support for poor producers (small and marginal farmers).

   As India successfully lobbied in favour of state-funded welfare schemes, the WTO inked a landmark pact in the form of a historical deal, Peace Clause, designed to ease movement of goods across countries and allow developing nations more options to feed their poor.

6. The Peace Clause is a temporary measure. The final agreement has not been reached yet. Some countries continue to raise questions on India's current price support arguing that these are exceeding the minimum level. India has to counter these in WTO panels by invoking the Peace Clause. For example, for rice and wheat (which are our main staple grains), the AMS in recent years has been claimed by some other countries as +26%, whereas if the base period is updated/adjusted for inflation since 1986–88, the AMS is only around 3% (below the minimum threshold). In this connection, it needs to be noted that the Article 18.4 of AoA allows some consideration to inflation in calculation of AMS, but it is not a unilateral right of a country. It depends on the discretion of other members of WTO during the review process. Further, the definition of excessive inflation is not clearly defined under AoA.

FREE/REGIONAL TRADE AGREEMENTS (FTAs/RTAs)

FTAs are arrangements between two or more countries or trading blocks that primarily agree to reduce or eliminate customs tariffs and non tariff barriers (NTBs) on their imports and exports. Since, 2000s, India's FTAs have expanded. According to WTO-RTA information system, globally 303 RTAs are in force, and those active for India are 16. Some of the India's active FTAs are as follows:

- 1972 (renewed from time to time) India–Bhutan
- 2004 India–Sri Lanka
- 2005 India–Singapore (CECA)
- 2006 SAARC Countries: SAFTA Pact
- 2009 India–Nepal (Treaty of Trade)
- 2010 India–South Korea (CEPA)
- 2010–11 India–ASEAN (for goods)
- 2011 India–Japan (CEPA)
- 2011 India–Malaysia (CECA)
- 2015 India–ASEAN (for services)

After FTA's, India's imports increased more than the exports. India's trade deficit widened with ASEAN, South Korea and Japan.

Till recently (up to August, 2019), India was negotiating a multilateral trade agreement known as RCEP (Regional Comprehensive Economic Partnership) which has 15 other members (10 member of ASEAN, Australia, China, Japan, New Zealand, and South Korea). Such agreements are referred as new age mega RTAs because of their size as well as ambitious coverage of issues, which are often termed as WTO-plus issues. The RCEP reflects the emerging trade and economic architecture globally. However, India withdrew from the negotiations because India wanted (a) a safeguard mechanism to allow it to raise tariffs in case of surge in imports of dairy products, specially from Australia and New Zealand; and (b) increased market access for India's services. As these conditions were not agreed, India had no option but to withdraw. India's dairy sector is the livelihood for 73 million small milk producers whereas in other countries only a few thousand large farmers are engaged in commercial dairy farming. If New Zealand exports 5% of its value added dairy products (VADPs) to India, it may capture 33% of India's domestic market for VADPs.

TARIFF RATE QUOTA (TRQ)

TRQ is established as a part of trade agreement between countries. Around 1200 TRQs are operated each year by WTO members, including US, Canada, Japan, and EU. A TRQ allows a set quantity of specific products to be imported at a low or zero rate of duty. TRQ does not limit the quantity of import of a product. It only specifies that once imports up to the TRQ commitment has been reached, a higher rate of duty will be applicable on further imports. In that sense, TRQ is a two-tiered tariff instrument. The use of this instrument is quite prevalent. The TRQs dovetail and protect the interests of both domestic consumers and producers. TRQs are being used on a wide range of products, mostly in the agricultural groups like cereals, meats, fruits, vegetables, dairy products and sugar.

IMPORTANT TERMS RELATED TO EXTERNAL TRADE

1. Quantitative Restrictions (QRs)

Quantitative restrictions are specific limits on the quantity or value of goods that can be imported or exported during a specific time period. Quantitative Restrictions are prohibited under GATT discipline. India has been maintaining QRs on imports of 825 agricultural products as on 1st April, 1997. These QRs have since been gradually removed from almost all products in phases. The Quantitative restrictions can easily be replaced with high import tariffs in case there is need to restrict import of these commodities for ensuring welfare of the farmers. Therefore, ability to restrict import of any commodity is not constrained in any manner by the provisions of GATT.

2. Tariff (Duty)

A tariff/duty is a tax levied on imports and less often on exports as they cross the borders into other countries. A specific tariff is imposed on each unit of an imported good. Ad-valorem tariff is a duty levied as a percentage on the price of the good to the importer. The imposition of tariff results in higher prices to domestic purchasers as the tariff is generally passed forward on resale.

3. Quota

A Quota in international trade is a type of trade barrier that nations place on the physical amount of imports or exports of specific kinds of goods. A quota differs from a tariff which is a schedule of taxes or duties placed on imports that does not categorically place limitations on the amount of goods that may be imported. Both tariff and quotas are detrimental to the concept of free trade and the GATT/WTO works to reduce such barriers.

4. Tariff Quota

A quota that allows for import of a commodity at less than the general applied rate is tariff quota. For example, if a country applies a general tariff of 100% on a particular commodity and then allows a limited quantity, say 20,000 tonnes, at a lower rate of say 20% is called tariff quota.

5. Dumping

Dumping is selling goods in a foreign country at a price which local producers regard as unfairly low. It means selling of goods at less than the long-run average cost of production plus transport cost, i.e. charging a lower price in export market than is charged for comparable goods in home markets; or simply selling at a price with which producers in the importing country cannot compete. Dumping is considered as unfair trade practice which can have a distortive effect on international trade.

6. Anti-dumping

Anti-dumping is a measure to rectify the situation arising out of the dumping of goods and its trade distortive effect. Anti-dumping duties are tariffs imposed in response to alleged dumping. The purpose of anti-dumping duty is thus to rectify the trade distortive effect of dumping. Anti-dumping duty as an instrument of fair competition is permitted by WTO. Anti-dumping is an instrument for ensuring fair trade and is not a measure of protection per se for the domestic industry. It provides relief to the domestic industry against the injury caused by dumping.

7. Green Box Policies

Green box policies is the term used to describe domestic support policies that are not subject to reduction commitments under the Agreement on Agriculture (AoA) under WTO. These policies are assumed to affect trade minimally and include support such as research, extension, foods security stocks, disaster payments and structural adjustment programmes.

8. Blue Box Policies

Blue box policies refer to direct payments to producers like decoupled income supports; payments not linked to production, structural adjustment assistance provided through investment aids to compensate for the structural disadvantage through resource retirement programmes, and through producer retirement programmes, government financial participation in income insurance and income safety net programmes. These relate to income and not to either the level of production or to prices.

9. Amber Box Policies

These are those policies which are trade distorting and are covered under reduction commitment in WTO.

10. Cairns Group

It is the group of free trading agricultural exporting nations which met in 1987 at Cairns (Australia) and agreed to present their common interests and concerns in the agricultural negotiations of the Uruguay Round. This group, comprising of countries viz., Argentina, Australia, Brazil, Canada, Chile, Colombia, Fiji, Hungry, Indonesia, Malaysia, New Zealand, Philippines, Thailand and Uruguay, emerged as a major negotiating group for agricultural trade negotiations and accounts for around 20% of world agricultural exports.

11. Most Favoured Nation (MFN) Status

All nations belonging to GATT have agreed to the most favoured nation principle as a condition of membership. The most favoured nation (MFN) status is a provision in the commercial treaty that grants each signatory the automatic right to any tariff reduction that may be negotiated by one of them with a third country. For example, if the United State were to negotiate a tariff reduction on automobiles with Japan, it would also be committed to such reductions with all its other trading partners to whom it has granted the MFN status.

12. Sanitary and Phyto-sanitary (SPS) Measures

The Uruguay Round had evolved a detailed discipline that a member country may apply trade restrictive measures for the protection of human life or its health and of plant or animal life or their health. These measures are contained in the Agreement on the Application of Sanitary and Phyto-Sanitary (SPS) Measures. SPS measures are applied to protect human life (health) or animal life (health) from risks arising from:

1. The additives, contaminants, toxins or diseases causing organisms in foods, beverages or food stuffs;
2. The entry of or spread of pests, diseases, disease carrying organisms or disease-causing organisms; and
3. The diseases carried by animals, plants or their products.

13. Aggregate Measure of Support (AMS)

This is a method of quantifying the aggregate value of domestic support or subsidy given to each category of agricultural products. Each WTO member country has made calculations to determine its AMS level wherever applicable. AMS includes non-product specific and product specific support.

14. TRIPS Agreement

The agreement on the Trade Related Aspects of Intellectual Property Rights (TRIPS) is a part of the Uruguay Round and covers seven categories of intellectual property namely:

(a) Copy rights and related rights;
(b) Trade marks;
(c) Geographical indications;
(d) Industrial designs;
(e) Integrated circuits;
(f) Trade secrets; and
(g) Patents (includes plant varieties).

The TRIPS Agreement inter alia prescribes the minimum standards to be adopted by members in respect of the above categories of intellectual property.

15. Balance of Payments (BOP)

The balance of payments of a country refers to the balance between the payments that are owed to the outside world and that are owned by the country. It is a recording of the value of the transactions across borders and comparison of in country transactions with outgoings. The difference of inflows and outflows shows the extent of balance of payments and the capacity or drain on foreign exchange of a country.

16. Tale Quale

Tale quale is a term used by UK based Grain and Feed Trade Association (GAFTA) with more than 900 members in 80 countries, including India. Tale quale means that the buyer agrees to accept the goods as they come provided they are shipped initially in good condition.

17. Rye Terms

Rye terms is also a term used by GAFTA, which means that the condition of goods on arrival is guaranteed by the seller.

18. Symmetric Coupling

Symmetric coupling is a situation when different components of manufactured products are split and produced in several geographic locations across national boundaries.

19. Technical Barriers to Trade (TBT)

These include measures used by different countries to restrict trade like labeling, packaging and nutritional attribute claims. At one time, the USA had 26 TBTs in place.

Appendices

COMMON ABBREVIATIONS IN AGRICULTURAL MARKETING

AAP	Agricultural and Allied Products
ADB	Asian Development Bank
AEZ	Agri-Export Zone
AFMA	Association of Food Marketing Agencies in Asia and the Pacific
AGMARK	Symbol of Agricultural Marketing on Graded Products
AOA	Agreement on Agriculture
APC	Agricultural Prices Commission
APEDA	Agricultural and Processed Food Products Export Development Authority, New Delhi
APL	Above Poverty Line
APMC	Agricultural Produce Market Committee
ASAC	Asian Standards Advisory Committee
ASI	Annual Survey of Industries/Agricultural Situation in India
AVORD	Association of Voluntary Organizations for Rural Development
BE	Budget Estimates
BICP	Bureau of Industrial Costs and Prices
BIS	Bureau of Indian Standards
CBs	Commodity Boards
BoP	Balance of Payments
BoT	Balance of Trade
BPL	Below Poverty Line
B2B	Busines to business
B2C	Busines to consumer
CACP	Commission for Agricultural Costs and Prices
CAM	Centre for Agricultural Marketing
CAIT	Confederation of All India Traders
CCB	Central Cooperative Bank
CCI	Cotton Corporation of India
CECA	Comprehensive Economic Cooperation Agreement
CEPA	Comprehensive Economic Partnership Agreement

CERC	Consumer Education and Research Centre
CFB	Corrugated Fibre Board
CGIAR	Consultative Group on International Agricultural Research
CIF Price	Cost, Insurance and Freight Price
CIP	Central Issue Price
CMIE	Centre for Monitoring Indian Economy
COSAMB	Council of State Agricultural Marketing Boards
CPI	Consumer Price Index
CPRs	Common Property Resources
CSC	Central Seeds Committee
CSCB	Central Seed Certification Board
CSO	Central Statistical Organization
CWC	Central Warehousing Corporation
DBT	Direct Benefit Transfer
DES	Directorate of Economics and Statistics
DFI	Doubling Farmers Income
DGCIS	Director General of Commercial Intelligence and Statistics
DMI	Directorate of Marketing and Inspection
DPIIT	Department of Promotion of Industry and Internal Trade
EAS	Electronic Auctioning System
ECAFE	Economic Commission for Asia and Far East
ECGC	Export Credit Guarantee Corportion
ENAM	Electronic National Agriculture Market
ENWR	Electronic Negotiable Warehouse Receipt
EOU	Export Oriented Unit
EXIM	Export-Import
FAO	Food and Agriculture Organization
FAQ	Fair Average Quality
FCI	Food Corporation of India
FDI	Foreign Direct Investment
FHP	Farm Harvest Price
FICCI	Federation of Indian Chamber of Commerce and Industry
FMC	Forward Markets Commission
FOB	Free on Board
FOR	Free on Rail
FPC	Farmer Producer Company
FPO	Farmer Producer Organization
FPS	Fair Price Shops
FSSA	Food Safety and Standards Act/Authority
FTZ	Free Trade Zone
FY	Financial Year
GAFTA	Grain and Feed Trade Association. (UK based)
GATT	General Agreement on Trade and Tariff
GCARD	Global Conference on Agricultural Research for Development
GDP	Gross Domestic Product
GM	Genetically Modified
GNP	Gross National Product
GST	Goods and Service Tax

HACCP	Hazard Analysis and Critical Control Point
HDPE	High Density Poly Ethylene
HP	Harvest Price. (Harvest Season Price)
HPMC	Himachal Pradesh Horticultural Processing and Marketing Corporation
ICAR	Indian Council of Agricultural Research
IFPRI	International Food Policy Research Institute
IGSI	Indian Grain Storage Research Institute, Hapur
IIFT	Indian Institute of Foreign Trade
IIP	Indian Institute of Packaging, Mumbai
IJAM	Indian Journal of Agricultural Marketing
ISAM	Indian Society of Agricultural Marketing
ISI	Indian Standards Institution
JCI	Jute Corporation of India
JNPT	Jawaharlal Nehru Port Trust/Terminal
KALIA	Krushak Assistance for Livelihood and Income Augmentation
KCC	Kisan Credit Card
KIAM	Karnataka Institute of Agricultural Marketing
KUMS	Krishi Upaj Mandi Samiti
LOOP	Law of One Price
MCs	Market Committees
MEP	Minimum Export Price
MIS	Market Intervention Scheme
MMTC	Minerals and Metals Trading Corporation
MNC	Multi National Corporation
MOAFW	Ministry of Agriculture and Farmers Welfare
MoFPI	Ministry of Food Processing Industries
MPDC	Market Planning and Design Centre
MPEDA	Marine Products Export Development Authority
MRTP	Monopolies and Restrictive Trade Practices
MS	Marketable/Marketed Surplus
MSP	Minimum Support Price
NAAS	National Academy of Agricultural Sciences
NABARD	National Bank for Agricultural and Rural Development
NABI	National Agrifood Biotechnology Institute
NAFED	National Agricultural Cooperative Marketing Federation
NAIP	National Agriculture Innovation Project
NBS	Nutrient Based Subsidy
NCAP	National Centre for Agricultural Economics and Policy Research
NCDC	National Cooperative Development Corporation
NDDB	National Dairy Development Board
NERAMAC	North Eastern Regional Agricultural Marketing Corporation
NHB	National Horticulture Board
NIAM	National Institute of Agricultural Marketing
NIAP	National Institute of Agricultural Economics and Policy Research
NITI	National Institution for Transforming India
NSC	National Seed Corporation
NSSO	National Sample Survey Organization

NWR	Negotiable Warehouse Receipt
OGL	Open General Licence
PBM	Parity Bound Models
PBP	Pay Back Period
PDS	Public Distribution System
PHT	Post Harvest Technology
PMFBY	Prime Minister Fasal Beema Yojana
PPV&FR	Protection of Plant Varietier and Farmers Rights
QRs	Quantitative Restrictions
RAJFED	Rajasthan State Cooperative Marketing Federation
RBI	Reserve Bank of India, Bombay
RCDF	Rajasthan Cooperative Dairy Federation
RCEP	Regional Comprehensive Economic Partnership
REC	Rural Electrification Corporation
RFID	Radio Frequency Identification
RKVY	Rashtriya Krishi Vikas Yojana
RPDS	Revamped Public Distribution System
RRBs	Regional Rural Banks
RSAMB	Rajasthan State Agricultural Marketing Board
RSWC	Rajasthan State Warehousing Corporation
RTE	Ready-to-Eat
SAMB	State Agricultural Marketing Board
SFCI	State Farms Corporation of Indian
SMP	Statutory Minimum Price
SPS	Sanitary and Phyto Sanitary
STC	State Trading Corporation
SSCs	State Seeds Corporations
SWC	State Warehousing Corporation
SWM	Secondary Wholesale Market
ToT	Terms of Trade
TBT	Technical Barriers to Trade
TPDS	Targeted Public Distribution System
TRIFED	Tribal Cooperative Marketing Development Federation of India
TRIPS	Trade Related Intellectual Property Rights
UBI	Universal Basic Income
UNCTAD	United Nations Conferences on Trade and Development
WBCIS	Weather Based Crop Insurance Scheme
WDRA	Warehousing Development and Regulation Act
WPI	Wholesale Price Index
WTC	World Trade Centre
WTO	World Trade Organization
XGS	Export of Goods and Services

SOME IMPORTANT INSTITUTIONS RELATING TO AGRICULTURAL MARKETING IN INDIA

1. Agricultural and Processed Food Products Export Development Authority, New Delhi.
2. Agro Economic Research Centres (various states).
3. Associated Chamber of Commerce and Industries, New Delhi.
4. Bureau of Indian Standards, New Delhi.
5. Central Food Technology Research Institute, Mysore.
6. Central Warehousing Corporation, New Delhi.
7. Coconut Development Board, Cochin.
8. Commission for Agricultural Costs and Prices, New Delhi.
9. Cotton Corporation of India, Mumbai.
10. Coffee Board, New Delhi.
11. Council of State Agricultural Marketing Boards, New Delhi.
12. Directorate of Economics and Statistics, Krishi Bhawan, New Delhi.
13. Directorate of Marketing and Inspection, Faridabad and Nagpur.
14. Directorate General of Foreign Trade, Kolkata.
15. Export Import Bank of India.
16. East India Cotton Association, Mumbai.
17. Fertiliser Association of India, New Delhi.
18. Food Corporation of India, New Delhi.
19. Federation of Indian Chambers of Commerce and Industry, New Delhi.
20. Food Safety and Standards Authority.
21. Forward Markets Commission, Mumbai.
22. Indian Grain Storage Management and Research Institute, Hapur.
23. Indian Institute of Foreign Trade, New Delhi.
24. Indian Institutes of Management
25. Indian Institute of Packaging, Mumbai.
26. Indian Society of Agricultural Marketing, Nagpur.
27. Indian Trade Promotion Organization, New Delhi.
28. Institute of Economic Growth, Delhi
29. Institute of Development Studies, Jaipur.
30. Institute of Rural Management, Anand
31. International Crops Research Institute for Semiarid Tropics, Hyderabad
32. Jute Corporation of India, Kolkata.
33. Karnataka Institute of Agricultural Marketing, Mysore.
34. Marine Products Export Development Authority (MPEDA).
35. Modern Food Industries (India) Limited, Delhi.
36. National Academy of Agricultural Sciences, New Delhi
37. National Bank for Agriculture and Rural Development, Mumbai.
38. National Institute of Agricultural Economics and Policy Research, New Delhi.
39. National Cooperative Development Corporation, Delhi.
40. National Cooperative Marketing Federation of India Limited, New Delhi.
41. National Dairy Development Board, Anand.
42. National Fertilizers Limited, New Delhi.
43. National Horticultural Board, Gurgaon.
44. National Institute of Agricultural Marketing, Jaipur.

45. National Institute of Crop Processing Technology, Thanjavour (TN).
46. National Institute of Food Technology Entreprenureship and Management, Sonepat (Haryana).
47. National Sample Survey Organization, New Delhi.
48. National Seeds Corporation Limited, New Delhi.
49. National Oilseeds and Vegetable Oils Development Board, Gurgaon.
50. Spices Board, Cochin.
51. Tea Board, Kolkata.
52. Tribal Cooperative Marketing Federation, New Delhi.
53. Tobacco Board, Guntur.
54. State Agricultural Marketing Boards, in the states at Hyderabad (Telangana), Guwahati (Assam), Patna (Bihar), Panchkula (Haryana), Shimla (Himachal Pradesh), Gandhi Nagar (Gujarat), Bangalore (Karnataka), Bhopal (Madhya Pradesh), Pune (Maharashtra), Shillong (Meghalaya), Bhubneshwar (Orissa), Chandigarh (Punjab), Jaipur (Rajasthan), Chennai (Tamil Nadu), Lucknow (Uttar Pradesh), Agartala (Tripura), Kolkata (West Bengal), New Delhi (Delhi), Kohima (Nagaland), Raipur (Chhatisgarh), Ranchi (Jharkhand) and Haldwani (Uttarakhand).
55. State Directorate of Agricultural Marketing, Hyderabad (Telangana), Chennai (TN), Kolkata (WB), Jaipur(Rajasthan), Lucknow (UP), Gandhi Nagar (Gujarat), Bangluru (Karnataka), Bhopal (Madhya Pradesh), Mumbai (Maharashtra) and Delhi.
56. Departments of Agricultural Economics, in all the state Agricultural Universities and many ICAR Institutes.
57. College/Institute of Agribusiness Management in state Agricultural Universities (for example at Pantnagar, Coimbatore, Navsari, Bikaner, Udaipur and Dharward).
58. ICSSR (Indian Council of Social Science Research) Institutes in various states

█ SELECTED BIBLIOGRAPHY ON AGRICULTURAL MARKETING

BOOKS

- Abbott, J.C., Marketing Problems and Improvement Programmes, Food and Agriculture Organization, Rome, 1958.
- Abbott, J.C., Agriculture and Food Marketing in Developing Countries — Selected Reading, Centre for Agricultural Bio-Sciences (CAB), 1993.
- Acharya, S.S.; Agricultural Production, Marketing and Price Policy—A Study of Pulses; Mittal Publications, Delhi-110035, 1988.
- Acharya, S.S.; Agricultural Marketing, study of Indian Farmers, Vol. 17, Ministry of Agriculture, Government of India, Academic Foundation, New Delhi, 2004.
- Acharya, S.S.; Agricultural Marketing and Rural Credit for Strengthening Indian Agricultural, INRM Policy Brief Ne 3, Asian Development Bank, New Delhi, 2006
- Acharya, S.S. and D.P. Chaudhri (Ed.), Indian Agricultural Policy at Cross-Roads, Rawat Publications, Jaipur, 2001.
- Acharya, S.S., M.S. Rathore and P.R. Sharma, Agricultural Marketing — A National Dialogue, Institute of Development Studies, Jaipur, 1998.
- Acharya, S.S. and N.L. Agarwal, Agricultural Prices—Analysis and Policy, Oxford and IBH, Publishing Co. Pvt. Ltd., New Delhi, 1994.
- Acharya, S.S. and R.K. Yadav, Production and Marketing of Milk and Milk Products, Mittal Publications, Delhi-110035, 1992.
- Acharya, S.S., Surjit Singh and Vidya Sagar (Ed.), Sustainable Agriculture, Poverty and Food Security, Asian Society of Agricultural Economics, Rawat Publications, Two Volumes, Jaipur, 2002.
- Agarwal, N.L., Agricultural Prices and Marketing in India—An Analytical Case Study of Rajasthan, Mittal Publications, Delhi-110035, 1986.
- Agarwal, N.L., Bharatiya Krishi ka Arthatantra, Rajasthan Hindi Granth Academy, Jaipur, 2008.
- Agarwal, N.L., Training Manual on Agricultural Marketing for Indian Farmers, Indian Society of Agricultural Marketing, Nagpur, 2001.
- Agarwal, V.K., Marketing of Dairy Products in Western U.P., Himalaya Publishing House, Mumbai, 1988.
- Amarchand, D. and B. Vardharajan, An Introduction to Marketing, Vikas Publishing House Private Ltd., New Delhi, 1980.
- Bhattacharya, S., Marketing Management, National Publishing House, Jaipur, 2004.
- Brunk, M.E. and L.B. Darrah, Marketing of Agricultural Products, The Ronald Press Company, New York, 1955.
- Cumming (Jr.), R.E., Pricing Efficiency in the Indian Wheat Market, Impex India, New Delhi, 1969.
- Desai, B.M., Banking Credit for Farm Inputs Marketing Business, Kaveri Book Service, New Delhi.
- Ferroni, Marco (Ed.), Transforming Indian Agriculture—India 2040, Sage Publications, 2013.
- Government of Madhya Pradesh; Reinventing the Mandi: Market Yards in a Globalizing World, 2008.
- Gupta, A.P., Marketing of Agricultural Produce in India, Vora and Co. Publishers Pvt. Limited, Mumbai, 1975.

- Indian Society of Agricultural Economics, Foodgrains Buffer Stocks in India, Seminar Series VIII, Mumbai, 1969.
- Indian Society of Agricultural Economics, Indian Agricultural Economy under Liberalized Regime, 1991–2015, Academic Foundation, New Delhi, 2016.
- Jain, Manohar Lal, Marketing of Agricultural Inputs, Himalaya Publishing Company, Mumbai, 1988.
- Jain, S,C., Principles and Practices of Agricultural Marketing and Prices, Vora and Co. Publishers Pvt. Limited, Mumbai, 1971.
- Jagdish Prasad, Encyclopedia of Agricultural Marketing, Vol. I to viii, 2000, Mittal Publications, Delhi-110035.
- Jasdanwalla, Z.Y., Marketing Efficiency in Indian Agriculture, Allied Publishers Pvt. Limited, Mumbai, 1966.
- Kahlon, A.S. and M.V. George, Agricultural Marketing and Price Policies, Allied Publishers Private Limited, New Delhi, 1985.
- Khusro, A.M., The Economics of Buffer Stocks and Storage of Major Foodgrains in India, Institute of Economic Growth, Delhi-110007, 1969.
- Kohls, R.L. and J.N. Uhl, Marketing of Agricultural Products, Macmillan Publishing Company Inc., New York, 5th Edition, 1980.
- Kotler, P., Marketing Management Analysis, Planning and Control, Prentice-Hall of India Pvt. Limited, New Delhi, 1972.
- Kumar, Shalander; Commercialization of Goat Farming and Marketing of Goats in India, Central Institute for Research on Goats, Mathura, 2007.
- Lele, Uma, Foodgrains Marketing in India—Private Performance and Public Policy, Popular Parkashan, Mumbai, 1973.
- Meena, R.K. and J.S. Yadav, Horticulture Marketing and Post-Harvest Management, Pointer Publications, Jaipur, 2001.
- Moore, J.R., S.S. Johl and A.M. Khusro, Indian Foodgrain Marketing, Prentice-Hall of India Private Limited, New Delhi, 1973.
- National Academy of Agricultural Sciences; State of Indian Agriculture, New Delhi, 2009.
- Narain, Dharam, Distribution of the Marketed Surplus of Agricultural Produce by Size-Level of Holdings in India, 1950–51, Asia Publishing House, New Delhi.
- Muzumdar, N.A. and Uma Kapila; Indian Agriculture in the New Millennium, Academic Foundation, New Delhi, 2006.
- Nayyar, H. and P. Ramaswamy, Globalisation and Agricultural Marketing, Rawat Publications, Jaipur, 1995.
- Neelamegham, S., Marketing Management and the Indian Economy, Vikas Publishing House Pvt. Ltd., New Delhi, 1978.
- Paul H. and S. Ricarda, Hungry Corporations: Transnational Corporations Colonise the Food Chain, Zed Books, London/New York, 2003.
- Prabhakar Rao, Marketing Efficiency in Agricultural Production, Himalaya Publishing House, Mumbai, 1985.
- Prasad, A. Shivarama, Agricultural Marketing in India, Mittal Publications, Delhi-110035.
- Rajagopal, Marketing In Peasant Economy—History and Trend, Manas Publication, Delhi, 1988.
- Rajagopal, Rural Marketing in India (Empirical Studies from Madhya Pradesh), Rennaissance Publishing House, Delhi, 1989.

- Rajagopal, Rural Marketing Administration in India, Kaveri Book Service, New Delhi.
- Rajagopal, Development of Agricultural Marketing in India, Printwell, Jaipur, 1996.
- Rajagopal, Micro Level Planning for Agricultural Marketing, Rennaissance Publishing House, Delhi, 1988.
- Rajagopal, Agri-Business and Entrepreneurship, Anmol Publications, New Delhi, 1990.
- Rajagopal, Entrepreneurship and Rural Markets, Rawat Publications, Jaipur, 1993.
- Rajagopal, Indian Rural Marketing, Rawat Publications, Jaipur, 1993.
- Ramasamy C. and K.R. Ashok, Vicissitudes of Agriculture in Fast Growing Indian Economy: Challenges, Strategy and Way Forward, Indian Society of Agricultural Economics and Academic Foundation, 2018.
- Ramesh Chand, Agricultural Development, Price Policy and Marketed Surplus in India (Study of Green Revolution Region), Concept Publishing Company, New Delhi-110059, 1991.
- Ramesh Chand and V.P.S. Arora (Ed.), Agricultural Input and Output Market Reforms in India, Advance Publishing Concept, New Delhi-110064, 2000.
- Ramphal; WTO and Indian Agriculture, Global Research Publication, New Delhi, 2010.
- Sarveshwara, Rao and G. Parthasarthy, Regulated Markets and Public Policy, Booklinks Corporation, Hyderabad.
- Seetharaman, S.P., Agricultural Input Marketing, Oxford and IBH Publishing Co., New Delhi.
- Shepherd, G.S., Marketing of Farm Products, Iowa State University Press, Ames, Iowa, USA, 1972.
- Sherlekar, S.A., Marketing Management, Himalaya Publishing House, Mumbai, 1981.
- Singh, J.P., Rural Godowns and Storage System in India — A Critical Review, National Institute of Agricultural Extension Management, Hyderabad, 2003.
- Singhal, A.K., Agricultural Marketing in India, Anmol Publications, New Delhi, 1989.
- Singh, D.V., Production and Marketing of Off-season Vegetables, Mittal Publications, Delhi-110035, 1990.
- Singh, Rajkumar, Agricultural Price Policy in India, Printwell Publishers, Jaipur.
- Sinha, J.C., Principles of Marketing and Salesmanship, R. Chand and Co., Delhi, 1976.
- Southworth, Herman, Some Studies on Fresh Fruits and Vegetable Marketing in Asia, Agricultural Development Council, Bangkok, Thailand, 1974.
- Subba Rao, K., Agricultural Marketing and Credit, ICSSR, New Delhi, 1985.
- Swaroop, R. and B.K. Sikka, Production and Marketing of Apples—An Economic Study in Himachal Pradesh, Mittal Publications, Delhi-110035, 1987.
- Talukdar, K.C. and B.C. Bhowmik (Ed.); Marketing of Perishable Products, B.R. Publishing Corporation, New Delhi, 1993.
- Tyagi, D.S., Managing India's Food Economy—Problems and Alternatives, Sage Publications India Pvt. Ltd., New Delhi, 1990.
- Thomsen, F.L., Agricultural Marketing, McGraw-Hill Book Company, New York, 1951.
- Waugh, F.W., Readings on Agricultural Marketing, Iowa State University Press, Ames, Iowa, USA, 1954.

REPORTS

(A) Indian Society of Agricultural Economics, Mumbai

(a) Seminar Series V, Seminar on Marketing of Agricultural Commodities, 1965.
(b) Seminar Series VIII, Seminar on Foodgrains Buffer Stocks in India, 1969.
(c) Seminar Series X, Seminar on Emerging Problems of Marketing of Agricultural Commodities, 1972.

(B) Directorate of Marketing and Inspection, Ministry of Agriculture, Government of India, Faridabad and Nagpur

(a) COMMODITY SURVEY REPORTS

- Foodgrains—Wheat, Rice, Barley, Maize, Millets, Gram and Pulses.
- Fruits and Vegetables—Bananas, Citrus fruits, Grapes, Guava, Papaya, Lichi, Mangoes, Pineapple, Stone fruits, Small fruits, Temperate fruits, Potatoes, Cole crops and root vegetables, Green peas, French beans, Onions, Tomatoes and Tapioca.
- Edible nuts—Cashew nuts and Walnuts.
- Spices—Chillies, Ginger, Turmeric, Cardamom and minor spices.
- Oilseeds and Oils—Castor, Groundnut, Linseed, Rapeseed—Mustard, Sesamum, Niger seed, Safflower seed, Coconut and Coconut products.
- Fibres—Aloe and Sisal, Cotton, Palmyra fibres and Sannhemp.
- Fish and Fish Products—Fish.
- Milk and Milk Products—Milk, Ghee and other milk products.
- Livestock and Livestock Products other than Milk—Cattle, Sheep and Goats, Poultry, Meat, Bones and Bonemeal, Hides, Skin, Wool, Goat hair, Bristles, Eggs, Animal Fats and By-products.
- Plantation Crops—Coffee.
- Essential Oils—Lemon grass oil and Sandalwood oil.
- Miscellaneous Crops—Tobacco, Lac, Henna, Isabgol, Myrobalans, Sugar, Arecanuts and Betalnuts.

(b) SPECIAL STUDIES/REPORTS

Market Research and Planning Cell (MRPC) have brought out the following reports:
1. Marketing of Wheat in Punjab and Haryana.
2. Marketable Surplus and Post-Harvest Losses of Paddy in India.
3. Marketing of Lemongrass Oil in India.
4. Price-Spread Studies of Onion in India.
5. Price-Spread Studies of F.C.V. Tobacco in India.
6. Marketing of Cotton in Karnataka.
7. Reasons for Decline of Sannhemp Export from India.
8. A Study on Trade Practices in the Sale of Banana in India.
9. Fairs, Markets and Produce Exchanges: Wholesale Agricultural Assembling Markets, Regulated Markets and Cooperative Marketing of Agricultural Produce.
10. Cold Storage—Cold storage for fruits and vegetables.
11. Grading—Grading instructions for different agricultural commodities.
12. Sampling and Testing—Methods of sampling and testing vegetable oils and fats under AGMARK.
13. Acts and Orders—Agricultural Produce (Grading and Marking) Act, Fruit Products Order, Cold Storage Order, Meat Food Products Order.
14. Marketing Costs and Margins of Agricultural Commodities in India, 1985.

15. Production, Utilization, Marketable and Marketed Surplus of Wheat, Rice and Maize, 1995.
16. Agmark Grading Statistics, 1995–96; 1997; 1998–99.
17. Infrastructural Facilities Provided by the APMC and their Utilization, MRPC-25, 1999.

(C) Ministry of Agriculture/Food, Government of India

1. Report of the High Power Committee on Agricultural Marketing (Guru), 1992.
2. Report of the Expert Group on Agricultural Marketing (S.S. Acharya), 1999.
3. Report of the Working Group on Agricultural Marketing Infrastructure (S.S. Acharya) as a part of the Expert Committee Report on Strengthening and Development of Agricultural Marketing.
4. Report of the Expert Committee on Strengthening and Developing of Agricultural Marketing (S.L. Guru), June, 2001.
5. Report of the High Level Committee on Re-orienting the Role and Re-structuring of Food Corporation of India (Shanta Kumar), 2015.
6. Report of the Committee to Examine Methodological Issues in Fixing Minimum Support Prices (Ramesh Chand), March 2015.
7. Assessment of Marketed and Marketable Surplus of Major Foodgrains in India (by V.P. Sharma and Harsh Wardhan of CMA, IIMA), Report for Ministry of Agriculture, 2015.

(D) Publications of National Institute of Agricultural Marketing, Jaipur

1. Flow Pattern of Fruits and Vegetables and Related Issues in the North West Hills Region, August, 1989.
2. A Report on Arrival Pattern and Related Issues on Fruits and Vegetables in Jaipur, May, 1991.
3. Training Programme on Agricultural Marketing in Tribal Areas, March, 1993.
4. A Case Study of Hadaspur Model of Direct Marketing, Pune, J.P. Mishra, 1992.
5. Systems of Sale in Azadpur Market of Delhi, 1993.
6. Country Level Action Plan for Market Development, 1994.
7. Agricultural Marketing Statistical Abstracts; 1999; 2000; 2001; 2003; 2005.
8. Gramin Bhandaran Yojana (Rural Godown Scheme), Training Manual for Training of Farmers, 2002.
9. Gramin Bhandaran Yojana (Rural Godown Scheme) Training Manual for Entrepreneurs, 2002.
10. Gramin Bhandaran Yojana (Rural Godown Scheme) Training Manual for Training of Trainers.
11. Proceedings of National Workshop on Marketing Extension, 19–20 July, 2001.
12. Proceedings of National Workshop on Imperatives of Quality Assurance, Grading and Standardization for Farmers, 23–24, January, 2002.
13. Proceedings of National Seminar on Rural godown, 12–13, April, 2002.
14. Marketing of Wheat in North India, 2001
15. Marketing of Pulses in Eastern India, 2001
16. Marketing of Sugarcane in South India, 2001
17. Marketing of Oilseeds in North India, 2001
18. Marketing of Rice in Eastern India, 2001
19. Marketing of Vegetables in Southern India, 2001

(E) Agricultural Universities

1. Chauhan, K.K.S. and R.V., Singh. Marketing of Wheat in Rajasthan, University of Udaipur, 1973.
2. Acharya, S.S. and N.L. Agarwal. Organization, Functioning and Benefit-Cost Analysis of Regulated Markets in Rajasthan, University of Udaipur, Udaipur, 1976.
3. Kahlon, A.S. and S.S. Grewal. A Study of Marketing of Wheat in the Punjab, Punjab Agricultural University, Ludhiana, 1964–65.
4. Kahlon, A.S. and Balwinder Singh. Marketing of Groundnut in the Punjab, Punjab Agricultural University, Ludhiana, 1968.
5. Kahlon, A.S. and Balwinder Singh. Orientation of Pome and Stone Fruit Market Structure to the Expanding Volume of Fruit Production in Kulu and Parvati Valleys, Punjab Agricultural University, Ludhiana.
6. Gill, K.S. and S.S. Johl. Marketing of Gram in Punjab, Punjab Agricultural University, Ludhiana, 1970.
7. Gill, K.S. Wheat Market Behaviour in Punjab and Haryana. Post-Harvest Period 1968–69 to 1970–71, Punjab Agricultural University, Ludhiana.
8. Chatha, I.S. and D.S. Sidhu. Production and Marketing of Potato in the Punjab State, Punjab Agricultural University, Ludhiana, 1980.
9. A Study of Impact of Changing Conditions on Grain Marketing Institutions and Structure of Grain Markets in the Punjab, 1966–67, Punjab Agricultural University, Ludhiana.
10. Seminar Report of First All India Agricultural Marketing Conference, G.B. Pant University of Agriculture and Technology, Pant Nagar, February, 1969.
11. Seminar Report of Second All India Agricultural Marketing Conference, J.N. Krishi Viswa Vidyalaya, Jabalpur, 1970.
12. Shah, S.L. and V.K. Pandey, Study of Marketable Surplus of Wheat in Critical Areas of India, G.B. Pant University of Agriculture and Technology, Pantnagar, 1976.
13. Singh, I.J. and K.N. Rai. Economic Aspects of Production and Marketing in Haryana, Haryana Agricultural University, Hisar, 1976.
14. Singh, I.J. and Kapil Chaudhary. Economic Analysis of Potato Production and Marketing in Haryana, Haryana Agricultural University, Hisar, 1977.
15. Singh, I.J, Kapil Chaudhary and R.C. Goel. Dynamics of Cotton Production and Marketing in Haryana, Haryana Agricultural University, Hisar, 1979.
16. Singh, I.J, and K.N. Rai. Production and Marketing of Sugarcane in Haryana, Haryana Agricultural University, Hisar, 1982.
17. Gangwar, A.C., K.N. Rai and Shri Niwas. Production and Marketing of Gram in Haryana, Haryana Agricultural University, Hisar, 1983.
18. Gangwar, A.C., K.N. Rai, and Shri Niwas. Production and Marketing of Rice in Haryana, Haryana Agricultural University, 1985.
19. Malik, H.S., Shri Niwas and A.C. Gangwar. Production and Marketing of Wheat in Haryana, Haryana Agricultural University, Hisar, 1988.
20. Burark, S.S., and Latika Sharma, Cost of Cultivation of Principal Crops in Rajasthan, MPUAT, Udaipur, 2017.

(F) Agro-Economic Research Centres (AERCs)

1. Green Revolution and Problems of Marketing — A Study of Production and Marketing of Bajra in Three Districts of Gujarat, Agro-Economic Research Centre, Vallab Vidya Nagar, 1972.

2. Pace and Pattern of Marketing of Groundnut in Saurashtra Region of Gujarat, Agro-Economic Research Centre, Vallab Vidya Nagar, 1969.
3. Relationship betweenn Wholesale Prices, Retail Prices and Details of contributing Factors for Price Difference of Onion in Gujarat, AERC Report 156, Vallabh Vidya Nagar, 2015
4. Marketed and Marketable Surplus of Major Foodgrains in Rajasthan, by V.D. Shah and M. Makwana, Working Paper 01, AERC, Vallabh Vidyanagar, 2016.

(G) National Council of Applied Economic Research (NCAER), New Delhi

Market Towns and Spatial Development in India, 1965.

(H) International Crops Research Institute for the Semi-Arid Tropics (ICRISAT), Hyderabad

1. Market Channels of SAT Crops and Comparative Market Efficiency in India—A Study Proposal, 1975.
2. The Effects of Inter-Regional Trade and Market Infrastructure on Aggregate Productivity of Agriculture.
3. Markets for ICRISAT Crops in Andhra Pradesh, June, 1976.
4. Agricultural Marketing in the Semi-Arid Tropics. A Research Plan, September, 1975.
5. Agricultural Marketing. A Preliminary Bibliography, August, 1975.
6. Marketed Surplus of Farm Products in India (Literature Review), November, 1976.
7. ICRISAT Strategic Plan to 2020: Inclusive Market-Oriented Development for Smallholder Farmers in the Tropical Drylands, 2010.
8. The Jewels of ICRISAT, Hyderabad, 2012.

(I) Ford Foundation, New Delhi

Kiehl, Elmer R. Agricultural Marketing in India — Role, Strategies and Implications, January, 1969.

(J) Agricultural Development Council, Bangkok/New York

PAPERS
1. No. 4, Some Building Blocks for an Agricultural Marketing Research Programme in Korea, June, 1971.
2. No.13, Food Shortages and Surpluses—A Marketing Trap for the Developing Countries, February, 1972.
3. No. 15, Myths about Agricultural Marketing, March, 1972.
4. No. 16, Improving Teaching of Agricultural Marketing in Asian Universities, May, 1972.
5. No. 21, Distributional Efficiency and Agricultural Price Policy—Foodgrain Marketing in India, October, 1972.
6. No. 22, The Basis for Agricultural Price Policy, November, 1972.
7. No. 28, Marketing Efficiency in Theory and Practice, April, 1973.

TEACHING AND RESEARCH FORUM
8. No. 12, A Comparative Study of Fertilizer Marketing System in Asia, November, 1977.

9. No. 18, Marketing of Rice in India: An Analysis of the Impact of Producer's Prices on Small Farmers, May, 1979.
10. No. 24, Needed Information and Economic Analysis for Fertilizer Policy Decisions, September, 1980.

RTN SEMINAR REPORT

11. No. 5, Marketing Problems Associated with Small Farm Agriculture, November, 1974.
12. Marketing Institutions and Service for Developing Agriculture, July, 1975.

(K) Commission for Agricultural Costs and Prices

The CACP's reports are published annually. These contain an incisive analysis of marketing problems and possible solutions.

(L) Ministry of Civil Supplies, Consumer Affairs and Public Distribution, Government of India

1. Report of the Committee on Forward Markets (Kabra Committee), 1994.
2. Government Intervention in Foodgrain Markets in the New Context, Report prepared by Ramesh Chand, National Centre for Agricultural Economics and Policy Research, New Delhi, September, 2002.
3. Report of the High Level Committee on Long Term Grain Policy, July, 2002. (Abhijit Sen).
4. Report of the National Advisory Council on Food Security Bill, 2010 (with Sonia Gandhi as chairperson)

(M) Planning Commission and Other Ministries of Government of India

1. Report of the Sub Group (chaired by S.S. Acharya) on Agricultural Economics, Marketing and Agribusiness for XI Five Year Plan, Planning Commission, 2006.
2. Report of the Working Group (chaired by S.S. Acharya) on Agricultural Marketing Infrastructure and External Trade Policy for XI Five Year Plan, Planning Commission, 2007.
3. Report of the High Powered Committee (with S.S. Acharya as member) for Krishi Vigyan Kendras, Ministry of Agriculture, 2006.
4. Report of the High Powered Committee on Cooperatives, Ministry of Agriculture, 2009.
5. Report of the Committee (Dalwai Committee) on Doubling Farmers Income, Post Production Agri-logistics: Maximising Gains for Farmers, Vol. III, MOAFW, August 2017, p. 1–165.
6. Report of the Committee (Dalwai Committee) on Doubling Farmers Income, Post Production Interventions: Agricultural Marketing, Vol. IV, MOAFW, August 2017, p. 1–109.
7. Report of the Committee (R.S. Paroda Committee) on Policies and Action Plans for Secure and Sustainable Agriculture, submitted to Principal Scientific Advisor to Government of India, August 2018, p. 1–198.

(N) World Bank (Washington) and Asian Productivity Organization (Tokyo)

1. India: Food Grain Marketing Policies, World Bank, Aug. 1999.
2. Marketing Systems for Agricultural Products, APO, 1997.

3. Enabling the Business of Agriculture, 2019, www.worldbank.org, Pp. 7 + 131.
4. Trading for Development in the Age of Global Value Chains, 2019, www.worldbank.org, Pp. 19 + 265.

(O) Institute of Development Studies, Jaipur

1. Acharya, S.S. and R.L Jogi, Minimum Support Prices in India: Some Issues, Working Paper No. 132, May, 2003.
2. Acharya S.S. and R.L Jogi; Farm Input Subsidies in Indian Agriculture, Working Paper No. 140, April 2004.

(P) Indian Society of Agricultural Marketing, Nagpur (India)

1. Vistas in Agricultural Marketing (1987–96), Vol. 1, 1996.
2. Vistas in Agricultural Marketing (1997–2005), Vol. 2, 2005.
3. Agribusiness Potential of Maharashtra, 2011
4. Agribusiness Potential of Gujarat, 2011.
5. Vistas in Agricultural Marketing (2006–12), Vol.3, 2012.
6. Status, Issues and Challenges of Agricultural Marketing in Mountain States of the Country, 2012.
7. Vistas in Agricultural Marketing, Vol. 4 (2012–18), 2018.
8. Women in Agricultural Production and Marketing, 2019.
9. Agribusiness Potential of Mizoram, 2020.

(Q) National Institute of Agricultural Economics and Policy Research (NIAP), New Delhi

1. Chand, Ramesh, Government Intervention in Foodgrain Markets in the New Context, Report Prepared for Ministry of Consumer Affairs, Food and Public Distribution, Government of India, March 2003.
2. Quinquennial Review Report 2006–10 (QRT Chaired by S.S. Acharya), 2012.
3. Saxena R. and Ramesh Chand, Understanding the Recurring Onion Price Shocks, Revelations from Production–Trade–Price Linkages, Policy Paper 33, 2017.

(R) Food and Agriculture Organization, Rome/Bangkok

1. Agricultural Marketing Training Improvement in Asia, 1979.
2. Regional Seminar on Applied Food Marketing Research and Related Training, 12–18 September, 1988, Kualalumpur, Malaysia.
3. Regional Workshop on Fruits and Vegetable Marketing and Processing, New Delhi, 7–15, December, 1992.
4. Agricultural Marketing in Asia and the Pacific: Issues and Priorities, A Theme Paper by S.S. Acharya, RAP, Bangkok, August, 2001.
5. Roundtable on Agricultural Marketing and Food Security in Asia, Proceedings, Conclusions and Recommendations, (S.S. Acharya) RAP-Bangkok, November, 2001.
6. Rice Price Policies in South-East Asia and Implications for East Timor by S.S. Acharya, September, 2001.
7. Horticultural Marketing, FAO Agricultural Service Bulletin No. 76, 1989.
8. Manual on Meat Cold Store Operation and Management, Animal Production and Health, Paper 92, 1991.
9. Market Integration and Price Transmission in India: A Case of Rice and Wheat with special reference to the World Food Crises of 2007–08, by S.S. Acharya, Ramesh Chand, P.S. Birthal, Shiv Kumar and others, www.fao.org/docrap/01pdf.

(S) IFPRI—International Food Policy Research Institute, Washington

1. Roy D., P.K. Joshi and Raj Chandra, Pulses for Nutrition in India, 2017.
2. Global Food Policy Report, 2018.

(T) Institute of Rural Management, Anand

1. Prasad S.C and G. Pareek, Farming Futures: An Annotated Bibliography on Farmer Producer Organizations in India, Working paper 290, April 2019.
2. Patel S.S., R. Pandey and H. Mishra, An Optimization Model for a Dairy Cooperative for Promoting Sustainable Operations for Milk Collection, Working Paper 293, May 2019.

JOURNALS/MAGAZINES

(a) Agricultural Marketing and Prices

1. Agricultural Marketing – Quarterly, Directorate of Marketing and Inspection, Ministry of Agriculture, Government of India, Faridabad/Nagpur.
2. Bulletin on Prices – Weekly, Directorate of Economics and Statistics, Ministry of Agriculture, Government of India, New Delhi.
3. NAFED News—Quarterly, National Agricultural Cooperative Marketing Federation of India Limited, NAFED House, Ashram Chowk, New Delhi-110014.
4. Fertilizer Marketing News – Monthly, The Fertilizer Association of India, Near Jawaharlal Nehru University, New Delhi-110067.
5. Indian Journal of Agricultural Marketing. Thrice a year, Indian Society of Agricultural Marketing, Hyderabad (Telangana).

(b) Agricultural Economics Journals and Others

1. Indian Journal of Agricultural Economics—Quarterly, The Indian Society of Agricultural Economics, A.K. Vaidya Marg, Goregaon (East), Mumbai-400063.
2. Agricultural Situation in India—Monthly, Directorate of Economics and Statistics, Ministry of Agriculture, Government of India, New Delhi.
3. Indian Agriculture in Brief—Annual, Directorate of Economics and Statistics, Ministry of Agriculture, Government of India, New Delhi.
4. Yojana—Monthly, Ministry of Information and Broadcasting, Government of India, New Delhi.
5. Economic and Political Weekly—Weekly, Bombay.
6. Bulletin on Food Statistics—Annual, Directorate of Economics and Statistics, Ministry of Agriculture, Government of India, New Delhi.
7. Asia–Pacific Journal of Rural Development, Half-yearly, CIRDAP, Dhaka (Bangladesh).
8. American Journal of Agricultural Economics – Quarterly.
9. The Asian Economic Review—Quarterly, Federation House, 11-6-841, Red Hills, Hyderabad 500004.
10. Economic Development and Cultural Change.
11. Economic Journal—Quarterly.
12. Artha Vijnana—Quarterly, Pune.
13. Anvesak, Half yearly, SPISER, Ahemedabad.
14. Arthaniti.
15. Artha Vikas, Vallabh Vidyanagar, Anand (Gujarat).

16. International Journal of Agricultural Economics.
17. Agricultural Economics Research Review, Half yearly, New Delhi.
18. Journal of Research, State Agricultural Universities.
19. Eastern Economist, New Delhi.
20. Margin—Quarterly, New Delhi.
21. Sections on Business and Economy in all Daily Newspapers.
22. Agricultural Statistics–At a Glance, Directorate of Economics and Statistics, Ministry of Agriculture, Government of India, New Delhi.
23. Economic Survey—Annual, Ministry of Finance, Government of India, New Delhi.
24. Journal of Agricultural Development and Policy, Half Yearly, Indian Society for Agricultural Development and Policy, Ludhiana (Punjab).
25. Journal of Rural Development, Quarterly, National Institute of Rural Development, Hyderabad, 500030.
26. Journal of Social and Economic Development—Half yearly, Institute of Social and Economic Change, Bangluru-560072.
27. Chinese Agricultural Economic Review, Chinese Agricultural University, Emerald (UK).
28. Asian Journal of Agriculture and Development, Southeast Asia Regional Centre for Graduate Study and Research in Agriculture (SEARCA), Laguna, Philippines.
29. Rural Development Statistics (Annual), National Institute of Rural Development, Hyderabad.
30. Fertilizer Statistics (Annual), Fertilizer Association of India.

Index